Intelligence, heredity, and environment

Intelligence, heredity, and environment

Edited by

ROBERT J. STERNBERG
Yale University

ELENA L. GRIGORENKO
Yale University
Moscow State University

CAMBRIDGE
UNIVERSITY PRESS

PUBLISHED BY THE PRESS SYNDICATE OF THE UNIVERSITY OF CAMBRIDGE
The Pitt Building, Trumpington Street, Cambridge, United Kingdom

CAMBRIDGE UNIVERSITY PRESS
The Edinburgh Building, Cambridge CB2 2RU, UK   http://www.cup.cam.ac.uk
40 West 20th Street, New York, NY 10011-4211, USA   http://www.cup.org
10 Stamford Road, Oakleigh, Melbourne 3166, Australia
Ruiz de Alarcón 13, 28014 Madrid, Spain

First published 1997
Reprinted 1999

Printed in the United States of America

Typeset in Times

Library of Congress Cataloging-in-Publication Data
Intelligence, heredity, and enmvironment / edited by Robert J.
  Sternberg, Elena Grigorenko.
     p.   cm.
  Includes indexes.
    ISBN 0-521-46489-7. – ISBN 0-521-46904-X (pbk.)
    1. Intellect. 2. Nature and nurture. I. Sternberg, Robert J.
II. Grigorenko, Elena.
  BF431.I534   1996
  155.7 – dc20   95-25931
                                                        CIP

A catalog record for this book is available from the British Library.

ISBN 0 521 46904 X Paperback

Contents

Contributors

Sheridan Bartholomew
Institute for Behavior Genetics
Campus Box 447
University of Colorado
Boulder, CO 80309–0345

Thomas R. Bidell
Department of Counseling
Campion 239D
Boston College
Chestnut Hill, MA 02167

Thomas Bouchard, Jr.
Department of Psychology
University of Minnesota
Elliot Hall – 75 East River Road
Minneapolis, MN 55455

Eddy de Bruyn
Human Development and Family
  Studies
Martha Van Rensselaer Hall
Cornell University
Ithaca, NY 14853

Stephen J. Ceci
Human Development &
Family Studies
Martha Van Rensselaer Hall
Cornell University
Ithaca, NY 14853

Stacey Cherney
University of Colorado at Boulder
Institute for Behavior Genetics
Campus Box 447
Boulder, CO 80309–0447

Yrjo Engeström
University of California, San Diego
Department of Communications
9500 Gilman Drive
La Jolla, CA 92093–0503

Ritva Engeström
University of California, San Diego
Department of Communications
9500 Gilman Drive
La Jolla, CA 92093–0503

Hans J. Eysenck
Department of Psychology
University of London
Bethlehem Royal &
  Maudley Hospitals
Denmark Hill, London SE5 8AF
England

Kurt W. Fischer
Harvard University
Larsen Hall – Appian Way
Cambridge, MA 02138

David Fulker
University of Colorado at Boulder
Institute for Behavior Genetics
Campus Box 447
Boulder, CO 80309–0447

Howard Gardner
Graduate School of Education
Harvard University
323 Longfellow Hall – Appian Way
Cambridge, MA 02138–3752

Edmund W. Gordon
3 Cooper Morris Drive
Pomona, NY 10970

Gilbert Gottlieb
Department of Psychology
University of North Carolina
Greensboro, NC 27412–5001

Elena Grigorenko
Department of Psychology
Yale University
P.O. Box 208205
New Haven, CT 06520–8205

Thomas Hatch
Graduate School of Education
Harvard University
323 Longfellow Hall – Appian Way
Cambridge, MA 02138–3752

John Hewitt
University of Colorado at Boulder
Institute for Behavior Genetics
Campus Box 447
Boulder, CO 80309–0447

Joseph M. Horn
Department of Psychology
University of Texas
Austin, TX 78712

Earl Hunt
Department of Psychology
NI-25
Guthrie Hall
University of Washington
Seattle, WA 98195

Arthur R. Jensen
School of Education
University of California
4511 Tolman Hall
Berkeley, CA 94720

Merja Kärkkäinen
Department of Education
University of Helsinki
Bulevardi 18
00120 Helsinki, Finland

Tatayana Kornilova
Moscow State University
20B Okhotnyi Road
Moscow 103009
Russia

Donald Y. Lee
Human Development and Family
  Studies
Martha Van Rensselaer Hall
Cornell University
Ithaca, NY 14853

Melissa P. Lemons
434 Napa Street
Sausalito, CA 94965

John C. Loehlin
Department of Psychology
University of Texas
Austin, TX 78712

Joan Miller
Department of Psychology
Yale University
Box 208205
New Haven, CT 06520–8205

Sandra Pipp-Siegel
Department of Psychology
University of Colorado
Boulder, CO 80309–0345

Robert Plomin
Institute of Psychiatry
113 Denmark Hill
London SE5 8A
England

Steven J. Reznick
Department of Psychology
Yale University
Box 208205
New Haven, CT 06520–8205

JoAnn Robinson
Institute for Behavior Genetics
Campus Box 447
University of Colorado at Boulder
Boulder, CO 80309–0345

Tina Rosenblum
Human Development and Family
  Studies
Martha Van Rensselaer Hall
Cornell University
Ithaca, NY 14853

Sandra Scarr, CEO
Kindercare Learning Centers, Inc.
2400 Presidents Drive
P.O. Box 2151
Montgomery, AL 36102–2151

S. J. Schoenthaler
Department of Sociology
California State University, Stanislaus
801 West Monte Vista Avenue
Turlock, CA 94380

Robert J. Sternberg
Department of Psychology
Yale University
P.O. Box 208205
New Haven, CT 06520–8205

Bruce Torff
Department of Psychology
Yale University
Box 208205
New Haven, CT 06520–8205

Douglas Wahlsten
University of Alberta
Department of Psychology
P-220 Biological Sciences Building
Edmonton, Alberta T6G 2E9
Canada

Irwin Waldman
Psychology Department
Emory University
532 Kilgo Circle
Atlanta, GA 30322

Lee Willerman
Department of Psychology
University of Texas
Austin, TX 78712

Preface

It is reassuring to know that virtually all researchers in a given field agree on something. In the field of intelligence, there are three facts about the transmission of intelligence that virtually everyone seems to accept.

1. Both heredity and environment contribute to intelligence.
2. Heredity and environment interact in various ways.
3. Extremely poor as well as highly enriched environments can interfere with the realization of a person's intelligence, regardless of the person's heredity.

Beyond these three facts, there is much divergence of opinion. We have edited this book in order to explore this divergence. Although there are certainly other books on the topic, we believe that, as of the book's publication date, no other volume is as up-to-date, balanced in its coverage of alternative views, and graced by the contributions of such a distinguished cast of authors, all of whom have been leading contributors to the field.

This book is for everyone who is interested in the question of how intelligence is transmitted: students, scholars, and curious laypersons. Authors were asked to contribute chapters that were technically sound and rigorous but that would nevertheless be interpretable to a wide variety of readers. We believe that the contributors fulfilled this mission.

Although we, as editors, have our own opinions and as authors have expressed these opinions, we have scrupulously tried to avoid favoring any one point of view. This field, like many others, finds itself with distinguished scholars representing a diversity of opinions, and our hope was that this would be the book people would seek if they wanted not only quality, but diversity, fairness, and balance, rather than the buzzing noise of an axe being ground.

The book is divided into a preface, 18 principal chapters, and two final concluding chapters that integrate the contributions presented in the volume. The chapters themselves are divided into four parts.

Part I, "The Nature–Nurture Question: New Advances in Behavior-Genetic Research on Intelligence," contains chapters that deal with the behavior genetics of intelligence at the most general level. There are five chapters in this part.

Chapter 1, by Sandra Scarr, is entitled "Behavior-Genetic and

Socialization Theories of Intelligence: Truce and Reconciliation." In her chapter, Scarr argues that socialization theories of intelligence have not stood up to the data, which are better fit by theories emphasizing the role of behavior genetics. In particular, Scarr concludes that (1) theorists believing in socialization should concentrate more on finding surprising relationships; (2) more attention needs to be paid to within-family, nongenetic variation; (3) socialization theorists should pay more attention to genotype–environment correlations, and in particular to how particular genotypes can lead people to create certain environments for themselves; (4) socialization researchers should not be content with saying that intelligence and environment interact, which merely restates what is obvious to all; (5) socialization theorists should take social-psychological research on social perception and constructivism more seriously; and (6) socialization theory needs to be tested scientifically and be freed from the demands of various kinds of advocacy.

Chapter 2, by Arthur R. Jensen, deals with "The Puzzle of Nongenetic Variance." An interesting feature of this chapter is that Jensen, who is most well known for his emphasis on the hereditary aspects of intelligence, here looks at environmental events, most occurring shortly after conception. In the chapter, Jensen concludes that the evolutionary process has ensured that the overwhelming majority of the members of a species will develop in common ways that enable them to adapt to the environment, while at the same time ensuring that there is enough genetic diversity to allow for adaptation to changes in environmental conditions. Jensen further suggests that shortly after conception, a series of very small and largely random events starts to influence neural development, resulting in differences in abilities, even between identical twins. Because the deviations result from very small and independent events, they are normally distributed in the population. In addition to these small and random events, there are also larger and generally negative environmental events that can affect abilities, usually negatively. The latter include, for example, what might be viewed as a critical mass of small negative events that then begin to snowball and lead to even more negative events, and thus lowered abilities.

Chapter 3 is by Robert Plomin and is entitled "Identifying Genes for Cognitive Abilities and Disabilities." In this chapter, Plomin suggests that most of what we know about cognitive abilities and disabilities derives from quantitative genetic research that documents levels of heritability – for example, from studies of twins and adoption. Plomin suggests that we now need to concentrate more on quantitative techniques that can go beyond merely specifying heritability coefficients. For examples, we need to try to track the developmental course of genetic contributions and to identify genetic links among various cognitive abilities. We also need to go beyond quantitative techniques to identify the specific genes that contribute to genetic variance in cognitive abilities. Plomin's own research has made some exciting starts in this direction.

Chapter 4, by John C. Loehlin, Joseph M. Horn, and Lee Willerman, deals with "Heredity, Environment, and IQ in the Texas Adoption Project." The authors conclude that the major contributor to familial resemblance is in the genes. Shared family environment has a considerable effect on IQ but only when children are young. By the time children enter later adolescence, the effect of shared family environment has become quite minor. In general, genetic effects increase with age rather than decrease, as people traditionally have expected. Intriguingly, the researchers found that there is a weak *negative* environmental association between the IQs of mothers and children: Birth mothers' IQs correlate more highly with the IQs of their children if they have no contact with the children since near birth than do the IQs of adoptive parents actually living with their own biological children.

Chapter 5, "IQ Similarity in Twins Reared Apart: Findings and Responses to Critics," is by Thomas J. Bouchard. In his chapter, Bouchard considers the findings of his and others' twin studies and also critiques of these studies. Bouchard concludes that there is no plausible alternative to genetic influence for explaining the similarities in IQ in monozygotic twins reared apart. Bouchard notes that results from these studies are consistent with the results of other kinds of behavior-genetic studies, such as of adult kinships, and also concludes that genetics predominates over environment in the transmission of human intelligence, at least in modern Western societies.

Part II of the book is entitled "Novel Theoretical Perspectives on the Genes and Culture Controversy." The six chapters in this part consider and evaluate some of the assumptions of behavior-genetic research.

Chapter 6 is by Douglas Wahlsten and Gilbert Gottlieb and is entitled "The Invalid Separation of Effects of Nature and Nurture: Lessons from Animal Experimentation." In this chapter, the authors suggest that analyses of the development of abilities need to look at levels of explanation. They suggest as plausible levels culture, society, immediate social and physical environments, anatomy, physiology, hormones, cytoplasm, and genes. The authors also critique some of the work in behavior genetics, suggesting among other things that the adoption method has certain limitations not clearly delineated by its users, that genetic and environmental effects are interdependent, and that different genes do not act in isolation.

Chapter 7, by Thomas R. Bidell and Kurt W. Fischer, is entitled "Between Nature and Nurture: The Role of Human Agency in the Epigenesis of Intelligence." The authors conclude that there has been little consensus in the nature–nurture debate because the framework in which the debate is conducted does not allow for satisfactory answers. In particular, they believe, it is a mistake to partition variance into two mutually exclusive sources – heredity and environment. Bidell and Fischer propose instead a framework in which genetics and environment are viewed not as separate, independent forces or sources of variation but rather as intrinsically related, integrative systems that take part in and are parts of the developing person. Changes in these systems are products of self-organizing activity; as a result, the development of intelligence is a constructive rather than a predetermined process. Genes and environment do not either single or jointly determine cognitive outcomes: Rather, we, as human agents, determine our own cognitive outcomes as we actively make sense of the world and build up the skills needed for participation in the world. Analysis of these skills and of differences in them needs to take place at multiple levels: biological, cognitive-behavioral, and sociocultural.

Chapter 8, by Howard Gardner, Thomas Hatch, and Bruce Torff, presents "A Third Perspective: The Symbol Systems Approach." Gardner, Hatch, and Torff note early on the paradox that as data accumulate, evidence for the role of both nature and nurture accumulates and gains plausibility. This fact suggests the need for a third perspective that integrates what have formerly been seen as incompatible perspectives. The authors introduce such a third perspective – namely, a symbol systems approach – which provides such an integration. They use the domains of musical thinking and spatial reasoning in order to elucidate this approach, which argues that the development of abilities is best understood in terms of the development of symbol systems upon which abilities operate.

Chapter 9, by Joan G. Miller, is entitled "A Cultural Psychology Perspective on Intelligence." Miller argues that theory and research drawn from the perspective of cultural psychology challenges many of the assumptions underlying research in the behavior-genetic tradition. Miller further suggests that is impossible to construct a measure of intelligence that is culture-free. Indeed, the effects of prior knowledge cannot be separated from IQ; furthermore, individuals who are successful in their lives do not necessarily perform well on tests designed to measure intelligence. Ultimately, the development of intelligence is an open and culturally variable process.

Chapter 10, by Stephen J. Ceci, Tina Rosenblum, Eddy de Bruyn, and Donald Y. Lee, presents "A Bio-Ecological Model of Intellectual Development: Moving Beyond h^2." The chapter makes several key points. First, even when heritability is extremely high, as for height, the environment can have a powerful effect on development. Indeed, heights have increased over the past several generations. Second, heritability coefficients can be misleading, because they can only refer to the proportion of actualized genetic variance. We have no way of knowing how much unactualized genetic potential exists. Third, estimates of heritability fail to take into account factors of the environment that can have a substantial effect on intelligence. The authors introduce the notion of *proximal processes*, which are reciprocal interactions between the developing child and other persons, objects, and symbols in the child's immediate setting that can influence intellectual development.

Chapter 11, by Edmund W. Gordon and Melissa P. Lemons, presents "An Interactionist Perspective on the Genesis of Intelligence." In this chapter, Gordon and Lemons propose that many researchers have overestimated the contribution of genetics, independent of environment, to intellectual development. They suggest that the debate between hereditarians and environmentalists has been unproductive and that an interactionist perspective tells us more than either of the other two more extreme perspectives tells us in isolation. They believe that both genetics and environment contribute to all behavior but that questions of percentages are invariably misleading. Overall, intelligence is both adaptive and transformative, and it continually emerges as the development of a person unfolds.

Part III is entitled "Specific Issues in the Nature–Nurture Controversy." The seven chapters in this part deal with specific questions that arise from the debate about the origins of intelligence.

Chapter 12, by Robert J. Sternberg, discusses the topic of "Educating Intelligence: Infusing the Triarchic Theory into School Instruction." In this chapter, Sternberg deals with the question of whether we can simultaneously teach subject matter and develop abilities in a way such that teaching for thinking results in teaching thinking. In particular, Sternberg describes the model and the program developed in a 5-year study to infuse his triarchic theory of human intelligence into the identification, instruction, and assessment of high-school students. Students were identified as being particularly strong in either analytic, creative, or practical abilities or as being balanced among the three. They were then placed in an advanced-placement psychology course that emphasized either analytic instruction, creative instruction, practical instruction, or memory. They were then assessed for analytic, creative, practical, and memorial achievements. Sternberg and his colleagues found that matching students' abilities to the form of instruction they received improved their performance in the course. In other words, conventional teaching may underestimate what students can do, because this teaching typically makes little or no attempt to match abilities to instruction and assessment.

Chapter 13, by H. J. Eysenck and S. J. Schoenthaler, also deals with the development of abilities, in this case, with "Raising IQ Level by Vitamin and Mineral Supplementation." The authors conclude that inadequate levels of vitamins and minerals in the bloodstream reduce children's levels of measured intelligence and that supplementation of diet by vitamins and minerals can raise children's nonverbal IQs. Supplementation appears to affect only fluid intelligence, not crystallized intelligence, and the effects are greater for younger than for older children. The effects are observable only if the children have had inadequate levels of vitamins and minerals and seem to be greater for vitamins than for minerals. The authors suggest that as many as 20% of American children could be helped by supplementation and that the effect would be greater in inner-city than in other environments.

Chapter 14 is by Elena L. Grigorenko and Tatiana V. Kornilova and is entitled "The Resolution of the Nature–Nurture Controversy by Russian Psychology: Culturally Biased or Culturally Specific?" In this chapter, Grigorenko and Kornilova argue that cultural and political biases exist in all of the available resolutions of the nature–nurture controversy. Indeed, the way in which the nature–nurture issue is resolved by a given theory is influenced by the cultural context in which the theory appears. The authors review Russian/Soviet theories and show how the cultural context affected them. But the problem is general to all theories, not just to some of them. In addition, the authors suggest that the mechanisms of cognitive development are highly flexible and that social forces play a major role in their various manifestations.

Chapter 15, by Yrjö Engeström, Ritva Engeström, and Merja Kärkkäinen, deals with "The Emerging Horizontal Dimension of Practical Intelligence: Polycontextuality and Boundary Crossing in Complex Work Activities." In their chapter, the authors focus on the practical dimension of intelligence. They introduce two key theoretical concepts: polycontextuality and boundary crossing. *Polycontextuality* refers to the fact that at the level of activity systems, experts are engaged in multiple simultaneous tasks and task-specific participation frameworks within one and the same activity. They are further involved in multiple communities of practice. Thus, the work of experts cannot be viewed adequately in a framework of expertise that looks only at the individual outside the contexts in which that individual participates. *Boundary crossing* refers to going beyond the boundaries that typically define a single type of expertise and thinking in an innovative way that merges what typically would be seen as the thinking of people with different specializations.

Chapter 16, by Stacey S. Cherny, David W. Fulker, and John K. Hewitt, analyzes "Cognitive Development from Infancy to Middle Childhood." In their chapter, the authors show that the nonshared environment that is uniquely experienced by individuals does not drive cognitive development. Indeed, the influences of this environment are transitory. Interestingly, shared environmental influences do not appear to be driving forces either; rather, the authors believe that genes are directing the developmental process. New genetic variation appears at each age, and the variation as it appears persists into later ages. Thus, consistent with the arguments of Loehlin and his colleagues, the influence of genetics appears to increase rather than decrease with increasing age.

Chapter 17, by J. Steven Reznick, examines "Intelligence, Language, Nature, and Nurture in Young Twins." Reznick uses data from the MacArthur Longitudinal Twin Study for twins assessed at 14, 20, and 24 months of age to address two questions about language and intelligence in the second year of life: First, what is the relation of language to intelligence? Second, are the effects of nature and nurture comparable for expressive and receptive language? Reznick finds that there is a relation between language and intellectual development for both receptive and expressive language and, moreover, that this relation increases during the second year.

Chapter 18, by Sandra Pipp-Siegel, JoAnn L. Robinson, Dana Bridges, and Sheridan Bartholomew, looks at "Sources of Individual Differences in Infant Social Cognition: Cognitive and Affective Aspects of Self and Other." The authors conclude that twins are faced with different developmental challenges from those that are encountered by nontwins. As a result, the relation between measures of self–other differentiation and other abilities (such as cognitive and language abilities) may be different for twins versus nontwins. As children grow older, the nature of their self-concept becomes more detailed and also more abstract. These chances may be reflected in twins' relations with each other. In particular, when twins are in grade school, they interact very heavily with each other, but as they enter adolescence, they start separating more and more, becoming more independent and even more likely to enter into conflict with each other.

The fourth and final part of this book, "Integration and Conclusions," contains two chapters, the concluding papers of Earl Hunt and Irwin D. Waldman. They pick up where we leave off in this preface, and so it is at this point that we state our hope that our readers will enjoy and profit from this book and come away with a better appreciation of the diversity of points of view found in people who study intelligence, heredity, and environment.

Part I

The nature–nurture question: New advances in behavior-genetic research on intelligence

1 Behavior-Genetic and Socialization theories of intelligence: Truce and reconciliation

Sandra Scarr

Theories should compete. It keeps them fit and trim. Left unchallenged, theories, like people, grow fat and lazy, and they eventually decay into shapeless blobs. Unchallenged theories petrify as "common wisdom," immobilizing critical faculties with intellectually paralyzing assumptions. A challenged theory is a creative network of conscious and considered ideas, casting its nomological net over new observations and predicting where to look for the next catch.

Psychological theories are rarely made to compete, but they can, and they should, be exercised in this way. Competing theories about determinants of intellectual differences offer a challenging opportunity to test theoretically required predictions with available observations. Predictions about crucial research results are generated by Socialization Theory and by Behavior-Genetic Theory, and those predictions are quite different. The critical observations to test the adequacy of predictions from these competing theories have been made. Yet, there has been little, direct theoretical confrontation. That the theories have not been tested with existing data in a systematic fashion is either a sign of mutual ignorance or an aversion to being challenged.

This chapter presents Socialization and Behavior-Genetic theories of intelligence and evaluates the adequacy of the theories' predictions to account for existing observations. The adequacy of the theories to generate new research is also assessed. By posing a stark contrast between theories, I hope to sharpen the debate and to modify developmental theory to fit existing observations about intellectual resemblance in families and to be more productive in future research.

Predictions from both Socialization and Behavior-Genetic theories can be partially correct, with one faring better than the other, or, less likely, one may be the overall winner by excluding the other. The test is whether each theory generates predictions that fit the available observations, and whether each can generate specific hypotheses that can be tested with additional research.

3

Most observations about determinants of intellectual differences can be fit to both Socialization and Behavior-Genetic theories and, thus, do not provide a critical test between them. Studies in the tradition of Socialization Theory typically sample one child per family and include only biologically related families, thus confounding genetic with social transmission of behavioral traits from parents to children (Scarr, 1992, 1993). Studies of biological families *only*, even if they include siblings, are not theoretically informative, because the observations of modest similarities among family members and correlations of family environments with children's intellectual differences can be fit as well, or better, to other theories. Other observations, mostly from behavior-genetic studies, do provide tests of different predictions that must be made by the two theories.

Definition of terms

Four definitions are essential at the outset for the terms *observation, intelligence, socialization,* and *behavior-genetic*.

Observations are scientifically credible (reliable, valid) measurements of phenomena that are the subject matter of a theory. Under the term *observation*, I include all kinds of measurements, such as tests, ratings, self-reports, and behavioral observations. Both Socialization and Behavior-Genetic Theory focus on processes that transmit behaviors from one generation to the next, so that observations of family influences are crucial to both theories. What is observed are co-occurrences or correlations between one event and another.[1] Observations are shaped into facts by theories, by making sense of the observation in terms of the theory (Scarr, 1985). Many observations can be made into different facts by different theories.

Intelligence. For the purposes of this chapter, intelligence is defined as scores on cognitive tests, including standard intelligence tests and factors from tests of specific cognitive abilities. Most often, intelligence will mean *general* intelligence, or *g*. It is assumed that intelligence develops through learning culturally valued knowledge and skills in a human, social environment (Scarr, 1993).

Western, especially U.S., ideas about intelligence refer to the individual learner, who acquires knowledge and skills incidentally from universal schooling, and from interactions with others. European psychologists refer more often to social supports for children's learning (e.g., Vygotsky), but it is the individual who becomes more or less intelligent. Bright children acquire learning sets, or executive functions, that guide their knowledge acquisition. Less bright children must be taught segmented information and skills that bright children acquire spontaneously.

Socialization Theory. For the past 50 years, thousands of articles and books have been written in psychology about the effects of parents as socializing agents on their children (see Bandura, 1977; Baumrind, 1993). In the context of Social Learning Theory, socialization researchers stress the many ways that parental behaviors directly affect their children's development, through modeling, social reinforcement, and interactions that promote or discourage learning. Socialization Theory is at root a description of parental behaviors and their association with child outcomes.

In this chapter, parental influences will include the many characteristics of schools, neighborhoods, and communities that parents expose children to as part of the correlated family environment.

Socialization appeals primarily to environmental differences *between families* – differences in the environments parents provide, including the parents themselves. Differences between families are hypothesized to make siblings alike (because they are reared by the same parents in the same household) and different from children reared in other families. Social disadvantage, a frequently evoked concept, is clearly a difference between low income and/or minority families and more affluent and/or majority-group families. Social class is a between-family difference, as are all family characteristics that siblings share but that differ among families.

Parental rearing styles would seem to be primarily between-family differences, because siblings share the same parents. Socialization research has rarely included siblings, so that characterizations of parental rearing styles have been based on parent reports and behaviors with one child per family and, thus, are de facto between-family differences. It is not clear whether parents' basic style is expected to vary much from one child to another. The theory is moot on this issue. Increasingly, socialization researchers have written about within-family environmental differences (e.g., Hoffman, 1991), but their studies have not included siblings or twins until very recent collaborations with behavior geneticists (e.g., Plomin et al., 1993; Plomin, Reiss, Hetherington, & Howe, 1994).

Behavior-Genetic Theory, derived from evolutionary theory, is focused on causes of individual variation in intelligence and other characteristics in populations (see Plomin, DeFries, & McClearn, 1990). Behavior genetics is not the study of species invariants or of gene action pathways that affect behaviors; rather, behavior genetics studies sources of variation in populations. It makes predictions about parental influences on children from both genetic and environmental (both biological and social) transmission (see Rowe, 1994). Environmental transmission includes both socialization differences between families and other, often unknown environmental differences between siblings in the same family. Behavior-genetic studies aim to sort out which influences have which effects in what populations.

In truth, generalizations from both socialization and behavior-genetic research are limited to the populations, measures, and methods sampled in studies. This limitation applies to all behavioral research, although it is often ignored in socialization studies.

Behavior-genetic explanations appeal to four major sources of variation to explain individual differences in intelligence.

1. Additive genetic effects refers to the combined effect of many genes, each contributing a small amount to phenotypic diversity. It is likely that human intelligence relies on many gene loci, as more than 100 genes have been identified, where abnormal variants can interfere with normal intellectual development. Roughly half of the genetic variation in a population occurs within families (except for identical twins, first-degree relatives share about half of their genes), and half occurs between families. An assortative mating correlation between spouses of about .3 for intelligence slightly reduces genetic variation within families and increases it between families.

2. Nonadditive genetic effects include dominance effects (where the heterozygote phenotypically resembles one of the homozygotes) and other major gene effects, genotype–environment correlations, genotype–environment interactions, and epistasis [the nonadditive effects of certain gene combinations, which Lykken, McGue, Tellegen, and Bouchard (1992) called *emergenesis*, because sometimes one individual in a family appears with an unusual characteristic, such as extraordinary musical talent, which is not shared with other family members; see Jensen, this volume]. Genotype–environment correlations are ubiquitous, nonrandom associations between one's personal characteristics and one's environment (Scarr, 1992; Scarr & McCartney, 1983). Gene–environment interactions occur when one genotype develops better in one environment, whereas a second genotype develops better in a different environment.

3. Between-family nongenetic effects make siblings more similar than children reared in different families. Nongenetic effects are usually thought of as psychosocial environmental differences between one family versus another. Social class and parental differences in rearing style can be considered between-family differences. But such differences may be prenatal and biological as well (see Jensen, chapter 2, this volume). Between-family environmental differences can be thought of as differences in *opportunities*. Children from lower social classes are said to have fewer opportunities to develop higher intelligence than those from higher social class families, not only in the home itself but in the correlated schools, neighborhoods, and communities in which they live.

4. Within-family nongenetic variance refers to those aspects of the environment, biological and social, that make siblings in the same family different from one another and that make genetically identical twins reared in the same home less than identical in all respects [although they are nearly as similar in intelligence as the same person tested twice (the limitation of test–retest reliability)]. Within-family, nongenetic differences include prenatal and biological environmental, as well as psychosocial environmental, events that affect one sibling in a different way from another (see Jensen, chapter 2, this volume).

Behavior genetic theory includes these four sources of potential intellectual variation. To make the contrast with Socialization Theory sharp and clear, I will stress the role of *genetic transmission* to account for individual differences in intelligence.

Why do individuals vary in intelligence?

The observation of differences in intelligence has many explanations, only a few of which we refer to in any one discussion. For example, if we ask why one person scores higher on an intelligence test than another, psychologists are unlikely to think first of differences in serotonin levels, or room temperature, or what each person had for breakfast, although in some circumstances all these could cause a difference in test performance. We are more likely to look for previous learning opportunities and motivation to perform or at how intelligent their parents are. Hesslow (1987) noted that when a house catches fire, we do not normally refer to the presence of oxygen as the cause of choice, although we would all accept the idea that oxygen plays a role in fires. Rather, we would refer to a lighted match as the cause of the fire.

How do we choose causes? The answer to this question will explain a lot about differences between Socialization and Behavior-Genetic theories. There must be a connection between the event and the putative cause, but a single cause is selected from an almost infinite variety of possible causes. Primarily, the selection of causes depends on the question asked; for example, if I ask why Down's syndrome children have lower than average IQ scores, I am not likely to refer to their diet, although I know that adequate nutrition plays some role in mental development. I am more likely to appeal to the chromosomal cause of their retardation. Theoretical and practical preferences also play important roles in causal selection, so that an educational researcher might focus on unmet stimulation needs of Down's syndrome children, whereas a genetic counselor might search for familial causes.

John Stuart Mill was the first philosopher to recognize that conditions which causally determine an event usually far outnumber the conditions mentioned in a causal explanation. Since the 18th century, many philosophers have pondered causal selection (Hesslow, 1987). There are six major reasons for selecting one cause over others.

1. *Unexpected or abnormal conditions.* Some conditions are already known and need not be mentioned (such as the presence of oxygen in a fire or adequate nutrition in retarded children). We do not need explanation of things that behave according to expectations. We ask "Why?" when something unexpected happens. Abnormal conditions that seem to make a difference between normal or expected functioning and the abnormal conditions that brought the event to our notice are prime candidates for causal selection. Lighted matches in a fire and a chromosomal defect in Down's Syndrome fall into this category.

2. *Precipitating causes.* conditions that immediately precede an event are more often selected than permanent or background causes, which may indeed be causes that have more impact than the precipitating events, as I will later show.

3. *Deviation from a theoretical ideal.* Ideals of how things are and how they work guide causal selection and promote as causes conditions that deviate from the ideal. When we observe children in economically deprived circumstances not doing well on IQ tests, we conclude that poverty must be the cause of poor test performance, because poverty deviates from our theoretical ideal of child-rearing circumstances. When we find a child with a background of poverty doing well on IQ tests and in school, we do not attribute good performance to poverty; rather, we look for other causes, such as resilience, that fit our theory about poverty. This causal selection can also be called *responsibility*: Conditions that deviate from what is good or right can be selected as causes of bad events, because on moral grounds they should not be allowed to occur. Poverty is associated with many poor social outcomes. Is poverty the cause?

4. *Predictive value.* We prefer those causes that would enable us to predict events. These are causes that are necessary, and perhaps sufficient, for the event to occur.

5. *Manipulability.* We have strong preferences in psychology to identify as causes those conditions that will allow us to bring about a desired goal or prevent an undesirable one. Hesslow (1987) proposes that predictability and manipulability are most often selected in science – predictability, because in science we want to control events, and manipulability (in more applied sciences such as medicine and clinical psychology), because fixing bad outcomes is highly valued.

6. *Theoretical or practical interest.* We also tend to select as causes those conditions that we know more about or know how to fix or deal with. With child intellectual failure, pediatricians see medical causes, psychologists see family or peer problems, and educators see poor instruction. All, some, or none may be causes of the event to be explained.

Hesslow (1987) proposes two principles to explain causal selection: the *difference* between objects with respect to a certain property, and *explanatory relevance*. The example he offered illustrates these principles very well. Causes selected to explain different wing lengths of fruit flies, shown in Figure 1.1, vary depending upon the relevance of the cause to the observed difference.

There is always an implied, if not explicit, comparison in any causal explanation, and the appropriateness of the choice depends upon the comparison. In Figure 1.1, mutant and normal flies are reared under three temperature conditions. The mutant has shorter wings under conditions of lower temperatures. If we ask the question "Why does M1 have short wings?", the causal explanation will depend *entirely* on which fly is the relevant comparison. If M1 is compared to N1, then genetic difference is the only possible explanation, as they were reared in the same temperature. If M1 is compared to M3 or M2, the explanation is entirely the temperature differences, as they have the same genotypes. If M1 is

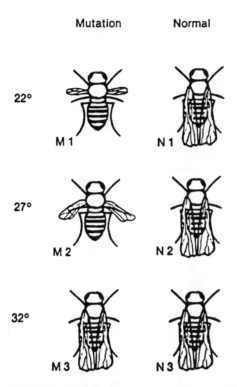

Figure 1.1. Wing length in mutant and normal flies reared in three temperatures (Hesslow, 1987).

Table 1.1. *Causes of variation in reading skills for two groups of children in three instructional conditions*

	Reading Skills	
Reading Instruction	Group A	Group B
1. Excellent	High	Low
2. Mediocre	Moderate	Low
3. Poor	Low	Low

compared to N3, the cause of the differences must be both genetic and environmental.[2]

An example from the educational domain can be invented with similar parameters (see Table 1.1). If we ask why children in Group A have high reading skills, the answer depends on which other group of children is chosen for comparison. If we compare Group A1 to B1 children in the same instructional conditions, the explanation has to be differences between personal characteristics (such as genetic differences) of the children. If the comparison is between Groups A1 and A2, similar children with excellent versus mediocre instruction, the explanation must lie in instructional differences.

Usually, both personal and instructional differences, or genetic and environmental differences, are involved. Complex causation requires causal parsing in order to determine *quantitatively* how much effect is caused by each of several factors. The theoretical and methodological abilities to parse causation is the major challenge of Behavior-Genetic Theory to Socialization Theory.

Explanatory power versus manipulability

I want to add another criterion for causal selection to Hesslow's criteria of difference and explanatory relevance. *We should prefer as explanations those conditions that account for more of the variation in the phenomenon to be explained, causes that have more **explanatory power**, even if they are not presently manipulable.*[3] In only the most applied, practical sciences is manipulability the most appropriate cause to be preferred above others. From a theoretical perspective, more important causes are those that give a more complete account of, or explain more about, the phenomenon. This proposal is intended to make human developmental science a more thoroughly scientific and multidisciplinary enterprise rather than being a front for advocacy of humane causes and social justice.

By contrast, Baumrind (1993) advocates causal selection based on manipulability. Her tone is that of a moral crusade against identifying and referring to causes that are not presently manipulable.

For psychologists, as for medical researchers, the purpose of identifying undesirable predispositions of individuals *should be* to devise more effective health-promoting interventions, not to discourage such attempts on the supposition that these predispositions are *genetically based and therefore intractable*. . . . Given limited resources, many socialization researchers choose to focus on those factors that are most susceptible to change – on how parents and educators can provide optimum environments for optimum development or optimum human behaviors [p. 1313, emphasis added].

Socialization researchers appeal almost exclusively to causes that can potentially, if not actually, be manipulated. Parents *should* be taught better parenting; teachers *should* become better instructors; children *should* be given safer and healthier environments in which to grow up. I applaud social advocacy and, as a citizen, participate in such movements. As a scientist, however, my role is to explain human development. In that context, causes that explain the phenomena more fully, even if they are not presently manipulable, are preferred. *Confusion between the legitimate but different goals of science and social advocacy is dangerous to both* (Marshall, 1992; Ernhart, Scarr, & Geneson, 1993; Scarr, in press).

Competing theories

Important aspects of Socialization and Behavior-Genetic theories can be contrasted and tested. Figure 1.2 shows statements of general theory, middle-level theory, and some specific predictions derived from theory. At the general theoretical level, Socialization Theory states that parental child-rearing styles or techniques are powerful determinants of children's intellectual development. Brighter children have parents who treat them in specific ways that are not common to parents who produce less bright children. At a middle theoretical level, more specific hypotheses are generated. To illustrate, Figure 1.2 shows two theoretical statements about socialization effects: (1) that parenting practices can be characterized as authoritative, authoritarian, permissive, and uninvolved; and (2) that specific child-rearing practices and proximal interactions produce more and less intelligent children.

From these middle-level statements, specific predictions can be made about what will be observed in family studies. By family studies, I do not mean the study or ordinary biological families only, because, taken alone, they are not at all informative about the putative effects of socialization practices (see Rowe, 1994; Scarr, 1993). A requirement for theory-relevant observations is that they must allow a critical test of the theory; to wit, if socialization practices are critical to children's intellectual development, then genetic transmission of intelligence from parent to child must be controlled or varied in order to rule out the major competing hypotheses from Behavior-Genetic Theory.

Thus, the specific predictions that differentiate Socialization Theory from Behavior-Genetic Theory are that (1) children reared by different parents, in different homes, will not resemble each other, even if they are genetically related, unless their parents use similar rearing practices, (2) genetic relatives reared in different homes will not be similar in intelligence unless their parents use similar rearing practices, and (3) authoritarian rearing style causes children to have low intelligence and poor academic achievement, regardless of parents' intelligence.

Behavior-Genetic Theory, at a general level, states that genetic differences are a major source of intellectual differences among children, regardless of parental rearing styles, unless parents are abusive or seriously neglectful (Scarr, 1992, 1993). At a middle level, the theory proposes that in societies such as Western democracies, where opportunities for learning

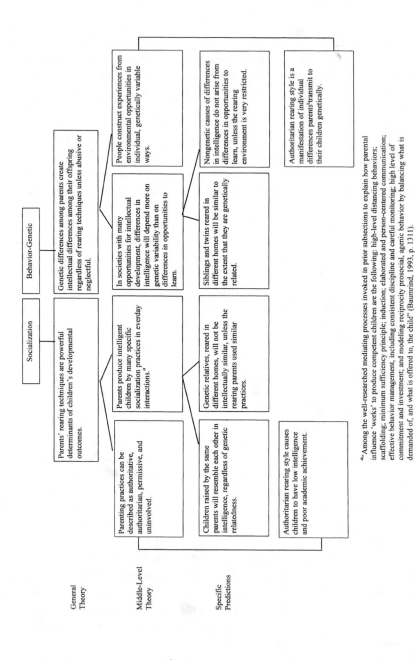

General
Theory

Middle-Level
Theory

Specific
Predictions

Socialization

Parents' rearing techniques are powerful determinants of children's developmental outcomes.

Parents produce intelligent children by many specific socialization practices in everday interactions.[a]

Parenting practices can be described as authoritative, authoritarian, permissive, and uninvolved.

Genetic relatives, reared in different homes, will not be intellectually similar, unless the rearing parents used similar practices.

Children raised by the same parents will resemble each other in intelligence, regardless of genetic relatedness.

Authoritarian rearing style causes children to have low intelligence and poor academic achievement.

Behavior-Genetic

Genetic differences among parents create intellectual differences among their offspring regardless of rearing techniques unless abusive or neglectful.

In societies with many opportunities for intellectual development, differences in intelligence will depend more on genetic variability than on differences in opportunities to learn.

People construct experiences from environmental opportunities in individual, genetically variable ways.

Siblings and twins reared in different homes will be similar to the extent that they are genetically related.

Nongenetic causes of differences in intelligence do not arise from differences in opportunities to learn, unless the rearing environment is very restricted.

Authoritarian rearing style is a manifestation of individual differences parents transmit to their children genetically.

[a]"Among the well-researched mediating processes invoked in prior subsections to explain how parental influence 'works' to produce competent children are the following: high-level distancing behaviors; scaffolding; minimum sufficiency principle; induction; elaborated and person-centered communication; effective behavior management, including consistent discipline and careful monitoring; high level of commitment and investment; and modeling reciprocity prosocial, agentic behavior by balancing what is demanded of, and what is offered to, the child" (Baumrind, 1993, p. 1311).

Figure 1.2. Competing predictions from socialization and behavior-genetic theories.

intellectual skills and knowledge are many and varied, individual difference in intelligence will result more from genetic differences than from differences in opportunities. A second middle-level proposition is that given a variety of opportunities, people construct their own experiences (Scarr & McCartney, 1983; Scarr & Weinberg, 1983), and those constructions are to a large extend expressions of their individual personalities, interests, and talents.

Specific predictions from Behavior-Genetic theory focus on observations of intellectual resemblance among family members. Intellectual similarity is predicted to the extent that parents and children, and siblings, are genetically related, with little or no similarity in intelligence to people genetically unrelated though reared in the same homes. Parental rearing style is predicted to be unimportant as a determinant of intelligence; rather, authoritarian rearing style is predicted to be a manifestation of parental intelligence and personality that are transmitted to children, in part, genetically. Observations of families with different degrees of genetic and environmental relatedness provide critical tests of Socialization and Behavior-Genetic theories by testing specific predictions made from the theories.

Competing models

To make entirely explicit the tests of Socialization and Behavior-Genetic theories that are to be made, Figure 1.3 shows models derived from the two theories. Models are shown for biologically related families, adoptive families, and identical (MZ) twins reared by different families. Three kinds of relationships are shown in the models: a solid line with one arrow for a causal path, a solid line with two arrows for a correlation that exists in nature apart from other elements in the model, and a dashed line with two arrows to indicate correlations created by the causal relations of other elements in the model.

First, let us look at the Socialization and Behavior-Genetic models for biological families. In the socialization model, parental rearing style and parental intelligence are correlated (in nature). Although socialization researchers seldom mention that authoritative parents are more intelligent than authoritarian ones, there are two reasons this must be so. First, authoritarianism is substantially, negatively correlated with IQ (−.5 in most studies). Second, to explain the observed correlation between parent and child intelligence in hundreds of studies, the only potential path is through parental rearing styles. Thus, socialization theory must predict that the parent–child IQ correlation is produced by the correlation between paren-

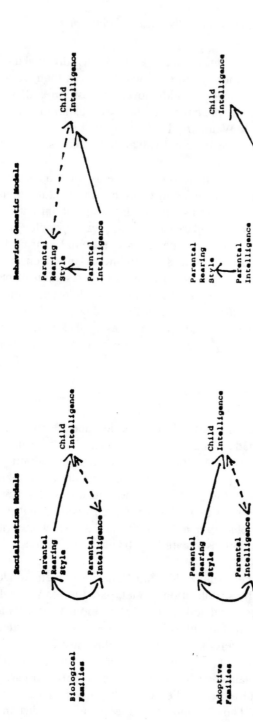

Figure 1.3. Competing models of parental effects on children's intelligence.

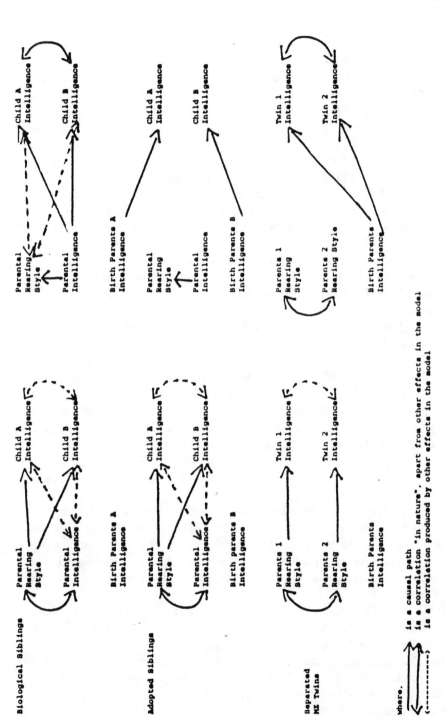

Figure 1.3. *Continued*

tal intelligence and rearing styles. In the Socialization model, parental rearing styles have the causal effect on child intelligence.

Although Behavior-Genetic Theory allows for both environmental and genetic transmission of intelligence from parents to children, the following models stress genetic transmission to maximize their contrast with those derived from Socialization Theory. In the Behavior-Genetic model for biological families, the causal paths are from parental to child intelligence and from parental intelligence to parental rearing styles. In this model, parental rearing styles are correlated with child intelligence as a product of the two causal paths and thus indicated by a dashed line with two arrows. In Behavior-Genetic Theory, parents of different levels of intelligence are predicted to use different rearing styles and predicted to have children of different intelligence levels, on average, so that rearing styles are *associated with* but do not cause differences in children's intelligence.

Note that observations of the relationships among parental rearing styles, parental intelligence, and children's intelligence *in biological families* cannot distinguish between Socialization and Behavior-Genetic models. Although the models are causally different, they specify the same observed correlations. Data that fit one will fit both.

Thus, let us turn to competing models for adoptive families. Here, based on what is observed, we can choose between the Socialization and Behavior-Genetic models. Despite occasional ad hoc muttering about adoptive families being weird (Baumrind, 1993), socialization researchers must predict the same observed correlations in adoptive as in biological families, because theirs is a social learning theory, whose effects are not dependent on genetic relatedness. Birth parents' intelligence should have no effect on adopted children's intelligence when there is no social contact.

By contrast, the predicted observations about genetic transmission of parent–child effects from Behavior-Genetic Theory are entirely different for adoptive and biological families. Neither adoptive parental intelligence nor rearing styles are predicted to affect adopted children's intelligence, but birth parent intelligence must be predicted to affect the rank order of adopted children's IQ scores.

Similarly, biological sibling studies alone are not informative for theory testing, because the two theories predict the same observed correlations, albeit for different theoretical reasons. Adopted sibling studies are informative, because the two theories predict different observations. Again, the Socialization Theory model for biological and adopted siblings must be the same, and adopted children's birth parents' intelligence must be irrelevant in predicting the adopted child's intelligence.

Behavior-Genetic Theory must predict a genetic effect for birth parents' intelligence on their offspring, even though the child was adopted away. The model does predict substantial effects for parental intelligence, because the adoptive parents are not genetically related to the child, or for adoptive parents' rearing style, which is only a manifestation of other parental characteristics.

An additional test of the competing models is shown for identical (MZ) twins reared in different adoptive homes. Here, for simplicity, adoptive parental intelligence has been left out of the model, because neither theory would predict any causal effect on child intelligence. Socialization Theory predicts that the rearing style of each set of parents has an effect on the child they rear, and there is no effect for birth parents' intelligence. To the extent that the rearing families are similar in rearing style, the twins will resemble each other in intelligence; if their rearing styles are not correlated, the twins will not have similar intelligence.

Predictions from Behavior-Genetic Theory are quite different: Birth parents' intelligence is predicted to have an effect on the twins' intelligence, and the twins' genetic correlation is predicted to make their IQs similar, despite being reared in different homes. Parental rearing styles are predicted to have no causal effect on the twins' intelligence.

In science, theories must subsume observations and remain internally consistent; that is, assumptions and predictions cannot contradict each other. A theory cannot predict all possible outcomes, or it is not testable, nor can it be constantly altered post hoc to accommodate unexpected observations that contradict its predictions. Socialization theory has been particularly guilty of post hoc revisionism, as shown in Hoffman's (1991) apologia for the observed lack of similarity among children reared in the same home, and in Baumrind's (1993) and Jackson's (1993) attempts to explain away incompatible family results. The predictions from the long history of socialization research are clear: Parents have important effects on their children through proximal interactions with them.

Observations from Socialization studies

Observations to be explained

Parental rearing styles, observed and reported by family members, have often been found to correlate with child outcomes, including intelligence and school achievement. Two recent studies serve as exemplars of the genre: one by the Patterson research group (DeBaryshe, Patterson, & Capaldi; 1993) and one from the Steinberg research group (Steinberg, Lanborn, Dornbusch, & Darling, 1992). I examine these well-designed studies in detail, because their problems with causal inference, in the face of

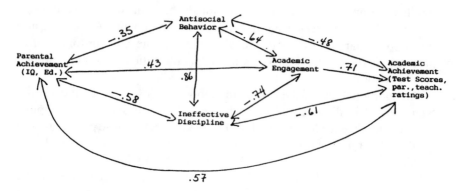

Figure 1.4. Estimated correlations among latent variables (DeBaryshe, Patterson, & Capaldi, 1993, Table 5).

their own, competently conducted, causal analyses, is acute, but no more so than problems in any other study inspired by Socialization Theory.

Patterson study

This study predicted academic achievement among 206 youths at risk for conduct disorder with parental rearing styles, parental IQ and education, and earlier measurements of the youths' own behavior problems and academic achievement. The authors proposed seven latent variable models to be tested for best fit. The report made clear that the preferred a priori model, which focused on parental discipline and conduct disorder, was not the best model fit to predict academic achievement differences among these boys. In these respects, and in its careful measurements, the study was exemplary.[4] Figure 1.4 shows estimated correlations among the latent variables in their models.

Correlations. In Figure 1.4, correlations among the latent variables in their best-fit model are given.[5] *These are the observations to be modeled*, so inspecting and understanding them is important. Note first the large correlations among some of the variables. Antisocial behavior was correlated .86 with ineffective parental discipline, which implies that they are very closely related constructs. Of the 10 correlations in this study, 7 exceed .50, and 9 > .40, implying excellent measurement of the manifest variables that are summarized in these latent variables. How should they be modeled to make theoretical sense of these observations?

Authors' models. Figure 1.5a shows the authors' hypothesized model, for which coefficients were not provided, but which was not a good fit to the observed correlations (Figure 1.4). Figure 1.5b is the best of the seven

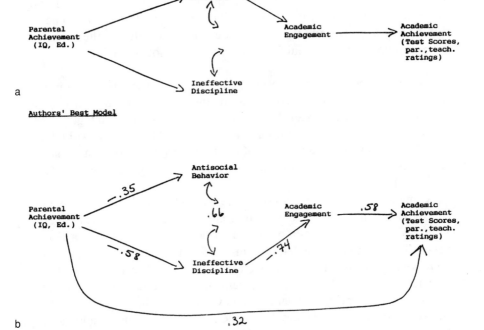

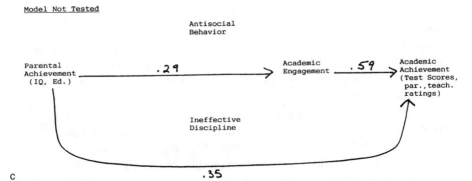

Figure 1.5. Tests of competing models of academic achievement (DeBaryshe, Patterson, & Capaldi, 1993).

models they tried, whose fits they tested competitively with chi-square, but seven does not exhaust the number of possible models for these observations, of course. *Their best-fit model was the only one that included a direct path from parents' IQ and education to youths' academic achievements.* The other predictive path in the best-fit model was from youths' academic engagement (effort at school) to their own achievement. Even this model was cluttered with (1) antisocial behavior, which did not contribute directly or indirectly to the prediction of academic achievement, and (2) ineffective discipline, whose indirect effect was through academic engagement. In no model did they drop both antisocial behavior and ineffective discipline to test for a best fit. Despite the authors' best modeling efforts, the evidence for causal effects from socialization practices was not compelling.

Models not tested.   More importantly, no direct path was tested from parental IQ and education to youths' academic engagement in their best-fit model or in any other model. The not-tested model is shown in Figure 1.5c. Based on the correlations shown in Figure 1.4, a still better fit was obtained with the very simple causal model of the transmission of parental achievements (IQ, Education) to adolescents' Academic Engagement and Academic Achievement.

At my request, the first author of the study, Deborah DeBaryshe, graciously calculated the fit of this alternative model and several others.[6] Whereas the authors' best-fit model had a chi-square of 89.45 and a Goodness of Fit Index (GFI) of .94, my simple parental achievement model has a chi-square of only 18.26 and a GFI of .97. One direct effect and one indirect effect (via Academic Engagement) of parental intelligence and education on offspring Academic Achievement are shown. One could, of course, go farther and predict Academic Achievement and Academic Engagement simply from Parental Achievement, with a correlation between the two adolescent variables; this would imply that smart, achieving parents have smart, achieving offspring. What else is new?

By eliminating parental discipline and conduct disorder from my more parsimonious model, I have removed the core Socialization Theory variables that were supposed to explain the correlations between parental and child characteristics (in these biological families).

A dispassionate reading of the Patterson group's report reveals that they tried very hard, with seven different models, to avoid the inference that higher-IQ parents have higher-achieving offspring, even among youths at risk for conduct disorder. The observation they ultimately could not avoid, and they included *in just one of the seven models*, was that the highest correlate of youths' academic test scores was parental IQ ($r = .424$).

Steinberg study

In a study of 6,400 adolescents, reports were obtained from the youths about their parents' child-rearing styles (authoritative or not), parents' school encouragement and involvement, and the adolescents' school achievement in 2 successive years. Socialization Theory predicted that parental Authoritativeness would cause parents to be more encouraging and more involved with their adolescents' schooling, which would cause higher academic achievement. Socialization Theory drove the data collection (although it is remarkable that the adolescents were trusted to supply all of the information) and the data analyses. Keep in mind that the research report was entitled, "Impact of parenting practices on adolescent achievement."

Reporting of correlations and explicit causal modeling were exemplary in this study as well. (Such good practices permit competing hypotheses to be tested and public discourse to further the aims of science.) First let us look at the one model they presented, as shown in Figure 1.6a; no competing models were tested.

The autocorrelations of School Performance and School Engagement across 2 years suggest considerable stability at adolescence. The core theoretical idea concerns the impact of parenting practices on adolescent achievement, for which the evidence is slim indeed. Even though the adolescent participants themselves supplied the information about parenting practices, the only reliable path coefficient that leads from parenting practices to adolescent achievement is the .073 impact of parent school involvement! The best parenting measure accounted for 0.53% of the variance in school achievement, which was reliably different from 0.00% of the variance because $N = 6,400$. Parental School Engagement had no impact on adolescents' School Engagement or on School Performance; parental School Involvement had no impact on adolescents School Engagement.

Observations to be modeled. In the second, more complete, model, shown in Figure 1.6b, I included other correlations, which the authors provided in the text and in a table,[7] to give a more complete picture of the data. Double-headed arrows signify correlations. *The correlations are the **observations** to be explained by a theoretical model.* Beginning our inspection of the observations in Figure 1.6b at the far left, we can see that the correlation between adolescent-reported parental authoritativeness and contemporaneous (1987) school performance was .27 and between parental authoritativeness and 1987 school engagement was .23, with parental authoritativeness explaining about 4–6% of the variation, before controlling for any other influences on 1987 school performance and engagement. The 1988 scores on the same school performance and engagement variables were to be controlled longitudinally for 1987 school performance and engagement (autocorrelations), so that one would surely have predicted that the already small coefficients at the right of this model between 1987 parental practices

Authors' Model (Figure 1)

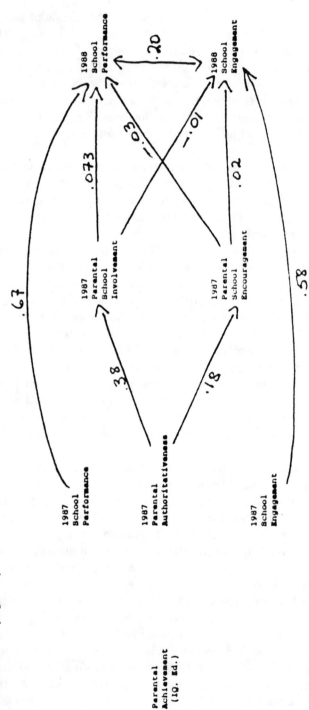

Figure 1.6. Over-time impact of parental authoritativeness on adolescent school performance and school engagement, mediated by parental involvement in school and parental encouragement of school success (Steinberg, Lamborn, Dornbusch, & Darling, 1992).

a

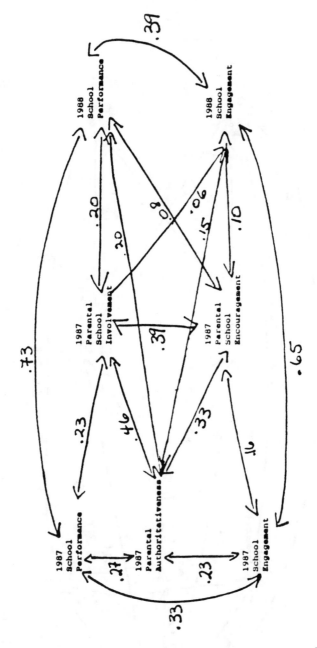

Figure 1.6. *Continued*

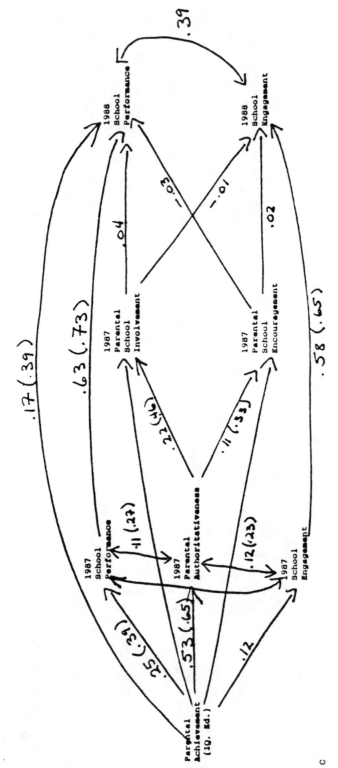

Figure 1.6. Continued

and 1988 adolescent school performance and engagement would decrease considerably over time, once autocorrelations were in the model.

Indeed, that is what happened, if one compares the first and the second models (Figure 1.6a and b). Although the observed correlation between 1987 parental school involvement and 1988 school performance was .20, the partial path coefficient is only .07, because 1987 school performance is a powerful predictor of the next year's school performance. Other longitudinal predictions from parental practices to adolescent outcomes were similarly reduced.

Direct and mediated effects.   It is noteworthy that the authors' model has no direct path between parental authoritativeness and school performance or engagement, yet the correlations were given. The 1987 parental authoritativeness had only modest correlations with contemporaneous school performance (.27) and school engagement (.23). Predictably, their explanatory power over time diminished, with the addition of the autocorrelations to the model, so that the observed relationships between 1987 parental authoritativeness and 1988 school performance and engagement were .20 and .06, respectively. A direct-effects model of parental authoritativeness could not explain more than <1–4% of the variation in school outcomes, once the stability of adolescents' behaviors was controlled.

Instead of a direct-effects model, the authors chose a mediated model, as stated in the title to their figure: "Over-time impact of parental authoritativeness on adolescent school performance and school engagement, mediated by parental involvement in school and parental encouragement of school success" (Steinberg et al., 1992, Figure 1). How promising were the mediated effects of parental authoritativeness? Adolescent reports of their parents' 1987 authoritativeness were correlated .33 and .46 with 1987 reported parental school encouragement and involvement, respectively. But the correlations of reported parental school encouragement in 1987 with 1987 adolescent outcomes were trivial (.16 for engagement, .05 for performance), and the correlations of parental school involvement only a bit higher (both .23). Thus, there was evidence from the observed correlations of contemporaneous measures that a mediated causal model would not explain much of the variation in adolescent school performance and engagement.

Indeed, the observed correlations of 1987 reported parental school involvement and encouragement with 1988 adolescent school performance and engagement ranged from .06 to .20, and were reduced to −.03 to .07 with covariates in the model. If parental authoritativeness was to be shown to have any effect on adolescent school outcomes, it was not likely to be mediated through these parent practices.

Models not tested.   I have a more serious objection to the study's theoretical frame: Parents who (are reported by their teenagers to) use authoritative rearing styles are *different* from those who use other styles, and their children are also different, for genetic reasons. Behind parental authoritativeness or authoritarianism stand IQ and education, especially the former (Bolger & Scarr, in press; Scarr, 1981; Scarr & Weinberg, 1978). High authoritarianism is moderately associated with low IQ ($r = -.50$) and with punitive discipline (Scarr, 1985, $r = .52$, controlling for education). Look back at Figure 1.5, with correlations from the Patterson study: The correlation between parental IQ/Education and Ineffective (nonauthoritative) Discipline is $-.58$.

Behind differences in adolescent school performance stand parental IQ and education, especially the former, as Patterson's research showed (adolescent School Achievement with parental education, $r = .27$; with parental IQ, $r = .42$) and as behavior-genetic research has confirmed many times (e.g., Scarr & Yee, 1980; Thompson, Detterman, & Plomin, 1993). Our research on *adoptive* and biologically related families showed that the links between parental intellectual achievements and adolescents' school achievements are almost entirely due to genetic transmission of intelligence (Scarr & Weinberg, in press; Scarr & Yee, 1980).

Variables added to Steinberg's model, based on correlations provided in their report and on our family research, present quite a different picture from that which Socialization Theory would predict. This model is shown in Figure 1.6c. In this third model, I included known relationships between parent practices and parental IQ and educational levels (Parental Achievements) from our research on educational and occupational achievements of adopted and biological adolescents and young adults (Scarr & Weinberg, 1978, in press). I used the biological family correlations.

On the left of Figure 1.5c are estimated path coefficients between Parental Authoritativeness and the 1987 adolescent and parental measures, based on their correlations, and Parental IQ and Education, as they related to parental rearing styles and to Adolescent School Performance in our studies. Correlations are given in parentheses. I will not belabor the model comparisons nor claim that the coefficients are correct to the third decimal place. They are not, as I only iterated the coefficients twice by hand. The major point is that if one includes measures of parental intelligence and social class (achievements), one can predict a lot more about *both* adolescent academic achievements and engagements *and* parent rearing styles. The *lack* of association between parenting practices and adolescent academic achievements that was so evident in the original article, despite its title, is amplified by a more critical theoretical review.

Observations from behavior-genetic studies

Table 1.2 provides a summary of family correlations on IQ tests and the first Principal Component from batteries of tests of cognitive abilities for adolescents and adults; the results are equivalent, so I will not dwell on possible distinctions. These data have been summarized and discussed so often in textbooks, review articles, and other books that it is not necessary to repeat here what can be found easily elsewhere (e.g., Bouchard, Lykken, Tellegen, & McGue, in press; Plomin, 1990; Plomin & Daniels, 1987; Scarr, 1992, 1993).

The picture is clear: Relatives resemble each other intellectually to the extent that they are genetically related, and genetic relatives reared in different homes are somewhat less similar intellectually than those reared in the same home. From adolescence to old age, identical twins reared in different homes have IQ correlations of .74 to .78 based on five studies, separated in time by 60 years, which sampled Europe and North America and included twins reared in working-class to professional-class homes. Analyses have corrected for small effects of correlated rearing in homes of the co-twins (see Bouchard, Lykken, McGue, Segal, & Tellegen, 1990; Thompson et al., 1993). The correlation of MZ twins reared apart *is* a direct estimate of heritability; from these studies, the nongenetic portion of IQ variance (including error) is about 24%, but there is no common environmental effect in these studies. By subtracting the MZA correlation from that of MZ twins reared together (.86), the estimate for the effects of shared environment is 10%.

In contrast to the MZA studies, four studies of genetically unrelated siblings adopted and reared from infancy to adulthood in the same homes show sibling IQ correlations that range from .05 to −.03, with a median of about 0.0 (Loehlin, chapter 4, this volume; Loehlin, Horn, & Willerman, 1989; Kent, 1985; Scarr & Weinberg, 1978; Teasdale & Owen, 1982).[8] The correlation of unrelated siblings *is* a direct estimate of the effects-shared environments (see Loehlin, chapter 4, this volume). Adopted adolescent and adult siblings, reared together since early infancy, are no more similar in intelligence than randomly chosen people in the same Euro-American population. Comparisons of biologically related to unrelated siblings reared in the same homes also yield heritability estimates of .70 or above (Loehlin, chapter 4, this volume; Scarr & Weinberg, 1978, in press). Although estimates of shared nongenetic effects on individual differences in IQ from twin and adoption studies vary from 0 to 10%, one can say that they appear to be rather small in adult populations in Western societies.

There is evidence in North American and Western European

Table 1.2. *Intelligence test correlations of siblings from behavior-genetic studies of biological and adoptive families and twins (adolescents and adults)*

Genetic r	Relationship	Same Home?	IQ Correlation	Number of Pairs
1.00	Same person, tested twice	yes	.90	—
1.00	Identical twins	yes	.86	4,672
1.00	Identical twins	no	.76	158
0.50	Fraternal twins	yes	.55	8,600
0.50	Fraternal twins	no	.35	112
0.50	Biological siblings	yes	.47	26,473
0.50	Biological siblings	no	.24	203
0.00	Adopted siblings	yes	.02	385

Sources: Loehlin, Horn, & Willerman (1989), Bouchard, Lykken, Tellegen, & McGue (in press); Pedersen, Plomin, Nesselroade, & McClearn (1992); Scarr & Weinberg (1978, in press); Teasdale & Owen (1985).

populations for quite high heritability of IQ past childhood (about 70%), with small effects of differences between families, which would include most parental socialization differences. The unexpected finding is that most of the environmental variation is found in unique, individual experiences that siblings do not share (that is, environments that make siblings dissimilar). Analyses of these family data in simple heritability and in multivariate models can be found in many sources (e.g., Bouchard et al., in press; Loehlin, chapter 4, this volume; Plomin, 1990; Scarr, 1992) and will not be repeated here.

Socialization Theory, focused on parental rearing practices, which presumably vary mostly *between* families, accounts for no more than 10% of the variation in intelligence in these populations. As Jensen (chapter 2, this volume) argues, even these "environmental" effects may not be the psychosocial variables that psychologists primarily think of when they consider sources of nongenetic variance. Effects of shared environments are larger in infancy and early childhood, as shown in twin studies (McCartney, Harris, & Bernieri, 1990) and in studies of adoptive families (Loehlin, chapter 4, this volume; Scarr & Weinberg, 1978), but the effects of rearing environments seem to decline across development, as predicted by our theory (Scarr & McCartney, 1983).

Social class

A favorite index of differences among families is social class. Social scientists have consistently claimed that social class differences in intelligence result from difference in opportunities to acquire the knowledge and skills valued by the majority culture. Behavior-genetic research offers strong evidence that differences in intelligence by social class arise, in large part, from genetic variability and not from differences in opportunities to learn, *in families that are not abusive, neglectful or culturally different*. Figures 1.7 and 1.8 show the IQ distribution of parents and children in biologically related and adoptive families of adolescents and young adults. By averaging the IQ test scores of mothers and fathers, a better estimate of the intellectual climate of the home is obtained in both kinds of families, as well as a better estimate of the genetic background of the adolescents and young adults in the case of biological families. By averaging the IQ scores of two young people, we get a more reliable estimate of the offspring value for each family.

First, notice that the range of midparent and midchild scores is restricted, with none having average IQ scores below 90 or above 135. Despite this attenuation, the regression coefficient for biological families is .76, and for adoptive families, .13, even with selective placement that increases intellectual resemblance in adoptive families Second, note that in adoptive families the lack of effect of parental intelligence on adopted children's intelligence means that some quite bright parents have low-IQ adopted children, and some low–average-IQ parents have quite bright adopted children. Home environments provided by low–average-IQ adoptive parents are nearly as advantageous for intellectual development as those provided by superior IQ adoptive parents.

By contrast, in the biological families, there is a close association between the average of children's IQ scores and those of their parents. These are the usual relationships observed when only biological families are studied, and the inference is usually made that brighter parents provide a more stimulating intellectual environment and, thus, have brighter children. That this cannot be the explanation for differences in intelligence among the adolescents and young adults is seen in the adoptive family graph, where a similar range of IQ differences among the parents produced little effect.

It must be that genetic differences among higher- and lower-IQ parents are transmitted to biological offspring, a fact corroborated by the similar correlation of birth parents' educational levels to their biological offspring's IQ scores among adopted children. Similar graphs can be shown for the regression of biological and adopted children on their parents' educational and occupational levels. The Texas Adoption Project (Loehlin, chapter 4, this volume) has reported similar results.

Thus, the unpalatable fact, generated by Behavior-Genetic Theory from the observation of parent–child IQ correlations in adoptive and biologically related families, is that in any one generation adults in different social-class

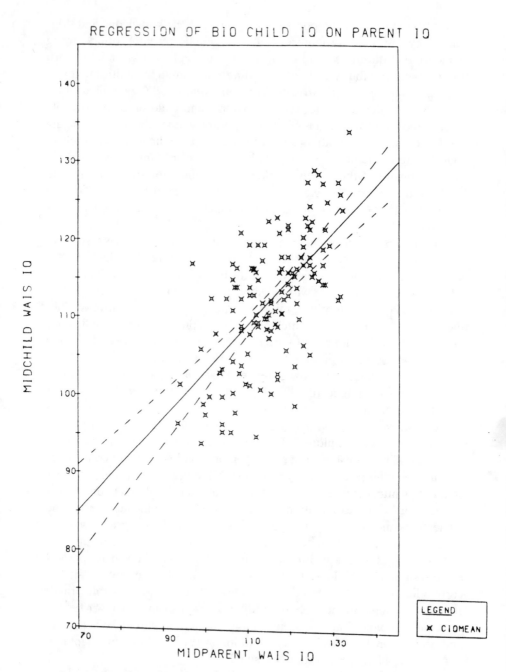

Figure 1.7. Regression of midchild on midparent IQ scores for adolescents in biologically related families (Scarr & Weinberg, 1978).

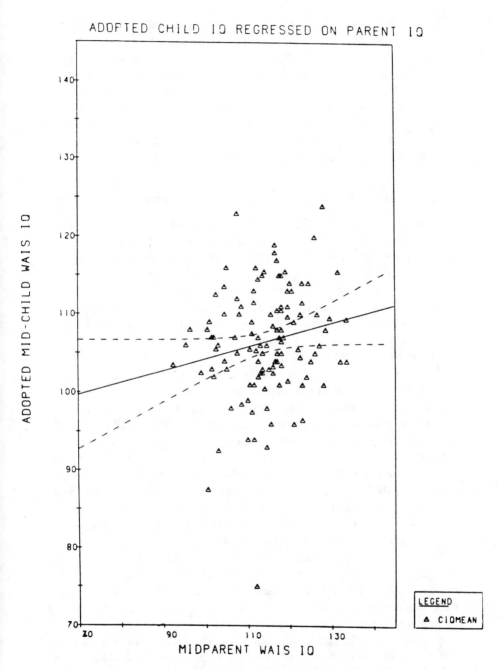

Figure 1.8. Regression of midchild on midparent IQ scores for adolescents in adoptive families (Scarr & Weinberg, 1978).

groups are not genetically equal, nor do children from different social class groups, on average, have equal intellectual potential. Intergenerational mobility, based in part on intellectual and educational achievements, guarantees that some children from lower social-class groups will rise in status and some from higher groups will decline in status from their parents' levels. Thus, in each generation, intellectual inequalities among social-class groups are reestablished. Because between-family environmental effects have been shown over and over again to be quite small, social-class differences must be largely genetic and to some extent heritable.

Genetic effects on intelligence: The larger political context

One cannot ignore the political context for presenting and interpreting research on genetic and environmental variability in intelligence. As has been said so many times, *Science is not value-free*, and it operates in a context of disputes about moral/ethical issues of distributive justice and a just society. How unequal should people be in the power and resources they have? Intelligence in pertinent to these disputes because it causes differences directly in educational achievements, directly and indirectly in occupational achievements, and indirectly in income (Scarr & Weinberg, in press; Taubman, 1976).

Behavioral scientists of all theoretical persuasions observe that intelligence is closely linked to educational success, which in turn is associated with eventual occupational status, which bears some relationship to earnings. Achievements and socioeconomic rewards are associated with intelligence, which makes some behavioral scientists very uncomfortable with the concept of intelligence itself. The observation of links between intelligence and socioeconomic achievements is not disputed, but the link is ascribed to societal injustice and unequal opportunities for people who come from advantaged versus disadvantaged families. The ubiquitous correlation between IQ and social class is ascribed to social-class differences in opportunities afforded to children during development.

The idea that differences in intelligence are associated with genetic differences is especially distressing to many behavioral scientists, because it implies necessarily that social-class differences in children's eventual achievements are also caused, in part, by genetic differences among parents at different social-class levels (Herrnstein, 1973; Teasdale & Owen, 1986). How can children have an equal opportunity to achieve if some are better endowed genetically with intellectual potential? It has been most convenient to deny the possibility and to excoriate those who dared to suggest it.

John Rawls (1971) dealt explicitly with the need to compensate for genetic inequalities as well as environmental ones. He argued that distributive justice depends on recognizing individual differences in talents that cause differences in social and economic achievements, and to compensate for them in ways that create a more just society than would occur through unchecked market forces. I agree that this may be the most equitable way to compensate for intellectual differences by detaching, in part, social and economic rewards from the fruits of intellectual differences. Of course, we cannot make considered judgments about these issues until we are willing to face the observation of genetic variability in intelligence and deal with our values about how much socioeconomic inequality is acceptable in a just society. Then, and only then, can we draw a rational plan for compensatory social justice (Scarr, 1994).

Values in the scientific community

As with most members of democratic societies, behavioral scientists believe that individual differences in intelligence and achievements should be under the control of individual effort and personal will. It is all right that some people are more intelligent and achieve more because they work harder than others, but it is not acceptable that some people are more favorably endowed. Any genetic hypothesis about group differences in behavior is anathema: social-class, race, and gender differences in behavior *must* result from environmental differences. The results of behavior-genetic studies are seen as contrary to the values of many social scientists, so these scientific results are buried, distorted, and ignored.

Science and social justice are not incompatible, just incommensurate. They are different realms of discourse, with different values, assumptions, and methods. Those who speak the language of advocacy for social justice often do not understand those who speak science, believing them to be socially insensitive and unsympathetic to the need for social change. Science is indeed imbued with values from its own historical times and cultural places (Scarr, 1985), but its methods that are open to critical review and replication make it fundamentally different from advocacy. Respect for the principles and rules of science is essential if Socialization Theory is to be a serious research enterprise.

Future research directions: Truce and reconcilliation

Hypothesis testing is the hallmark of science. Theory-driven research into human intellectual variation cannot be primarily experimental. Thus, we must look for quasi-experimental situations in which to test competing hypotheses. This is exactly what Behavior-Genetic Theory directs us to:

Quasi-experiments that vary both genetic and environmental relatedness in families. By adopting the research designs of behavior genetics, socialization investigators can test and refine their theories in ways that are not now possible with their antiquated, confounded designs.

Genetics creeps into Socialization studies

After a very critical review of Socialization Theory's principles and predictions, and the observations they have produced, it may seem unlikely that a truce and reconciliation can be arranged with the Behavior-Genetic forces. On the contrary, I believe that more and more socialization researchers recognize (1) the ubiquitous influence of individual differences in intelligence and personality on the socialization variables that they study, and (2) the need to take genetic variability into account.

The need to incorporate environmental measures in behavior-genetic studies has been recognized (e.g., Plomin & McClearn, in press; Scarr, 1993), but the need to include genetically informative designs in socialization studies is less often mentioned. The omission of genetically informative families in socialization studies has produced only uninterpretable results. With only these data, there is no way to test competing theories about sources of individual differences, as was shown in Figure 1.3. Moreover, it is not at all clear how to measure family environments without including genetic differences among parents and offspring (Plomin & Neiderhiser, 1992). Whatever modest degree of similarity in intelligence, personality, and interests exists among biological family members is not primarily due to growing up or living in the same household (Hoffman, 1991; Loehlin, 1992; Plomin & Deniels, 1987; Scarr & Grajek, 1982; Scarr, Webber, Weinberg, & Witting, 1981).

Socialization Theory has been moot on possible effects of genetic variability on intellectual development, and it ignores the confounding of genetic and environmental variability reported for the most common measures of family environment (Chipuer, Plomin, Pedersen, McClearn, & Nesselroade, 1993; Plomin & Bergeman, 1991; Plomin & Daniels, 1987; Plomin, Loehlin, & DeFries, 1985; Plomin, McClearn, Pedersen, Nesselroade, & Bergeman, 1989a,b; Rowe, 1981, 1983, 1994). Because family members report *perceptions* of the family environment, and because individual differences in social perception are partly genetic, measures of family environments, such as the *Family Environment Scales* (Moos & Moos, 1981) and the *HOME* scales (Caldwell & Bradley, 1984), are partly heritable. In addition, *observations* of family interactions yield moderate to high heritabilities for parent–child and sibling relationships (Hetherington, Reiss, & Plomin, 1994; Plomin, 1994).

Parenting child-rearing styles, such as Baumrind's *authoritative, authoritarian, permissive,* and *uninvolved* classification are certainly genetically variable. Parental styles *are,* after all, parental behaviors, for which there is a great deal of evidence for heritability. Parent behaviors reflect intellectual and other personal differences among people, and the proportion of genetic variance in those characteristics has been repeatedly found to be 50–70%, with much of the remaining nongenetic variance due to individual experiences. The new collaboration with behavior genetics can help to revise Socialization Theory to accommodate previously incompatible observations about effect of family environments.

An important unresolved issue is the nature of nonshared environments: Are there systematic effects that can be studied scientifically, or are the effects due to individually idiosyncratic, random events that differ unsystematically for each person? Scientists wish to find systematic nonshared environmental effects in the social environment (see Hetherington et al., 1994) or in the biological environment (Jensen, chapter 2, this volume). It may be, however, that truly random environmental events affect developmental outcomes (Bandura, 1982), in which case there cannot be a general theory about them (McCall, 1993). If one person is devastated by the death of a pet and another relieved at the loss of responsibility (Gottesman, 1990), and most people are unaffected because they lack pets, it is difficult to formulate a scientific theory about the infinite variety of such events. The study of nonshared environments is so new that it is not possible to conclude whether or not a systematic scientific theory of nonshared environments will account for much of the nonshared, nongenetic variance in intelligence.

One source of nonshared environment could be *differential* parental treatment of two children in the same family. That is, parental rearing styles may vary systematically *within* families and thereby account for some nonshared environmental influences. This promising lead has proved a dead end so far: In two recent studies, differential parental discipline was not found to be associated with differential parental perceptions of their children's personalities or behavior problems (Deater-Deckard, 1994). Although differential discipline and differential perceptions of siblings were measured reliably, and parents agreed in their differential perceptions of their children (.40–.50) and agreed about differential discipline (.30–.40), there were no systematic effects of differential perceptions of siblings' characteristics (except age) on parents' treatment of them. In the next decade there will be many more such studies of within-family environmental differences. Whether systematic effects will be found, or whether the Heisenberg Principle applies within the walls of the home, nonshared environments will attract a great deal of research attention.

Suggestions for revisions in Socialization theory

1. Socialization scientists should look for incongruent or surprising relationships, because they are often the most informative for theory testing and can have important implications for intervention efforts (Scarr, 1989). For example, if a vocabulary score explains differences in parental discipline practices as well or better than parental education does (Pinkerton & Scarr, under review; Scarr, 1985), one might infer that brighter parents both had more education and used more appropriate discipline techniques. It is not plausible to infer that educational differences mean simply *knowledge* differences about child rearing (see Scarr, 1985, 1989). The inference that parent-education programs can easily remedy bad parenting by providing information alone is questionable; rather, parent education programs can improve bad parenting to the extent that they work around intellectual differences among parents who use different practices naturally. Parent-education programs that ignore or deny the connection between intelligence and parenting practices are likely to be ineffectual in changing parenting behaviors.

2. There must be more attention focused on *within-family nongenetic variation* (Jensen, chapter 2, this volume; Rowe, 1994). Current Socialization Theory *describes* the interplay between parents and children; it does not predict or explain the low co-variation between parental and child behaviors. Within-family sources of nongenetic variation are mysterious; there are no good theories about them. Perhaps they are random small events that initiate a developmental trajectory with its own pathway and inertia. Perhaps, they are idiosyncratic convergences of personal characteristics and exposures to environments that amplify original developmental trends. Perhaps, the process of emergenesis (Lykken et al., 1992) is more common than previously thought. What is needed is thoughtful approaches to describing the phenomenon and testing competing hypotheses about, in addition to genetic differences, how siblings grow up to be so different.

3. Socialization Theory should look at genotype–(or person–)environment correlations, which describe the evocative and active roles that people play in creating their own environments (Plomin, DeFries, & Loehlin, 1977; Scarr, 1992, 1993; Scarr & McCartney, 1983). The environment is not primarily a random set of opportunities, of advantages and disadvantages, that happen to just anyone. The environment does not have the same meaning to all people.

4. Socialization researchers should abandon an empty *interactionism* (which is better termed *transactionism*) that merely states the obvious: Organisms have environments with which they transact and in which they

develop. There is no profound theoretical insight here, because there is no plausible, alternative model. Organisms do not develop without environments, and environments have no effects on organisms that are not there.

Specifying the exact *nature* of transactions between organisms and environments can be theoretically productive, and many European developmentalists test hypotheses about the active roles of persons in perceiving their environments and in shaping their own experiences (e.g., Magnusson & Torestad, 1992). Biological differences among persons in intelligence, personality, motivation, and emotion are hypothesized to result in different experiences. Thus, they can test for the effects of persons on environments (transactions), of environments on persons (transactions), and the differential effects of person–environment combinations (true interactions).

If more socialization researchers would take a biosocial approach to transactions, they would stop being so partial to $g \times e$ interactions over g–e correlations (e.g., Baumrind, 1993). $G \times e$ interactions are nonlinear effects of different environments on different genotypes. To espouse interactions, over other processes, is to deny main effects of genes and of environments, to deny additive effects of linear scales from better to worse, or more to less facilitative. Although I gave two illustrations of $g \times e$ interactions earlier in the chapter (flies and children), these are experimental designs where the environments to which the organism is exposed are arbitrary with respect to their own characteristics. The everyday world for most people consists of choices about what to listen to and look at, what to ignore, where to be and with whom. People expose themselves differentially to opportunities for experiences. These are g–e correlations, not $g \times e$ interactions.

5. Socialization Theory (not to mention contemporary philosophy) should take the large social-psychological research literature on social perception and constructivism more seriously. Whatever can be observed and measured about objective aspects of the environment is given different meanings by different people in developmentally changing and individually different ways (Scarr, 1992, 1993). More important than objectifying environments is to know how individuals construct their experiences from those environments to which they are exposed, or more often expose themselves, and to understand how they integrate what they encounter with what they are.

6. Socialization Theory Revised deserves to be tested scientifically. To do this, it needs to be free of advocacy demands. Socialization researchers, as with other citizens, can be advocates for what they believe to be right and just; some of what they advocate may result from scientific studies of social problems. But their science must meet the criteria of scientific discourse. At

present, Socialization Theory does not meet those tests, but it can be revised to do so.

Notes

1. Experimental studies of the core processes posited by both theories are not possible in human populations. Laboratory simulations can be experimentally arranged, but the phenomena studied are trivial reflections of the powerful processes posited by the theories to occur in everyday life. Parents have pervasive, longitudinal effects on child development that cannot be manipulated easily in short-term, artificial experiments on social learning. Nor can artificial selection experiments be used to test genetic hypotheses.
2. In fact, this example is a gene–environment interaction, where the mutant's wing length depends on the temperature, whereas the normal genotype develops normal wing length under all three temperature conditions.
3. What is manipulable changes with scientific invention, of course. In the near future, gene therapy promises to fix many conditions that have heretofore been considered intractable or amenable to only symptomatic relief, such as sickle cell anaemia, cystic fibrosis, and metabolic disorders, many of which affect brain development and intelligence.
4. In addition, the first author, Deborah DeBaryshe, has commented on the first draft of this chapter and provided additional analyses. The authors' scientific conduct is also exemplary.
5. DeBaryshe provided correlations among the latent variables based on a confirmatory factor analysis; she agrees that they are very similar to the published ones, so I have not changed them here.
6. DeBaryshe ran two other models with the same paths as my three-latent-variable model, with and without a path from ineffective discipline to academic engagement but including the correlations between discipline and antisocial behavior. Neither model that includes the socialization variables and paths from parental achievement to adolescent engagement and achievement fits as well as the authors' best model or as my three-variable model.
7. I estimated correlations from those provided for subgroups of the sample in their Table 5. Averaging was done by conversion of r to Fisher's z and weighing by sample size.
8. Results from the Minnesota Transracial Adoption Study (Scarr, Weinberg, & Waldman, 1992) show the predicted decline in IQ correlations of genetically unrelated siblings from an average age of 7.5 years to 18.5 years (.31 to .19), but the black and interracial adoptees' IQ correlation in late adolescence was statistically greater than 0.0. IQ correlations of biological siblings in the same families did not decline; they were .48 and .44 in childhood and late adolescence, respectively. Although heritability estimates for the transracial adoptees' IQ scores were about .50 at both assessments, the effect of shared environment still accounted for 19% of the IQ variance in late adolescence.

References

Bandura, A. (1977). *Social learning theory*. Englewood Cliffs, NJ: Prentice-Hall.

Bandura, A. (1982). The psychology of chance encounters and life paths. *American Psychologist, 37*, 747–55.

Baumrind, D. (1993). The average expectable environment is not good enough: A response to Scarr. *Child Development, 64*, 1299–317.

Bolger, K. E., & Scarr, S. (in press). Not so far from home: How family characteristics predict child care quality. *Early Development and Parenting*.

Bouchard, T. J., Jr., Lykken, D. T., Tellegen, A., & McGue, M. (in press). Genes, drives, environment and experience: EPD theory – revised. In C. Benbow & D. Lubinski (Eds.), *From psychometrics to giftedness: Essays in honor of Julian Stanley*. Baltimore: Johns

Hopkins University Press.

Bouchard, T. J., Jr., Lykken, D. T., McGue, M., Segal, N. L., & Tellgen, A. (1990). Sources of human psychological difference: The Minnesota study of twins reared apart. *Science, 250*, 223–28.

Caldwell, B. M., & Bradley, R. H. (1984). *Home observation for the measurement of the environment.* Little Rock, AK: University of Arkansas Press.

Chipuer, H. M., Plomin, R., Pedersen, N. L., McClearn, G. E., & Nesselroade, J. R. (1993). Genetic influence on family environment: The role of personality. *Developmental Psychology, 29*, 110–18.

Deater-Deckard, K. (1994). *Differential parental discipline and differential parental perception of siblings' personality and behavior problems.* Unpublished doctoral disertation, University of Virginia, Charlottesville.

DeBaryshe, B. D., Patterson, G. R., & Capaldi, D. M. (1993). A performance model for academic achievement in early adolescent boys. *Developmental Psychology, 29*, 795–804.

Ernhart, C. B., Scarr, S., & Geneson, D. F. (1993). On being a whistleblower: The Needleman case. *Ethics and Behavior, 3*, 73–93.

Gottesman, I. I. (1990). Personal communication.

Herrnstein, R. (1973). *IQ in the meritocracy.* Boston: Atlantic Monthly Press.

Hesslow, G. (1987). The problem of causal selection. In D. Hilton (Ed.), *Contemporary science and natural explanation: Commonsense conceptions of causality.* New York: Harvester Press.

Hetherington, E. M., Reiss, D., & Plomin, R. (Eds.) (1994). *Separate social worlds of siblings.* Hillsdale, NJ: Erlbaum.

Hoffman, L. W. (1991). The influence of family environments on personality: Accounding for sibling differences. *Psychological Bulletin, 110*, 187–203.

Jackson, J. F. (1993). Human behavioral genetics, Scarr's theory, and her views on interventions: A critical review and commentary on their implications for African American children. *Child Development, 64*, 1318–32.

Loehlin, J. C. (1992). *Genetics and personality.* Thousand Oaks, CA: Sage.

Loehlin, J. C., Horn, J. M., & Willerman, L. (1989). Modeling IQ change: Evidence from the Texas Adoption Project. *Child Development, 60*, 993–1004.

Lykken, D. T., McGue, M., Tellegen, A., & Bouchard, T. J., Jr. (1992). Emergenesis. *American Psychologist, 47*, 1565–77.

Magnusson, D., & Torestad, B. (1992). The individual as an interactive agent with the environment. In W. B. Walsh, K. H. Craik, & R. H. Price (Eds.), *Person–environment psychology: Models and perspectives.* Hillsdale, NJ: Erlbaum.

Marshall, E. (1992). When does intellectual passion become conflict of interest? *Science, 257*, 620–1.

McCall, R. B. (1993). Environment effects on intelligence: The forgotten realm of discontinuous nonshared within-family effects. *Child Development, 54*, 408–15.

McCartney, K., Harris, M. J., & Bernieri, F. (1990). Growing up and growing apart: A development meta-analysis of twin studies. *Psycholgical Bulletin, 107*, 226–37.

Moos, R. H., & Moos, B. S. (1981). *Family Environment Scales Manual.* Palo Alto, CA: Consulting Psychologists Press.

Pedersen, N. L., Plomin, R., Nesselroade, J. R., & McClearn, G. E. (1992). A quantitative genetic analysis of cognitive abilities during the second half of the lifespan. *Psychological Science, 3*, 346–53.

Plomin, R. (1990). The role of inheritance in behavior, *Science, 248*, 183–8.

Plomin, R. (1994). *Genetics and experience: The interplay between nature and nurture.* Thousand Oaks, CA: Sage.

Plomin, R., & Daniels, D. (1987). Why are children in the same family so different from one another? *Behavioral and Brain Sciences, 10*, 1–60.

Plomin, R., DeFries, J. C., & McClearn, G. E. (1990). *Behavioral genetics: A primer.* New

York: Freeman.

Plomin, R., Emde, R. N., Braungart, J. M., Campos, J., Corley, R., Fulker, D. W., Kagan J., Reznick, J. S., Robinson, J., Zahn-Waxler, C., & DeFries, J. C. (1993). Genetic change and continuity from fourteen to twenty months: The MacArthur Longitudinal Twin Study. *Child Development, 64*, 1354–76.

Plomin, R., Loehlin, J. C., & DeFries, J. C. (1985). Genetic and environmental components of "environmental" influences. *Developmental Psychology, 21*, 391–402.

Plomin, R., & McClearn, G. E. (Eds.) (in press). *Nature, nurture, and psychology*. Washington, DC: American Psychological Association Press.

Plomin, R., McClearn, G. E., Pedersen, N. L., Nesselroade, J. R., & Bergeman, C. S. (1989a). Genetic influence on childhood family environment perceived retrospectively from the last half of the lifespan. *Developmental Psychology, 24*, 738–45.

Plomin, R., McClearn, G. E., Pedersen, N. L., Nesselroade, J. R., & Bergeman, C. S. (1989b). Genetic influences on adults' ratings of their current environment. *Journal of Marriage and the Family, 51*, 791–803.

Plomin, R., & Neiderhiser, J. M. (1992). Genetics and experience. *Current Directions in Psychological Science, 1*, 160–4.

Plomin, R., Reiss, D., Hetherington, E. M., & Howe, G. W. (1994). Nature and nurture: Genetic contributions to measures of the family environment. *Developmental Psychology, 30*, 32–43.

Rawls, J. (1971). *Inequalities and social justice*. Cambridge, MA: Harvard University Press.

Rowe, D. C. (1981). Environmental and genetic influences on dimensions of perceived parenting: A twin study. *Developmental Psychology, 17*, 203–8.

Rowe, D. C. (1983). A biometrical analysis of perceptions of family environment: A study of twin and singleton sibling kinships. *Child Development, 54*, 416–23.

Rowe, D. (1994). *The myth of family influences*. New York: Guilford.

Scarr, S. (1981). *Race, social class and individual differences in IQ: New studies of old issues*. Hillsdale, NJ: Erlbaum.

Scarr, S. (1985). Constructing psychology: Making facts and fables for our times. *American Psychologist, 40*, 499–512.

Scarr, S. (1989). Protecting general intelligence: Constructs and consequences for interventions. In R. L. Linn (Ed.), *Intelligence: Measurement, theory, and public policy*. Urbana: University of Illinois Press.

Scarr, S. (1992). Developmental theories for the 1990's: Development and individual differences: *Child Development, 63*, 1–19.

Scarr, S. (1993). Biological and cultural diversity: The legacy of Darwin for development. *Child Development, 64*, 1333–53.

Scarr, S. (1994, March 1). *Individuality and community: The moral role of the State in family life*. Paper presented at the Conference on Child Welfare Policy, Ministry of Children and Families, Oslo, Norway.

Scarr, S. (in press). Psychology in the public arena: Four cases of dubious influence. *Scandinavian Journal of Psychology*.

Scarr, S., & Grajek, S. (1982). Similarities and differences among siblings. In M. E. Lamb & B. Sutton-Smith (Eds.), *Sibling relationships*. Hillsdale, NJ: Erlbaum.

Scarr, S., & McCartney, K. (1983). How people make their own environments: A theory of genotype–environment effects. *Child Development, 54*, 424–35.

Scarr, S., & Yee, D. (1980). Heritability and educational policy: Genetic and environmental effects on IQ, aptitude, and achievement. *Educational Psychologist, 15*, 1–22.

Scarr, S., & Weinberg, R. A. (1978). The influence of "family background" on intellectual attainment. *American Sociological Review, 43*, 674–92.

Scarr, S., & Weinberg, R. A. (1983). The Minnesota adoption studies: Genetic differences and malleability. *Child Development, 54*, 260–7.

Scarr, S., & Weinberg, R. A. (1994). Educational and occupational achievements of brothers and sisters in adoptive and biologically related families. *Behavior Genetics 24*(4), 301–25.

Scarr, S., Weinberg, R. A., & Waldman, I. D. (1992). IQ correlations in transracial adoptive families. *Intelligence, 17*, 541–55.

Steinberg, L., Lanborn, S. D., Dornbusch, S. M., & Darling, N. (1992). Impact of parenting practices on adolescent achievement: Authoritative parenting, school involvement, and encouragement to succeed. *Child Development, 63*, 1266–81.

Taubman, P. (1976). The determinants of earnings: Genetics, family, and other environment; a study of white, male twins. *American Economic Review, 66*, 858–70.

Teasdale, T. W., & Owen, D. R. (1985). Heredity and familial environment in intelligence and educational level – a sibling study. *Nature, 309*, 620–2.

Teasdale, T.W., & Owen, D. R. (1986). The influence of paternal social class on intelligence and educational level in male adoptees and non-adoptees. *British Journal of Educational Psychology, 56*, 3–12.

Thompson, L. A., Detterman, D. K., & Plomin, R. (1993). Cognitive abilities and scholastic achievement: Genetic overlap but environmental differences. *Psychological Science, 3*, 158–65.

Arthur R. Jensen

Within the past decade, empirical findings in behavior genetics have importantly changed how researchers in this field think heredity and environment affect individual differences in mental ability. These insights are hardly new. Some are even found in the writings of Sir Francis Galton (1822–1911), the father of behavior genetics. But the recognition, formalization, and empirical support given to them by behavior geneticists in recent years can be considered significant advances. The most surprising findings, only conjectured by Galton, concern the role of environment in the development of mental ability. The present picture is quite different from the beliefs generally held only a decade ago.

Genotype–environment covariance

One such idea is that the perceptible environment is like a cafeteria. People make different selections according to their genetic makeup, or *genotype*. The environment is not a "given" but is largely the person's own creation. This becomes increasingly true as persons develop from infancy to maturity.

Behavioral differences between persons that result from their self-selected and self-fashioned environments are the phenotypic expression of *genotype–environment covariance*.[1] It accounts for more of the total variance (i.e., individual differences) in abilities and achievements than was formerly thought. Genotype–environment (GE) covariance is neither a strictly genetic nor a strictly environmental component of phenotypic variance but reflects the genetically driven differential selection of experiences from the available environment. It also includes the effects of differential treatment by parents, teachers, and peers, because their responses are largely evoked by the person's distinctive genotypic characteristics.

Environmental forces peculiarly accommodate people's genotypic propensities. People seek out different environments, including friends and activities, that are congenial to their nature. The wider the variety of genotypes in a population, and the more varied the environment, the larger is the GE covariance component of the total phenotypic variance. The familiar phrase "nature *and* nurture" is now replaced by "nurture *via* nature." This is not just a subtle distinction; it proves theoretically crucial for understand-

42

ing some essential data of behavior genetics. Researchers Robert Plomin and Sandra Scarr have most prominently furthered this idea (Plomin, 1986; Plomin & Bergeman, 1991; Plomin, DeFries, & Loehlin, 1977; Plomin & Neiderhiser, 1992; Scarr, 1985; Scarr & Carter-Saltzman, 1982; Scarr & McCartney, 1983; also see Bouchard & Segal, 1985; Rowe, 1987; Rowe & Plomin, 1981; Willerman, 1979).

Epistasis and emergenesis

Another new focus is on the genetic mechanism called *epistasis*. Epistasis is the interaction of two or more genes at different chromosomal loci to produce a distinct phenotypic effect, which cannot be explained by the additive effects of multiple genes. The idea of epistasis has been broadened to include interactions between polygenic systems that affect distinct phenotypic traits. Termed *emergenesis* in this context, it is essential for understanding the occurrence of conspicuous phenotypic differences between close relatives in certain traits (Lykken, McGue, Tellegen, & Bouchard, 1992).

Occasionally, one sees a remarkable difference between one family member and all the others. The person's exceptional talent or trait seems *too* exceptional to be explained in terms of the usual *additive* effects of polygenes or differences in upbringing. For instance, Beethoven's brothers also had music lessons but showed only mediocre talent; the parents and siblings of the mathematical genius Ramanujan showed neither a mathematical bent nor any other intellectual distinction; and the parents, siblings, and children of the great conductor Toscanini had no outstanding musical talent. One could list countless other examples showing that most geniuses seem to just "come out of the blue."

According to emergenesis, the unusual development of certain abilities and talents depends on some rare combination of genes or polygenic systems that simultaneously influence several different abilities and traits. Only if this critical combination is present does the talent appear, given an appropriate environment. For example, Galton suggested that a higher than average level of general mental ability, energy, and persistence are involved in most outstanding achievements. Each parent's genotype may carry only some part of this combination. As parents pass on a random half of their genes to each of their children, there is some very small probability that any one child will get the particular combination of genes needed for the talent. It is like getting a royal flush in poker. All five of the critical cards must be in the shuffled deck when the game begins, but a habitual poker player's chances of getting them in his hand all at once are so slight that it rarely or never happens in his lifetime.

Geniuses rarely pass on their extraordinary emergenic gift. Like anyone else, they transmit but a random half of their genes to each of their offspring. That this random half will include the genius's particular rare combination of genes is very unlikely. Hardly anyone questions the conclusion that the extraordinary achievements of genius exemplify both emergenesis and GE correlation. John B. Watson notwithstanding, there is no evidence that any special kind of environmental influences, if applied to a random sample of healthy infants, would be at all likely to produce the equivalent of a Shakespeare, a Beethoven, a Newton, a Michelangelo, a Gandhi, or a Babe Ruth.

In such examples, the importance of the environment is often overrated. Many parents who have hoped their seemingly talented child would become a great musician have done everything they could for their child to achieve this goal. Yet exceedingly few ever become famous musicians, even with unusual ambition and efforts of parents and child. In contrast, when Leonard Bernstein was a child, his parents even went so far as to get rid of the piano in their home, because their young son showed such intense devotion to practicing on it that they feared he might one day think of becoming a professional musician, a possibility his father extremely wished to preclude. Years later, in Carnegie Hall, after one of Leonard Bernstein's concerts with the New York Philharmonic, a family friend chided Bernstein's father for having tried early on to discourage his famous son's passionate interest in music. The elder Bernstein pleaded, "How was I to know he would become Leonard Bernstein?"

How do genetic researchers discover that a particular trait is emergenic? If, for the given trait, one finds a very high correlation (say, .75) between monozygotic (MZ) twins reared apart, and a very low correlation (say, .15) between dizygotic (DZ) twins reared together, one suspects emergenesis. Although DZ twins (and ordinary full siblings) have, on average, about half their genes in common, very rarely do they both have the same unique *combination* of genes associated with the emergenic trait. Also, if there is a high correlation between MZ twins reared apart, a very low correlation found between DZ twins reared together is not likely to be a result of environmental differences. Lykken et al. (1992) give many examples of emergenesis identified by the twin method (i.e., DZ correlation significantly less than one-half of the MZ correlation). Most of the psychological examples are in the realm of personality traits (e.g., extraversion), attitudes (e.g., religiosity), and interests (e.g., hunting and fishing, gambling). There are many examples of emergenesis in the realm of physical traits, where individual differences in some traits (e.g., a beautiful or handsome face) often depend on an ideal *configuration* of features that are genetically independent. So there is a very low probability that the relatives of an exceptionally good-looking person will inherit the same combination of genes that make for an ideal configuration. A Greta Garbo or a Clark Gable typically has quite ordinary-looking parents and siblings.

It is noteworthy that general intelligence, as represented by IQ or by psychometric g (i.e., the general factor common to a diverse battery of cognitive tests), does not behave as an emergenic trait in genetic analyses. However, an above average level of g is often a critical condition for the development of an emergenic talent when the expression of the talent itself depends on the possession of certain complex cognitive skills. For example, acquiring the knowledge and technical skills needed to express an emergenic talent in mathematics or musical composition usually requires a superior level of general intelligence. It is exceedingly unlikely that there was ever a great mathematician or composer who was not above average in general intelligence. (Being above a threshold level in some trait for the manifestation of emergenesis, however, does not apply to every emergenic trait, particularly those in the personality domain.)

Sources of environmental variance

The most startling discovery in recent years concerns the locus of environmental effects on general intelligence. It was once believed that the most potent sources of environmental variance in IQ are conditions that differ between families. These are variables such as socioeconomic status (SES), cultural background, parents' education and occupation, style of child rearing, number of books in the home, and the like. In the last decade, we have seen the results of several large-scale studies of adopted children, such as the Texas, the Colorado, and the Minnesota adoption studies. They show, to everyone's astonishment, that these environmental differences between families account for little or none of the variation in the IQs of adolescents and adults, although these shared environmental factors account for about half of the total environmental variance in preadolescent children. Also, in childhood, the proportion of shared environmental variance among relatives is directly related to their degree of genotypic similarity, which decreases going from twins to siblings to parents-offspring to cousins (Chipuer, Rovine, & Plomin, 1990). But adoption studies based on adolescents and young adults show that the effects of shared (or between-families) environment have diminished to almost zero, with little change in the proportion of nonshared (or within-families) environmental variance and a marked increase in the proportion of genetic variance. Yet the adoptive families in these postadolescent studies range widely in SES and other variables on which many families typically differ from one another. Yet such differences scarcely contribute to the variance in IQ after childhood.

This remarkable fact could have been discovered only by studying adopted children. In children reared by their biological parents, the effects of heredity and environment are completely confounded. The children's IQs and the quality of the environment are both correlated with the parents' genotypes. Countless studies of children reared by their biological parents report large correlations between IQ and environmental assessments. However, these correlations are unable to prove anything about the importance of environmental factors for individual differences in IQ. This is because the observed IQ–environment correlation reflects more than just the *direct* effect of environment on the person's mental development. It includes also the effect of the parents' genotypes on the environment, plus the parent–offspring correlation due to parents and their offspring having about half their genetic variance in common.

Past investigations of the effects of the home environment on IQ have too often overlooked the influence of the genetic correlation between parents and offspring. A careful study (Longstreth et al., 1981) that took this factor into account correlated children's IQs with ratings of the home environment (based on a 2-hour interview with the parents) on those aspects commonly believed to affect children's intellectual development. It demonstrated, as have many such studies, a significant and substantial correlation between the environmental measures and children's IQs. The correlation dropped to nonsignificance, however, when the mothers' IQs were partialed out.

Such outcomes are easily understood from the path model in Figure 2.1, which shows the causal effects of heredity and environment on the child's mental development (here indicated by IQ) for nonadoptive and adoptive children. As is evident in Figure 2.1, the crucial advantage of an adoption study is that it eliminates the effect of the parent–child genetic correlation from the connection between environment and IQ (or any other trait). Therefore, any significant correlation between adoptive children's IQs and the typical environmental differences *between* adoptive families must be due solely to environmental effects. (This is true, of course, only if the adoptees and their adoptive parents are not genetically related, and if adoptees are not selectively placed according to their supposed genotypes for intellectual development.)

Empirically, it turns out that this between-families source of environmental variance constitutes almost half of the total environmental variance in IQ in childhood but is practically nil in adolescents and adults (McGue, Bouchard, Iacono, & Lykken, 1993). By late adolescence, almost none of the environmental component of IQ variance results from differences *between* family environments – that is, those aspects of the environment that are *shared* by children reared together in the same family but that differ between one family and another. Most of the strictly environmental, or

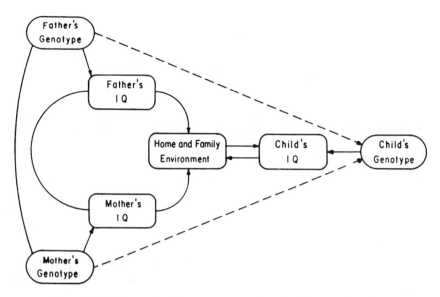

Figure 2.1. Diagram showing causal connections (straight arrows) and correlations (curved lines) between the genetic and the systematic (i.e., nonrandom) environmental factors that influence a child's IQ when the child is reared by its biological parents. In the case of an adopted child whose biological parents are unrelated to its adoptive parents, the two dashed arrows are deleted.

nongenetic, variance exists *within* families. It comprises those environmental effects that are *unshared* or *specific* to each child in a family. To the exent that adult family members resemble each other in intelligence, they do so almost entirely because of their genetic similarity. Apparently, as individuals progress from childhood to adulthood and encounter an ever-increasing range of experiences, they discover and select from their widening environment those aspects that are most compatible with their own genotypic proclivities. Therefore, with increasing maturity, the individual's genotype is increasingly expressed in the individual's phenotypic characteristics, reflected by the diminishing proportion of environmental variance and the increasing proportion of genetic variance (broad heritability).

This amazing fact, which contradicts popular belief, is one of the major discoveries of behavior genetics in the past decade. And it poses an extremely important puzzle – the puzzle of nongenetic variance. To understand it, we need to review a few technical matters.

Variance components in behavior genetics

The total variance in a metric trait, such as IQ, can be partitioned into several components, as shown in Figure 2.2. Each main component, or

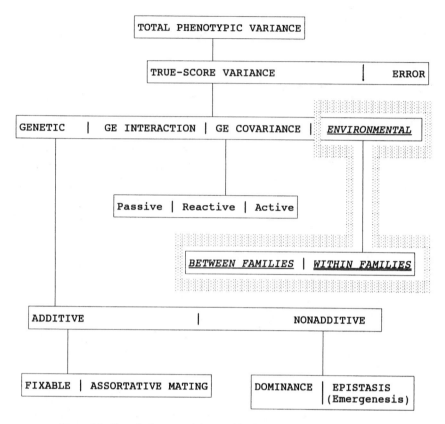

Figure 2.2. Branch diagram of the partitioning of the total phenotypic variance into the components (i.e., sources of variance) that can be estimated by the techniques of quantitative genetics. The main focus of the present article is on the environmental components (particularly *within-families*), shown highlighted.

source of variance, can be subdivided into more specific components. The genetic variance, for example, is analyzed into additive and nonadditive components, and each of these is analyzed into two components. Explanation of every component in Figure 2.2, and of how they are estimated by the methods of quantitative genetics, is beyond the scope of this chapter. It is covered in most textbooks of behavior genetics.

When the total phenotypic variance (V_P) is standardized (i.e., $V_P = 1$), the variance of each component then becomes a decimal fraction, or proportion, of V_P. Empirical studies usually report standardized values of the variance components.

This chapter focuses on the environmental variance (V_E). It is analyzable into two components, called Between Families (BF) and Within Families (WF), with variances V_{BF} and V_{WF}.

It should be noted that some writers use other terms, as follows, for the BF and WF components of the environmental variance. They all have the same meaning.

Between Families (BF)	Within Families (WF)
Common Environment	Specific Environment
Shared Environment	Nonshared Environment
Systematic Environment	Random Environment
E_2	E_1

The BF environment, by definition, is any environmental influence on a trait that causes two or more persons who were reared together to be more alike, on average, than persons who were not reared together. The BF component is the variance of the means of each of many sets of persons who were reared together (e.g., sets of siblings or pairs of twins). A simple rule: The variance *between* families is the covariance *within* families.[2]

The WF environment, by definition, is the environmental influence on a trait that causes persons who were reared together to differ from each other. The WF variance component is the environmental variance specific to each person.

A coefficient of correlation (r) between persons is also a variance component. (This correlation should not be squared to represent a proportion of variance; it is itself the proportion of common or shared variance.) Correlation coefficients based on different classes of persons are used to estimate the variance components shown in Figure 2.2. For example, if persons take the same test twice (a few days apart), or take two equivalent forms of a test, the coefficient of correlation between the scores obtained on test and retest (or between equivalent forms) is the proportion of *true-score* variance. This correlation is also known as the test's *reliability coefficient* (r_{xx}). The proportion of error variance, therefore, is the complement of the reliability, or $1 - r_{xx}$.

The variance components of particular interest in this article are obtained from correlations based on the kinds of data shown in the accompanying chart. Regarding monozygotic twins reared apart (MZA), the genetic component V_g estimated by r_{MZA} includes some fraction of the GE covariance found in MZ twins reared together (MZT), so r_{MZA} actually estimates $V_g + kV_{ge}$, where k is some fraction of the V_{ge} of MZT. Also, the correlation between unrelated persons reared together (r_{UT}) excludes the genetic component only if the adopted children have not been selectively placed according to their supposed genotypes. All of the components listed in the table are attenuated by measurement error. The correction for attenuation, which eliminates the effect of measurement error, is to divide each correlation or variance component by the reliability (r_{xx}) of the test used to obtain it. The r_{xx} of IQ tests in an unrestricted sample is typically about .90. When possible, estimated variance components intended for theoretical interpretation should be corrected for attenuation.[3]

Correlation Between	Variance Components
Text–retest on same persons	True-score (V_{TS})
Monozygotic twins reared apart (r_{MZA})	Genetic (V_g)
Monozygotic twins reared together (r_{MZT})	Genetic (V_g), GE interaction (V_i) GE covariance (V_{ge}) BF environment (V_{BF})
Unrelated persons reared together (r_{UT})	BF environment (V_{BF})

Other variance components that cannot be measured directly are estimated by subtracting one empirically obtained component from another. Certain components can be estimated by several different kinds of data.

The BF environmental component (V_{BF}), for example, is estimated directly by r_{UT} and indirectly by the formula $r_{MZT} - r_{MZA}$. But this formula may underestimate V_{BF}, because the GE covariance (V_{ge}) is likely to be smaller in MZA than in MZT. MZ twins, of course, have identical genotypes, but those who are reared together usually have a more similar environment than those reared apart. The greater similarity in environment makes the GE interaction and GE covariance larger in MZT than in MZA. Therefore, the difference, $r_{MZT} - r_{MZA}$, comprises not only V_{BF} but some fraction of $(V_i + V_{ge})$, and thus overestimates V_{BF}. The preferred estimate of V_{BF} is the correlation between unrelated persons who were reared together (r_{UT}). Another possible estimate of V_{BF}, though less compelling than r_{UT}, is the correlation ($r_{P_A C_A}$) between adoptive parents (P_A) and their adopted children (C_A).

Estimating the WF environmental variance (V_{WF}) allows three options. The simplest is based on only one correlation and is therefore less liable to error than formulas based on two or more correlations. Many estimates of V_{WF} in the literature are based on $1 - r_{MZT}$, but this is spuriously inflated by variance due to measurement error. If one has a good estimate of the reliability of measurement (r_{xx}), the better estimate, corrected for attenuation, is $1 - r_{MZT}/r_{xx}$.

Another estimate of V_{WF} is $V_{xx} - r_{MZA} - r_{UT}$. However, this method is proper only if the samples of MZA and UT are of about the same age. The relative sizes of the genetic component of IQ and the BF and WF environmental components all change between childhood and maturity (Plomin, 1986, chapter 14).

Still another estimate of V_{WF} is $r_{xx} - r_{MZA} - r_{P_A C_A}$. As in the previous formula, the MZA and C_A samples should be of similar age. Using correla-

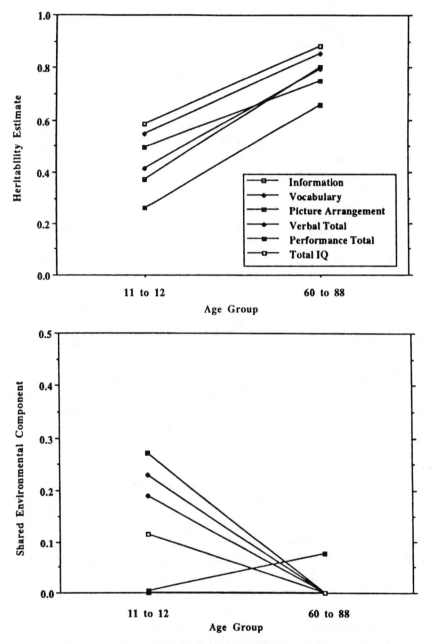

Figure 2.3. Proportion of variance in Wechsler test performance associated with heritability estimates (top panel) and shared environmental components (bottom panel) derived from the ongoing University of Minnesota cross-sectional study of reared-together twins. (From McGue et al., 1993, p. 72, with permission of the authors and the American Psychological Association.)

tions based on different age groups in one and the same formula may result in seriously inconsistent estimates of the variance components.

Estimates of all these components are made more accurate when corrected for attenuation (by dividing each of the correlation coefficients on which they are based by r_{xx}, or the equivalent, by dividing the final variance component by r_{xx}), assuming, of course, that r_{xx} is itself quite accurate.

MZ twins reared together from birth, despite having identical genotypes, do differ significantly in many personal characteristics, including IQ. Although these differences are typically much smaller than the differences between dizygotic (DZ) twins or ordinary siblings reared together, they are nevertheless real. The true-score differences between MZT afford probably the least ambiguous estimate of the WF environmental variance ($1 - r_{MZT}/r_{xx}$). This WF component becomes the large part of the nongenetic true-score variance in IQ after early adolescence. From childhood to maturity, the BF environmental component dwindles almost to nonexistence. By adulthood, virtually the *only* nongenetic variance in IQ is the WF component. The marked increase in heritability and decrease in the between-families (or shared) environmental variance is well illustrated in Figure 2.3, which is based on data from the Wechsler intelligence scales obtained from two different age groups of reared-together MZ and DZ twins.

The term *nongenetic* here seems preferable to *environmental*. In a psychological context, most people think of *environment* as only the psychosocial-cultural milieu. But the main causes of WF variance are still uncertain. They could be more directly biological than psychosocial-cultural.[4] The more neutral term *nongenetic*, therefore, is less apt to prejudice possible conceptions of the nature of WF variance.

Empirical estimates of BF and WF variance components

Reviews (Bouchard & McGue, 1981; Bouchard, Lykken, McGue, Segal, & Tellegen, 1991; Plomin, 1986, 1988; Plomin & Daniels, 1987) of studies of the kinship correlations used in the genetic analysis of human mental ability provide the evidence for the following conclusions. The most telling are studies of genetically unrelated persons who were adopted in infancy, reared together in the same family, and tested in late adolescence or adulthood (Scarr & Weinberg, 1978; Teasdale & Owen, 1984; Willerman, 1987). They show much smaller correlations (close to zero) than those obtained with adoptees tested in childhood. The N-weighted mean correlations (obtained via Fisher's Z transformation) based on all the available studies are probably the best estimates of the correlations one can obtain (where N is the sample size in each study). Because the correlations used in the follow-

Table 2.1. *Weighted mean correlations used for WF variance estimates*

Relationship	Reared	Symbol	Number[a]	Correlation	Corrected[b]
MZ twins	Together	MZT	4,672	.86	.95
MZ twins	Apart	MZA	162	.75	.83
Childhood[c]					
Unrelated	Together	UT	570	.29	.32
Postadolescent[d]					
Unrelated	Together	UT	385	.0025	.003
Parent–child[e]	Adopted	$P_A C_A$	1,397	.19	.21

[a] Number of pairs.
[b] Correction for attenuation based on reliability coefficient of .90.
[c] Mean age of 9 years.
[d] Mean age of 17 years.
[e] Adoptive parents and adopted child are genetically unrelated.

ing analysis are the N-weighted means of all the correlations reported in the published studies of each type of kinship data, it amounts to a meta-analysis of the estimated variance components.

Basic correlations. The N-weighted means of all the available correlations used for estimating WF variance are shown in Table 2.1.

Estimates of BF environmental variance. Four different estimates of the BF variance (with correction for attenuation in parentheses) are as follows.

1. $V_{BF} = r_{MZT} - r_{MZA} = .86 - .75 = .11$ (.12)
2. Childhood $V_{BF} = r_{UT} = .29$ (.32)
3. Postadolescent $V_{BF} = r_{UT} = .0025$ (.003)
4. $V_{BF} = r_{P_A C_A} = .19$ (.21)

Estimates of WF nongenetic variance.

a. $V_{WF} = r_{xx} - r_{MZT} = .90 - .86 = .04$ (.044)
b. $V_{WF} = r_{xx} - r_{MZA} - r_{UT} = .90 - .75 - .29 = -.14$ (−.15)
c. $V_{WF} = r_{xx} - r_{MZA} - r_{UT} = .90 - .75 - .0025 = .15$ (.16)
d. $V_{WF} = r_{xx} - r_{MZA} - r_{P_A C_A} = .90 - .75 - .19 = .05$ (.06)

The observed differences between the various estimates of the nominally same component, whether BF or WF, call for some explanation. These differences are almost entirely a result of the fact that, for IQ, the relative sizes of the components of genetic variance, GE covariance, and BF and WF environmental variances systematically change with age. Genetic variance (V_g) and GE covariance (V_{ge}) gradually increase from infancy to maturity. V_{BF} increases from early childhood to puberty, then decreases markedly to late adolescence and maturity. V_{WF} decreases from early childhood to midchildhood and then remains nearly constant to maturity. From early childhood to maturity, the major trade-off is between the increasing $(V_g + V_{ge})$ and the decreasing V_{BF}. Therefore, the discrepancies in the estimates of V_{BF} and V_{WF} are mostly attributable to these age changes and the fact that the correlations for MZT, MZA, and UT are based on different age groups. It so happens that studies of MZT are mostly based on school-age children, while MZA studies are nearly all postadolescents and adults. UT studies are based both on children and on postadolescents; the N-weighted average correlations obtained separately within each age group are used here. In the Texas adoption study, for example, pairs of unrelated adopted subjects reared together were tested for IQ as children (average age 10 years), and showed a correlation of .26. When they were tested again in late adolescence (average age 18 years), they showed a correlation of only .02 (Willerman, 1987). In view of these facts, several of the alternate estimates of the BF and WF components, as identified by the numbers or letters used earlier, call for comment.

1. Because MZT have more GE covariance in common than do MZA, the difference between r_{MZT} and r_{MZA} is a slightly inflated estimate of V_{BF}, which, by this estimate, really consists of $V_{BF} + kV_{ge}$, where $k < 1$. The value of k is not precisely known but is probably greater than 1/2. Therefore, a reasonable guesstimate of V_{BF} (corrected for attenuation) would be about .06.

4. This estimate of V_{BF} is based on the $r_{P_A C_A}$ for children, but because the adoptive parents are adults, they share less of the BF environment with their adopted children than is shared by two children of similar age reared together. Therefore, we should expect $r_{P_A C_A} < r_{UT}$, and this is what is found (i.e., .19 < .29).

a. Because MZT as children share more of the BF environment than MZT as adults, and studies of MZT are based mostly on school-age children, they have larger V_{BF} than adults. The formula $r_{xx} - r_{MZT}$, therefore, probably underestimates adult V_{BF} to some degree. The estimate obtained in (c), based entirely on postadolescent data, is predictably larger (.15 > .04).

b. This estimate is clearly anomalous, as it results in negative variance, which is impossible. The reason for the anomalous estimate is that the formula includes one correlation based on adults (i.e., $r_{MZA} = .75$) and one based on children (i.e., $r_{UT} = .29$). Note that $r_{xx} - r_{MZA}$ therefore estimates $V_{BF} + V_{WF}$ for adults, and if we subtract from it r_{UT} (= V_{BF}) based on children, a *negative* value is obtained, because V_{BF} is larger for children than for adults. Therefore the V_{WF} obtained in (c), which is based entirely on postadolescent data, is probably a good estimate.

d. Because $r_{P_A C_A}$ is based on children, it overestimates adult V_{BF}; therefore, when it is subtracted from $r_{xx} - r_{MZA}$, it gives an underestimate of V_{WF}.

An exact estimate of the WF variance, however, is not crucial to the present argument. What we do know with reasonable certainty is that, beyond childhood, there is almost zero BF environmental variance in IQ. Whatever environmental variance exists is WF (.10 to .20 of the total true-score variance). The rest of the reliable IQ variance consists mainly of genetic variance (.60 to .70) and GE covariance (.10 to .20).

There is no significant evidence of a GE interaction component for IQ. One statistical test of GE interaction is the correlation (r_{md}) between the *means* of MZ pairs and the absolute *differences* between individuals in each pair (Jinks & Fulker, 1970). In the pooled 69 pairs of MZA for whom scores are available in the literature, $r_{md} = -.09$, $p = .22$. The weighted average of r_{md} in all studies of MZT (totaling 1,435 pairs) is $-.04$, $p = .06$. But this is at best a weak test, which assumes that the genes controlling sensitivity to the environment are the same as those that affect the average expression of the trait in MZ twins. Because other factors besides GE interaction, such as skewness of the score distribution, can cause r_{md} to differ from zero, this test may exclude the presence of GE interaction if the null hypothesis ($r_{md} = 0$) cannot be rejected but cannot prove the existence of GE interaction if the null hypothesis is rejected. The null hypothesis cannot be rejected on the basis of the existing studies of MZA. (The technical problems of detecting GE interactions are well discussed by Neale and Cardon [1992, pp. 22–3].)

Distinction between IQ and psychometric *g* for genetic analysis

Genetic models often make a distinction between the *phenotype* of interest and some particular *index* of the phenotype. *Intelligence*, as a psychological construct, and IQ, as a standardized score on a particular mental test, are examples of a phenotype and its index. There is not necessarily a perfect correlation between the true phenotype (if it were measurable) and an index of it. Therefore, analysis of the index variance into genetic and nongenetic components does not necessarily yield precisely the same proportional values of the components as would be obtained from a parallel analysis of the true phenotype.

For *intelligence*, however, this proposition cannot be examined, because there is no generally agreed upon meaning of *intelligence*. Undefined terms are unsuitable phenotypes for behavior-genetic analysis. So we are left with only an index, an IQ score based on a particular test, which is highly correlated but not perfectly correlated (even when corrected for attenuation), with other IQ scores based on different tests.

The fact that IQ tests are all quite highly correlated (about .80) with each other, however, means that they measure some factor in common, whatever that factor may be. Many analyses have shown that this factor is the same one that is common to individual differences in performance on virtually all cognitive tests and other manifestations of mental ability, however diverse these may be in the specific information content and particular skills involved (Jensen, 1992). This *general factor* is called Spearman's *g*, or psychometric *g*, or just *g*. It can be estimated with varying degrees of accuracy by factor-analyzing large and diverse batteries of cognitive tests – the larger the battery and the more diverse the tests, the better the estimate of *g*.

Typical IQ tests, when factor-analyzed with a large and varied assortment of other cognitive tests, have large *g* loadings. Some 75–85% of the reliable variance in IQ consists of *g* variance. This distinction between IQ and *g* should be kept in mind in genetic analyses of IQ, because the results are slightly different for the same analyses applied to *g* factor scores. The difference is theoretically important. IQ variance has a smaller genetic component than *g*, even though IQ may also reflect the genetic component of other ability factors besides *g*, such as verbal and spatial ability (Bouchard, Lykken, McGue, Segal, & Tellegen, 1990, 1991).

Most but not all of the genetic variance in a battery of diverse tests is contained in the *g* factor, while the environmental variance resides mainly in the group factors and the variance specific to each test (Cardon, Fulker, & Plomin, 1992; Luo, Petrill, & Thompson, 1994). However, verbal and spatial group factors show some slight heritability independent of *g*. Various mental tests differ markedly in heritability (i.e., the proportion of genetic variance in test scores), and the tests' heritability coefficients are positively correlated to a high degree (.6 to .8) with the tests' *g* loadings (Jensen, 1987; Pedersen, Plomin, Nesselroade, & McClearn, 1992). In other words, the more a test reflects genetic variance, the larger its *g* loading. Factor analysis was developed by Pearson and Spearman at the turn of the century without any thought of genetics behind it. Yet the process of extracting the general factor, or *g*, from a number of diverse cognitive tests by means of factor analysis filters out, so to speak, much of the environmental variance (not including GE covariance).

In computer terms, the *g* factor reflects mostly individual differences in the genetically conditioned "hardware" of information processes, while individual differences in the "software" arise from environmental influences, learning, and experience. Individual differences in IQ based on any particular test typically reflect more of the "software" or experiential component of variance than does the *g* factor. Even so, the best estimate of the heritability of IQ as measured in adults by a single IQ test is about .75, which if corrected for attenuation would be about .80.

The nature of the nongenetic WF variance in IQ

The puzzle of nongenetic variance is this: By late adolescence, the *between-*families (BF) environmental variance in IQ has diminished to near zero, and the only remaining source of nongenetic variance is within families (WF). What, then, are the kinds of environmental effects that could be the source of the WF variance? This is a puzzling question, because psychologists have generally believed that the main environmental effects on IQ exist between families as differences in the psychological-educational-socioeconomic-cultural environment (PESC). Also, psychologists have generally believed that BF differences in these PESC effects are a much greater source of IQ variance than the more subtle differences in the psychological environment that cause differences between persons who were reared together. If these beliefs were true, why should the BF environmental variance in IQ diminish to almost zero from childhood to maturity, while the WF variance remains nearly constant throughout this period? Evidently, BF environmental differences, or PESC effects, do not have a strong or lasting influence on mental development, at least as it is indexed by IQ.

This important conclusion, however, should not be generalized to include individual differences in how effectively people have used their general mental ability in various attainments. Educational and occupational achievements reflect considerably more than mental ability as indexed by IQ (Jensen, 1993). Many other variables are involved, such as opportunities, interests, values, energy level, ambition, persistence, work habits, lifestyle, and other aspects of character and personality. Some of these personal variables involve genetic factors, although to a probably lesser degree than IQ, and they may be more influenced by the PESC aspects of the BF environment. But this is a separate issue and beyond the scope of this chapter.

To understand the nature of the predominant environmental effects on the distribution of IQ, at least within the typical range of the PESC environment in our population, we must focus on the WF environmental variance. One way to do this is to propose a working hypothesis and seek relevant evidence. As a working hypothesis, which is not yet tested and implies no theoretical commitment, it focuses examination of a class of nongenetic variables that has been peculiarly slighted in research on individual differences in IQ.

The physical microenvironment as a cause of WF variance. To account for nongenetic variance that shows up in genetic analyses even of very highly heritable physical characteristics, such as height, Sir Ronald Fisher (1918) hypothesized what he termed the *random somatic effects* of the environment. The causes of these random somatic effects can be prenatal, perinatal, or postnatal. Each such effect can be so slight as not to be individually detectable. Because the single effects are many and random, however, their net effect may differ considerably between persons. For the same reasons, according to the law of errors, individual differences in the net effects would have a normal distribution in the population.

The innumerable causes of these effects can be called the *microenvironment*. We can hypothesize that the microenvironment affects both physical and mental development and that the nonshared or WF nongenetic variance in IQ mainly reflects microenvironmental effects.

This is illustrated by the following analogy. Suppose that the microenvironment is represented by a huge stack of cards. Each card bears a single integer number ranging, say, from -3 to $+3$, where the negative and positive numbers indicate the degree of unfavorable or favorable effect of a single microenvironmental factor. The numbers occur with equal frequencies in the stack. The cards are shuffled, and 10 cards are dealt at random to each of 10,000 persons. Each person's net score is the sum of the numbers on the 10 cards received. The total range of scores, therefore, would have an approximately normal, or bell-shaped, distribution extending from -30 (very bad luck) to 0 (average luck) to $+30$ (very good luck). This normal distribution of luck would have a mean of 0 and a standard deviation (*SD*) of approximately 10. (The particular numerical values in this analogy are of course wholly arbitrary.) If we aggregate scores into many small groups (analogous to families), the variance of the means of these random net effects (analogous to the between-families variance) will be much smaller than the variance of individual scores. (The total variance between individuals is the sum of the between-groups variance and the within-groups variance. The between-groups variance is $1/n$th of the individual variance, where n is the average number of individuals in a group.) Therefore, the small between-groups variance may be swamped by other, larger sources of trait variance (e.g., genetic). Behavior-genetic analyses intended to estimate BF environmental variance, such as the correlation between unrelated children reared together, will scarcely reflect the microenvironmental effects responsible for the WF environmental variance.

This random model of the microenvironment is essentially the same as the genetic model for the inheritance of polygenic traits.[5] The approximately normal distribution of a polygenic trait (e.g., height) in all the offspring of the same parents results from the net effect of a different random assignment to each offspring of one-half of each parents' genes. Each gene produces a small positive or negative effect on the phenotype, with a different net effect for each offspring (unless they are MZ twins).

Differences between full siblings in their genetic endowments in any polygenic trait are like a random lottery. Some individuals have better luck than others.

Geneticists also recognize certain rare or mutant genes that singly can have a large phenotypic effect (usually deleterious), which overrides the normal polygenic determinants. (These are called *major gene* effects.) By analogy, a similar feature can be incorporated in our model of the microenvironment. Some single, rare environmental factors, such as accidental trauma or disease, can have a large, overriding phenotypic effect. These *macro*environmental effects are added to the normal distribution of microenvironmental effects in the population. One or both tails of the resulting composite distribution, therefore, would deviate from a normal curve. The amount of deviation would reflect the proportion of the total WF environmental variance contributed by macroenvironmental effects. These effects can be incorporated in the cards analogy by including in the stack of cards a small proportion bearing large numbers (e.g., ranging from ±20 to ±30).

The idea of randomness, or luck, as a source of important behavioral difference has been neglected in psychology. It is not a new idea. Galton (1908/1974) may have been the first to suggest its explanatory value. More recently, Paul Meehl (1978) has invoked a *random walk* hypothesis to explain the discordance of MZ twins for the development of psychiatric illnesses that have a strong genetic component, such as schizophrenia. The random walk hypothesis is attractive because research has failed to confirm hypotheses about MZ twin discordance in highly heritable traits that posit only a few categorical or systematic environmental variables, each with a big effect.

It has also been hypothesized that random epigenetic effects, or developmental noise, is an intrinsic phenomenon in all complex biological systems and may occur even under conditions of identical genotypes and uniform environment. This phenomenon is even regarded by Molenaar, Boomsma, and Dolan (1993) as a *third* source of variance, distinct from and in addition to genetic and environmental variance. These geneticists cite much relevant literature on what might be termed *autonomous chaos* in the developmental process, and they mention examples of it in physical traits studied in isogenic strains of animals raised under uniform conditions. It would as likely affect the structural and functional variance in neural networks as it does other anatomic features. They suggest, therefore, that some part of the variance classified as nonshared (or within-family) environmental variance cannot be traced to any exogenous effects. They write: "In our opinion, an important reason why the sources of these [nonshared environment] influences are still unknown is because a significant part of nonshared environmental influences may not be due to environmental differences at all, but result from intrinsic variability in the output of deterministic, self-organizing developmental processes" (p. 523). A similar view has been expressed by the Nobel laureate biochemist Gerald M. Edelman (1987):

As a result of the dynamic character of this model [of neural development], vast amounts of connectional variability will be found at all places in the nervous system, but particularly at the level of axonal and dendridic arbors in their finest ramifications. This insures individuality – while identical twins may have closer neuroanatomic structures than outbred individuals, it is predicted that they will nonetheless be found to have functionally significant variant wiring. (p. 323)

Our problem, then, is to try to get an empirical handle on the microenvironmental and epigenetic component of the WF nongenetic variance in IQ.

Intrapair IQ differences in MZT. Intrapair differences between monozygotic twins reared together (MZT) are probably the most direct measure of the WF environment. Any other kinship differences necessarily include genetic, GE covariance, and GE interaction effects, which together swamp the WF environmental effects. Because microenvironmental effects are hypothesized to be random, they cannot contribute to GE covariance, and their contribution to IQ variance therefore must be entirely nongenetic. Intrapair MZT differences consist exclusively of true WF nongenetic effects plus measurement error (e). The e must be taken into account, because it is a considerable part of the average intrapair difference, and we are really interested in the true-score differences.

The one possible disadvantage of using twins to measure environmental effects is that twins share the same uterine environment during gestation; this may present unique and unequal biological hazards, possibly increasing the intrapair differences in critical ways (Bulmer, 1970). Such effects, if unique to twins, would not add to the nongenetic differences between single-born children. The Oxford cytogeneticist C. D. Darlington (1954) argued that MZ twin differences overestimate environmental effects, because some of the difference is due to unequal division of the fertilized ovum, creating what Darlington terms *cytoplasmic discordance* and *asymmetry*.

These differences in the epigenetic landscape thus occur at the earliest stage of development. Their enduring differential effects on the twins are not genetic and are not really environmental but are probably best viewed as random biological noise – the "random somatic effects" mentioned by R. A. Fisher. Sometimes, there is even a marked inequality in placental blood supply, a condition peculiar to MZ twins. It also causes differences in development. The authors of a well-known twin study stated, "Such differences are neither genetic, in the ordinary sense, nor environmentally in-duced. In comparing the variability of identical [MZ] and fraternal [DZ] twins, therefore, it is not proper to consider all differences in identical twins reared together as environmentally determined" (Newman, Freeman, & Holzinger, 1937, p. 51).

The use of MZT for estimating the WF environmental variance, there-fore, might overestimate the WF environmental variance in the population of singletons. However, these uniquely biasing prenatal factors in twin differences might be offset by the fact that MZ twins both share the same prenatal conditions and early environment on many variables, such as the mother's age, compatibility (or incompatibility) with the mother's blood group, health, parity, mother's medication before or during childbirth, and similarity of infant and early childhood experiences. Such variables prob-ably contribute to the *unshared* environmental effects, or WFE variance, in the population of singletons.

Analysis of MZT intrapair differences in IQ. In the total literature on MZT, I have found ten studies in which all of the twins were tested on the Stanford-Binet, and the individual IQs were reported. They total 368 pairs, all school-age children (ranging in age from 5 to 16 years, with an average age of 10 years). Because of sampling differences, it would be statistically undesirable to pool all of these studies. Therefore, every sample was com-pared with every other sample to find out whether the means and variances of the IQ distributions differed significantly at the .05 level between sam-ples, using *t* tests of the mean differences and Bartlett's test for homogene-ity of variances.

Six samples (from studies by Hirsch, 1930; Merriman, 1924; Stocks, 1930, 1933; Wingfield & Sandiford, 1928) did not differ significantly ($p > .05$) from each other and therefore can be treated statistically as samples from the same population. They comprise 180 pairs tested on the 1916 revision of the Stanford-Binet.[6] The pooled sample (with total $N = 360$) has a mean IQ of 96.91 and *SD* of 15.78. (The mean IQ of twins is typically a few points below the population mean of approximately 100.)

The intraclass correlation (r_i) between the twins is .878, which differs little from the average $r_i = .88$ of the ten studies (totaling 661 pairs) of MZT based on Stanford-Binet IQs or the *N*-weighted average $r_i = .86$ of all existing studies of MZT (totaling 4,672 pairs [Bouchard & McGue, 1981]). The r_i is also the proportion of total shared variance, which, for MZT, consists of genetic variance, GE covariance and interaction, and shared, or BF, environmental influences. All the rest (i.e., $1 - r_i$) is WF environmental variance and error variance.

Table 2.2. *Distribution of MZT intrapair absolute differences in Stanford-Binet IQ*

$|D|$	f	cf	cIQ	cIQH	cIQL	cVIQ	cVIQH	cVIQL
0	18	180	96.91	99.76	94.06	249.12	242.78	239.22
1	24	162	97.34	100.51	94.17	260.06	249.26	250.80
2	14	138	96.70	100.33	93.07	263.12	253.07	246.81
3	20	124	97.09	101.02	93.16	274.68	260.44	258.07
4	17	104	96.95	101.35	92.56	254.97	237.86	233.46
5	19	87	97.16	102.02	92.30	271.84	249.63	246.76
6	15	68	97.83	103.35	92.31	286.70	253.99	258.42
7	9	53	98.90	105.13	92.66	290.09	241.47	260.94
8	8	44	99.07	105.86	92.27	330.06	271.03	296.74
9	4	36	97.39	104.81	89.97	343.13	283.88	292.36
10	3	32	97.63	105.41	89.84	345.23	278.37	291.01
11	2	29	97.10	105.17	89.03	360.27	291.38	298.93
12	3	27	96.48	104.74	88.22	379.03	310.19	311.43
13	5	24	97.08	105.63	88.54	361.95	285.23	292.75
14	0	19	96.18	105.26	87.11	370.52	287.98	288.20
15	4	19	96.18	105.26	87.11	370.52	287.98	288.20
16	1	15	98.17	107.67	88.67	337.07	238.76	254.89
17	3	14	97.96	107.57	88.36	355.96	255.67	271.66
18	1	11	96.09	106.00	86.18	319.81	216.73	226.51
19	5	10	95.50	105.50	85.50	339.85	235.65	244.05
20	2	5	92.90	103.40	82.40	283.69	170.24	176.64
21	1	3	94.17	105.00	83.33	369.47	244.67	259.56
22	2	2	83.00	94.00	72.00	125.00	4.00	4.00

$|D|$: absolute intrapair difference
  f: frequency
  cf: cumulative frequency
 cIQ: cumulative mean IQ ($N =$ cf)
cIQH: cumulative mean IQ of higher-scoring twins
cIQL: cumulative mean IQ of lower-scoring twins
cVIQ: cumulative variance of IQ ($N =$ cf)
cVIQH: cumulative IQ variance of higher-scoring twins
cVIQL: cumulative IQ variance of lower-scoring twins

Table 2.2 shows the frequency distribution of the absolute (i.e., unsigned) intrapair differences ($|D|$) in IQ, and the cumulative frequencies (*cf*) (going from the largest $|D|$ of 22 IQ points to a $|D|$ of zero), with the corresponding cumulative means and variances. Figure 2.4 shows the bivariate frequency distribution of the twins' IQs grouped in the class intervals 50–59, 60–69, and so on. What further information can we obtain from the statistics in Table 2.2 and Figure 2.3?

1. Given the total phenotypic variance $V_P = 249.12$ of IQs in the twin sample and the twin intraclass correlation $r_i = .878$, the *total* WF (i.e., within-

Higher-IQ Twin

	IQ	60	70	80	90	100	110	120	130	140	150
	150										1
	140										
	130								1		
	120							3	1	1	
Lower-IQ Twin	110						10	6			
	100					23	12	4			
	90				25	24	3				
	80			23	11	5					
	70		6	7	3			r_i = +.878			
	60	5	2	2							
	50	2									
	Σ	7	8	32	39	52	25	13	2	1	1

Figure 2.4. Bivariate frequency distribution of lower-IQ and higher-IQ twins.

Note: The *intraclass* correlation, r_i = +.878, is the correlation between the twins regardless of their classification as higher or lower in IQ [or any other basis for classification]. The intraclass correlation is not the same as the Pearson correlation [or *interclass* correlation], r, between lower- and higher-IQ twins. The Pearson correlation between the lower-IQ and higher-IQ twins in the present sample is r = +.941. [Both interclass correlation and intraclass correlation are clearly explicated by R. A. Fisher (1970, pp. 213–49), the inventor of the intraclass correlation.]

pair) variance (including measurement error) is calculated as $V_{WFt} = V_P (1 - r_i) = 249.12 (1 - .878) = 30.4$. The *SD* of the WF IQs, then, is $(30.4)^{(0.5)} = 5.5$ IQ points. (This includes true-score environmental effects plus measurement error.)

2. The best estimate of the Stanford-Binet equivalent forms reliability in the age range and IQ level of the present sample is r_{xx} = .93 (McNemar, 1942, chapter 6). The twin correlation corrected for attenuation, then, is .878/.93 = .944. With the measurement error removed, we can estimate the WF environmental variance as $.93 \times 249.12(1 - .944) = 12.95$. So the *SD* of WF environmental effects is $(12.95)^{(0.5)} = 3.6$ IQ points. This may be compared with the total true-score *SD* of IQ, or $(.93 \times 249.12)^{(0.5)} = 15.2$ IQ points. All this can be most clearly presented in the typical form of an analysis of variance, as shown in Table 2.3.

The error variance is larger than the true-score WF variance in this sample. This is generally true in all of the studies of MZT reported in the

Table 2.3. *Analysis of variance of MZT IQs*

Source[a]	Uncorrected			Corrected for attenuation		
	Variance	Percent	*SD*	Variance	Percent	*SD*
BF	218.73	87.8	14.8	218.73	94.4	14.8
WFE	12.95	5.2	3.6	12.95	5.6	3.6
Error	17.44	7.0	4.2	0	0	0
Total	249.12	100.0	15.8	231.68	100.0	15.2

[a] BF = between families (i.e., twin pairs).
WFE = within-families environment.
Note: Variances are additive; *SD*s are not additive.

literature. Given the weighted mean correlation of .86 based on all of the 34 published studies of MZT (totaling 4,672 pairs), and assuming that the average test reliability is .90, then the proportion of within-families environmental variance (V_{WFE}) is .04. If the population variance of IQ is $16^2 = 256$ (as on the Stanford-Binet), the V_{WFE} would be 10.24, and the *SD* of WF environmental effects would be $(10.24)^{(0.5)} = 3.2$ IQ points. By comparison, the *SD* of measurement errors would be $[(1 - .90)256]^{(0.5)} = 5.1$ IQ points. Again, measurement errors are larger than WFE differences. The MZT intrapair differences in IQ attributable to WFE effects are almost swamped by measurement error. Unfortunately, for any given twin pair, we have no way of knowing how much of the intrapair IQ difference consists of measurement error and how much of true-score WFE effects. Yet we can make some theoretically informative inferences about the nature of the WFE from some further quantitative analyses of the total frequency distribution of intrapair differences (shown in the first two columns of Table 2.2).

Mean absolute difference ($|\bar{D}|$). The model of random nongenetic effects with which the MZ twin data are to be compared posits that each twin's nongenetic deviation from the twin pair's common genotypic value is a normal random deviate, as would be expected for the distribution of many small and independent random effects. Therefore, the MZ twin differences, if they reflect random nongenetic effects and therefore conform to this model, should approximate the same distribution that would obtain for differences between pairs of values taken at random from a normal distribution. In the normal, or Gaussian, distribution there is an exact relationship between the σ and the mean absolute difference ($|\bar{\Delta}|$) between every pair of values in the distribution taken at random. The formula usually called *Gini's mean difference*, as given by the statistician Acardo Gini in 1914 (see Kendall & Stuart, 1977, pp. 48–9, 257), is $|\overline{\Delta}| = 2\sigma/\sqrt{\pi} = 1.1284\sigma$. Its standard deviation is $\sigma_{|\Delta|} \approx 0.8068\sigma$. The sample value of $|\overline{\Delta}|$ is signified by $|\overline{D}|$ and its *SD* by $SD_{|D|}$.

If, as hypothesized, all WF effects (i.e., both WFE and error) on IQ are random and therefore normally distributed, the theoretical values of $|\overline{\Delta}|$ and $\sigma_{|\Delta|}$ can be calculated from the *SDs* for WFE and Error shown in Table 2.3, using Gini's formulas. These theoretical values are as follows:

	Uncorrected		*Corrected*	
	$|\overline{\Delta}|$	$\sigma_{|\Delta|}$	$|\overline{\Delta}|$	$\sigma_{|\Delta|}$
WFE	4.1	2.9	4.1	2.9
Error	4.7	3.4	0	0
WFE + Error	6.2	4.4	4.1	2.9

Because we cannot separate WFE effects from measurement error for each intrapair difference, we can only compare the uncorrected theoretical values of WFE + Error ($|\overline{\Delta}| = 6.2$, $\sigma_{|\Delta|} = 4.4$, CV [coefficient of variation] $= \sigma/|\overline{\Delta}|$ $= .7150$) with the corresponding obtained values of all the twin differences in IQ, which are $|\overline{D}| = 5.70$, $SD_{|D|} = 5.37$, CV $= .9421$. These obtained values, calculated directly from all of the intrapair IQ differences, differ from the theoretical values based on Gini's formulas for $|\overline{\Delta}|$ and $\sigma_{|\Delta|}$ applied to the uncorrected *SD* of the combined WFE + Error $= 5.51$ [i.e., from Table 2.3: $(12.95 + 17.44)^{(0.5)} = 5.51$]. The standard deviations differ significantly ($F_{179,179}$ $= (5.37/4.44)^2 = 1.46$, $p < .05$), and the CVs differ significantly ($t = 2.80$, $df = 179$, $p < .01$ [formula for standard error of the CV in Kendall & Stuart, 1977, p. 248]). This can only mean that the distribution of the obtained differences ($|D|$) is *not* consistent with the proposed working hypothesis that the differences result from random and normally distributed effects. And, as measurement errors are conventionally considered random and normally distributed, with $\mu = 0$, $\sigma_e = \sigma_x \sqrt{1 - r_{xx}}$, we must infer that a nonnormal distribution of WFE true-score effects is what causes the departure of the composite WFE plus error distribution from a normal distribution. The obtained distribution of $|D|$ has a lower mean and a larger *SD* than theoretically expected because, compared to the theoretical distribution, there is an excess of very small $|D|$ values having quite large frequencies (which lowers the overall mean of $|D|$) and an excess of large $|D|$ values having relatively small frequencies (which increases the *SD* more than the mean). The nonnormal frequency distribution of $|D|$ is most logically regarded as a composite of the normal distribution of measurement error and a nonnormal distribution of WFE effects. To infer the nature of the distribution of WFE effects on IQ, one has to try to read through the noise of measurement error. This requires a more detailed examination of the distribution of $|D|$.

The delta |Δ| distribution.   The MZ twin intrapair difference |D| is the sum of each twin's deviation (*d*) from some value that both have in common. If that common value is the average of both twins, one twin's deviation is positive (+*d*), the other's is negative (−*d*), and their absolute values, |*d*|, are identical. Then, obviously, 2|*d*|, equals |D|, that is, the intrapair difference. In a group of twins, the mean of these intrapair deviations of course will be zero. (If the positive and negative deviations are normally distributed around the mean, their σ_d is given by a rearrangement of Gini's formula, viz., $\sigma_d = \frac{1}{2}|\overline{\Delta}|\sqrt{\pi}$.)

But the symmetry of +*d* and −*d* is merely a formalism, without any heuristic theoretical value or empirically testable implications. It would be theoretically more interesting to hypothesize that each twin's IQ is a deviation, not from the mean of both twins' IQs, but from their common genotypic value, whatever that may be. Unfortunately, we have no way of measuring any individual's (or any twin pair's) genotypic value. Over many twins, however, the means of each pair of twins are, on average, probably closer to their genotypic values than to any other values that could be directly calculated from the twin data. (Geneticists theoretically define *genotypic value* as the mean phenotypic value of all individuals with the same genotype.)

For any given pair of twins, the deviation of each twin's true-score deviation from the genotypic value, attributable to WFE effects, may be unequal. The higher-IQ twin, for example, could be less deviant from the genotypic value than is the lower-IQ twin (or vice versa). If this inequality were true more often than not in the twin population, the total distribution of twin deviations around their genotypic values would be nonsymmetrical and, ipso facto, nonnormal. Such asymmetry cannot be seen by direct examination of the distribution of |D| but must be inferred indirectly from certain statistics of the separate distributions of the higher-IQ and the lower-IQ twins from each pair. Certain other departures from normality, however, can be observed directly from a proper graph of the cumulative frequency distribution of |D|.

The absolute differences (|Δ|) between all possible pairs of variate values in a unit normal curve ($\mu = 0$, $\sigma = 1$, z = a standardized deviation from μ) are distributed as the delta (|Δ|) distribution. Its frequency distribution resembles the right-hand side of one-half of the normal curve, but it has different parameters.[7] Some of the parameters of the |Δ| distribution are:

Range:	0 to +∞
Mean:	$\mu_{|\Delta|} = |\overline{\Delta}| + 2/\sqrt{\pi} = +1.1284z$
Standard deviation:	$\sigma_{|\Delta|} \approx .8068z$
Coefficient of variation:	$CV = \sigma/\mu = .7150$

These parameters are useful for comparison with the corresponding statistics of an empirical distribution of |D|, to determine its resemblance to a |Δ|

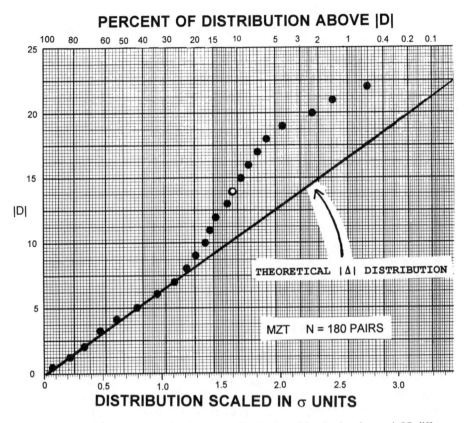

Figure 2.5. A cumulative frequency distribution of the absolute intrapair IQ differences ($|D|$) in 180 pairs of MZ twins, plotted on normal probability paper. (The one open data point for $|D| = 14$ is an interpolation, as no twins in this sample have an intrapair difference of 14 IQ points. With this one exception, the values of $|D|$ range continuously from 0 to 22 IQ points. Any single value of $|D|$ is properly interpreted as falling within the class interval $|D| \pm 5$.)

distribution. If the distribution of all the intrapair twin differences in IQ very closely resembles a $|\Delta|$ distribution, it would be consistent with our working hypothesis that the WFE effects are normally distributed and therefore most likely the result of many small, random positive and negative environmental influences on IQ.

Figure 2.5 shows the percentiles of the cumulative frequency distribution of $|D|$ (based on the column labeled cf in Table 2.2), plotted on a normal probability grid. The reason for this kind of plot is that if the distribution of effects (e.g., WFE + Error), from which the values of $|D|$ arose, were randomly and normally distributed, all of the plotted data points should fall along a straight line, as would the theoretical $|\Delta|$ distribution. Any systematic and significant departure from the straight line indicates that the $|D|$

values could not have been generated solely from random differences between normally distributed values. This outcome is apparent in Figure 2.5. The values of $|D|$ fit a straight line in the range from 0 to 9, which comprises about 80% of the twin pairs. However, values of $|D|$ larger than 9, which comprise about 20% of the twin pairs, depart significantly and systematically from a straight line. They are considerably larger than the $|D|$ values predicted from the hypothesis that all values of $|D|$ arise from normally distributed nongenetic effects on IQ.

The possibility that some same-sex DZ twins have been misclassified as MZ cannot be dismissed. (For DZ twins, $|\overline{D}| = 11.4$ IQ points.) This type of misclassification, however, has been found to be not more than about 3–4% with the method of zygosity diagnosis used in these early studies, which would amount to 6 or 7 DZ pairs included in the present sample of 180 pairs that were misdiagnosed as MZ.

To what extent is this departure of values of $|D| > 9$ from their expected values attributable to environmental effects and to errors of measurement? Errors are presumed to be normally distributed, so they should not deviate significantly from the straight line. McNemar (1942, chapter 6) has shown that the distribution of children's IQ (and mental age) differences between equivalent forms L and M of the Stanford-Binet conform almost perfectly to the theoretical $|\Delta|$ distribution throughout the full range of IQ. Deviations of the obtained values from the theoretical values are extremely small. When the properties of the $|\Delta|$ distribution were used to predict the equivalent-forms test-reliability coefficients based on the actual test–retest correlation, the predicted reliability coefficients have a mean absolute deviation of only .005 from the obtained reliability coefficients.

From this fact, it seems reasonable to infer that the distribution of twin differences, plotted as $|D|$ in Figure 2.5, deviates from the hypothetical $|\Delta|$ distribution for $|D| > 9$, not because of measurement errors but because of the nonnormality of WFE effects. The nonnormality involves only the most extreme 20% of the twin pairs' $|D|$ values, and this 20% could be evenly divided between the left and right tails of the nonnormal distribution (each tail with 10%), or it could be divided asymmetrically. A high degree of symmetry would mean that the extreme WFE effects are as frequently positive as negative; that is, large environmental effects would raise IQ as often as they lower it. Is this in fact what happens?

Before examining this question, we should take another look at the distribution of measurement errors for individuals who were tested with equivalent forms of the Stanford-Binet when the test–retest interval is a considerable period. Changes in IQ then reflect not only measurement error in the strict sense (or the complement of the internal consistency reliability) but also true-score developmental variation. In the course of mental development, as in physical development, children show lags and spurts in growth, which are partly environmental but also, we know, partly genetic, because MZ twins show higher concordance than DZ twins in the pattern of lags and spurts in their cognitive development (Wilson, 1974, 1983).

Thorndike, Fleming, Hildreth, and Stranger (1940) provide ideal data on which we can check the fit of Stanford-Binet IQ changes over a minimum test–retest interval of 2.5 years in 1,167 elementary-school children. The distribution of test–retest differences has a mean absolute difference ($|\overline{D}|$) of 10.5 IQ points, and the test–retest correlation is +.65. I have plotted this $|D|$ distribution on a normal probability grid as in Figure 2.5. All of the $|D|$ values in Thorndike et al.'s test–retest data fall on a straight line, without the least suggestion of the kind of deviation from linearity seen in Figure 2.5.

This suggests that whatever environmental and genetic effects are reflected in IQ changes over a period of more than 2.5 years are normally distributed and could have resulted from many small, randomly distributed effects. The twin differences greater than 9 IQ points, on the other hand, do not fit this model. Some proportion of them deviate more than would be expected from the normal distribution of many small random effects. Because developmental changes in IQ during the elementary-school years do not show nonnormal effects, as Thorndike et al.'s data seem to indicate, it is a likely hypothesis that some part of the large nonnormal twin differences originated prenatally or before school age.

Another test of nonnormality of the twins' IQ deviations is to look at the distribution of IQ deviations (d) of each twin from the mean of each pair. This distribution of d, which ranges from -11 to $+11$, has a mean = 0, $SD =$ 3.91. The distribution is, of course, necessarily symmetric about the mean, so it cannot be informative about skewness. It can, however, be informative about another possible index of nonnormality – namely, *kurtosis*. Kurtosis refers to the degree of peakedness or flatness of the distribution and is indexed by the ratio of the 2nd and 4th moments (μ_2 and μ_4) of the distribution, the measure of kurtosis being Pearson's $\beta_2 = \mu_4/\mu_2^2$. For the normal distribution, $\beta_2 = 3$. A $\beta_2 < 3$ indicates a *platykurtic* distribution; $\beta_2 > 3$ indicates a *leptokurtic* distribution. In our twin sample's distribution of d, $\beta_2 = 3.93$, which is very significantly ($p < .001$) greater than 3. So this d distribution is decidedly nonnormal. It is leptokurtic, which means that, compared to the normal curve, it has an excess of small absolute deviations ($|d| < 3$) and also an excess of large deviations ($|d| > 5$).

The small deviations are scarcely larger than would be expected from measurement error alone, when the reliability of the IQ is .93. For the whole distribution of $|D|$ (as shown in Figure 2.5), the mean difference, $|\overline{D}|$, is 5.70 (SD = 5.37). But assuming a test–retest reliability of .93 (with error variance = $1 - .93 = .07$), the mean difference between repeated measures on the same persons (with the same total IQ variance [249.12] as in the present twin sample) would be 4.71 (SD = 3.37). The intraclass correlation (r_i) between the 80% of twins with $|D| < 10$ is .85; for twins with $|D| \geq 10$, the r_i = .63. Because the SDs of the IQs on which these two correlations are based differ considerably (15.11 and 18.50, respectively), the correlations should be corrected for this difference. When the correlations are thus corrected to a common SD equal to that of the whole sample (15.78), the r_i for twins with $|D| < 10$ is .86, and the r_i for twins with $|D| \geq 10$ is .57. This difference (.86 – .57 = .29) in the twin correlations implies that most of the WFE effects that are visibly larger than measurement error occur in only about 20% of the twin pairs – that is, those with $|D| \geq 10$ IQ points. The WFE effects in the 80% of twin pairs with $|D| < 10$ must be quite small.

It should be noted that the picture shown in Figure 2.5 is not peculiar to this set of MZ twin data. When other sets of MZT IQ data (Osborne, 1980; Rosanoff, Handy, & Plesset, 1937) and three combined studies (totaling 69 pairs) of MZ twins reared apart (Juel-Nielsen, 1965; Newman et al., 1937; Shields, 1962) were each plotted in the same fashion as in Figure 2.5, the same distinctive features of the plot shown in Figure 2.5 are seen in each of the other sets of twin data.[8]

Asymmetry of WFE effects. Are the 20% of twin differences that are large (i.e., $|D| \geq 10$) attributable to WFE effects that enhance IQ or depress IQ by equal amounts, on average, for the higher- and lower-IQ twins in each pair? Or do the higher- and lower-scoring twins reflect unequal effects of the WFE? The answer to this question may be revealed by looking for systematic differences between certain features of the IQ distribution of the higher-IQ (HIQ) members and of the lower-IQ (LIQ) members of each twin pair.

Michael Bailey and Joseph Horn (1986) were probably the first researchers to apply this strategy to MZ twin data. They reported a larger IQ variance for the LIQ than for HIQ twins, which led them to conclude that in MZ twin pairs, the LIQ twin reflects disadvantageous nongenetic effects. That is, the LIQ twin's phenotype deviates, on average, further below the pair's common genotypic value than the HIQ twin's phenotype deviates above it. Their finding at least suggests that the IQ distributions of the HIQ and the LIQ twins differ in ways other than their defining mean difference in IQ. Their distributions are in some way asymmetrical.

The difference in variances reported by Bailey and Horn, however, is subtle at best, and several other MZ twin studies do not consistently show the variance of IQL > variance of IQH. The Bailey and Horn variance ratios (i.e., $F =$ LIQ variance/HIQ variance) based on five well-known MZ twin studies, are all larger than 1 (averaging 1.14), but only two of the five are significant. In our present twin data, including all levels of $|D|$, the $F = 0.99$, which seems not to replicate the Bailey and Horn finding. However, looking at the cumulative variances (cVIQH and cVIQL) of the HIQ and LIQ twins in the last two columns of Table 2.2, we see that in 18 out of 23 comparisons the VIQL > VIQH. Going from twin differences of 22 to 6 IQ points, the cumulative variances are consistently larger for the LIQ twins, but this trend reverses markedly when we add in the twins who differ by only 5 to 0 points. For twin pairs with $|D| \geq 6$, $F = 1.02$; for twins with $|D| \leq 5$, $F = 0.75$. Besides their defining difference in $|D|$, twins with small intrapair differences appear to differ also in other ways from twins with large intrapair differences. As seen in Figure 2.5, the demarcation between small and large differences falls at a $|D|$ of about 9 or 10 IQ points.

Bailey and Horn (1986) also noted that if there was a significant difference in the degree to which the IQs of the LIQ and the HIQ twins predicted $|D|$, it would suggest that one of the groups accounts for a larger part of the twin differences than the other group. They hypothesized that the IQs of the LIQ twins would show a larger correlation with $|D|$ than would the HIQ twins. If this hypothesis were borne out, it would mean that the WFE effects that cause MZ twins to differ in IQ are larger (in a negative direction) for the lower-scoring twins than for their higher-scoring co-twins (in a positive direction). Probably the best way to look at this is to compare the correlation between $|D|$ and the cumulative means of the HIQ twins with the correlation between $|D|$ and the cumulative means of the LIQ twins (columns cIQH and cIQL of Table 2.2). This correlation for the HIQ twins is $-.075$; for the LIQ twins, the correlation is $+.293$. There is no proper test for the significance of this difference, although the direction of the difference is consistent with the Bailey and Horn hypothesis.

But, as previously noted, we are really dealing with two distinct distributions demarcated by $|D| < 10$ and $|D| \geq 10$. So we should look at the Bailey and Horn hypothesis separately within each distribution. For $|D| < 10$, the correlation between $|D|$ and the cumulative means of the HIQ twins is $+.678$; the corresponding correlation for the LIQ twins is $+.035$. For $|D| \geq 10$, the correlation for HIQ twins is $-.45$; for LIQ twins, the correlation is $-.76$. (If the largest twin difference, $|D| = 22$, is regarded as an outlier and is omitted from the calculations, the correlations for $|D| \geq 10$ are $-.05$ for the HIQ twins and $-.87$ for the LIQ twins.)

The absolute size of these correlations is unimportant here. It is the *difference* between the correlations for the HIQ and LIQ twins that is most informative. It shows that the intrapair IQ differences are not symmetrical, for if they were symmetrical, these correlations should be nearly the same. But we see that for twin pairs with $|D| < 10$, the HIQ twins' IQs predict $|D|$ better than the LIQ twins' IQs do. And for twin pairs with $|D| \geq 10$, the LIQ twins' IQs twins' IQs predict $|D|$ much better than the HIQ twins' IQs do.

Finally, to look at this phenomenon with greater statistical power than is afforded by the 180 twin pairs in the above analyses, I have analyzed MZT data from several studies totaling 1,435 twin pairs (studies by Hirsch, 1930; Merriman, 1924; Newman, Freeman, & Holzinger, 1937; Osborne, 1980; Rosanoff, Handy, & Plesset, 1937; Stocks, 1930, 1933; Wingfield & Sandiford, 1928.)

Because these studies used different IQ tests and the various samples are heterogeneous in means and SDs, it was necessary to standardize the twins' IQs and the intrapair differences within each study, scaling both variates as $(X - \overline{X})/SD = z$ scores, where X is an individual IQ (or an intrapair $|D|$) and $\overline{X}$ is the sample mean IQ (or the sample $|\overline{D}|$), and SD is the sample standard deviation of each variate. With the twin data in each sample separately transformed from IQ to z_{IQ} and from $|D|$ to $z_{|D|}$, the total of 1,435 MZT pairs in these samples then could be pooled. The z scores for IQ (z_{IQ}) were regressed on the z scores for intrapair difference ($z_{|D|}$), separately for the HIQ and LIQ twins. (The regression coefficient for z scores is identical to r, the Pearson correlation coefficient.) If, at each level of $z_{|D|}$, the HIQ and LIQ twins' standardized IQs (z_{IQ}) in each pair differed, on average, equally (but in opposite directions) from the grand mean $\overline{z}_{IQ} = 0$ of all the twins, then the regression coefficients (or correlations) of the HIQ and LIQ twins' z_{IQ} on $z_{|D|}$ should not differ significantly in absolute size. (Of course, they necessarily have opposite signs.)

As it turns out, however, the correlations are +.14 for the HIQ twins and −.26 for the LIQ twins, as depicted in Figure 2.6. The difference between the two absolute values of r is highly significant ($t > 4$, $df = 1,434$, $p < .001$). The crucial point is that r_L is a significantly larger correlation than than r_H. This result can be interpreted as showing that whatever specific, or unshared, or within-family nongenetic effects (WFE) cause MZ twins to differ in IQ, these effects are more strongly negative than positive. That is, the lower-IQ twins are more disadvantaged than the higher-IQ twins are advantaged by whatever nongenetic factors make MZ twins differ in IQ. Apparently, the nongenetic influences on mental development are more frequently deleterious than they are advantageous. However, in the general population (excluding MZ twins), this possibility remains only an untested hypothesis. It would be exceedingly difficult to test this hypothesis in samples composed entirely of persons who differ in genotypes, including

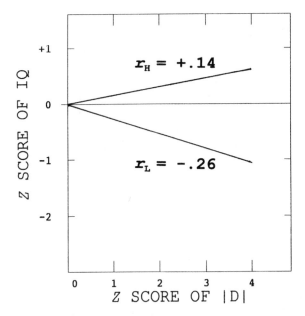

Figure 2.6. The linear regression (or correlation, *r*) of the IQs of the higher-(H) and lower-(L) scoring twins in each pair on the absolute twin differences, |*D*|. Both IQ and |*D*| were transformed to *z* scores, with the H and L intercepts both set at 0.

dizygotic (DZ) twins, because of positive genotype × environment covariance in IQ, which is a component of intrapair differences in DZ twins (and single-born siblings). Even if both genetic and environmental effects were perfectly symmetrical, on average, for DZ twins or siblings, the effects of GE covariance would not necessarily be symmetrical. Persons with different genotypes elicit and seek different environmental conditions, which may interact nonadditively with genetic effects.

Thus, differences due to GE covariance and differences due to WFE are confounded in DZ twin and sibling data. And for differences between genetically unrelated children reared together, there is no way to distinguish WFE variance from genetic variance or GE covariance.

On the other hand, there is little reason to believe that WFE effects are peculiar to MZ twins. Singletons are probably even more subject to such WFE effects than are MZ twins, because single-born siblings are exposed prenatally to differences in mother's age, parity, blood antigens, health, and other conditions that have an impact on development. Such conditions are the same for MZ twins but may differ markedly for singletons.

In summary, the results of these analyses of MZ twin differences in IQ are consistent with the hypothesis that for the vast majority (about 80%) of twins (and probably singletons), the WFE variance results from many small and randomly distributed microenvironmental factors whose net effects in individuals are normally distributed. A minority of MZ twins (about 20%), however, show larger nongenetic deviations in IQ than can be assumed under this model of small, random microenvironmental effects. What could account for these more deviant individuals, of whom the downwardly deviant are the more affected? There are two likely possibilities, and both are probably true. The first is that in a minority of twins, one member of the pair encounters some exceptionally strong, or macroenvironmental, factor that has an impact on mental development. The second possibility is that an unusually unlucky combination of random microenvironmental factors may exceed a critical threshold in their phenotypic consequences such that a nonrandom stochastic process – a snowball effect – alters the trajectory of mental growth for better or (more often) worse.

Effects of the physical microenvironment on IQ

There is enough evidence of physical microenvironmental effects on mental development that this source of IQ variance cannot be ignored. The main reason that these effects have remained far in the background of research on intelligence is that, typically, each effect alone is so small as to be statistically insignificant and unrecognized, except in huge samples, which are rare. Also, in most studies, these effects are confounded with the much greater proportion of genetic variance and GE covariance.

Yet the physical microenvironment may well account for most of the specific or WF variance. Compared with the WF environmental variance to be explained, which is but a small percent (probably not more than about 5%) of the total IQ variance in MZT twins, the part attributable to all physical microenvironmental effects is large. It should be realized that any individual is not affected by more than some very small proportion of the total population of microenvironmental effects. In large samples, the physical factors with the strongest effects on IQ are so infrequent in the population that, in a random sample, each factor contributes a barely detectable increment to a multiple correlation with IQ. The effect has little chance statistically of being replicated in studies based on smaller-sized samples. Thus, it is highly likely that a significant correlation between IQ and a single physical variable found in one small-sample study will not replicate in another small-sample study. The significant correlation in the first study then is discarded as sampling error, and the variable in question escapes further investigation.

One example is seen in the famous study of 19 pairs of MZ twins reared apart (MZA), by Newman, Freeman, and Holzinger (1937). They found a correlation of +.51 (*p* < .05) between intrapair absolute differences in Stanford-Binet IQ and intrapair absolute differences in fingerprint ridge count. The differences have to be nongenetic, of course, but whatever caused the intrapair differences in fingerprints must have occurred at some time during the first trimester of gestation, because fingerprints are fully developed by the fourth prenatal month. The set of twins (Gladys and Helen) with the largest intrapair IQ difference (24 points) showed by far the largest difference in fingerprints. Fingerprints usually differ between MZ twins no more than the fingerprints of a person's right and left hands. The larger MZ differences are due to epigenetic biological noise in embryonic development. Intrapair differences in MZ twins' palm-print ridge counts, attributable to developmental noise, are also significantly correlated with intrapair differences in certain personality traits measured by the MMPI (Rose, Reed, & Bogle, 1987).

On the other hand, the Minnesota Twin Study, with 48 MZA pairs, found a near-zero correlation between intrapair differences in fingerprints and IQs (Thomas Bouchard, personal communication). Small but significant effects that fail to replicate in small-sample studies are obviously hard to distinguish from Type I error. Feasible solutions are meta-analysis of the statistics from many studies, and analysis of mental measurements of MZ twins specially selected for much larger than the average intrapair differences either in IQ or in various physical characteristics.

In seeking clues to the nature and developmental timing of microenvironmental effects, MZ twin studies should correlate intrapair differences in a variety of physical variables with intrapair differences in IQ. Unfortunately, there have been few studies of this type. The results of one such study (Burks, 1940), based on 20 MZT twins, are shown in Table 2.4. The correlations are presented with their 95% confidence intervals; thus, three of the correlations in the first column are significant at *p* < .05. Note that twin intrapair differences in physical traits can be moderately correlated with intrapair differences in IQ, although the physical traits have little or no correlation with IQ (see the last column in Table 2.4). The intrapair differences in physical traits serve merely as signs of developmental noise.

Many physical conditions are correlated with IQ. Some of these were identified in the Collaborative Perinatal Project of the National Institute of Neurological Diseases and Stroke (Broman, Nichols, & Kennedy, 1975). This study, with nearly 27,000 subjects, reported the correlations of 169 prenatal, perinatal, and postnatal variables with the Stanford-Binet IQ at 4 years of age. Of the 169 variables, 32 are related to race, socioeconomic status, and family history. These contribute mainly to the between-families variance and usually involve genetic factors. Of the remaining 137 variables,

Table 2.4. *Correlation (and 95% confidence interval) of IQ and anthropometric measurements in monozygotic twins[a]*

Trait	Correlation between intrapair IQ difference and intrapair difference in physical trait[b]	Correlation between twins[c]	Correlation between IQ and physical trait[d]
IQ		.95 ± .06	
Height	.47 ± .35	.96 ± .06	.17
Weight	.12 ± .43	.98 ± .03	−.02
Leg length	.11 ± .43	.92 ± .09	.29
Trunk length	.40 ± .38	.98 ± .03	.08
Iliac[e]	.41 ± .38	.89 ± .12	−.11

[a] Prepared from information in Burks (1940, pp. 89–90).
[b] Based on average intrapair differences of 20 MZ pairs of both sexes.
[c] Based on 10 pairs of male MZ twins of ages 9 years 7 months to 10 years 6 months.
[d] Based on 21 males (members of 11 twin pairs), ages 9 years 7 months to 10 years 6 months. Confidence intervals not computed because of the high intrapair correlations on these traits.
[e] For a single measure of the iliac taken beyond age 12 years 6 months, the correlation with IQ drops to −.04 ± .46.

most are physical conditions that are environmental as far as the child is concerned; that is, they are not causally related to the child's genotype.

Some of these 137 conditions can be considered aspects of the WF environmental variance. I have classified them into five categories. Within each category, I have tabulated the total number of variables, the number significantly ($p < .001$) correlated with IQ at age 4, and the number of these that can be strictly regarded as a within-family environmental effect (WFE), in that it is unlikely to involve genetic factors and could differ between single-born siblings reared together. This tabulation is shown in Table 2.5. Also given is the variables' mean absolute (unsigned) zero-order correlations with IQ for correlations that are significant at $p < .001$.

The average correlations are quite small, but it should be remembered that each of these physical environmental variables alone affects only some fraction of the population; the increments or decrements in IQ could be considerable for the affected individuals. Also, different individuals are "hit" by different microenvironmental elements and some individuals are "hit" more or less often than others. Thus, even common but singly minute effects can accumulate randomly, with a substantial net effect on IQ for some individuals. When all these variables are combined in a multiple-regression equation to predict IQ, they account for about 4 percent of the

Table 2.5. *Tabulation of physical conditions correlated with IQ in the National Collaborative Prenatal and Perinatal Project and the mean absolute correlation (|r|, p < .001) in each category*

	Type of variable				
	Maternal	Prenatal	Labor and delivery	Neonatal	Infancy and childhood
Total	19	26	30	25	37
p < .001	13	17	14	13	28
WFE	8	3	2	0	2
Mean |r|	.070	.085	.064	.066	.030

Source: Based on Appendix 4, table 1, in a report by Broman, Nichols, & Kennedy, 1975.

total IQ variance. But this is scarcely less than the total WFE percent of the variance, about 5–6% as estimated from the disattenuated MZT correlation.

Some specific within-family microenvironmental effects

Maternal and prenatal factors. As the geneticist Geoffrey Ashton (1986) has stated, "A developing fetus is a special kind of graft in which the fetus is potentially incompatible with the maternal genotype at all polymorphic loci. . . . A reasonable biological hypothesis is that antigenic incompatibility exists at many loci and is expressed through subtle effects during brain development *in utero*. The more homozygous an individual is, the less developmental deficit is incurred" (p. 528). When the gene at one (or more) of the chromosomal loci that controls a particular physical characteristic has identical alleles,[9] such genes are called *homozygous*. Ashton's research, based on genetic markers (blood antigens) for 18 chromosomal loci, found that homozygosity is associated with higher scores on verbal and spatial tests. The genotype of a more homozygous fetus, having less intraindividual genetic variation, is less likely to be incompatible with the mother's genotype. Thus, the mother's greater immunological tolerance protects the more homozygous fetus from the developmental damage that could otherwise result from its antigenic incompatibility with the mother if the fetus were more heterozygous.

Effects of the Rhesus (Rh) blood antigen are well known. When the mother is Rh-negative and the fetus is Rh-positive (as happens when the father is Rh$^+$), the mother builds up antibodies that attack the Rh$^+$ fetus's red blood cells. The effects of the maternal antibodies are so subtle in the first pregnancy as to go undetected, but the antibodies continue to build up in subsequent pregnancies, with serious consequences for the developing fetus, sometimes including stillbirth. Some 7–8% of pregnancies of Rh-negative mothers are at risk for Rh incompatibility. The average (negative) effect on IQ in this group is about –6 IQ points; children born in later ordinal positions show the greater effects (Costiloe, 1969). Children of unknown blood type whose mothers are Rh-negative average about one-half an IQ point lower than children whose mothers are Rh-positive, a statistically significant effect (Mascie-Taylor, 1984). Fortunately, since the 1970s, it has been possible to make Rh-negative mothers immune to the Rh factor by vaccination with a blood extract called *Rh immune globulin*. Its widespread use and the gradual decrease in family size are probably among the many causes of the secular rise in IQ in industrialized countries in recent decades (Flynn, 1987).

But Rh is not the only blood antigen incompatibility that accounts for some fraction of the variance in IQ and other behavioral traits. The ABO blood groups of the mother are also correlated with IQ, although to a lesser degree than the Rh factor (Broman et al., 1975; Mascie-Taylor, 1984). These are only a few of the many blood antigens and other polymorphisms that could affect the developing brain because of mother–fetus incompatibility. Immunoreactive factors have also been invoked to explain the greater incidence of developmental disorders in the male fetal brain, as a male fetus is antigenically apt to be less compatible with the mother than a female fetus (Gaultieri & Hicks, 1985).

Other *prenatal* factors that are correlated (negatively) with IQ are the mother's age, parity (i.e., number of prior pregnancies), X-ray exposure during pregnancy, the mother's smoking or excessive use of alcohol or drugs during pregnancy, fever during pregnancy, a shorter or longer than normal period of gestation, maternal diabetes, and placental abnormalities (Broman et al., 1975; Mascie-Taylor, 1984). When a placental abnormality is present, the difference in IQ between MZ twins is related to whether they are monochorionic or dichorionic (Melnick, Myrianthopoulos, & Christian, 1978). (The chorion is the outer embryonic membrane.) Lower birth weight is related to lower IQ. That the effect is not genetic is shown by the fact that the MZ twin with the lower birth weight usually has a lower IQ at school age, a result explained by the twins' unequal sharing of nutrients during gestation (Churchill, Neff, & Caldwell, 1966; Scarr, 1969; Willerman & Churchill, 1967).

Several *perinatal* factors are correlated (negatively) with IQ. Anoxia at birth, usually a result of premature separation of the placenta from the uterus, can have a drastic effect on early psychomotor development, but the deficit in IQ generally diminishes throughout childhood, averaging about 3 IQ points by 7 years of age (Corah, Anthony, Painter, Stern, & Thurston, 1956). Mother's pelvic size and the fetus's head position during delivery (Willerman, 1970a,b) and breech delivery (Broman et al., 1975) are also related to the child's later IQ.

The purported negative relation between *birth order* and IQ has often been regarded as a clear-cut example of a within-family environmental effect. But even the existence of any birth order effect on IQ, when all the methodological artifacts in this research are controlled, has been critically questioned in the most thorough review of the evidence available (Ernst & Angst, 1983). The birth-order effect claimed by some researchers accounts for at most about 2–3% of the total variance in IQ. The only purely psychological theory of the birth-order effect on mental development is the so-called *confluence theory* of Robert Zajonc (1976). This model, however, has been found decisively faulty (Retherford & Sewell, 1991), and other explanations not involving causal factors of a psychological nature better explain the data (Page & Grandon, 1979).

What little effect birth order may have on IQ is perhaps best explained by the increasing probability in successive pregnancies of maternal immune attack on the fetal brain. For example, the relative frequency of type AB blood increases with birth order, suggesting an increasing mother–fetus incompatibility and spontaneous abortion of fetuses with the other blood types in the ABO system. At least 25% of all conceptuses are spontaneously aborted, usually in the early stage of pregnancy. Spontaneous abortion is a critical threshold on a continuum of causal factors, which at subthreshold levels may result in subtle forms of disadvantage that contribute a part of the WFE variance. A considerable body of evidence has been adduced in support of this immunoreactive theory of the effect of birth order on IQ (Foster & Archer, 1979). A decrease in immunoreactive effects because of the gradually decreasing family size in all industrialized countries over the last three generations might account for some part of the secular rise in IQ during this period (Flynn, 1987).

One of the most striking *postnatal* environmental variables found to affect IQ is whether the infant is given breast milk or a formula. In a large ($N = 300$) and methodologically exemplary study in Cambridge, England, children born preterm (under 1,850 g at birth) were fed by tube with either breast milk or a preterm formula. The neonates in both groups were well matched for birth weight, gestation, and other medical variables. The mothers' social class, education, family structure, and other potentially confounding factors were statistically controlled. The experiment continued in the hospital under professional supervision until the babies were discharged or had reached 2,000 g body weight. At 7.5 to 8 years of age, the children who had received breast milk scored, on average, 8.3 IQ points higher on the WISC than those who had received a formula, a difference significant beyond the .0001 level of confidence (Lucas, Morley, Cole, Lister, & Leeson-Payne, 1992). The authors explain this result in terms of nutritional factors that affect brain development and are present only in mothers' milk.

Many other physical health-related factors are probably correlated with IQ, but these have not yet been studied in detail. Lubinski and Humphreys (1992) found that medical and physical well-being are considerably above the norm in the mathematically gifted; they are more highly associated with giftedness even than extreme levels of socioeconomic privilege. It is also likely that common childhood diseases, such as whooping cough, measles, mumps, and chicken pox, could each take a toll on IQ, perhaps of one IQ point. Inoculation against these diseases would prevent this negative effect. As inoculation is a mild induction of the disease that stimulates the body's immune system, it might also favorably affect physical growth, including brain development. Some part of the gradual secular rise in IQ over the past three generations could be attributable to such factors, which, along with improved nutrition, have become widespread in industrialized countries.

Summary

From an evolutionary standpoint, the genetic inheritance, or innateness, of fitness characteristics is essential. The evolutionary process has ensured normal development to the vast majority of every species by biologically programming the ontogeny of their crucial characteristics, at the same time maintaining enough genetic diversity in certain traits for adaptation to changing environmental conditions. In humans, intelligence and the ability to learn are such characteristics. An overly plastic nervous system, with its functions shaped too easily by the environment, would put the organism's adaptive capacity at risk of being wafted this way or that by haphazard experiences. A half-century of research in physical anthropology and behavioral genetics supports the idea that general mental ability, or g, is a fitness trait with increasing cybernetic stability during its course of development. As argued by Moffitt, Caspi, Harkness, & Silva (1993), it is elastic rather than plastic in its temporary deviations from its biologically programmed trajectory. Genetic variance, genotype–environment covariance, and $G \times E$ interaction are the major components of g variance. The variance attributed to shared, or between-families, environmental factors, which is considerable throughout childhood, gradually shrinks to near-zero between early adolescence and maturity. During this period, most of the environmental variance is converted into genotype–environment covariance, as persons elicit, seek, select, and modify those elements of the available cognitive-social-cultural milieu that are most compatible with their genotypically conditioned proclivities.

After such sources of IQ variance have been accounted for, psychologists generally try to explain the one remaining source of variance – the specific, or within-family environment (WFE) – in wholly psychological terms of social learning and possible differences in opportunities and motivation that may exist among full siblings who are reared together. Environmental variables of a biological nature are most often slighted. Yet the physical-biological microenvironment might well contribute most of the WFE variance in g.

According to the microenvironmental theory of WFE, the neural basis of mental development is affected in each individual by a limited number of physical events beginning shortly after conception, each with a biologic effect usually too small to be detected individually. Their reliably detectable effects result from their aggregation in some individuals. These small biologic events are a random selection from among all such microenvironmental events that may affect development. Because they "hit" individuals more or less at random, they vary in both number and kind for different individuals. The net effects of these small, independent physical-environmental influences for individuals are deviations (positive and negative) of individuals' phenotypic IQs from their genotypic values. The statistical properties of these deviations can be inferred from the intrapair IQ differences between MZ twins. Because the net deviations have resulted from many small, independent events, they are normally distributed in the population.

Superimposed on this normal distribution of random environmental effects on IQ is a distribution resulting from a small number of comparatively large environmental effects, more often negative than positive, that "hit" only a fraction of the population. They are attributable to (1) a nonrandom, stochastic snowball effect on a few unlucky individuals who by chance have received a critical preponderance of unidirectional small effects, which increases the likelihood of incurring still more effects in the same direction; and (2) the occurrence of rare events with large effects that "hit" only a small fraction of the population. The composite of these two distributions of net environmental effects forms a population distribution that is leptokurtic, with excess frequencies in the two tails, especially in the tail on the negative side. This, then, is the form of the distribution of phenotypic IQ deviations that behavior-genetic models attribute to the specific, or within-family, environment.

A host of nongenetic but biologic factors – prenatal, perinatal, and early postnatal – are known to affect mental development, each factor alone having only a small effect. But the number of these presently known factors is probably only a fraction of all the biologic factors that affect mental growth. The additive and interactive effects of their random combinations probably accounts for most of the *g* variance ascribed to WFE. If so, it makes more understandable the notably unsuccessful efforts of researchers to produce any bona fide evidence that *g* can be significantly and lastingly raised by any purely psychological or educational means (Detterman & Sternberg, 1982; Jensen, 1989; Spitz, 1986). Because *g* reflects individual differences mainly in the neural mechanisms of information processing, it is more susceptible to biological than to psychological influences.

The secular increase of IQ in industrialized countries over the past three generations can be attributed in part to the widely increased availability of improved health care, nutrition, obstetrical advances, and other factors with biologic effects on mental growth. In First World countries, such benefits have almost universally minimized a significant portion of the micro-environmental factors that negatively affect mental development.

In the picture we see emerging from behavior-genetic analyses of mental abilities, the psychometric construct called *g* appears to be a biological phenomenon with many behavioral correlates, including performance on IQ tests. Some of these correlates are trivial, except as knowledge of them may help to advance understanding of the nature of *g*. However, *g* has correlates of great significance in their own right. The phenomenon represented by *g* is an undoubtedly crucial factor in understanding individual differences in many educationally, economically, and socially important variables.

Notes

1. It is important to distinguish between genotype–environment (GE) *covariance* (or correlation, which is simply the standardized covariance) and genotype–environment *interaction*. They are entirely different concepts, but each may account for some part of the phenotypic variance in a trait.

   GE *covariance* is the result of the nonrandom occurrence of different genotypes in different environments. In other words, genotypes and environments may be correlated. Persons whose genotype is favorable for the development of a certain trait (e.g., musical talent) are more likely than chance to grow up in an environment that is favorable to the development of the trait (e.g., parents with musical interests, opportunity for music lessons, etc.). The correlation of genotypes and environments for a given trait in the population increases the phenotypic variance over what it would be if the correlation were zero. Assuming for simplicity that there is no GE interaction, the total phenotypic (P) variance (V) is the sum of the genetic (G) variance and the environmental (E) variance *plus* twice the covariance (Cov) of G and E, or, as it is expressed in biometrical genetics, $V_P = V_G + V_E + 2\text{CovGE}$. (Regarding $\text{CovGE} = r_{GE}\sqrt{V_G}\sqrt{V_E}$, note that CovGE depends on there being substantial values of V_G and V_E; if either one is zero, there can be no GE covariance, or correlation either, because GE correlation depends on variance in both genetic and environmental effects.)

GE *interaction* is a component of the phenotypic variance that is due to different genotypes reacting differently to the same environmental condition. That is, an environmental condition that favors the phenotypic development of individuals who have genotype A may have no effect, or may even have an unfavorable effect, on individuals who have genotype B. A classic example is a condition known as *galactosemia*. Most infants thrive on milk, but a small number have a genotype that prevents their normally metabolizing milk, and the abnormal metabolites damage the infant's brain, resulting in severe mental retardation. Another example: A pair of orphaned monozygotic twins (hence, identical genotypes) separated in infancy, one reared by a very unmusical family, the other by an intensely musical family; neither twin even shows any sensitivity to music or develops any interest in it. Another pair of MZ twins in the identical circumstances shows a very different outcome: The twin reared in the unmusical family shows little sensitivity or interest in music, while the twin reared in the musical family turns out to be a highly accomplished musician. One set of twins (i.e., one genotype) is insensitive to the musical environment, and the other set (i.e., another genotype) is highly sensitive to a musical environment if exposed to music. The phenotypic variance in musicality among these two sets of twins would have a large component of GE *interaction*.

2. The statement – "the variance *between* families is the covariance *within* families" – is most easily explained in terms of the analysis of variance and its relation to the intraclass correlation. If we perform a simple one-way analysis of variance on a population of persons grouped in families, we arrive at three variances: the Between-Families variance (V_{BF}), the Within-Families variance (V_{WF}), and the Total variance ($V_T = V_{BF} + V_{WF}$). The correlation between persons *within* families is the *intraclass correlation*, which is $r_i = V_{BF}/V_T$. But a correlation coefficient is just a standardized covariance – that is, a covariance divided by the total variance. So if we multiply the intraclass *correlation* (between the persons within families) by the total variance, we have the *covariance* between family members, which is equal to V_{BF}. Therefore, the variance between families is the covariance within families.

3. One rarely sees corrections for attenuation in behavior-genetic literature, although it is often called for when the aim is to estimate components of variance that certainly comprise some variance due to measurement error, which is unique to the particular measuring instrument and is of no theoretical interest. Surely if error-free measurements of the variables of interest were available, investigators would prefer to use them. Although there are problems with the correction for attenuation, such as the reliability of the reliability coefficient itself, it is still possible to more closely approximate the true-score variance components by correction for attenuation than by not correcting at all; noncorrection, in effect, assumes perfect reliability, which we know is impossible. We cannot achieve perfection with the correction for attenuation, but we can come somewhat nearer to the error-free values of the correlations and variance components if we do make the correction, provided we have a reliable reliability coefficient. Therefore, I consider it preferable in the present analysis to correct for attenuation, using a "best estimate" of the measurement's reliability coefficient. The best estimate for intelligence tests, based on the standardization data of a variety of individual and group tests, is .90 (see Jensen, 1980, chapter 7). (Specifically for the Stanford-Binet, beyond age 6, the best estimate is .93.) The most relevant reliability is test–retest or equivalent forms reliability rather than internal consistency (Kuder-Richardson) reliability, although these two conceptually distinct types of reliability are usually of comparable magnitude.

4. By *psychosocial-cultural*, I mean environmental influences that arise from the individual's subjective waking experiences that involve personal interactions, identification with role

models, learning opportunities, language, customs, parental and peer demands, and values and interests acquired in the individual's environment. By *biological* or *physical* environmental influences, I refer to factors, both endogenous and exogenous, that directly impinge on and affect the individual's anatomy and physiology, or that directly affect the physical growth process of any organ system, particularly the nervous system. By *directly affect*, I mean that conditioning, learning, and awareness are not the agencies of the influence. Examples of physical influences on an individual are the mother's health and nutrition during the individual's prenatal development, perinatal anoxia, childhood diseases, malnutrition, brain injury, hormonal imbalance, sensory defects, and the like. Other physical environmental effects are mentioned later on in this chapter.

5. A *polygenic* trait is one whose genetic variance is contributed by genes at two or more chromosomal loci and for which all the genes have small and more or less equal effects, whether their alleles have additive effects, or are interactive within the same locus (i.e., dominance) or between different loci (i.e., epistasis). Intelligence is a polygenic trait; the number of genes involved in IQ variance has been variously estimated in the genetics literature at between 20 and 100, although these numbers are not taken very seriously, they represent reasonable limits within which the true value probably falls.

6. The question arises of whether a sample size of 180 MZ twin pairs can afford the statistical power needed for the analysis of WF variance, which is a small proportion (about .10) of the total phenotypic variance. Neale and Cardon (1992, chapter 9) point out that enormously larger samples of twins than that used here are needed to estimate certain small variance components (e.g., the shared environmental variance) in genetic analyses with a satisfactory degree of statistical confidence, such as $p < .05$. The main concern of the present analysis, however, is not with the estimation of the relative size of the BF and WF environmental components of variance (which in any case would not be possible using only MZ twins reared together) but with the form of the distribution of IQ differences between MZ twins and whether this distribution conforms to the distribution of effects that are predicted from a model of environmental effects that are purely random. The twin data show that certain statistics of the obtained distribution of twin differences depart from the corresponding parameters (σ and CV = σ/μ) of the theoretical distribution predicted by the random effects model at $p < .05$ and $p < .01$, respectively. Also, the highly significant ($p < .001$) asymmetry of the lower- and higher-scoring twins (Figure 2.6) contradicts the random-effects model.

7. The right-hand ($+z$) half of the normal curve has the following parameters, where z is a standardized deviate (x/σ):

$$\begin{aligned}
\text{Range:} \quad & 0 \text{ to } +\infty \\
\text{Mean:} \quad & \mu_{+z} = \sqrt{2}/\sqrt{\pi} = +.79788z \\
\text{Standard deviation:} \quad & \sigma_{+z} = \sqrt{1/2} = .7071z \\
\text{Coefficient of variation:} \quad & \text{CV} = \sigma/\mu = .8862
\end{aligned}$$

8. It is noteworthy that the same distinctive features seen in Figure 2.5 are also found when such a plot is performed on Burt's (1966) reputed data on 53 pairs of MZ twins reared apart (MZA), with their IQ correlation of .771. Since 1976, Burt's MZA results have been excluded from all meta-analyses in behavior genetics because of their questioned authenticity. Yet there is no evidence in Burt's publications that the idea of plotting twin differences in this fashion had ever occurred to him. If he had faked his MZA data, as alleged by his detractors, he would have to be credited with clairvoyant intuition.

9. Every normal person's 46 chromosomes come in 23 identifiable homologous pairs, one chromosome from each parent. The genes, each at different loci on a chromosome (like beads on a string) control the production of enzymes, which in turn affect the development of all of the body's physical structures and functions. Many genes (called *polymorphic* or *segregating genes*) have two or more forms, called *alleles* (or *allelomorphs*), which all have somewhat different developmental effects on the same system. When a gene at a particular

chromosomal locus has identical alleles (e.g., AA or aa, instead of Aa) in the two chromosomes, it is said to be *homozygous*. When the alleles of a gene at a particular chromosomal locus are different (e.g., Aa instead of AA or aa) in the two chromosomes, the gene is said to be *heterozygous*.

References

Ashton, G. C. (1986). Blood polymorphisms and cognitive abilities. *Behavior Genetics, 16*, 517–29.

Bailey, J. M., & Horn, J. M. (1986). A source of variance in IQ unique to the lower-scoring monozygotic (MZ) twin. *Behavior Genetics, 16*, 509–16.

Bouchard, T. J., Jr., Lykken, D. T., McGue, M., Segal, N., & Tellegen, A. (1990). Sources of human psychological difference: The Minnesota study of twins reared apart. *Science, 250*, 223–8.

Bouchard, T. J., Jr., Lykken, D. T., McGue, M., Segal, N. L., & Tellegen, A. (1991). IQ and heredity. *Science, 252*, 191–2.

Bouchard, T. J., Jr., & McGue, M. (1981). Familial studies of intelligence: A review. *Science, 212*, 1055–9.

Bouchard, T. J., Jr., & Segal, N. L. (1985). Environment and IQ. In B. B. Wolman (Ed.), *Handbook of intelligence* (pp. 391–464). New York: Wiley.

Broman, S. H., Nichols, P. L., & Kennedy, W. A. (1975). *Preschool IQ: Prenatal and early developmental correlates*. Hillsdale, NJ: Erlbaum.

Bulmer, M. G. (1970). *The biology of twinning in man*. Oxford: Clarendon.

Burks, B. S. (1940). Mental and physical developmental patterns of identical twins in relation of organismic growth theory. In G. W. Whipple (Ed.), *Intelligence: Its nature and nurture* (Part II, 399th Yearbook of the NSSE). Bloomington, IL: Public School Publishing Co.

Burt, C. (1966). The genetic determination of differences in intelligence: A study of monozygotic twins reared together and apart. *British Journal of Psychology, 57*, 137–53.

Cardon, L. R., Fulker, D. W., & Plomin, R. (1992). Multivariate genetic analysis of secific cognitive abilities in Colorado Adoption Project at age 7. *Inelligence, 16*, 383–400.

Chipuer, H. M., Rovine, M. J., & Plomin, R. (1990). LISREL modeling: Genetic and environmental influences on IQ revisited. *Intelligence, 14*, 11–29.

Churchill, J. A., Neff, J. W., & Caldwell, D. F. (1966). Birth weight and intelligence. *Obstetrics and Gynecology, 28*, 425–9.

Corah, N. L., Anthony, E. J., Painter, P., Stern, J. A., & Thurston, D. (1956). Effects of perinatal anoxia after seven years. *Psychological Monographs, 79*, No. 3, 1–34.

Costiloe, T. M. (1969). Ordinal position in sibship and mother's Rh status among psychological clinic patients. *American Journal of Mental Deficiency, 74*, 10–16.

Darlington, C. D. (1954). Heredity and environment. *Proceedings of the 9th International Congress of Genetics, 9*, 370–81.

Detterman, D. K., & Sternberg, R. J. (Eds.) (1982). *How and how much can intelligence be increased*. Norwood, NJ: Ablex.

Edelman, G. M. (1987). *Neural Darwinism: The theory of neuronal group selection*. New York: Basic Books.

Ernst, C., & Angst, J. (1983). *Birth order: Its influence on personality*. Berlin: Springer-Verlag.

Fisher, R. A. (1918). Correlation between relatives on the supposition of Mendelian inheritance. *Transactions of the Royal Society of Edinburgh, 52*, 399–433.

Fisher, R. A. (1970). *Statistical methods for research workers* (14th ed.). New York: Hafner.

Flynn, J. R. (1987). Massive IQ gains in 14 nations: What IQ tests really measure. *Psychological Bulletin, 101*, 171–91.

Foster, J. W., & Archer, S. J. (1979). Birth order and intelligence. *Perceptual and Motor Skills, 48*, 79–93.

86 JENSEN

Galton, F. (1908/1974). *Memories of my life*. New York: AMS Press.

Gaultieri, T., & Hicks, R. E. (1985). An immunoreactive theory of selective male affliction. *Behavioral and Brain Sciences, 8*, 427–41.

Hirsch, N. D. M. (1930). *Twins: Heredity and environment*. Cambridge: Harvard University Press.

Jensen, A. R. (1987). The g beyond factor analysis. In J. C. Connoly, J. A. Glover, & R. R. Ronning (Eds.), *The influence of cognitive psychology on testing and measurement*. Hillsdale, NJ: Erlbaum.

Jensen, A. R. (1989). Raising IQ without increasing g. *Developmental Review, 9*, 234–58.

Jensen, A. R. (1992). Commentary: Vehicles of g. *Psychological Science, 3*, 275–8.

Jensen, A. R. (1993). Psychometric g and achievement. In B. R. Gifford (Ed.), *Policy perspectives on educational testing* (pp. 117–227). Boston: Kluwer.

Jinks, J. L., & Fulker, D. W. (1970). Comparison of the biometrical, genetical, MAVA, and classical approaches to the analysis of human behavior. *Psychological Bulletin, 73*, 311–49.

Juel-Nielsen, N. (1965). Individual and environment: A psychiatric-psychological investigation of MZ twins reared apart. *Acta Psychiatrica Scandinavia*, Supplement 183. Cophenhagen: Munksgaard.

Kendall, M., & Stuart, A. (1977). *The advanced theory of statistics* (4th ed.) (Vol. 1). New York: Macmillan.

Longstreth, L. E., Davis, B., Carter, L., Flint, D., Owen, J., Richert, M., & Taylor, E. (1981). Separation of home intellectual environment and maternal IQ as determinants of child IQ. *Developmental Psychology, 17*, 532–41.

Lubinski, D., & Humphreys, L. G. (1992). Some bodily and medical correlates of mathematical giftedness and commensurate levels of socioeconomic status. *Intelligence, 16*, 99–115.

Lucas, A., Morley, R., Cole, T. J., Lister, G., & Leeson-Payne, C. (1992). Breast milk and subsequent intelligence quotient in children born preterm. *Lancet, 339*, 261–4.

Luo, D., Petrill, S. A., & Thompson, L. A. (1994). An exploration of genetic g: Hierarchical factor analysis of cognitive data from the Western Reserve Twin Project. *Intelligence, 18*, 335–47.

Lykken, D. T., McGue, M., Tellegen, A., & Bouchard, T. J., Jr. (1992). Emergenesis: Traits that may not run in families. *American Psychologist, 47*, 1565–77.

Mascie-Taylor, G. C. N. (1984). Biosocial correlates of IQ. In C. J. Turner & H. B. Miles (Eds.), *The biology of human intelligence*. Proceedings of the 20th Annual Symposium of the Eugenics Society, London.

McGue, M., Bkouchard, T. J., Iacono, W. G., & Lykken, D. T. (1993). Behavioral genetics of cognitive ability: A lifespan perspective. In R. Polomin & G. E. McClearn (Eds.), *Nature, nurture & psychology*. Washington, DC: American Psychological Association.

McNemar, Q. (1942). *The revision of the Stanford-Binet Scale*. Boston: Houghton Mifflin.

Meehl, P. E. (1978). Theoretical risks and tabular asterisks: Sir Karl, Sir Ronald, and the slow progress of soft psychology. *Journal of Consulting and Clinical Psychology, 46*, 806–34.

Melnick, M., Myrianthopoulos, N. C., & Christian, J. C. (1978). The effects of chorion type on variation in IQ in the NCPP twin population. *American Journal of Human Genetics, 30*, 425–433.

Merriman, C. (1924). The intellectual resemblance of twins. *Psychological Monographs, 33*, 1–49.

Moffitt, T. E., Caspi, A., Harkness, A. R., & Silva, P. A. (1993). The natural history of change in intellectual performance: Who changes? How much? Is it meaningful? *Journal of Child Psychology and Psychiatry, 14*, 455–506.

Molenaar, P. C. M., Boomsma, D. I., & Dolan, C. V. (1993). A third source of developmental differences. *Behavior Genetics, 23*, 519–24.

Neale, M. C., & Cardon, L. R. (1992). *Methodology for genetic studies of twins and families*. Boston: Kluwer.

Newman, H. H., Freeman, F. N., & Holzinger, K. J. (1937). *Twins: A study of heredity and environment.* Chicago: University of Chicago Press.

Osborne, R. T. (1980). *Twins: black and white.* Athens, GA: Foundation for Human Understanding.

Page, E. B., & Grandon, G. M. (1979). Family configuration and mental ability: Two theories contrasted with U.S. data. *American Educational Research Journal, 16,* 257–72.

Pedersen, N. L., Plomin, R., Nesselroade, J. R., & McClearn, G. E. (1992). A quantitative genetic analysis of cognitive abilities during the second half of the life span. *Psychological Science, 3,* 346–53.

Plomin, R. (1986). *Development, genetics, and psychology.* Hillsdale, NJ: Erlbaum.

Plomin, R. (1988). The nature and nurture of cognitive abilities. In R. J. Sternberg (Ed.), *Advances in the psychology of human intelligence* (Vol. 4). Hillsdale, NJ: Erlbaum.

Plomin, R., & Bergeman, C. S. (1991). The nature of nurture: Genetic influences on "environmental" measures. *Behavioral and Brain Sciences, 14,* 373–427.

Plomin, R., & Daniels, D. (1987). Why are children in the same family so different from one another? *Behavioral and Brain Sciences, 10,* 1–60.

Plomin, R., DeFreies, J. C., & Loehlin, J. C. (1977). Genotype–environment interaction and correlation in the analysis of human behavior. *Psychological Bulletin, 84,* 309–22.

Plomin, R., & Neiderhiser, J. M. (1992). Genetics and experience. *Current Directions in Psychological Science, 1,* 160–3.

Retherford, R. D., & Sewell, W. H. (1991). Birth order and intelligence: Further tests of the confluence model. *American Sociological Review, 56,* 141–58.

Rosanoff, A. J., Handy, L. M., & Plesset, I. R. (1937). The etiology of mental deficiency with special reference to its occurrence in twins. *Psychological Monographs, 48*(4).

Rose, R. J., Reed, T., & Bogle, A. (1987). Asymmetry of *a-b* ridge count and behavioral discordance on monozygotic twins. *Behavior Genetics, 17,* 125–40.

Rowe, D. C. (1987). Resolving the person–situation debate. *American Psychologist, 42,* 218–27.

Rowe, D. C., & Plomin, R. (1981). The importance of nonshared (E1) environmental influences in behavioral development. *Developmental Psychology, 17,* 517–31.

Scarr, S. (1969). Effects of birth weight on later intelligence. *Social Biology, 16,* 249–56.

Scarr, S. (1985). Constructing psychology: Facts and fables for our times. *American Psychologist, 40,* 499–512.

Scarr, S., & Carter-Saltzman, L. (1982). Genetics and intelligence. In R. J. Sternberg (Ed.), *Handbook of human intelligence.* Cambridge: Cambridge University Press.

Scarr, S., & McCartney, K. (1983). How people make their own environments: A theory of genotype–environment effects. *Child Development, 54,* 424–35.

Scarr, S., & Weinberg, R. A. (1978). The influence of "family background" on intellectual attainment. *American Sociological Review, 43,* 674–92.

Shields, J. (1962). *Monozygotic twins brought up apart and brought up together.* London: Oxford University Press.

Sitz, H. H. (1986). *The raising of intelligence: A selected history of attempts to raise retarded intelligence.* Hillsdale, NJ: Erlbaum.

Stocks, P. (1930, 1933). A biometric investigation of twins and their brothers and sisters. *Annals of Eugenics, 4,* 49–108, *5,* 23–5.

Teasdale, T. W., & Owen, D. R. (1984). Heredity and familial environment in intelligence and educational level – A sibling study. *Nature, 309,* 620–2.

Thorndike, R. L., Fleming, C. W., Hildreth, G., & Stranger, M. (1940). Retest changes in IQ in certain superior schools (pp. 351–61). In National Society for the Study of Education, *39th yearbook.*

Willerman, L. (1970a). Fetal head position during delivery, and intelligence. In C. R. Angle & E. A. Bering (Eds.), *Physical trauma as an etiological agent in mental retardation* (pp. 105–8). Washington, DC: U. S. Dept. of Health, Education, and Welfare.

Willerman, L. (1970b). Maternal pelvic size and neuropsychological outcome. In C. R. Angle & E. A. Bering (Eds.), *Physical trauma as an etiologic agent in mental retardation* (pp. 109–12). Washington, DC: U. S. Dept. of Health, Education, and Welfare.

Willerman, L. (1979). Effects of families on intellectual development. *American Psychologist, 34*, 923–9.

Willerman, L. (1987, April). *Where are the shared environmental influences on intelligence and personality?* Paper presented at the Society for Research on Child Development, Baltimore, MD.

Willerman, L., & Churchill, J. A. (1967). Intelligence and birth weight in identical twins. *Child Development, 38*, 623–9.

Wilson, R. S. (1974). Twins: Mental development in the preschool years. *Developmental Psychology, 10*, 580–8.

Wilson, R. S. (1983). The Louisville Twin Study: Developmental synchronies in behavior. *Child Development, 54*, 298–316.

Wingfield, A. H., & Sandiford, P. (1928). Twins and orphans. *Journal of Educational Psychology, 19*, 410–23.

Zajonc, R. (1976). Family configuration and intelligence. *Science, 192*, 227–36.

3 Identifying genes for cognitive abilities and disabilities

Robert Plomin

I predict that the next generation of psychologists will wonder what all the nature–nurture fuss was about. It will seem obvious that genetic differences as well as environmental differences contribute to individual differences in cognitive abilities and disabilities. This prediction relies on the hope that psychology will continue to be a science, because in science, data reign supreme, and the data are clear: Genetics is important. Converging evidence from family, twin, and adoption studies provides a better case for significant and substantial genetic contributions to cognitive abilities, especially general cognitive ability (intelligence), than for any other domain of behavior or medicine (Plomin, Owen, & McGuffin, 1994b). Few scientists seriously dispute any longer the conclusion that cognitive abilities show significant genetic influence. (See, chapters 4 and 5, this volume, for discussions of this issue.)

The magnitude of genetic influence is still not universally appreciated, however. For general cognitive ability, the world's literature suggests that about half of the total variance in IQ scores can be accounted for by genetic variance (Chipuer, Rovine, & Plomin, 1990; Loehlin, 1989). It should be noted that the total variance includes error variance. Correcting for unreliability of measurement, heritability would be higher. Also, the world's literature includes disproportionate numbers of young children. As discussed below, new evidence indicates that adults show greater heritability. Estimating the magnitude of the genetic effect is more difficult than determining its statistical significance, but regardless of the precise estimate of heritability (the genetic effect size), the point is that genetic influence on IQ test scores is not only significant; it is also very substantial.

The conclusion that genetic contributions to individual differences in cognitive abilities are significant and substantial is one of the most important facts that have been uncovered in research on cognitive abilities. However, even when these conclusions about *whether* and *how much* are fully accepted, we are much closer to the beginning than the end of the story of heredity and cognitive abilities (Plomin & Neiderhiser, 1991). The rest of

89

the story is the story beyond heritability. The purpose of this chapter is to peek at one of the story's later plots that is only now being written. It is the story of the *new genetics* that will revolutionize genetic research on cognitive abilities by identifying specific genes that contribute to genetic influence.

During the past decade, advances in molecular biology have led to the dawn of a new era for genetic research on cognitive abilities. The new genetics makes it possible to identify specific genes responsible for genetic influence on cognitive abilities. Finding genes that account for even a small portion of the genetic contribution will revolutionize quantitative genetic research by making it possible to identify relevant genotypes directly from a few drops of blood or a few cells scraped from the lining of the cheek rather than resorting to indirect inferences of genetic influence from twin and adoption studies.

Because DNA markers and linkage play a major role in the new genetics, we begin with a brief description of DNA markers and linkage.

DNA markers

Until the 1980s, classical single-gene genetic markers were limited to a few dozen gene products in the blood and urine, such as the ABO blood types. A key element of the new genetics was the discovery of thousands of DNA markers spread throughout our 23 pairs of chromosomes. DNA markers are genetic polymorphisms (differences among individuals) in DNA itself rather than in gene products. Although space does not permit a detailed description of these DNA markers, a brief history is warranted.

The first DNA markers were discovered in 1980 and are called *restriction fragment-length polymorphisms* (RFLP). Restriction enzymes are special enzymes that cut DNA at a particular sequence of nucleotide bases. DNA comprises four nucleotide bases – G, C, A, and T – with G always pairing with C, and A always pairing with T, as steps in the spiral staircase that is the double helix of DNA. Our total genome consists of more than 3 billion such pairs of nucleotide bases that form the genetic code. For example, one of the hundreds of restriction enzymes recognizes a nucleotide base sequence of GAATTC and severs the DNA molecule between the G and A bases whenever the enzyme encounters this sequence. If an individual happens to have a different nucleotide base anywhere in this sequence, the restriction enzyme will not cut that individual's DNA at that place. This results in a longer DNA fragment for that individual compared to other individuals whose DNA is cut at the step by the restriction enzyme. Thus, this type of DNA marker received its name because it is a DNA fragment-length polymorphism detected by restriction enzymes.

RFLP have two alleles that represent the presence or absence of a particular nucleotide base site recognized by a restriction enzyme. In the early 1980s, a type of DNA marker was discovered that has many alleles, which is better for linkage analysis (described in the next section). For unknown reasons, as much as one-third of the human genome consists of repetitive DNA sequences. Some sequences that are several hundred base pairs in length repeat many times; these are called *tandem repeats*. The number of repeats differs for individuals. When a restriction enzyme is used to cut a DNA fragment with one of these repeat elements in it, the resulting fragment will differ in length across individuals because individuals differ in the number of repeats. This type of DNA marker is called a *variable number tandem repeat* (VNTR) polymorphism or a *minisatellite repeat marker*.

In 1987, another type of repeat marker, called *microsatellite repeats*, or *short-sequence repeats* (SSR), was discovered that also has many alleles. Microsatellite repeat markers are polymorphisms involving DNA that, again for unknown reasons, repeat a sequence of two, three, or four nucleotide bases. The human genome is estimated to contain as many as 50,000 such microsatellite repeats; more than a thousand have already been identified. Moreover, genotyping microsatellite repeat markers can be automated. For these reasons, the human genome project has now switched to this type of marker to chart the human genome. In 1992, an evenly spaced set of 813 such markers was made available for linkage analysis (Weissenbach et al., 1992).

In summary, at the foundation of the new genetics are thousands of new genetic markers that are polymorphisms in DNA itself, not just in a gene product. The three major categories of DNA markers are RFLP, minisatellite repeats, and microsatellite repeats. These markers detect different types of DNA differences among individuals, but many DNA differences are not detected by any of them. The next generation of DNA markers will be determined by direct sequencing of nucleotide bases, which will detect every DNA difference.

Linkage. A few hundred of these DNA markers, especially highly polymorphic microsatellite repeat markers, spread at roughly equal intervals throughout the genome, can be used with traditional linkage analysis to locate the chromosome and then the spot on the chromosome where the gene responsible for a single-gene disorder resides. Linkage is a violation of Gregor Mendel's second law of inheritance, which states that genes assort independently. Genes do assort independently unless the genes happen to be close together on the same chromosome. If chromosomes were inherited intact, a particular pair of alleles on the same chromosome would always be inherited together. However, a process called *crossing-over* occurs in which members of a chromosome pair come into contact and exchange parts. For this reason, after many generations, crossing-over separates alleles for two genes on the same chromosome unless the two genes are very close together. In contrast, within a family, crossing-over will separate particular combinations of alleles for genes on the same chromosome only if the genes are far apart on the chromosome. Highly polymorphic DNA markers with many alleles such as microsatellite repeat markers are valuable because both chromosomes of both parents are likely to have different alleles, which makes it possible to track unambiguously the parental origin of both of the child's chromosomes. Linkage assesses cosegregation between a DNA marker and a single-gene trait within families. In this way, linkage can localize genes without a priori knowledge of gene products. This was first accomplished in 1983 using RFLP markers to determine that the gene for Huntington's disease, a single-gene disorder that leads to neural degeneration later in life, is near the tip of chromosome 4.

Once linkage obtains the general chromosomal address of a gene, finer-grained linkage analyses with more DNA markers in the region can zoom in on the gene itself. When markers are found on each side of the gene, the gene can be fished out and cloned – that is, inserted in bacterial DNA so that the gene can be replicated at will. Then the product of the gene can be studied in order to understand its pathological process.

Single genes and cognitive disabilities

Thousands of disorders have been identified that show simple Mendelian patterns of inheritance for which defects in a single gene are the necessary and sufficient cause of the disorder (McKusick, 1990). A single gene is necessary in the sense that the disorder occurs only when the particular defective form (allele) of the gene is present; it is sufficient in the sense that the allele produces the disorder independent of other genetic and environmental factors.

There are several examples of such single-gene causes of subtypes of mental retardation. For these, the new genetics guarantees that the responsible genes will be identified, and several have already been identified. The new frontier for molecular genetics lies with common and complex dimensions such as cognitive abilities. The challenge is to use molecular genetic techniques to identify genes involved in complex systems influenced by multiple genes as well as multiple nongenetic factors.

The goal of the rest of this chapter is to introduce molecular genetic techniques and to describe recent findings relevant to cognitive abilities and disabilities. The place to begin is single-gene causes of cognitive disabilities.

Most of these are rare, and, although often devastating for affected individuals, they contribute little to genetic variability in the normal range of
cognitive ability.

More than 100 rare single-gene disorders include mental retardation
among their symptoms (Walhsten, 1990). In this section, the classic example
of phenylketonuria and two newer examples are discussed – fragile X,
which is the most common single-gene cause of mental retardation, and
Alzheimer's disease, which is a common dementia among the elderly.

Phenylketonuria (PKU). In the early 1930s, a dentist in Norway had two
retarded children who exuded a peculiar odor that so aggravated his asth-
matic condition that he was unable to stay with them in a closed room. The
dentist had the children examined by Fölling, who began the search for the
cause of the odor by analyzing the children's urine. The search quickly paid
off in the isolation of excess phenylpyruvic acid in their urine. Fölling
postulated that the excess phenylpyruvic acid was due to a disturbance in
the metabolism of phenylalanine, an essential amino acid, and that this
somehow was related to the children's retardation. Because neither parent
showed this disorder, Fölling predicted that it was due to a single recessive
gene. That is, if both parents carried one copy (allele) of the recessive gene,
they would not themselves display the trait, but their children could inherit
a defective allele from each parent and thus display the trait.

The disorder became known as *Fölling's disease* or *phenylketonuria*
(PKU). Other retarded children with an excess of phenylpyruvic acid were
found. Although the incidence of PKU is low – about 1 in 10,000 births –
PKU accounted for about 1% of institutionalized severely mentally retarded individuals. Family studies confirmed the hypothesis that the disorder was inherited as a single recessive gene. Biochemical research revealed
that the metabolic problem is caused by the inactivity of a particular enzyme, phenylalanine hydroxylase. This enzyme converts phenylalanine to
tyrosine. If this conversion is blocked, phenylalanine levels increase in the
blood, and phenylpyruvic acid accumulates in the urine. The high level of
phenylalanine in the blood depresses the level of other amino acids, depriv-
ing the developing nervous system of needed nutrients.

This knowledge concerning the biochemical origin of PKU led to a
rational therapy. If PKU mental retardation is caused by a buildup of
phenylalanine, the amount of phenylalanine in the diet can be reduced.
Phenylalanine is found in a wide variety of foods, particularly meats. In
1953 a special diet was prepared that was very low in phenylalanine. Although it did not improve the cognitive ability of older PKU children, it
largely prevented retardation when the diet was administered to very young
children.

During the 1980s, the new genetics using DNA markers and a technique called *linkage analysis* showed that the gene for phenylalanine hydroxylase (PAH) is on chromosome 12 and that PKU is caused by mutations in the PAH gene. The PAH gene turned out to be huge, involving more than 100,000 nucleotide bases. Moreover, several different mutations in the PAH gene were found to cause PKU (Woo, 1991). Multiple markers are available that together can detect the majority of the PAH mutations that result in PKU. When more than 90% of the PKU alleles can be detected, screening can be implemented to identify carriers who have only one copy of a PKU allele and do not show PKU symptoms. Although PKU is rare, the PKU allele is not. Only about 1 in 10,000 individuals have two recessive PKU alleles, but about 1 in 50 individuals are carriers. This is typically the case for recessive disorders because most of the PKU alleles exist in carriers of a single PKU allele who do not show the disorder and for this reason are not selected out by natural selection via reduced reproduction. However, if PKU carriers mate, they have a 25% chance of producing an affected offspring with two copies of the PKU allele.

In summary, PKU is a distinct form of mental retardation caused by a single recessive gene that is necessary and sufficient to cause the disorder. DNA markers and linkage were used to identify the PAH gene responsible for the disorder.

Fragile X and mental retardation. In 1991, a single gene responsible for another distinct type of mental retardation, fragile X, was discovered (Verkerk et al., 1991). Fragile X involves a CGG microsatellite repeat in the FMR-1 gene on the X chromosome. Its incidence is 1 in 1,250 males and 1 in 2,500 females, making it the single most important cause of mental retardation after Down's syndrome, which is a chromosomal anomaly caused by inheriting an extra copy of chromosome 21. A surprise lay hidden in the FMR-1 gene for fragile X. Although the number of CGG repeats for a particular allele is heritable across generations, for unknown reasons, the number of CGG repeats sometimes increases greatly in a single generation, from a few dozen repeats to hundreds of repeats. This causes the X chromosome to be fragile at this site in the sense that the chromosome is easily broken when the chromosome is prepared in a certain way. Such unstable expansions of microsatellite repeats have been implicated in several other genetic disorders, including Huntington's disease. Another fragile site on the X chromosome has been linked to a less common form of mental retardation (Knight et al., 1993).

As with PKU, these fragile X varieties of mental retardation involve single genes that are necessary and sufficient to produce the disorder.

Single-gene linkage and Alzheimer's disease. Dementia is a common disorder marked by progressive memory loss and confusion. The most common dementia is called *Alzheimer's disease* (AD), which generally appears late in life. AD also includes a rare type (FAD), accounting for fewer than 1% of AD cases, that appears in middle adulthood and shows a dominant single-gene pattern of inheritance. DNA markers and linkage analysis discovered that mutations in the amyloid precursor protein gene on chromosome 21 are associated with a few cases of FAD (Goate et al., 1991). The majority of FAD cases have more recently been found to be linked to a gene on chromosome 14 (Schellenberg et al., 1992), and the gene has been identified (Sherrington et al., 1995).

One gene, one disorder? PKU, fragile X mental retardation, and the early-onset form of dementia called *FAD* are examples of single-gene disorders that are necessary and sufficient to cause the disorder. The traditional reductionistic approach of molecular biology assumes that all complex disorders such as mental retardation are a concatenation of several disorders, each caused by a different gene necessary and sufficient to cause the disorder, or at least a gene of major effect that largely accounts for genetic influence. This could be called the *one-gene, one-disorder* (OGOD) hypothesis (Plomin et al., 1994b).

Quantitative trait loci (QTL) and cognitive abilities and disabilities

In contrast to the OGOD perspective, quantitative genetics assumes that genetic influences on complex, common behavioral dimensions and disorders are due to multiple genes of varying effect size that contribute additively and interchangeably like risk factors to vulnerability. Any single gene in a multigene system is neither necessary nor sufficient to cause a disorder. From this perspective, genetic effects involve probabilistic propensities rather than predetermined programming. One implication of the assumption of a multigene system is that genotypes are distributed dimensionally regardless of whether traits are assessed using a dichotomous diagnosis. For this reason, genes that contribute to multigenic systems have been called *quantitative trait loci* (QTL) (Gelderman, 1975). The term *QTL* replaces the word *polygenic*, which literally means *multiple genes* but has come to connote many genes of such infinitesimal effect size that they are unidentifiable. QTL denote multiple genes of varying effect size. The hope is to be able to detect QTL of modest effect size. *Oligogenic* is another word that has been used as a substitute for *polygenic*, but it presupposes that only a few ("oligo") genes are involved.

Both the OGOD and QTL approaches are likely to contribute to complex behaviors. The critical unresolved issue is the extent to which genetic influences on complex behaviors are due to genes operating according to the OGOD or QTL hypotheses. Molecular biology traditionally favors the OGOD hypothesis, whereas quantitative genetics leans toward QTL. The distinction between OGOD and QTL has important implications for molecular genetics research strategies.

Linkage and allelic association. The most important implication of the distinction between the OGOD and QTL perspectives is that much greater statistical power is needed to detect QTL because their effects can be much smaller. OGOD effects segregate simply in families, and linked DNA markers cosegregate with the disorder. However, conventional linkage analysis of large pedigrees is unlikely to have sufficient power to detect QTL for a disorder unless a particular locus accounts for most of the genetic variance, in which case it is not really a QTL. Moreover, as mentioned earlier, QTL imply quantitative dimensions, and these are not easily handled by linkage approaches, which traditionally assume a dichotomous diagnosis that cosegregates with a single gene. Newer linkage designs such as affected relative designs are more powerful than traditional pedigree designs and make no assumptions about mode of inheritance (Risch, 1990). It is also possible to incorporate quantitative measures in linkage studies (Fulker et al., 1991; Fulker & Cardon, 1994). These newer linkage designs may be able to detect some of the largest QTL effects. The advantage of linkage approaches is that they can detect OGOD effects without a priori knowledge of pathological processes in a systematic search of the genome using a few hundred highly polymorphic DNA markers. Such systematic screens of the genome can exclude OGOD effects. However, they cannot exclude QTL effects because linkage is quite limited in its power to detect QTL of small effect size.

The flip side of the low power of traditional linkage designs is that if linkage is found, the locus must account for a large portion of the genetic influence on the disorder. This also applies to the newer linkage designs unless very large samples are employed. For example, linkage reported for reading disability, if true, would imply a major effect of a single gene. Reading disability has been reported to be linked to markers on chromosome 15 in a traditional linkage analysis (Smith, Kimberling, Pennington, & Lubs, 1983), although a later report with additional families showed less evidence for linkage (Smith, Pennington, Kimberling, & Ing, 1990). A sibpair linkage analysis of 161 sibling pairs in these same 19 families also suggested linkage to chromosome 15 and possibly to chromosome 6 (Smith,

Kimberling, & Pennington, 1991), as did a sib-pair linkage analysis that incorporated quantitative scores of reading ability (Fulker et al., 1991). An independent sample using a sib-pair design failed to replicate the chromosome 15 linkage but found strong evidence for linkage to chromosome 6 at the major histocompatibility locus (Cardon et al., 1994).

Although linkage remains the strategy of choice from the OGOD perspective and for skimming off the largest QTL effects, other strategies are needed to detect QTL of smaller effect size. Most likely, given the pace of discovery in molecular biology, new techniques will be developed to reach this goal. For the present, allelic association represents an increasingly used strategy that is complementary to linkage (Owen & McGuffin, 1993).

In contrast to linkage, allelic association refers to a correlation in the population between a phenotype and a particular allele, usually assessed as an allelic frequency difference between cases and controls. Loose linkage between two loci does not result in population associations because, as mentioned earlier, alleles on the same chromosome at all but the tightest linked loci are separated by crossing-over. However, if a DNA marker is very close to a QTL, alleles for the marker and the QTL may remain together on the same chromosome for thousands of years. For this reason, allelic association research uses DNA markers in or near candidate genes likely to affect the target trait.

Allelic association can occur for another reason: The DNA marker itself may be the QTL. That is, the DNA marker may code for a functional polymorphism that directly affects the phenotype. Use of such functional DNA markers greatly enhances the power of the allelic association approach to detect QTL (Sobell, Heston, & Sommer, 1992). Many new DNA markers are of this type.

The main limitation of allelic association is that a systematic genome search would require thousands of DNA markers (because the markers would have to be very close to the QTL), unlike linkage that can scan the genome for OGOD genes with only a few hundred markers. Until genotyping on such a massive scale is feasible, allelic association studies will be limited to screening functional polymorphisms or DNA markers in or near possible candidate genes. Despite its limitations, allelic association can at least provide the statistical power needed to detect QTL of small effect size, QTL much too small to be detected by linkage. Statistical power can be increased to detect small QTL associations by increasing sample sizes of relatively easy-to-obtain unrelated subjects.

Some encouraging examples of the success of the allelic association approach to QTL have been reported for complex traits including cognitive disabilities.

QTL association with Alzheimer's disease. The best QTL example for cognitive disability is the recently discovered allelic association between late-onset Alzheimer's disease (AD) and Apo-E4 (Corder et al., 1993). Unlike the rare early-appearing FAD, the prevalence of AD increases steeply with age from less than 1% at age 65 years to 15% in the ninth decade. The frequency of the Apo-E4 allele is 40% in AD individuals as compared to 15% in control populations. The odds ratio, or approximate relative risk, is 6.4 for individuals with one or two Apo-E4 alleles. However, some 40% of AD cases do not possess an Apo-E4 allele. In this sense, the Apo-E4 association is an example of a QTL effect because the Apo-E4 allele is neither necessary nor sufficient to develop the disorder. It has been estimated that Apo-E4 contributes approximately 17% to the population variance in liability to develop the disorder (Owen, Liddle, & McGuffin, 1994). Although this is a large effect from a QTL perspective, it is a small effect from an OGOD perspective. A linkage study using 32 small pedigrees found only relatively modest evidence of linkage for the Apo-E region of chromosome 19 (Pericak-Vance et al., 1991), and other studies were not able to replicate this linkage. A QTL of this magnitude may be near the lower limit of detection by linkage analysis with realistic sample sizes.

QTL associations with cognitive ability. Genes that account for disabilities are not necessarily important sources of variance for abilities. For example, although a greatly increased number of CGG repeats is related to fragile X mental retardation, CGG repeat length is not related to variation in cognitive ability in the normal range of ability (Daniels et al., 1994). Similarly, the Apo-E4 allele does not show a significant association with normal variation in cognitive ability (Plomin et al., 1995).

However, the genetic links between the normal and abnormal are an empirical issue. I predict that, despite relatively rare OGOD single-gene examples such as fragile X, QTL associated with normal variation in cognitive ability will also contribute to risk for mild mental retardation as well as high cognitive ability.

The normal range of general cognitive ability is a reasonable candidate for molecular genetic research because it is one of the most heritable dimensions of behavior. Several studies investigated associations between classical genetic markers such as blood groups and IQ without notable success (e.g., Ashton, 1986; Gibson, Harrison, Clarke, & Hiorns, 1973; Mascie-Taylor, Gibson, Hiorns, & Harrison, 1985). Some evidence suggests that carriers for recessive disorders such as PKU show slightly lowered IQ scores (Bessman, Williamson, & Koch, 1978; Propping, 1987).

In addition to research on cognitive disabilities, the new DNA markers have begun to be used in research on cognitive abilities (Plomin et al.,

1994a, 1995). An allelic association strategy is employed in this ongoing research that uses DNA markers that are in or near genes likely to be relevant to neurological functioning, such as genes for neuroreceptors. Allelic frequencies of these DNA markers are compared in groups differing in IQ. Comparing allelic frequencies in high- and low-IQ groups is more cost efficient than genotyping individuals throughout the IQ distribution.

In an original sample of Caucasian children from 6 to 12 years of age, three groups were examined: a low-IQ group ($N = 18$) with a mean IQ of 82, a middle-IQ group ($N = 21$) with a mean IQ of 105, and a high-IQ group ($N = 24$) with a mean IQ of 130. Independent samples of even lower- and higher-IQ children were included for purposes of replication: a low-IQ group ($N = 17$) with a mean IQ of 59, and a high-IQ group ($N = 27$) with a mean IQ of 142. These sample sizes provide statistical power to detect only relatively large allelic frequency differences, differences of about .20 or greater. QTL associations of this magnitude account for 2% or more of the population variance of IQ. From a linkage perspective, this is a very small effect size, but from a QTL perspective, it is a large effect. If the heritability of IQ is 50%, 25 genes of this effect size could account for it.

An analysis of 100 DNA markers have found several suggestive associations with IQ, but none have consistently replicated (Plomin et al., 1994a, 1995). For example, Apo-E4 showed a slightly greater frequency in the low-IQ group, as it did in a Dutch study of elderly men (Feskens et al., 1994). Interestingly, allele 3 of Apo-E yielded a larger difference, showing allelic frequencies in the low- and high-IQ groups of about 65% and 80%, respectively, in both the original and replication samples. Only one marker yielded a significant association in both the original and replication samples, with frequencies of about 75% in the low-IQ group and 100% in the high-IQ group. Although this might be a chance result because 100 markers were examined, the marker is interesting because it was found to be mitochondrial DNA rather than genomic DNA (Skuder et al., 1995). Mitochondrial DNA is inherited maternally because mitochondria are outside the nucleus of the cell in the cell's cytoplasm, which comes from the egg. Mitochondrial DNA might be involved in neurological disorders during childhood that affect cognitive development (Taylor, 1992). A second phase of the project has begun that will increase the power to detect QTL associated with IQ by obtaining even more extreme samples, larger samples, and additional replication groups. Regardless of the ultimate success of this first attempt to identify QTL associated with cognitive abilities, the project indicates the potential for harnessing the power of molecular genetics to identify QTL responsible for genetic influence on cognitive abilities.

The major implication of identifying QTL will be for basic research. From a basic science point of view, the way forward clearly is to begin to

identify some of the specific genes responsible for the substantial heritability of cognitive abilities. It is an empirical issue whether any of these genetic
effects are of sufficiently large effect size to be detected. In the QTL and
cognitive abilities project just described, if several hundred DNA markers
for neurologically relevant genes, especially the new generation of functional polymorphisms, yield a single replicable association, even if it accounts for less than 1% of the variance, it will begin to establish the
parameters of the task. Even a small handhold on the genetic contribution
to individual differences will help in the climb toward understanding how
genes interact with neural, psychological, and environmental processes in
cognitive development. Identifying specific genes, even QTL of small effect,
will revolutionize quantitative genetic research on cognitive abilities by
making it possible to identify genotypes directly rather than resorting to
indirect inferences from twin and adoption studies. Moreover, genes of
small effect on average in the population might have strong effects for some
individuals.

It should be noted that even when DNA markers are identified that
reliably differentiate high- and low-IQ groups, the markers are not likely to
be useful in predicting high- and low-IQ children in the general population.
Indeed, it is doubtful whether allelic associations in sufficient number and
strength will ever be identified that in combination can reach levels of
prediction that rival those that can be made at present on the basis of
parental IQ. The regression of offspring on midparent IQ (that is, the
average IQ of the mother and father) is about .60 in Caucasian populations.
In other words, parents' IQ can predict more than one-third of the total
variance in offspring IQ scores, which is more than two-thirds of the genetic
variance if heritability is .50. If QTL responsible for genetic variance in IQ
scores typically account for less than 1% of IQ variance in the population,
dozens of such QTL might be required to make a reasonable prediction of
children's IQ. Moreover, if many QTL individually account for far less than
1% of the variance, most of these QTL will never be identified; thus,
identified QTL will fall far short of predicting all of the genetic variance of
IQ.

As is the case with most important advances, identifying genes for cognitive abilities could raise new ethical issues as well (e.g., Wright, 1990). For
single-gene disorders, identifying genes has already led to concerns such as
employment and insurance discrimination (e.g., Bishop & Waldholz, 1990;
Nelkin & Tancredi, 1989). The heat that will be engendered in a debate on
these issues can be anticipated from the vehement responses to a 1988
editorial in *Science* concerning the human genome project, which asserted
that "we must step boldly and confidently across the threshold" (Koshland,
1989, p. 189). A 1993 commentary in *Nature* hyperbolized that "the isola-

tion of the first gene involved in determining 'intelligence' (whatever that is) will be a turning point in human history" (Müller-Hill, 1993, p. 492). However, this same commentary wisely argued:

Anticipating such conflicts, many may conclude that we do not need or want this genetic knowledge. I disagree. The knowledge will simply unveil reality, emphasizing the injustice of the world.... Laws are necessary to protect the genetically disadvantaged. Social justice has to recompense genetic injustice. (Müller-Hill, 1993, p. 492)

The potential of measured QTL for understanding the etiology of cognitive abilities and disabilities seems likely to far outweigh its potential abuses. Moreover, forewarned of problems and solutions that have arisen with single-gene disorders, we should be forearmed as well to prevent abuses.

Conclusions

Most of what is currently known about the genetics of cognitive abilities and disabilities comes from quantitative genetics research. Twin and adoption studies have documented significant and substantial genetic influence. More quantitative genetics research is needed that goes beyond heritability. New quantitative genetics techniques can track the developmental course of genetic contributions and identify genetic links among cognitive abilities. Candidate environmental factors need to be investigated in the context of genetically sensitive designs to follow up on the far-reaching findings of nonshared environment and genetic influences on experience, exploring the developmental processes of genotype–environment correlation by which genotypes become phenotypes.

Such quantitative genetics research will also guide attempts to identify specific genes that contribute to genetic variance in cognitive abilities. So far, only a few genes have been identified that affect cognitive disabilities. Most of these involve relatively rare disorders such as PKU and fragile X, for which a single gene is the necessary and sufficient cause of distinct mental retardation disorders, in line with the OGOD hypothesis. The allelic association between Apo-E4 and late-onset Alzheimer's disease is the best example of a cognitive QTL. The coming confluence of quantitative genetics and molecular genetics will be synergistic for the investigation of cognitive abilities. Research on the new genetics will add an exciting new chapter to the story of heredity and cognitive abilities and disabilities.

Acknowledgments

The research on QTL associations with cognitive ability is supported by Grant HD-27694 from the National Institute of Child Health and Human

Development. I am grateful to Hans Eysenck, Arthur Jensen, and John Loehlin for reviewing an earlier version of this chapter.

References

Ashton, G. C. (1986). Blood polymorphisms and cognitive abilities. *Behavior Genetics, 16*, 517–29.

Bessman, S. P., Williamson, M. L., & Koch, R. (1978). Diet, genetics, and mental retardation interaction between phenylketonuric heterozygous mother and fetus to produce nonspecific diminution of IQ: Evidence in support of the justification hypothesis. *Proceedings of the National Academy of Sciences, 78*, 1562–6.

Bishop, J. E., & Waldholz, M. (1990). *Genome.* New York: Simon and Schuster.

Cardon, L. R., Smith, S. D., Fulker, D. W., Kimberling, W. J., Pennington, B. F., & DeFries, J. C. (1994). Quantitative trait locus for reading disability on chromosome 6. *Science, 266*, 276–9.

Chipuer, H. M., Rovine, M., & Plomin, R. (1990). LISREL modelling: Genetic and environmental influences on IQ revisited. *Intelligence, 14*, 11–29.

Corder, B., Saunders, A. M., Strittmatter, W. J., Schmechel, D. E., Gaskell, P. C., & Small, G. N. (1993). Gene dose of apolipoprotein E type 4 allele and the risk of Alzheimer's disease in late onset families. *Science, 261*, 921–3.

Daniels, J., Owen, M. J., McGuffin, P., Thompson, L., Detterman, D. K., Chorney, M., Chorney, K., Smith, D. L., Skuder, P., Bignetti, S., McClearn, G. E., & Plomin, R. (1994). IQ and variation in the number of fragile X CGG repeats: No association in a normal sample. *Intelligence, 19*, 45–50.

Feskens, E. J. M., Havekes, L. M., Kalmijn, S., de Knijff, P., Launer, L. J., & Kromhout, D. (1994). Apolipoprotein E4 allele and cognitive decline in elderly men. *British Medical Journal, 309*, 1202–6.

Fulker, D. W., & Cardon, L. R. (1994). A sib-pair approach to interval mapping of quantitative trait loci. *American Journal of Human Genetics, 54*, 1092–103.

Fulker, D. W., Cardon, L. R., DeFries, J. C., Kimberling, W. J., Pennington, B. F., & Smith, S. D. (1991). Multiple regression analysis of sib-pair data on reading to detect quantitative trait loci. *Reading and Writing: An Interdisciplinary Journal, 3*, 299–313.

Gelderman, H. (1975). Investigations on inheritance of quantitative characters in animals by gene markers. I. Methods. *Theoretical and Applied Genetics, 46*, 319–30.

Gibson, J. B., Harrison, G. A., Clarke, V. A., & Hiorns, R. W. (1973). IQ and ABO blood groups. *Nature, 246*, 498–500.

Goate, A., Chartier-Harlin, M-C., Mullan, M., Brown, J., Crawford, F., Fidani, L., Giuffra, L., Haynes, A., Irving, N., James, L., Mant, R., Newton, P., Rooke, K., Roques, P., Talbot, C., Pericak-Vance, M., Roses, A., Williamson, R., Rossor, M., Owen, M. J., & Hardy, J. (1991). Segregation of a missense mutation in the amyloid precursor protein gene with familial Alzheimer's disease. *Nature, 349*, 704–6.

Knight, S. J. L., Flannery, A. V., Hirst, M. C., Campbell, L., Christodoulou, Z., Phelps, S. R., Pointon, J., Middleton-Price, H. R., Barnicoat, A., Pembrey, M. E., Holland, J., Oostra, B. A., Bobrow, M., & Davies, K. E. (1993). Trinucleotide repeat amplification and hypermethylation of a CpG island in FRAXE mental retardation. *Cell, 74*, 127–34.

Koshland, D. E., Jr. (1989). Sequences and consequences of the human genome. *Science, 246*, 189.

Loehlin, J. C. (1989). Partitioning environmental and genetic contributions to behavioral development. *American Psychologist, 44*, 1285–92.

Mascie-Taylor, C. G. N., Gibson, J. B., Hiorns, R. W., & Harrison, G. A. (1985). Associations between some polymorphic markers and variation in IQ and its components in Otmoor villagers. *Behavior Genetics, 15*, 371–83.

McKusick, V. A. (1990). *Mendelian inheritance in man* (9th ed.). Baltimore, MD: Johns Hopkins University Press.

Müller-Hill, B. (1993). The shadow of genetic injustice. *Nature, 362,* 491–2.

Nelkin, D., & Tancredi, L. (1989). *Dangerous diagnostics: The social power of biological information.* New York: Basic Books.

Owen, M., Liddle, M., & McGuffin, P. (1994). Apo E and Alzheimer's disease. *British Medical Journal, 308,* 672–3.

Owen, M. J., & McGuffin, P. (1993). Association and linkage: Complementary strategies for complex disorders. *Journal of Medical Genetics, 30,* 638–9.

Pericak-Vance, M. A., Bebout, J. L., Gaskell, P. C., Jr., Yamaoka, L. H., Hung, W. Y., Alberts, M. J., Walker, A. P., Bartlett, R. J., Haynes, C. A., Welsh, K. A., Earl, N. L., Heyman, A., Clarck, C. M., & Roses, A. D. (1991). Linkage studies in familial Alzheimer's disease: Evidence for chromosome 19 linkage. *American Journal of Human Genetics, 48,* 1934–50.

Plomin, R., McClearn, G. E., Smith, D. L., Vignetti, S., Chorney, M., Chorney, K., Kasarda, S., Thompson, L. A., Detterman, D. K., Daniels, J., Owen, M. J., & McGuffin, P. (1994a). DNA markers associated with high versus low IQ: The IQ QTL Project. *Behavior Genetics, 24,* 107–18.

Plomin, R., McClearn, G. E., Smith, D. L., Skuder, P., Vignetti, S., Chorney, M. J., Chorney, K., Kasarda, S., Thompson, L. A., Detterman, D. K., Petrill, S. A., Daniels, J., Owen, M. J., & McGuffin, P. (1995). Allelic associations between 100 DNA markers and high versus low IQ. *Intelligence, 21,* 31–48.

Plomin, R., & Neiderhiser, J. N. (1991). Quantitative genetics, molecular genetics, and intelligence. *Intelligence, 15,* 369–87.

Plomin, R., Owen, M. J., & McGuffin, P. (1994b). Genetics and complex human behaviors. *Science, 264,* 1733–9.

Propping, P. (1987). Single gene effects in psychiatric disorders. In F. Vogel & K. Sperling (Eds.), *Human genetics: Proceedings of the 7th international congress, Berlin* (pp. 452–7). New York: Springer.

Risch, N. (1990). Linkage strategies for genetically complex traits. II. The power of affected relative pairs. *American Journal of Human Genetics, 46,* 229–41.

Schellenberg, G. D., Bird, T. D., Wijsman, E. M., Orr, H. T., Anderson, L., Nemens, E., White, J. A., Bonnycastle, L., Weber, J. L., Alonso, M. E., Potter, H., Heston, L. L., & Martin, G. M. (1992). Genetic linkage evidence for a familial Alzheimer's disease locus on chromosome 14. *Science, 258,* 668–71.

Sherrington, R., Rogaev, E. L., Liang, Y., Rogaeva, E. A., Levesque, G., Ikeda, M., Chi, H., Lin, C., Li, G., Holman, K., et al. (1995). Cloning of a gene bearing missense mutations in early-onset familial Alzheimer's disease. *Nature, 375,* 754–60.

Skuder, P., Plomin, R., McClearn, G. E., Smith, D. L., Vignetti, S., Chorney, M. J., Chorney, K., Kasarda, S., Thompson, L. A., Detterman, D. K., Petrill, S. A., Daniels, J., Owen, M. J., & McGuffin, P. (1995). A polymorphism in mitochondrial DNA associated with IQ? *Intelligence, 21,* 1–11.

Smith, S. D., Kimberling, W. J., & Pennington, B. F. (1991). Screening for multiple genes influencing dyslexia. *Reading and Writing: An Interdisciplinary Journal, 3,* 285–98.

Smith, S. D., Kimberling, W. J., Pennington, B. F., & Lubs, H. A. (1983). Specific reading disability: Identification of an inherited form through linkage analysis. *Science, 219,* 1345–7.

Smith, S. D., Pennington, B. F., Kimberling, W. J., & Ing, P. S. (1990). Familial dyslexia: Use of genetic linkage data to define subtypes. *Journal of the American Academy of Child and Adolescent Psychiatry, 29,* 204–13.

Sobell, J. L., Heston, L. L., & Sommer, S. S. (1992). Delineation of the genetic predisposition to a multifactorial disease: A general approach on the threshold of feasibility. *Genomics, 12,* 1–6.

Taylor, R. (1992). Mitochondrial DNA may hold a key to human degenerative diseases.

Journal of NIH Research, 6, 62–6.

Verkerk, A. J., Pieretti, M., Sutcliffe, J. S., Fu, Y-H., Kuhl, D. P., Pizzuti, A., Reiner, O., Richards, S., Victoria, M. F., Zhang, F., Eussen, B. E., VanOmmen, G-J., Blonden, L. A., Riggins, G. J., Chastain, J. L., Kunst, C. B., Galjaard, H., Caskey, C. T., Nelson, D., Oostra, B., & Warren, S. T. (1991). Identification of a gene (FMR-1) containing a CGG repeat coincident with a breakpoint cluster region exhibiting length variation in fragile X syndrome. *Cell, 65,* 905–14.

Walhsten, J. (1990). Gene map of mental retardation. *Journal of Mental Deficiency Research, 34,* 11–27.

Weissenbach, J., Gyapay, G., Dib, C., Vignal, A., Morisette, J., Millasseau, P., Vaysseix, G., & Lathrop, M. (1992). A second-generation linkage map of the human genome project. *Nature, 359,* 794–801.

Woo, S. L. C. (1991). Molecular genetic analysis of phenylketonuria and mental retardation. In P. R. McHugh & V. A. McKusick (Eds.), *Genes, brain, and behavior* (pp. 193–203). New York: Raven Press.

Wright, A. (1990). Achilles' helix. *The New Republic,* July 9–16, 21–31.

4　Heredity, environment, and IQ in the Texas Adoption Project

John C. Loehlin, Joseph M. Horn, and Lee Willerman

Historically, by far the most common design in human behavior-genetic studies has been the comparison of the resemblance of identical and same-sex fraternal twin pairs. It remains a basic approach today. The twin design permits the investigator to compare pairs of individuals of the same age and family environment, but of different degrees of genetic relatedness, on some trait of interest. A greater resemblance for those genetically more similar, the identical twin pairs, is taken as evidence of the importance of genes in contributing to individual differences on the trait in question. A second important design is the study of adoptive families. An adoption study examines the resemblance of genetically unrelated individuals growing up together in the same family, as parents and children or as siblings. If they are less similar than corresponding members of ordinary biological families, the difference is taken as evidence of the effect of the genes in causing the resemblance in the biological families.

Both twin and adoption studies also permit inferences concerning the effect of the shared family environment on traits. In the twin study, such inferences are indirect: If the twins are more similar than would be expected from the estimated genetic effect, an effect of shared environment is deduced. In the adoption study, the inference is direct: Do genetically unrelated family members show any resemblance *at all*? If so, there is evidence of the effect of the family environment (or possibly, of selective placement of children by adoption agencies); if not, the effects of the family environment, if any, are unsystematic. Thus, the two kinds of studies can complement and support one another; results consistent across both are at less risk of being due to some idiosyncratic feature of the twin or adoption situation.

At the time we were planning the present study, there had been only a handful of adoption studies of IQ (e.g., Burks, 1928; Freeman, Holzinger, & Mitchell, 1928; Leahy, 1935; Skodak & Skeels, 1949; Snygg, 1938). None of these was recent, and there were some conflicts among their results. Since that time, new studies have appeared – for a review of several, see Loehlin (1980); see also Plomin, DeFries, and Fulker (1988). Those most like our study in terms of tests, ages, and populations are two done in Minnesota by Scarr and Weinberg (1977, 1978, 1983). We will compare our results to theirs at several points.

The Texas Adoption Project

The Texas Adoption Project began with 300 Texas families who had adopted one or more children through a church-related home for unwed mothers. The bulk of the adoptions occurred during the 1960s. For the cases included in our study, the placement in the adoptive home occurred within a short time of birth and resulted in permanent adoption. Most of the adoptive families as well as the families from which the birth mothers came were white and middle-class. While in residence at the home, as time permitted, the birth mothers had been given IQ tests, along with various measures of personality and interests. The IQ test most frequently administered was the *Revised Beta Examination*, a nonverbal pencil-and-paper IQ test; however, sometimes a Wechsler test, such as the Wechsler Adult Intelligence Scale (WAIS) or the Wechsler Intelligence Scale for Children (WISC), was used. The agency made the IQ test scores available to us from their files.

With the aid of the agency, families who had adopted a child born of a mother for whom an IQ score was on record were contacted and invited to participate in the study. In each case, arrangements were made for a nearby psychologist to administer an individual IQ test to members of the adoptive families. These family members often included, in addition to the parents and the adopted child, other adopted children, or one or more biological children of the adoptive parents. Several personality questionnaires were also given; the results from these have been reported elsewhere (e.g., Loehlin, Horn, & Willerman, 1981; Loehlin, Willerman, & Horn, 1982). The individual IQ tests given to the children were the Stanford-Binet, for those aged 3 or 4 years, the WISC, for those 5 to 15, and the WAIS, for those 16 or older. The median age of all the adopted children in the sample was about 8 years, and that of the natural children about 10, but both groups spanned a considerable range of ages. The adoptive parents were given both the WAIS and the Beta. More details on the IQ testing and the results from the original sample may be found in earlier publications (e.g., Horn, Loehlin, & Willerman, 1979, 1982).

About 10 years after the original testing, these families were recontacted, and 181 agreed to participate again. Only the children were tested this time. They were administered current versions of the Wechsler tests, the WAIS-R or WISC-R, depending on age. In addition, they were given the Beta, which their adoptive parents, and in many cases their birth mothers, had taken a decade or more earlier. Again, personality tests were also given, whose results have been reported elsewhere (Loehlin, Willerman, & Horn, 1987; Loehlin, Horn, & Willerman, 1990; Willerman, Loehlin, & Horn, 1992).

IQ test scoring and norms

All tests were scored according to their respective manuals, and IQs were obtained using the tabled norms. For correlational analyses, these IQs should be satisfactory (see the last paragraph of this section for a check on this). However, for comparisons of means, the IQs were adjusted for obsolescence of norms and for racial composition in the manner described by Flynn (1984). The norm adjustment allowed for an upward movement of U.S. IQ test performance of three-tenths of an IQ point per year. Thus, comparisons of IQ tests normed in different years or of performances on the same test given at different times could reasonably be made.

The Beta IQs present some special problems. The norming of this test was done on a stratified sample of male prisoners selected to match U.S. census data on age, education, and socioeconomic status (Lindner & Gurvitz, 1946) rather than on an actual sample of the U.S. population. Because cases were selected to match the 1940 census data, this was taken as the base year for the norm adjustment rather than the 1946 date of publication of the norms. As a check on the adequacy of the original age-norming of this test, a comparison was made between the use of the tabled norms and an empirical age-adjustment based on the present sample. The IQ was correlated with the sum of standardized raw subscale scores adjusted for a linear and a quadratic component of age. The correlations between IQ and the age-adjusted score composite were .98 for the fathers, .98 for the mothers, .99 for the birth mothers, and .98 for each of two groups of children – those who took the WISC-R, and those who took the WAIS-R. For comparison purposes, similar correlations were obtained for the children between IQs and age-adjusted scores, separately for the WISC-R and the WAIS-R. The correlations were .96 and .97, respectively. All of these correlations are high enough to suggest that deficiencies in age-norming per se should not present a problem for analyses of a correlational nature.

Likewise, the obsolescence of norms, although important for mean comparisons, should not have much effect on correlations. For two groups, each of which was given a single test on one occasion, the adjustments should have no effect at all on correlations, as they would involve making the same correction to every score in each group. Where individuals were tested at different times, as in the case of the birth mothers, or where different tests

Table 4.1. *Means and standard deviations of IQ (adjusted for age of norms) for various parent and child groups in the Texas Adoption Project, for those in first study only, and for those in both original and follow-up studies*

	In first only			In both studies					
				At first study			At follow-up		
Group & test	Mean	*SD*	*N*	Mean	*SD*	*N*	Mean	*SD*	*N*
Adoptive fathers									
WAIS	105.3	13.3	114	108.2	10.7	178			
Beta	103.4	8.3	115	105.7	6.9	181			
Adoptive mothers									
WAIS	101.7	11.6	114	106.3	10.2	175			
Beta	100.2	8.2	116	103.3	7.0	176			
Birth mothers									
Wechsler	100.5	12.0	26	101.6	14.4	18			
Beta	100.4	9.1	144	100.5	8.4	198			
Adopted children:									
with a birth mother									
IQ									
Wechsler	103.3	11.0	159	105.1	11.9	204	105.4	13.8	204
Beta							96.9	8.3	203
with no birth mother									
IQ									
Wechsler	99.6	12.8	50	104.5	9.4	52	103.9	12.9	52
Beta							98.5	7.1	48
Natural children									
Wechsler	104.2	11.3	71	103.6	11.4	92	106.0	14.3	92
Beta							97.4	8.8	89

Note: Beta = Revised Beta Examination; Wechsler = WISC or WAIS (or Stanford-Binet) in first study and WISC-R or WAIS-R in follow-up, depending on age. See text for details. *SD* = standard deviation. Adoptive fathers and mothers are classified by whether any child in the family participated in both studies; children (and birth mothers) by whether the child in question did. Three birth mothers tested with earlier Wechsler IQ tests were excluded here and in Table 4.2 but included elsewhere.

were given to different individuals, as in the case of the children, there in theory could be a small effect on correlations. To verify that this effect was in fact small, adjusted IQs were correlated with the original unadjusted IQs for a number of relevant subgroups. These correlations were in every instance above .99; indeed, nearly all were above .995. Thus, the norm adjustments can safely be ignored for correlations; the correlational analyses presented in this chapter are based for the most part on ordinary, unadjusted IQs.

IQ means and standard deviations

Table 4.1 gives means and standard deviations on IQ tests for individuals in several categories. The IQs have been adjusted as described in the preceding section. The designation "Wechsler" in the table refers to an age-appropriate individual IQ test taken by the subject, a test from the Wechsler series except for the Stanford-Binets given to the youngest children; IQs for each test were separately adjusted before combining. Several points are evident from the table. All of the subject groups tend to be slightly above average in IQ, ranging from IQs of 100.5 for the birth mothers (averaged across tests and subgroups) to 105.9 for the adoptive fathers. These average IQs are, by the nature of the adjustments, considerably lower than the uncorrected IQs published in our original reports (Horn et al., 1979, 1982) and in the studies of Scarr and Weinberg (1977, 1978). They are also lower than those for most other published IQ studies involving similar populations, where IQs are typically inflated by the use of dated norms (Flynn, 1984). The uncorrected IQs would run roughly 6 to 8 IQ points higher than those in Table 4.1 for the Wechsler tests in the first study, and 8 to 10 points for the Beta. In the follow-up study, because the revised Wechsler norms were newer and the Beta norms were older, the differences would be about 3 to 5 and 13 IQ points, respectively.

All the subject groups are restricted in range. As against a nominal 15 on the Wechsler tests, the observed standard deviations range from 9.4 to 14.4. The clientele of this adoption agency are a selected group and were probably further selected by their participation in our study. The standard deviations on the Beta, which has the same nominal standard deviation of 15 as the Wechsler tests, are even more restricted, lying between 6.9 and 9.1. This is partly a result of ceiling effects, which are apparent for a number of the Beta subscales. For example, the adoptive mothers have a mean raw score of 26.6 on the Digit Symbol subtest, where the maximum possible score is 30, and a mean of 21.8 on Identities, a test of clerical speed and accuracy, where the maximum possible score is 25. However (in an analysis not shown in the table), samples selected to avoid high IQs on either test (i.e., groups for whom differential ceiling effects should presumably be minimal) had standard deviations for the Beta that remained smaller than those on the Wechsler tests.

The Beta IQs also tend to be systematically a little lower than the Wechsler IQs. The difference between the tests may reflect the less adequate norming of the Beta, or, conceivably, an overcorrection in the adjustment process. The discrepancy was the greatest at the time of the follow-up, when the correction discrepancy was most extreme. The test was not included in Flynn's study, so that a direct comparison of norm changes cannot be made; however, the content of the Beta is generally similar to that of other IQ tests, and, if anything, Flynn found nonverbal scales to show more upward movement over time than did a verbal scale, Vocabulary (Flynn, 1984, p. 46).

Table 4.1 permits a number of meaningful comparisons. The left-hand and middle columns address the issue of selection in the follow-up study. The left-hand columns give the IQs for families from the original sample who elected not to participate in the follow-up study or were unreachable; the middle columns give the IQs from the first study for families that did participate in the follow-up. It appears that there was selection for higher IQ among the participating adoptive parents and adopted children. Although not shown in the table, the parents in the families who participated tended also to be younger (by about a year, on the average) and to have had more education. It is not very surprising that younger families, in which more of the children were still residing at home, were more likely to participate, or that the effectiveness of an appeal to an increase in knowledge was greater for the more intelligent and educated. The consequence, however, was a restriction in IQ range for the adoptive parents.

There are some theoretically interesting comparisons among the rows of Table 4.1, as well. Do the adoptive parents and birth mothers differ in average IQ? If so, which do the adopted children resemble? Does this resemblance increase or decrease with age? Do the children of the birth mothers have higher average IQs than their mothers did, suggesting a favorable effect on IQ of being reared in the adoptive home? Do the natural children have higher average IQs than the adopted children, suggesting a favorable effect of their parents' genes?

Let us consider these questions in turn. First, do the adoptive parents score higher on IQ tests than the birth mothers of the children they adopted? Mostly so, but by small amounts, ranging from −.2 to 6.6 IQ points, depending on which test and sample is being considered. The larger differences are for the Wechsler; however, the sample of birth mothers who took a Wechsler test is small. The average differences on the Beta between the adoptive parents and birth mothers is 2.7 IQ points. This small difference between levels of parental IQs permits only weak tests of genetic and environmental hypotheses based on the mean IQs of their children.

Granting this, we may still at least consider the remaining questions. First, which set of parents do the adopted children most resemble in their average level of IQ? On the Wechsler tests, the average for those children for whom a birth mother's IQ is available tends to lie between that of the adoptive parents and the birth mothers, closer to the adoptive parents; however, on the Beta, it is lower than either, which makes it closer to the birth mothers.

Do these resemblances change with age? The question can be answered only for the Wechsler tests – the Beta, designed for adults, was not given to the children in the first study. On the Wechsler tests, the average adjusted IQs of the adopted children stay about the same across the 10-year interval between the first and second testings; those for the natural children rise by a couple of IQ points (the increase is a statistically significant one), moving them closer to their parents.

Do the children of the birth mothers have higher average IQs than the birth mothers themselves did, suggesting a favorable effect on IQ of being reared in the adoptive home? The evidence is mixed. *Yes* if one looks at the Wechsler; *No* if one looks at the Beta – the size of the difference is about the same in each case.

Finally, do the natural children have higher average IQs than the adopted children, suggesting a favorable effect of their parents' genes? Not at the time of the initial study, when the two groups of children appear quite similar, overall. There is a slight trend in favor of the natural children by the follow-up study, but it is not very striking.

In general, these results are different from those obtained in the classic adoption studies and in the studies of Scarr and Weinberg. There, adopted children tend to show considerably lower IQs than biological children in the same or comparison families.

As previously noted, the difference in measured IQs between the adoptive parents and the birth mothers in our study is small, limiting the force of any conclusions that might be drawn from differences between overall means. Other studies have worked with less select groups of birth mothers, allowing stronger inferences from mean differences. What would happen if differences were created within the present sample by appropriate selection of subsamples? To see, we divided the birth mothers into those of lower and higher IQ and, as a control, the adoptive families into those of higher and lower socioeconomic status (SES). The division was at the median in each case. The Beta IQs of the children were compared, and the results are shown in Table 4.2. The level of the birth mothers' IQ had a significant effect on the average IQs of the adopted children, and the level of the adoptive families' SES did not, although there was about a 1-point average difference in favor of the higher SES homes. The effect of birth mothers' IQ was similar at the two SES levels. In short, when we look at selected subgroups of birth mothers, we see clear evidence for genetic effects on mean IQ.

Note that a breakdown of the type in Table 4.2 can provide information on both genotype–environment correlation, the tendency for genotypes and environments to be associated, and on genotype–environment interaction, the tendency of environmental variables to affect different genotypes

Table 4.2. *Mean Beta IQs of adopted children, by birth mother's Beta IQ and adoptive family SES*

	Lower-SES adoptive families	Higher-SES adoptive families
Higher-IQ birth mothers (Mean IQ = 107.0)	98.6 (43)	99.7 (55)
Lower-IQ birth mothers (Mean IQ = 94.1)	94.0 (56)	95.6 (44)

Note: Ns in parentheses. All IQ scores norm-corrected. Lower & higher = below or above median of group. Difference between rows significant by ANOVA ($p < .001$); column difference and interaction not significant.

Table 4.3. *IQ correlations between parents, tested in original study, and children tested both in original and follow-up studies*

IQ tests (parent/child)	Biologically unrelated		Biologically related		
	F-Ad	M-Ad	F-Na	M-Na	B-Ad
Parent and child at original testing					
Wechsler/Wechsler	.19	.13	.29	.04	.36
Beta/Wechsler	.08	.10	.09	.14	.23
Parent at original testing with child at follow-up					
Wechsler/Wechsler	.10	.05	.32	.14	.39
Beta/Wechsler	.07	−.02	.26	.19	.26
Wechsler/Beta	.15	.07	.16	−.01	.78
Beta/Beta	.08	−.02	.20	.21	.33
Number of pairs	248–257	242–248	90–93	87–90	*a*

Note: F, M = father, mother in adoptive family; Ad, Na = adopted, natural child in adoptive family; B = birth mother of adopted child. Number of pairs ≈ number of children, since parents may be multiply entered. Tests: Beta = Revised Beta Examination; Wechsler, in first study = Wechsler Adult Intelligence Scale (WAIS), Wechsler Intelligence Scale for Children (WISC), or Stanford-Binet, depending on age; Wechsler, in follow-up = WAIS-R or WISC-R, depending on age. *a* = 21 pairs for parent Wechsler, 199–200 pairs for parent Beta.

differently. There is some evidence of the former, in that the children of higher-IQ birth mothers were somewhat more often placed into higher-SES homes, and correspondingly for lower IQ and SES (i.e., the Ns are larger in the lower left- and upper right-hand cells of Table 4.2 than off the diagonal). However, the degree of correlation is quantitatively quite modest (phi = .12, not statistically significant).

Examples of hypotheses involving interaction might be that a relatively favorable environment would give genotypes more freedom to express themselves, or that genotypes for low intelligence are more susceptible to environmental effects than genotypes for high intelligence. The former hypothesis would be supported by a larger difference in the second column of Table 4.2 than in the first; the latter, by a larger difference in the bottom row than in the top. The fact that the interaction in Table 4.2 is not even close to statistically significant means that hypotheses like these gain little support from our data.

It should also be noted that our study does not provide a very powerful test of such hypotheses, because the adoption agency in its placements excluded obviously unfavorable environments. The adoptive families in the original sample averaged about a standard deviation above the general Texas population in occupational status (Horn et al., 1982). Although they did span a substantial range, few, if any, were from the lowest part of the socioeconomic spectrum.

In summary, when we look at selected subgroups of birth mothers, we see clear evidence for genetic effects on mean IQ, although not for the effects of socioeconomic status or for gene–environment correlation or interaction. Let us turn next to consider the evidence from individual differences in IQ (i.e., correlations).

IQ correlations

Table 4.3 shows IQ correlations between parents, tested at the time of the original study (or earlier, in the case of the birth mothers), and their children, tested both at the time of the original study and at the 10-year follow-up. This and all subsequent tables are restricted to the families who participated in both studies, in order to minimize the effects of differential selection in making comparisons. Table 4.3 shows on the left correlations between family members who are biologically unrelated, and whose resemblance must thus largely reflect environmental factors, and on the right correlations between family members who share on the average one-half their genes (parents and biological offspring). The rightmost column, the correlations between birth mothers and their adopted-away children, is especially interesting, because it reflects genetic resemblance in the absence of shared environment: These birth mothers had no contact with their children after the first few days of life; in fact, many of the infants went directly from the hospital to their adoptive families.

The correlations in the left-hand columns of Table 4.3 suggest a modest effect of family environmental factors, one that may be decreasing over time. The correlations between adoptive fathers and their adopted children average about .14 at the time of initial testing and about .10 at the time of follow-up; those for adoptive mothers, about .12 and .02, respectively. A decrease in the effect of family environment on the IQs of children as they grow older has also been a characteristic finding in twin studies (McCartney, Harris, & Bernieri, 1990).

On the right-hand side of Table 4.3, where genes are involved, the correlations are generally higher than those on the left, suggesting that genetic factors influence IQs. The correlations at the bottom of the table are on the average at least as large as those at the top, suggesting no drop-off with age in the influence of genetic factors. (Twin studies suggest some increase in IQ heritability with age.)

A particularly intriguing feature of the Table 4.3 correlations is that the biological correlations on the right seem to *decrease* with the amount of parent–child contact. Those involving birth mothers, where there has been essentially no postnatal contact, are highest; those with adoptive fathers are next highest; and those with adoptive mothers, with whom the child may be presumed to have had the most contact, are lowest. The same direction of difference holds on the left-hand side of the table as well: Correlations of the adopted child with the adoptive mother run a bit lower than correlations with the adoptive father.

The sample sizes are not so large that we can exclude sampling error as an explanation for these differences. Also, there is confounding with restriction of range – the birth mothers were least restricted, and the adoptive mothers were the most restricted (see standard deviations in Table 4.1). However, differences of the same kind were also present in Scarr and Weinberg's Minnesota adoption studies (1978, 1983). In a study with young children, parent–adoptee IQ correlations were .27 and .21 for fathers and mothers, respectively; in a study with late adolescents, they were .16 and .09. Thus, we may need to entertain the hypothesis that parents' interactions with their children *do* have an effect on the latter's IQs, but one that can tend to make them less, not more, like the parents.

We do not wish to exaggerate the uniformity of this result. A number of adoption studies have not obtained it. Among the early studies, Freeman et al. (1928) showed a difference in this direction, but Burks (1928) and Leahy (1935) did not. In a subsample in Scarr and Weinberg (1977) involving only early adoptions, the difference was the other way. In an adoption study with very young children (Plomin et al., 1988), differences were small and inconsistent from age to age. Even in our own study, such differences do not hold for all combinations of measures (see Horn et al., 1979). Thus, an effect of this kind should be regarded only as a possibility, not as a firm finding. But if it is real, it suggests that genetic and environmental factors may sometimes act in opposition to each other in affecting family resemblance. This could go some way toward explaining the puzzle noted by Plomin and

Table 4.4. *IQ correlations between biologically related and unrelated individuals reared as siblings (and tested in both original and follow-up studies)*

	Biologically unrelated		Biologically related
IQ tests	Ad-Ad	Na-Ad	Na-Na
At original testing			
WISC/WAIS/Binet	.11	.20	.27
At 10-year follow-up			
WISC-R/WAIS-R	−.09	.05	.24
Beta	−.02	−.03	.33
df within families	75	106	25
pairs (Beta)	75	107	27
Across testings			
WISC/WAIS/Binet to WISC-R/			
WAIS-R	−.06	.10	.22
df within families	162	222	62

Note: Ad, Na = adopted, natural child; df = degrees of freedom. Tests: WISC/WAIS = Wechsler Intelligence Scale for Children or Wechsler Adult Intelligence Scale, depending on age; −R = Revised; Binet = Stanford-Binet; Beta = Revised Beta Examination.

Loehlin (1989) – that estimates of IQ heritability made by so-called direct methods tend to yield higher values than those made by indirect methods. As it happens, the direct methods compare individuals who were *not* reared together in the same families (e.g., unwed mothers and their adopted-away children), whereas the indirect methods compare individuals who *were* reared in the same families (e.g., studies of reared-together twins).

Table 4.4 shows IQ correlations between siblings. Correlations are given on the left for two kinds of biologically unrelated pairs (two adopted children in a family and an adopted and a natural child), and on the right for biologically related pairs (two natural children of the adoptive parents). Correlations are shown across various combinations of tests and testings. Again, as for the parent–child relationship, the correlations of biologically related siblings tend to be higher, suggesting the influence of the genes.

Model fitting

The implications of sets of familial correlations such as these can be summarized by formalizing the relationships among them in a path model. Previous model fitting has been done with the data of the initial study (Loehlin,

1979; Horn et al., 1982), mostly using Beta IQs for the parental generation and Wechsler Performance IQs for the children. The basic model fitting for the initial data suggested that additive effects of the genes accounted for about 38% of the true-score IQ variance, that shared family environment accounted for about 19%, and that gene–environment correlation accounted for about 10%. Environment not shared by family members, nonadditive genetic effects, and gene–environment interaction were presumably responsible for the remaining 33%. A number of analyses with variations on the basic model or using various subdivisions of the sample assessed the sensitivity of the model fitting to various chance and systematic factors, and in general found the results to be reasonably stable. The particular quantitative estimates of the effects of the genes and of shared environment might differ somewhat from analysis to analysis, but the former typically exceeded the latter. Among the systematic variables investigated were socioeconomic status, child's gender, child's age, test reliability, and IQ levels of the birth mother and adoptive parents.

Another model-fitting analysis, which focused on IQ change, was carried out for the families participating in both studies, using Wechsler full-scale IQs for the adoptive family members, and Beta IQs for the biological mothers (Loehlin, Horn, & Willerman, 1989). This more selected sample of families yielded somewhat lower estimates of genetic variance than did the full sample: 26% rather than 38% at the time of the first study. The estimate rose to 37% at the second study, when the children were 10 years older. Based on the first testing, there was a small but statistically significant effect of shared environment, accounting for about 6% of the IQ variance; by 10 years later, this had shrunk to a nonsignificant 1%. The continuity from first to second testing appeared to be best conceptualized as a persistence of the developed IQ across the intervening years; the changes between testings appeared to represent (1) the introduction of new genetic variance and random environmental effects, and (2) the weakening of shared familial effects.

The preceding analyses were based on the data of the selected samples and thus most likely would represent underestimates of population heritabilities. They also were based on comparisons across different IQ tests and involving different ages – factors that would attenuate heritability estimates. With the testing of the children on the Revised Beta in the second study, it becomes possible to carry out a single analysis in which everyone receives the same IQ test – not quite all at the same ages but at least avoiding young children, who create the most serious problem of comparability with adult IQs.

Figure 4.1 shows the path model used. It is based on a strategy suggested by Coon, Carey, and Fulker (1990; see also Loehlin, 1992), which leads to

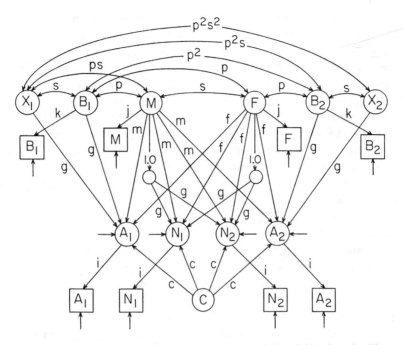

Figure 4.1. Path model. Circles represent theoretical variables (e.g., intelligence); squares represent measured variables (e.g., IQ). M, F = mother, father in adoptive family; N_1, N_2 = two natural children of M and F; A_1, A_2 = two adopted children of M and F; B_1, B_2 = birth mothers of A_1 and A_2; X_1, X_2 = birth fathers of A_1 and A_2; C = shared environmental variables (other than parental intelligence) that may influence the intelligence of the children in a family. Small unlabeled circles = phantom variables (see text). *Paths* (straight, directional arrows): g = genetic effect of parent on child; m,f = environmental effect of mother's, father's intelligence on child; c = effect of children's shared environment; i,j,k = square roots of reliabilities of child, adoptive parent, and birth mother IQ. *Correlations* (curved, bidirectional arrows): s = spouse correlation; p = placement correlation; others (p^2, ps, etc.) are derived from these – only representative examples are shown in the figure. *Residuals* (short, unlabeled arrows): effects not otherwise accounted for (e.g., errors of measurement, individual experiences, nonadditive genetic effects).

conceptually simple models that require fewer explicit assumptions about such matters as the basis of spouse resemblance or the nature of genetic transmission yet still permit meaningful comparisons of genetic and environmental effects.

The conventions of the figure are the usual ones of path analysis: Squares represent measured variables, circles represent inferred or latent variables, curved two-headed arrows represent correlations, and straight arrows represent causal relationships. The short, unlabeled arrows represent the residual causes of a given variable (i.e., "everything else"). The figure represents a hypothetical adoptive family of a mother and father, two adopted and two biological children, and the two sets of birth parents of the adopted children. The circles at the top of the figure represent the intelligences of the parent generation: The M and F in the center of the row are the adoptive mother and father, the B_1 and B_2 beside them represent the birth mothers of two adopted children in the family, and the Xs represent the two (unmeasured) birth fathers. Associated with the circles (except for the Xs) are squares representing the actually measured IQs of the corresponding individuals. The row of four circles lower down in the figure represents the intelligences of the children: The As are the two adopted children and the Ns are two natural biological children of M and F. Again, each has a corresponding square for his or her measured IQ. The C at the bottom of the figure represents common environmental factors that the siblings might share that are unrelated to the IQs of their parents.

The curved arrows at the top of Figure 4.1 represent correlations among the intelligences in the parental generation. Only a few representative ones are shown; however, all possible pairs are connected in the actual model. The two basic correlations are s, the resemblance between spouses, and p, the placement correlation produced by the adoption agency trying to match the birth mother's characteristics to those of the adoptive home. The various other correlations, such as the p^2 between the two birth mothers or the p's between birth father and adoptive parent, are derived as combinations of these two. In combining these, it is assumed that selective placement occurs equally with respect to both parents in the adoptive family and that it is independent of the birth father's IQ, except insofar as the latter is correlated with the birth mother's.

The labeled paths g, m, f, and c are the ones of principal interest. The paths labeled g represent genetic transmission from parent to child; m and f represent environmentally mediated influences of the mother's and father's intelligences on that of the child; and c represents the effect on intelligence of other aspects of the environment that the children share. These might be material, such as family nutrition or the quality of the neighborhood schools, or psychological, such as the personalities of the parents or their beliefs about child rearing. The residual arrows pointing to the circles representing children's intelligence remind us that there can be genetic and environmental factors that the children do *not* share – individual differences in their experiences, and the luck of the draw of parental gene combinations. Finally, the two small unlabeled circles below M and F, connected to them by paths labeled 1.0, are so-called *phantom variables*, merely a technical device for keeping genetic and environmental paths separate in the diagram.

The paths labeled i, j, and k from the circles to the corresponding squares are set to the square roots of the reliabilities of measurement of the respective variables. That is, we are assuming that the measured IQ is partly a reflection of the true scores (the circles) and partly of errors of measurement (the residual arrows pointing at the squares). The square of the path value represents the reliability, the proportion of variance of the observed score that is due to the true score. Internal consistency reliabilities were used; they were estimated separately for the children, the adoptive parents, and the birth mothers from the intercorrelations among the six Beta subscales. The reliability estimates for the three groups were .621, .709, and .678, so that $i = .788$, $j = .842$, and $k = .824$. The relatively low reliabilities are presumably a function of the substantial restriction of range in these samples. Applying a standard correction for restriction of range (Gulliksen, 1950, p. 124) raises the reliabilities to .88, .94, and .90, respectively, values quite comparable to the internal-consistency reliability in the standardization sample, which is given as .90 in the test manual (Kellogg, Morton, Lindner, & Gurvitz, 1957).

Table 4.5 shows 14 equations derived from the path diagram following the path-tracing rules of path analysis. Generally speaking, each correlation is expressed as a sum of paths connecting the two variables involved, with the value of a path being the product of the arrows along it. The part of the expression within the square brackets represents the correlation between the true-score variables; the multiplication outside it adjusts for the reliabilities. Within the square brackets, in the equations numbered 5 and higher, environmental paths are grouped together first, then genetic paths, and finally, paths involving both genes and environment (in the sibling equations 12–14). On the right in the table are the observed correlations of Revised Beta IQs for individuals in the specified relationship, as well as the number of pairings on which each is based. The equations and correlations were given to a computer program that solves sets of nonlinear simultaneous equations by an iterative procedure, and the parameter estimates of Table 4.6 were obtained as a result.

A weighted least squares criterion was used for the fitting, with the numbers of pairings used for the weights. The program fits to Fisher z-transforms of the observed correlations, but because the highest observed correlation is .333, this feature would have had very little effect. Under certain assumptions that include multivariate normality, the goodness-of-fit criterion at the point of solution can be interpreted as a chi square. Such an interpretation is only approximate for the present solutions, for at least two reasons. The numbers of pairings for some relationships are inflated due to multiple entry – for example, in a family with two adopted children, the father's score would be entered twice, once with each. This will tend to raise the chi square. Also, there is overlap between different relationships – for example, many of the same fathers are involved in pairings with both adopted and natural children. Such lack of independence will tend to lower

Table 4.5. *Equations for path modeling and correlations among Beta IQs*

Equation	Correlation	Pairs
1. $r_{MF} = [s]j^2$	.164	175
2. $r_{MB} = [p]jk$	.028	193
3. $r_{FB} = [p]jk$	.093	199
4. $r_{BB} = [p^2]k^2$	−.030	38
5. $r_{MN} = [m + sf + g + sg]ji$	.205	87
6. $r_{FN} = [f + sm + g + sg]ji$	.205	90
7. $r_{MA} = [m + sf + pg + psg]ji$	−.019	244
8. $r_{FA} = [f + sm + pg + psg]ji$	.082	252
9. $r_{B1A1} = [pm + pf + g + sg]ki$	.333	198
10. $r_{B1A2} = [pm + pf + p^2g + p^2sg]ki$	.021	109
11. $r_{BN} = [pm + pf + 2pg]ki$	.045	92
12. $r_{AN} = [c^2 + m^2 + f^2 + 2msf + 2g^2p(1 + s)$ $+ (g + gs + gp + gps)(m + f)]i^2$	−.031	107
13. $r_{AA} = [c^2 + m^2 + f^2 + 2msf + g^2p^2(1 + 2s + s^2)$ $+ (gp + gps)(m + f)]i^2$	−.017	75
14. $r_{NN} = [c^2 + m^2 + f^2 + 2msf + 2g^2(1 + s) + (2g + 2gs)(m + f)]i^2$	.328	27

Note: Subscripts: M, F = mother, father in adoptive family; A = child adopted by M and F; N = natural biological child of M and F; B, X = birth mother, father of A; 1 and 2 identify different individuals in a category. For path symbols, see Figure 4.1 and text. Pairs = number of pairings.

Table 4.6. *Parameter estimates and significance tests from model-fitting to Beta IQ correlations*

Parameter	Full model	Reduced model
g	.371[a]	.355[a]
f	.041	.000
m	−.093	.000
c	.000	.000
s	.221[a]	.225[a]
p	.099	.000
χ^2	3.469	6.486
df	8	12
p	>.90	>.80

[a] $p < .05$, based on chi-square difference test.

chi squares (McGue, Wette, & Rao, 1987). As a result, the chi squares and probability values should be regarded only as approximations.

The chi square for the fit of the full model of Figure 4.1 and Table 4.5 is 3.47, with 8 *df*. The probability associated with such a chi square is approximately .90, meaning that the model represents an excellent fit to the data, even allowing for some uncertainty as to the exact value of chi square.

The differences between models (when they are nested – i.e., the parameters of one are a subset of those of the other) can also be tested by chi square. Several successively simpler models were examined. The setting in turn of c, f, m, and p to zero did not significantly worsen the fit; the genetic path g and the spouse correlations s were, however, essential. The values for the reduced model containing just these two parameters are shown at the right in Table 4.6. The chi square for this model is 6.49, with 12 df, $p > .80$, still a very good fit to the data.

The negative path m in the full model from mother's to child's intelligence suggests that the environmental effect of mothers on their children is such as to make the children less, rather than more, like themselves. These data by themselves do not compel us to accept such a path – the model testing shows it not to be significantly different from zero. However, the fact that something of the sort seemed also to be present in the families who were tested only in the original study, and was observed in some other studies mentioned earlier, suggests that one might at least want to retain such a hypothesis for further investigation.

Heritability of IQ

How is one to translate parameters from the foregoing analysis into an estimate of heritability, the proportion of variance of IQ in the general population that is due to the genes? The fact that these are selected families complicates such estimates. However, we can use twice the (genetic) regression of child on parent IQ to yield an approximate heritability estimate that will be relatively independent of the effects of selection on the parents (Falconer, 1981, chapter 10). The parameter g estimates such a regression, but it is in standard-score terms, and we need a raw-score regression. We may obtain this by multiplying the standard-score value by the ratio of the offspring to the parental standard deviations. Because the path estimate is for true scores, we use true-score standard deviations for this purpose (obtained by multiplying raw-score variances by reliabilities and taking the square root, pooling subgroups as necessary). These true-score standard deviations come out to be 6.24 and 6.53 for the parental and children's generation, respectively, yielding an estimated raw-score genetic regression of .39. Twice this is an estimate of heritability in the population – namely, .78. This is an estimate of heritability in the narrow sense – that is, of the proportion of IQ variance due to genetic effects transmissible from parent to child, the so-called additive effects of genes. It should be emphasized that several approximations have entered into this estimate of heritability and that it should not be considered to represent sacred truth. Most notably, we are extrapolating from results in a fairly restricted sample. Nevertheless, this estimate seems reasonably consistent with other data on adult IQs, if genetic effects on IQ are largely additive. For example, the correlation of twins reared apart gives a direct estimate of broad-sense heritability – the total effect, additive and nonadditive, of the genes. An average for this IQ correlation from four studies is about .74 (Bouchard, Lykken, McGue, Segal, & Tellegen, 1990, table 2). A recent Swedish study yielded .78 for the same correlation based on the first principal component of a battery of cognitive tests, for older twins (average age 65) who had been separated early in life (Pedersen, Plomin, Nesselroade, & McClearn, 1992). A comparison of male Norwegian identical and fraternal twins, based on an intelligence test taken upon entering compulsory military service, yielded heritability estimates of .74 and .69 in two groups, each involving over 1,000 twin pairs (Tambs, Sundet, Magnus, & Berg, 1989). Obviously, exact equivalences are not expected, but, with allowance for the effects of measurement error in the other studies, the various estimates suggest that something over three-fourths of the reliable variance of adult IQ in these populations reflects the influence of the genes.

It will be recalled that in earlier model-fitting analyses based on the data from the first sample, genotype–environment covariance contributed an appreciable amount to IQ variance (about 10%). What happens to this in the present analysis? For practical purposes, it vanishes. The contribution of this covariance may be estimated as twice the product of the genetic path, the environmental path, and the correlation between genes and environment. As any one of these terms approaches zero, so does their product. Thus, in the reduced model of Table 4.6, in which m and f are zero, the contribution of genotype–environment covariance to the IQ variance of the children's generation will be zero. For the full model, where these paths are small, it will be close to zero. The genotype–environment covariance in question here is the so-called *passive* variety (see Plomin, DeFries, & Loehlin, 1977), which results from the fact that in biological families, parents may contribute to their offspring both genes and environmental factors affecting intelligence; or, in the case of adopted children, genes and environments may be associated via selective placement. Our study provides no estimates of the other two forms of genotype–environment covariance, reactive and active, which depend on the reaction of other people to the trait, or the trait's active role in selecting environments that affect it.

Summary and conclusions

The results on IQ from the Texas Adoption Project are generally consistent with the results from other behavior-genetic methods, such as the comparison of identical and fraternal twins, or the study of twins reared apart. The major contributor to familial resemblance is the genes. Shared family environment has an appreciable effect on IQ when children are small, but this becomes minor by the time they are late adolescents. However, we found in our data some tantalizing suggestions that the full story of family effects may prove to be more complicated than this, in a weak *negative* environmental association between mothers' and childrens' IQs. A particularly striking manifestation was that the birth mothers showed, if anything, higher IQ correlations with the children they had had no contact with since near birth than the adoptive parents did with their own biological children with whom they had lived all their lives.

The genetic effects on IQ increased with age. We estimated the heritability (for true scores in the population) to be about .78 for the Revised Beta test at the time of the second study, when the children averaged about 17 years old. This is a figure consistent with results of adult studies of identical and fraternal twins and of separated identical twins. (It should be noted that the majority of twin studies of IQ have been done with children and that these tend to result in lower heritabilities. This accounts for the overall average figure of around .50 often mentioned in the literature – e.g., by Scarr, 1978; Henderson, 1982; Loehlin, 1989; and Plomin, DeFries, & McClearn, 1990).

It seems to us that our results point to two especially interesting lines for further research on the genetic and environmental determinants of intellectual abilities. One is on the possible differentiating effects of family environments on the intellectual development of their members. The other is on how genetic and environmental influences vary across the life span. In following both of these lines of research, it seems to us that it should also be profitable to pursue them past general intelligence, IQ, to more specialized abilities. How do the genes and family influences affect the distinctive aspects of intellectual skills, apart from the common core that they share? Do these processes differ at different stages in life? And if so, how?

References

Bouchard, T. J., Jr., Lykken, D. T., McGue, M., Segal, N. L., & Tellegen, A. (1990). Sources of human psychological differences: The Minnesota Study of Twins Reared Apart. *Science, 250*, 223–8.

Burks, B. S. (1928). The relative influence of nature and nurture upon mental development. *The Twenty-Seventh Yearbook of the National Society for the Study of Education*, Part I, 219–316.

Coon, H., Carey, G., & Fulker, D. W. (1990). A simple method of model fitting for adoption data. *Behavior Genetics, 20*, 385–404.

Falconer, D. S. (1981). *Introduction to quantitative genetics* (2nd ed.). New York: Wiley.

Flynn, J. R. (1984). The mean IQ of Americans: Massive gains 1932 to 1978. *Psychological Bulletin, 95*, 29–51.

Freeman, F. N., Holzinger, K. J., & Mitchell, B. C. (1928). The influence of environment on the intelligence, school achievement, and conduct of foster children. *The Twenty-Seventh Yearbook of the National Society for the Study of Education*, Part I, 103–217.

Gulliksen, H. (1950). *Theory of mental tests*. New York: Wiley.

Henderson, N. D. (1982). Human behavior genetics. *Annual Review of Psychology, 33*, 403–40.

Horn, J. M., Loehlin, J. C., & Willerman, L. (1979). Intellectual resemblance among adoptive and biological relatives: The Texas Adoption Project. *Behavior Genetics, 9*, 177–207.

Horn, J. M., Loehlin, J. C., & Willerman, L. (1982). Aspects of the inheritance of intellectual abilities. *Behavior Genetics, 12*, 479–516.

Kellogg, C. E., Morton, N. W., Lindner, R. M. & Gurvitz, M. (1957). *Revised Beta Examination manual*. New York: Psychological Corporation.

Leahy, A. M. (1935). Nature–nurture and intelligence. *Genetic Psychology Monographs, 17*, 235–308.

Lindner, R. M., & Gurvitz, M. (1946). Restandardization of the Revised Beta Examination to yield the Wechsler type of IQ. *Journal of Applied Psychology, 30*, 649–58.

Loehlin, J. C. (1979). Combining data from different groups in human behavior genetics. In J. R. Royce & L. P. Mos (Eds.), *Theoretical advances in behavior genetics* (pp. 303–34). Alphen aan den Rijn, The Netherlands: Sijthoff & Noordhoff.

Loehlin, J. C. (1980). Recent adoption studies of IQ. *Human Genetics, 55*, 297–302.

Loehlin, J. C. (1989). Partitioning environmental and genetic contributions to behavioral development. *American Psychologist, 44*, 1285–92.

Loehlin, J. C. (1992). Using EQS for a simple analysis of the Colorado Adoption Project data on height and intelligence. *Behavior Genetics, 22*, 239–45.

Loehlin, J. C., Horn, J. M., & Willerman, L. (1981). Personality resemblance in adoptive families. *Behavior Genetics, 11*, 309–30.

Loehlin, J. C., Horn, J. M., & Willerman, L. (1989). Modeling IQ change: Evidence from the Texas Adoption Project. *Child Development, 60*, 993–1004.

Loehlin, J. C., Horn, J. M., & Willerman, L. (1990). Heredity, environment, and personality change: Evidence from the Texas Adoption Project. *Journal of Personality, 58*, 221–43.

Loehlin, J. C., Willerman, L., & Horn, J. M. (1982). Personality resemblances between unwed mothers and their adopted-away offspring. *Journal of Personality and Social Psychology, 42*, 1089–99.

Loehlin, J. C., Willerman, L., & Horn, J. M. (1987). Personality resemblance in adoptive families: A 10-year follow-up. *Journal of Personality and Social Psychology, 53*, 961–9.

McCartney, K., Harris, M. J., & Bernieri, F. (1990). Growing up and growing apart: A developmental meta-analysis of twin studies. *Psychological Bulletin, 107*, 226–37.

McGue, M., Wette, R., & Rao, D. C. (1987). A Monte Carlo evaluation of three statistical methods used in path analysis. *Genetic Epidemiology, 4*, 129–55.

Pedersen, N. L., Plomin, R., Nesselroade, J. R., & McClearn, G. E. (1992). A quantitative genetic analysis of cognitive abilities during the second half of the life span. *Psychological Science, 3*, 346–53.

Plomin, R., DeFries, J. C., & Fulker, D. W. (1988). *Nature and nurture during infancy and early childhood*. New York: Cambridge University Press.

Plomin, R., DeFries, J. C., & Loehlin, J. C. (1977). Genotype–environment interaction and correlation in the analysis of human behavior. *Psychological Bulletin, 84*, 309–22.

Plomin, R., DeFries, J. C., & McClearn (1990). *Behavioral genetics: A primer* (2nd ed.). New York: Freeman.

Plomin, R., & Loehlin, J. C. (1989). Direct and indirect IQ heritability estimates: A puzzle. *Behavior Genetics, 19*, 331–42.

Scarr, S. (1978). From evolution to Larry P., or what shall we do about IQ tests? *Intelligence, 2*, 325–42.

Scarr, S., & Weinberg, R. A. (1977). Intellectual similarities within families of both adopted and biological children. *Intelligence, 1,* 170–91.

Scarr, S., & Weinberg, R. A. (1978). The influence of "family background" on intellectual attainment. *American Sociological Review, 43,* 674–92.

Scarr, S., & Weinberg, R. A. (1983). The Minnesota Adoption Studies: Genetic differences and malleability. *Child Development, 54,* 260–67.

Skodak, M., & Skeels, H. M. (1949). A final follow-up study of one hundred adopted children. *Journal of Genetic Psychology, 75,* 85–125.

Snygg, D. (1938). The relation between the intelligence of mothers and their children living in foster homes. *Journal of Genetic Psychology, 52,* 401–6.

Tambs, K., Sundet, J. M., Magnus, P., & Berg, K. (1989). No recruitment bias for questionnaire data related to IQ in classical twin studies. *Personality and Individual Differences, 10,* 269–71.

Willerman, L., Loehlin, J. C., & Horn, J. M. (1992). An adoption and a cross-fostering study of the Minnesota Multiphasic Personality Inventory (MMPI) Psychopathic Deviate scale. *Behavior Genetics, 22,* 515–29.

5 IQ similarity in twins reared apart: Findings and responses to critics

Thomas J. Bouchard, Jr.

Research on genetic influence on intelligence has a long and contentious history (Brand, 1993; Fancher, 1985; Kamin, 1974). Both the idea of a general factor of cognitive ability, Spearman's *g*, and the idea that genetic factors might be an important source of variance in cognitive ability have been continuously debated since they were first systematically expounded by Galton (1869, 1876). Reviews of Galton's books published in the *London Times* at the time of their appearance could, if slight changes were made, be published today. The debate on the nature of mental abilities and the influence of heredity on such abilities (as well as most other psychological traits) initiated by Galton continues unabated.

The current status of *g*

There should be no doubt that the issues of the measurability of IQ and its usefulness are still controversial issues. Consider the following recommendation regarding the measurement of abilities and other psychological traits:

Make explicit to everyone (pupils, parents, public and professionals of all kinds) that a person's abilities, activities, and attitudes cannot be measured. The public, especially, misperceive that hard data exist, and that test scores constitute these data. The public does not realize how quickly the point is reached where we do not know how to discriminate validly among people, but where data mislead us to think we do. This is what is meant by the myth of measurability [Tyler & White, 1979, p. 376].

This recommendation is from a report published by the National Institute of Education in the United States. I would be inclined to argue that such a claim (that abilities, activities, and attitudes cannot be measured) reflects an abysmal level of ignorance about psychometrics and the accomplishments of social and behavioral scientists over the last 100 years (Bollen, 1989; Ghiselli, Campbell, & Zedeck, 1981; Robinson, Shaver, & Wrightsman, 1991). However, R. W. Tyler, the lead author of the report, is a senior scholar with a distinguished career in education. He cannot be

126

unfamiliar with the evidence. The reason for such a ludicrous claim, consequently, must lie outside the body of empirical evidence generated by social scientists. Indeed, if such a claim is true, then the physical sciences must also be in a dismal state, as there is good evidence that measurement in the social sciences is not nearly as poor, in comparison with that of the physical sciences, as many people believe (Hedges, 1987). The measurement of IQ is more precise and has been more fully explored for sources of artifact than any other construct in psychology (Barrett & Depinet, 1991; Gottfredson, 1986; Gottfredson & Sharf, 1988; Hartigan & Wigdor, 1989; Humphreys, 1992; Jensen, 1980).

Here is a second example:

Because intelligence is not the objectively defined explanatory concept it is often assumed to be, it is more an obstacle than an aid to understanding abilities [Howe, 1990, p. 100].

This quote is by a distinguished professor of educational psychology at Exeter University in England. His claims, in my opinion, are also unsupportable by the evidence (Detterman, 1993; Matarazzo, 1992; Vernon, 1993).

A last example demonstrates that derisive, but unsupportable, comments are not the exclusive domain of educational psychologists:

Spearman's g is not an ineluctable entity; it represents one mathematical solution among many equivalent alternatives. The chimerical nature of g is the rotten core of Jensen's edifice, and of the entire hereditarian school [Gould, 1981, p. 320].

These views can be contrasted with those of equally eminent scholars. John Carroll (1993), for example, has recently completed the most comprehensive survey of the factor-analytic literature ever published; he is confident that there is a factor of general intelligence and that it is influenced by genetic factors. More importantly he argues:

In *The Abilities of Man*, Spearman (1927) developed what was probably the first formal theory of cognitive abilities, the so-called two-factor theory whereby any cognitive test was conceived to be "saturated" with a general factor g and a specific factor s unique to that test. . . . In the main, I accept Spearman's concept of g, at least to the extent of accepting for serious consideration his notions about the basic processes measured by g – the apprehension of experience (what might now be called metacognition) and the eduction of relations and correlates [pp. 636–7].

Nathan Brody (1992) has recently reviewed the substantive research on both theories and correlates of IQ measures and draws a similar conclusion:

The first systematic theory of intelligence presented by Spearman in 1904 is alive and well. At the center of Spearman's paper of 1904 is a belief that links exist between abstract reasoning ability, basic information-processing abilities, and academic performance. Contemporary knowledge is congruent with this belief [p. 349].

Neither of these reviewers believes that the "problem of intelligence" is completely solved, but they do agree that considerable progress has been made. I agree with this evaluation. Stated in different terms I believe the descriptive problem – namely, the answer to the question, "At the level of phenotypic test scores derived from cognitive tests, what is the structure of human cognitive abilities?" – has been largely answered. Technical arguments about the proper method of rotation and so forth are simply irrelevant distractions. In assessing g, it simply makes little difference what method is used (Jensen, 1994). This is not to deny that more research will clear up many details. Second, I believe and will argue that genetic factors play a profound role in the determination of an individual's ultimate level of cognitive ability when that individual is reared under a normal range of circumstances (Scarr, 1992, 1993). The mechanisms or processes that control these outcomes remain largely a mystery, although we do have some clues and tentative theories (Bouchard, Lykken, Tellegen, & McGue, in press; Byrne & Whiten, 1988; Reed, 1984, 1990).

Why is there so much controversy over the construct of intelligence or g? Because it is one of the most important and powerful constructs in the armamentarium of psychology (Miller, 1984), and taking it seriously has immense repercussions. This argument has been brought to the fore with a vengeance by Herrnstein and Murray in their recent book *The Bell Curve* (1994). With very few exceptions, virtually any dimension of behavior scaled from the less valued end to the more valued end correlates positively with IQ (Jensen, 1980, chapter 8; Matarazzo, 1972, chapter 12). The correlations are modest, but they are seldom zero, and they are almost never negative. None of the correlations are high enough to allow one to conclude that g is an overall measure of goodness or human worth or anything else of the sort. The correlations are modest enough so that one can easily find "bad people" with high IQs – a common complaint against IQ tests. Such cases, no matter how often they are cited, do not constitute evidence sufficient to refute stable statistical trends. As I will show, one of the major blocks to advancing our understanding of these issues, both in psychology as a profession and in the public at large, is the abysmally low level of quantitative understanding in both populations. Verbal sophistry – bolstered by anecdotes, linked to emotional appeals, and buttressed by claims of evil intent – masquerade as explanations of embarrassing findings even though they cannot withstand the most elementary quantitative scrutiny. It does not seem to be very widely understood that virtually all these wordy arguments can, if they are sensible, be reformulated into quantitative arguments and evaluated. In order for them to be taken seriously, numbers must be attached! As I show in this chapter, when numbers are properly attached, the explanatory power of most of these arguments evaporates.

Genetic influence on mental ability: Current status

Psychology no less than other human endeavors is subject to fads and fashions. The view that heredity is an important source of human individual differences has waxed and waned over the years (Degler, 1991; Richards, 1987), and these changes were often unrelated to the amount and quality of evidence available. It now seems clear that part of the problem was that psychology (and many other social sciences as well) (1) was wedded to the concept of testing the null hypothesis and testing for statistical significance, (2) lacked a systematic means of integrating data from multiple studies of different kin, and (3) failed to put the evidence and arguments into systematic quantitative form. These problems have now been largely solved. The null hypothesis is known, to put it lightly, to be "bunk" and the testing of statistical significance downright misleading in almost all instances where it is used (Cohen, 1994; Lykken, 1968; Meehl, 1990; Schmidt, 1994). Formal testing of substantive hypotheses via model fitting has now become the norm in behavior-genetic research (Neale & Cardon, 1992) and is likely to become widespread throughout psychology in the future (Schmidt, 1993). Model fitting can, like any other methodology, be abused. This approach to data analysis does, however, force researchers and critics alike to state their claims in testable forms. The failure to specify a model underlying a verbal claim reveals the claim for what it often is: an unsubstantiated assertion disguised as knowledge.

The most recent round in the long running debate about the importance and validity of IQ measures, as well as the debate about the influence of heredity on IQ, was launched by Arthur Jensen in a now famous article entitled, "How much can we boost IQ and scholastic achievement?" (1969). In this article, Jensen claimed on the basis of his review of the evidence that compensatory education has been tried and it apparently has failed. . . . Why has there been such uniform failure of compensatory programs wherever they have been tried? What has gone wrong? In other fields, when bridges do not stand, when aircraft do not fly, when machines do not work, when treatments do not cure despite all conscientious efforts on the part of many persons to make them do so, one begins to question the basic assumptions, principles, theories, and hypotheses that guide one's efforts. Is it time to follow suit in education [p. 2]?

Jensen then presented a systematic body of evidence to show that what had failed were two theories that continue to permeate American social science – namely, the *average-child concept* and the *social-deprivation hypothesis*. The average-child concept encompasses the belief that all children are basically equivalent in their capacity to learn and develop. Observed differences are due to their upbringing (socioeconomic status) and to other general social and/or idiosyncratic influences. The social-deprivation hypothesis is ancillary to the average-child hypothesis and asserts that children in minority groups and children of the poor are invariably less capable only because of the environmental deprivations that they experience as excluded groups.

After demonstrating that this theory had failed to explain the observed differences in IQ, Jensen argued that we should replace it with a Genetic-Diversity Theory of Individual Differences and its natural complement on the environmental side, a diversity of learning opportunities. Specifically, he asserted that

if diversity of mental abilities, as of most other human characteristics, is a basic fact of nature, as the evidence indicates, and if the idea of universal education is to be successfully pursued, it seems a reasonable conclusion that schools and society must provide a range and diversity of educational methods, programs, and goals, and occupational opportunities, just as wide as the range of human abilities [p. 117].

The Genetic-Diversity Theory of Individual Differences proposed by Jensen is a continuation of the Galtonian model. It asserts that there are fundamental differences between human beings in their capacity to develop intellectual skills (IQ, special mental abilities) and most other characteristics. In Darwin's (1871) words:

So in regard to mental qualities, their transmission is manifest in our dogs, horses and other domestic animals. Besides special tastes and habits, general intelligence, courage, bad and good tempers, etc., are certainly transmitted.

As with Galton, who had concerned himself with, among other things, individual differences, social class differences and race differences in ability, Jensen also addressed social-class differences and race differences. I forego discussion of race differences here as it is not germane to this chapter. Jensen argued that social-class differences were in part genetic in origin. Jensen's work set off a storm of protest (Hirsch, 1975), and since then a great deal of research has been carried out with the goal of refuting his claims. As I will show in this chapter, using the MZA (monozygotic, or identical, twins reared apart) data, and as I have shown elsewhere using the entire array of kinship data (Bouchard, 1993b; McGue, Bouchard, Iacono, & Lykken, 1993), the evidence for a large degree of genetic influence on individual differences in intelligence, as measured by IQ test scores, is now irrefutable. The evidence for genetic influences on SES (socioeconomic status) differences in IQ has also grown (Bouchard, 1976; Bouchard, Lykken, McGue, Segal, & Tellegen, 1990a). Jensen drew on the work of Barbara Burks (1938) among others. Scarr and Weinberg (1978) report, on the basis of their adoption study: "Burks estimated that genetic differences among the occupational classes account for about .67 to .75 of the average IQ differences among children born into those classes. Our studies support that conclusion" (p. 689). This interpretation of the evidence is reasonable even though there are studies that demonstrate SES effects in the context of extreme placement – namely, the French cross-fostering study (Capron & Duyme, 1989) and the French adoption study (Schiff & Lewontin, 1986). It is crucial to keep in mind that the average age of the children in the Capron and Duyme study is 14 years. As I will show shortly, common family environmental influences appear to attenuate to near zero as adulthood is reached (McGue et al., 1993). As McGue (1989) has pointed out, it will be interesting to see the results of follow-ups of the French adoption studies.

The relevance of Jensen's work to this chapter is that Jensen relied heavily on the studies of Sir Cyril Burt of identical twins reared apart. Jensen's work evoked a scathing review of the IQ literature by Leon Kamin (1974). Kamin was especially critical of Burt and is credited with exposing Burt as a fraud. The case against Burt has, however, weakened considerably in recent years (Aldhous, 1992; Fletcher, 1991, 1993; Joynson, 1990). In collaboration with like-mined colleagues, Kamin eventually went on to criticize the entire enterprise of behavior genetics (Lewontin, Rose, & Kamin, 1984). I will address here only the issue of IQ. In his 1974 book, Kamin concluded that "there exists no data which should lead a prudent man to accept the hypothesis that IQ test scores are in any degree heritable" (p. 1). This conclusion and his criticisms, particularly the criticism of the MZA studies, over the last 20 years have been widely repeated in introductory psychology texts and elsewhere. The most recent publication in which he is repeatedly quoted, demonstrating that he has not changed his mind, is the attack on behavior genetics published by *Scientific American* (Horgan, 1993).

It should be noted that, Kamin's criticisms not withstanding, there is a strong consensus among experts regarding the findings in this domain. Snyderman and Rothman (1987, 1988) carried out a survey of expert opinion about IQ tests, their meaning, and the nature–nurture controversy. The results show that experts agree with the conclusions drawn by Carroll and by Brody, cited earlier, regarding *g* and the findings regarding the heritability of IQ discussed in this chapter.

A brief historical background on the study of MZAs

The origin of the study of twins adopted early in life and reared apart – the twin-reared-apart (TRA) method – is unknown. Although Francis Galton can be credited with introducing the twin method and the adoption method (Bouchard, 1993c), he never mentioned the TRA method, even though after the publication of his now famous paper on twins (Galton, 1876), one of his correspondents mentioned the existence of one such pair (Townsend, 1874–5).

The first systematic collection of quantitative data from a pair of monozygotic twins reared apart (Bessie and Jessie) was carried out by H. J. Muller who followed up on a more discursive treatment of the same pair by Popenoe (1922). It is of considerable interest that while Muller published his paper on this pair of MZA twins in the biologically oriented *Journal of Heredity* (1925), the study dealt primarily with psychological, not physical or medical, variables and was subtitled, "The extent to which mental traits are independent of heredity, as tested in a case of identical twins reared apart." Muller understood the value of such cases and articulated the logic of this experiment of nature even before the necessary statistical methods were developed to analyze properly the data collected from a series of such cases.

Cases are required in which the identical twins are reared apart, under environments differing as much as those *commonly met with do, in order that we may gain an idea of the amount of effect of such environmental differences as distinguish separate families in a community* [italics added]. Not one such case has heretofore been systematically investigated by modern methods, as such cases are very rare. Each such case is extremely valuable, however, since in any one such case, if a mental trait is found which shows marked similarity in the two members of the pair, and wide diversity in other individuals, in spite of the fact that the environments of the twin members differed considerably in such features as would be most likely to influence the trait, it may be pretty safely concluded that the trait in question, when measured by the method used, is genetically narrowly determined, and is reliable as a genetic indicator; where on the other hand, great differences appear, it is highly probable that the latitude of genetic indetermination is great, that the character differences so indicated are largely non-genetic, and that some other method of observation or testing must be used for estimating the genes which may be concerned with such characters. The results, then, may indicate not only the amount of variation caused by environment in the trait measured, but also the reliability of the method of measurement used, for indicating genetic facts [p. 434].

Muller made a number of important methodological points that are apparently not widely understood 70 years later and still warrant discussion. I will return to them after I discuss the analytic methods currently applied to data gathered on such twins and summarize the previous findings.

The quantitative analysis of twins-reared-apart data

The simplest way to conceptualize the quantitative evaluation of TRA data is via path analysis. Figure 5.1 shows three path diagrams, one for identical twins reared together, one for identical twins reared apart, and one for unrelated individuals reared together. For simplicity I have left out sources of variance that do not contribute to similarity. Path diagrams allow us to quantify our intuitive notions of influence and correlation. The notations are as follows: items in circles indicate underlying (latent) variables; items in boxes indicate measured phenotypes (scores) for the kinship indicated

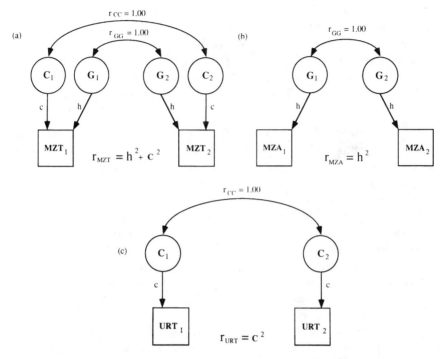

Figure 5.1. Path diagrams for (a) monozygotic twins reared together, (b) monozygotic twins reared apart, and (c) unrelated individuals reared together.

(e.g., MZT_1 is the score, on the trait under consideration, for the first member of a twin pair raised together); G = genotype; C = common (shared) environment; and h and c equal, respectively, genetic and shared-environmental path coefficients. Single-headed arrows denote causal influences, with the lower-case letters representing the degree to which the phenotypic standard deviation is a function of the variability in the latent causal entities. Double-headed arrows indicate correlations. Figure 5.1a diagrams the model for monozygotic twins reared together. We see that G_1 and G_2 are correlated 1.00, reflecting the identical genotypes of monozygotic twins – for DZ (dizygotic, or fraternal) twins, the correlation would be .5, thereby showing what we know from genetic theory, that on average they share 50% of their segregating genes. Both twins are influenced by G (genes cause similarity), and the magnitude of influence is indexed by the path coefficient h. Because we are discussing correlations, not covariances, the path coefficients in these models are standardized. We see that common environment, which by definition is correlated 1.00, also influences (causes similarity between) the twins, and its influence is indexed by the term c. The rules of path analysis allow us to estimate the influence of an underlying latent trait by multiplying the terms of the path. There are two paths in the case of MZ twins reared together (MZT). The first path is through the genes shown by the term ($h*1.00*h$), or more simply h^2. The second path is through the common environment ($c*1.00*c$), or c^2. The correlation between MZT twins is the sum of these influences, or:

$$r_{MZT} = h^2 + c^2$$

This equation formalizes our intuitive notion that MZ twins are alike because we recognize that two different factors can be the cause of their similarity – heredity and common environmental influences. For DZ twins, we can state on the basis of genetic theory that under certain assumptions (no nonadditive genetic variance and no assortative mating), the genetic influence is half that of MZ twins. These equations show that genetic and environmental factors are confounded when relatives are reared together (Bouchard & Segal, 1985; Scarr, chapter 1, this volume). The most common criticism of the TRA design is the argument that the design assumes that MZ and DZ twins experience similar, common family environmental influences – the so-called *equal environment assumption*. Consider the following quote from Lewontin et al. (1984):

There are also some obvious environmental reasons to expect higher correlations among MZ than among DZ twins, especially when one realizes the degree to which an MZ pair creates or attracts a far more similar environment than that experienced by other people. Because of their striking physical similarity, parents, teachers, and friends tend to treat them much alike and often even confuse them for one another.... *There is no great imagination required to see how such a difference between MZs and DZs might produce the reported difference in IQ correlations* [italics added]. It is entirely clear that the environmental experiences of MZs are much more similar than those of DZs [pp. 115–16].

It is certainly true that MZ twins experience more similar environments than do DZ twins, but it is also true, if perhaps surprising, that no one has been able to show that such imposed similarities in treatment are trait-relevant. The critical assumption being made, when this argument is brought forward, is the trait relevance of the treatment. Loehlin and Nichols (1976) studied this problem using very large samples of twins. They related differences within pairs of twins to differences in treatments as reported by the twins' mothers. Consider the dressing-alike argument; it is often claimed that because MZ twins dress alike much more than do DZ twins, they are made more similar. Measured differences in dress were related to differences on the 18 California Psychological Inventory (CPI) Scales for 451 MZ twins. The average correlation was .004. The corresponding correlation with a composite measure of differential experiences (as reported by the mothers) was .056. These effects are obviously trivial. Statistically sophisticated readers may note that difference scores are notoriously unreliable and discount these findings on that basis. However, similar difference scores have been shown to be sufficiently reliable to capture artifacts in our own MZA analyses and yield correlations in the .60 range. In addition, the quantitative findings are replicable using alternate methods (Bouchard & McGue, 1990). A large number of studies have now been carried out on this problem (DeFries & Plomin, 1978; Kendler, Neale, Kessler, Heath, & Eaves, 1993; McCartney, Harris, & Bernieri, 1990; Rose, 1981; Rose, Kaprio, Williams, Viken, & Obremski, 1990; Rowe & Clapp, 1977; Rowe, Clapp, & Wallis, 1987; Scarr, 1968; Scarr & Carter-Saltzman, 1979; Scarr, Scarf, & Weinberg, 1980). Most of this evidence was available and had been brought to their attention prior to the time Lewontin et al. wrote their book. Nevertheless, the only citation on the subject they provide demonstrates that there are treatment differences, not that such differences are trait-relevant.

The TRA design largely overcomes the objection of a highly similar common rearing environment. If the twins are not subject to placement bias, a testable proposition I will discuss later, then they no longer share a common environmental source of similarity so that (see Figure 5.1b):

$$r_{\text{MZA}} = h^2$$

Those with a psychometric background will recognize that this model is of the same form as the true-score model for test–retest and (more pertinently) parallel form reliability (Hayes, 1973). The unsquared correlation between the two forms represents the proportion of variance explained by the true scores. The MZA correlation similarly represents the variance explained by genetic influences (Bouchard, Lykken, McGue, Segal, & Tellegen, 1990b; Jensen, 1971; Miller & Levine, 1973).

The correlation between MZA twins estimates the *broad heritability* of a trait as opposed to the *narrow heritability*. The broad heritability includes all genetic factors that make MZA twins alike. These include nonadditive genetic factors (dominance, epistasis), which while genetic in origin are nontransmissible from parents to offspring (Lykken, McGue, Tellegen, & Bouchard, 1992). Methods that estimate the narrow heritability of IQ find a somewhat lower figure than methods that estimate the broad heritability, suggesting that nonadditive variance may be important for this trait (Pedersen et al., 1992). This distinction gives rise to the common practice of claiming that the heritability of IQ is between .4 and .8 (Herrnstein & Murray, 1994). For the kinds of samples ordinarily studied, the narrow heritability is probably between .4 and .6, and the broad heritability is, as I show in this chapter, around .75.

In other sciences, investigators go through a great deal of trouble to create efficient model systems for investigating a phenomenon. The goal is to create a system that gives the most direct and clearest answer to a question. In animal behavior genetics, the most obvious example is the widespread use of inbred strains of animals. The comparison of random samples of strains allows the investigator to hold heredity constant in order to allow investigation of the influence of various environmental manipulations. Conversely, different strains exposed to identical environments are compared to detect genetic influences. Of course, if animals respond to selective breeding for a behavioral trait, the evidence is even more conclusive (DeFries, Gervais, & Thomas, 1978). In human behavior genetics, monozygotic twins are the closest we can come to inbred strains of animals. Monozygotic twins reared apart combine an experiment of nature (twins) and an experiment of nurture (adoption). The intraclass correlation between MZA twin members is the most powerful and most direct way to estimate the broad heritability of a trait (Plomin, DeFries, & McClearn, 1990). The statistical power of this design is remarkable. For a trait with a heritability of about .50 (an estimate close to that found for many psychological characteristics), 50 pairs of MZA twins have roughly the same statistical power as 1,000 pairs (500 MZ and 500 DZ) of twins reared together – the heritability estimates have the same 95% confidence interval (Lykken, Geisser, & Tellegen, 1978).

An efficient design to detect both the broad heritability and the influence of common family environment is one that contains equal proportions of MZA and MZT twins (Eaves, 1970). A simple, powerful, and direct design for estimating common family environmental influences is the study of unrelated individuals reared together as siblings (URT); the path diagram for this design is shown in Figure 5.1c. As with the MZA correlation, the URT correlation is also a direct estimate of a parameter – in this instance common (shared) environmental influence. While it is understandable that, because of their rarity, MZA twins have been infrequently studied, it is a mystery why URTs have been studied so seldom (Scarr & Weinberg, 1994). Compared to twins reared apart, URTs are relatively common. It almost as though psychologists did not wish to collect data using a sample that would refute their favorite hypotheses.

IQ findings from twins reared apart

Table 5.1 summarizes the entire world literature on the IQ correlations between twins reared apart, including recent data from the Minnesota Study of Twins Reared Apart (MISTRA) (Bouchard et al., 1990a) and the Swedish Adoption Study of Aging (SATSA) (Pedersen et al., 1992).

As shown by the path model for MZA twins, introduced in Figure 5.1b, the MZA intraclass correlation gives us what Muller called "an idea of the

Table 5.1. *Intraclass correlations, confidence intervals, sample sizes, and tests utilized for IQ in five studies of MZA twins*

Study and Test used (Primary/Secondary/ Tertiary)	N for each Test	Primary Test	Secondary Test	Tertiary Test	Mean of Multiple Tests
Newman, Freeman, & Holzinger (1937) (Stanford-Binet/Otis)	19/19	.68 ± .12	.74 ± .10		.71
Juel-Nielsen (1980) (Wechsler-Bellevue/ Raven)	12/12	.64 ± .17	.73 ± .13		.69
Shields (1962) (Mill-Hill/Dominoes)	38/37	.74 ± .07	.76 ± .07		.75
Bouchard, Lykken, McGue, Segal, & Tellegen (1990a) WAIS/Raven-Mill-Hill First principal component	48/42/43	.69 ± .07	.78 ± .07	.78 ± .07	.75
Pedersen, Plomin, Nesselroade, & McClearn (1992) First principal component	45	.78 ± .06			.78
Weighted Average					.75

amount of effect of such environmental differences as distinguish separate families in a community." Kinship studies, like any kind of scientific study, allow us to generalize only to populations similar to the one sampled, in this case to the range of environments found in the community from which the MZA twins had been sampled. This well-known restriction has a long history, and Galton similarly restricted the range of his generalizations when he asserted:

There is no escape from the conclusion that nature prevails enormously over nurture *when the differences of nurture do not exceed what is commonly to be found among persons of the same rank of society and in the same country* [italics added] [Galton, 1876, p. 576].

The goal of determining the magnitude of genetic and environmental influence on IQ should be clearly distinguished from the goal of determining the full reaction range of a trait (Turkheimer, 1991). I discuss the concept of reaction range in the next section.

The Minnesota Study of Twins Reared Apart has been very explicit regarding the range of environments to which its conclusions can be generalized.

The IQs of the adult MZA twins assessed with various instruments in four independent studies correlate about 0.70, indicating that about 70% of the observed variance in IQ in this population can be attributed to genetic variation. Since only a few of these MZA twins were reared in real poverty or by illiterate parents and none were retarded, this heritability estimate should not be extrapolated to the extremes of environmental disadvantages still encountered in society. Moreover, these findings do not imply that traits like IQ cannot be enhanced [Bouchard et al., 1990a, p. 227].

It is often argued that the heritability statistic is uninformative because like any statistic it may vary from population to population and from one set of circumstances to another. I find this argument nonsensical because it directly implies that we should do away with all descriptive statistics. Furthermore, the implied claim that heritability varies greatly is an empirical one and can only be answered by obtaining estimates of this statistic in a variety of settings. We have reason to believe that genetic estimates are more generalizable than often claimed (Rushton, 1989), and the data in Table 5.1 confirm this conjecture for IQ. A second argument against its use is that heritability is misunderstood and that its use furthers that misunderstanding. I would simply argue that what is needed is more education, not less, and certainly not the suppression of statistics. If there is one important point being made in this chapter, it is that the implications of a wide variety of verbal arguments and claims regarding IQ are quantifiable and testable. The only way that we can resolve these disagreements is via statistical evidence, as the phenomena are inherently probabilistic (Bouchard, 1993c).

As Table 5.1 indicates, the weighted average of the MZA IQ correlations is .75. This figure should be compared to the reliability of the types of tests used in these studies, a figure that is unlikely to be above .90 (Parker, Hanson, & Hunsley, 1988). Clearly, the MZA method tells us that a very significant portion of the reliable variance in measured IQ is genetic in origin. How dependable and meaningful are these findings? I turn to these questions next.

Reaction range – Genotype × environment interaction

It always bears repeating that IQ is a phenotype and the genotype is a biochemical code. Measurement of a phenotype presumes a previous developmental process. If that developmental process is nonlinearly related to the genotype, there may be complex interactions. The possibility of such interactions has been repeatedly put forward as a reason for not computing heritabilities (Feldman & Lewontin, 1975; Layzer, 1974; Lewontin, 1974; Wahlsten, 1990; and numerous commentators).

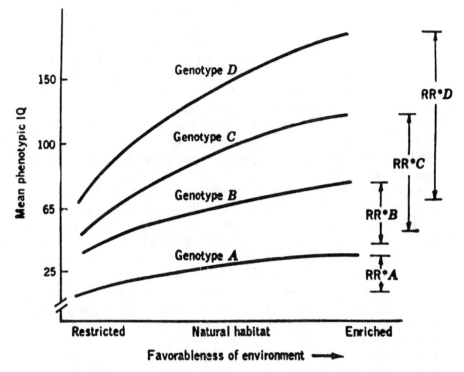

Figure 5.2. Scheme of the reaction-range concept for four hypothetical genotypes. Note: Marked deviation from the natural habitat has a low probability of occurrence. RR signifies reaction range in phenotypic IQ. (From Gottesman, 1963, p. 255.)

The concept of reaction range was initially introduced into the discussion of genetic influence on IQ by I. I. Gottesman (1963, 1968; Turkheimer & Gottesman, 1991). Gottesman's classic hypothetical reaction-range curves for IQ are shown in Figure 5.2.

Unlike the heritability statistic, which summarizes the bottom-line outcome for a population of individuals exposed to a range of environments, the reaction-range curve attempts to illustrate the degree to which genotypes have variable expression in different environments. The hypothetical curves in Figure 5.2 convey a number of important ideas. First, there is a strong genetic main effect and no disordinal interaction (i.e., a genotype keeps the same rank order under all environments, and none of the curves cross). There is an ordinal interaction (a fan-shaped spread): Genotype D responds much better to the enriched environment than Genotype A (the reaction range for each genotype is given at the far right). *One of the features of a reaction-range curve is that it specifically attempts to character-*

ize the degree of expression of a trait for different genotypes under varying environments. Gottesman's curves have the added advantage of pointing out that there is a range of environments which we tend to characterize as the *natural habitat* for a particular organism. Gottesman's range of environments could, of course, be extended to include what might be called, using the plant analogy, *hothouse environments* – environments in which every conceivable effort is made to enhance the trait of interest. It is possible, for example, under very special conditions to make a tomato plant grow into the size of a tree and produce enormous tomatoes. There is a lower tail to the environmental dimension as well. Without the right conditions, no organism will survive.

An examination of the extremes of the favorableness dimension is actually very informative. It quickly becomes clear that a single dimension of favorableness is misleading. At the low end, a tomato plant can die from drought, heat, excessive dampness, or frost. At the high end, a giant tomato plant needs support for its branches lest they break off; it also needs protection from the wind or else it will be blown over and uprooted; and any of these events can quickly lead to infection and death. This brings us to the issue of natural selection for organismic characteristics. One of the most important limiting selective forces that shape a species are the extremes of environment that it faces. Thus, a species of trees lives in a natural habitat where the most extreme winds experienced over long periods of time do not destroy all the trees, or the ground does not flood sufficiently frequently to drown the roots of all members of the species. Organisms in the natural world colonize ranges or niches that are close to the ones in which they evolved. There are interesting examples of organisms changing the environment to fit their needs. Eucalyptus trees in California pull enough moisture out of the fog with their leaves to make up for the lack of rain. The occasional freezes are, however, a continual threat to their existence and limit their spread.

Can we think of comparable examples for the trait of IQ? Clearly, there are environments so bad that they are incompatible with life. There are also environments so intellectually impoverished that mental growth is stunted, and there is a cumulative deficit in IQ (Jensen, 1977). If there are many individuals living in such environments, and they are sampled properly in an MZA study, the computed heritability will reflect this fact (h^2 will be higher if they are not sampled). It is widely agreed that such environments are undersampled by MZA studies (Bouchard et al., 1990a, p. 227), and in that sense, the figures overestimate the degree of genetic influence in the entire population. As I have pointed out elsewhere, however, in the Minnesota study, 90% of the population have IQs in the range we studied (Bouchard, Lykken, McGue, Segal, & Tellegen, 1991). What does not seem to be adequately appreciated, and is a direct implication of this rendition of the reaction-range curve, is that as environments get better, genetic differences generally become a more important source of individual differences than environment (h^2 becomes larger). It is also important not to underestimate what can be done at the highly favorable end of the environmental continuum. Unlike the dangers that exist for a tomato plant grown in a hothouse, and despite widespread beliefs to the contrary, there is no reason to believe that accelerating the educational progress of gifted children is detrimental to their well-being (Stanley, 1973). Nor is there any doubt that the opportunity for extensive practice is important in the development of high-level skills (Ericsson & Charness, 1994; Ericsson, Krampe, & Heizman, 1993).

If the reaction-range curves in Figure 5.2 were largely disordinal – that is, if different genotypes performed very differently as they moved along the environmental continuum and repeatedly crossed each other – this would complicate the computation of heritabilities. There are those who argue that this is likely to be the case, although they seldom refer to human IQ data. One particular example is often presented by Lewontin (Lewontin, 1982; Lewontin, 1975; Schiff & Lewontin, 1986, p. 172) and others (Byne, 1994). It involves demonstrating that it is possible to find phenotypic features of some organisms, in this case the Achillea plant, that interact in a disordinal manner with the environments in which they are raised. In their example, seven different genotypes, rank-ordered according to their height when grown at a low elevation (the Stanford University Botanical Garden), are then shown to differ in their rank order when grown at medium (California Foothills) and high elevations (Mountains of the High Sierra). After detailed scrutiny of the original source (Clausen, Keck, & Hiesey, 1940), I have been unable to locate the precise figure or set of data used by Lewontin. I am not implying bias here; he could easily and legitimately have combined data from a variety of tables and figures presented in the book. The problem is one of incomplete reporting. The figure most similar to Lewontin's figure is figure 122 (p. 310). That figure and the legend are reproduced here as Figure 5.3.

I would argue that this figure also shows, as Clausen et al. indicate in the legend, that there are ecotypes and ecospecies of the California Achilleas, and that the type chosen from a particular environment generally does best in its natural environment or one close to it. In addition, extremes kill. The Maritime plant (*A. borealis arenicola*) picked near sea level does best at sea level but quite well at medium elevation and does not even flower at Timberline. The Mid-Sierran race (1315–1) actually does better at sea level than Mather (where it was picked) but does not flower at Timberline. The High Sierran form (2459–1) appears to do well at all altitudes, but the

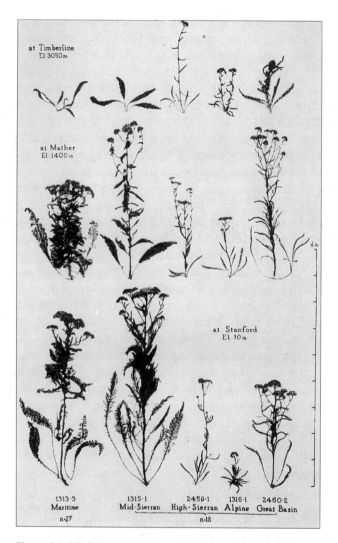

at Timberline
El 3050 m

at Mather
El 1400 m

at Stanford
El 30 m

1313-3	1315-1	2459-1	1316-1	2460-2
Maritime	Mid-Sierran	High-Sierran	Alpine	Great Basin
n-27		n-18		

Figure 5.3. Modifications at three transplant stations in five clones representing altitudinal ecotypes and ecospecies of the California Achilleas. The lowest row consists of specimens grown at Stanford, the middle row at Mather, and the top row at Timberline. (From Clausen, Keck, & Hiesey, 1940, p. 310.)

Alpine form (1316–1) appears to do poorly at Stanford. The Great Basin Form appears to do best at its own altitude. We might be able to claim that forms selected for extremes do well in milder environments, but, of course, we have no idea if they would survive against competitors over time. In this study, natural selection has been removed. We can, of course, say that two of the types do not survive in the Timberline environment as they do not flower. Clausen et al. also show that plants that do flower in one season sometimes simply do not survive because in a subsequent season the frost occurs on an earlier date (cf. their figure 123). Interestingly, Clausen et al. point out that Achilleas has "pronounced individual differences in earliness (time of flowering) within the same ecotype" (p. 314). The fact is, however, that the kinds of disordinal effects emphasized by Lewontin in this plant are exhibited at the extremes (many do not survive more than a few years at Timberline), and we do not know what would happen under natural circumstances. The meaningful interactions may indeed be slightly ordinal. This is not to argue that this approach to the matter is uninformative. As Turkheimer and Gottesman (1991) point out, "In some contexts, it is perfectly reasonable to ask how individuals in the natural environment come to vary as they do; in others, it is reasonable to ask how they might vary if the environments were to be altered radically" (p. 19). An excellent discussion of this issue can also be found in Haldane (1946).

Another widely cited complex interaction should also be dealt with at this point. Cooper and Zubeck (1958) demonstrated that if two strains of rats (Bright and Dull) were raised in three different environments (enriched, normal for laboratory rats, and restricted) their performance (error rate) would yield a strong interaction. The primary finding is that both strains do poorly in deprived environments; they differ in the environment under which they were selected (as they should) but do not differ in enriched environments. Unfortunately, this article is extremely misleading, as the authors themselves admit that the results may be due to an artifact – namely, "the ceiling of the test may have been too low to differentiate the animals, that the problems may not have been sufficiently difficult to tax the ability of the brighter rats" (p. 162). It must also be mentioned that studies of environmental influence using inbred strains of animals, while of great interest theoretically, create serious problems with regard to generalizability to hybrid organisms, which almost all species of animals, including human beings, are. Inbred strains of animals are hardly representative of their own species (most originating lines do not survive the process of inbreeding) and are unduly sensitive to most environmental variation relative to hybrids. Hybrids are probably buffered from environmental influences (Hyde, 1973). These issues have been discussed in detail in the technical literature (Bouchard, 1993b, pp. 72–3; Crow, 1990; Falconer, 1990; Henderson, 1990; Hyde, 1974).

While the possible existence of complex interactions may make the analysis of the main effects of genes and environment a futile exercise (Feldman & Lewontin, 1975), it must be kept in mind that there is very little evidence for such effects on IQ even though a great deal of work has been carried out on the problem (Eaves, Last, Martin, & Jinks, 1977a; Eaves, Last, Young, & Martin, 1977b; Jinks & Fulker, 1970). The fact is that "everything in the world can be explained by factors about which we know nothing" (Urbach, 1974, p. 253). More to the scientific point, as Rao, Morton, and Yee (1974) have argued, "since armchair examples of significant interactions in the absence of an additive effect are pathological and have never been demonstrated in real populations, we need not be unduly concerned about interaction effects. The investigator with a different view should publish any worthwhile results he may obtain" (p. 357).

The misuse of environment as an explanation of MZA similarity in IQ: Trait relevance and the partialling fallacy

Another common argument against the MZA method is that few of the twins are reared in extremely different environments. This argument is not relevant if one is attempting to describe the source of variation found in a specific community. An additional flaw that often accompanies this argument is that the environment can be characterized along one dimension (e.g., good ↔ bad). In point of fact, it is virtually certain that different environments are relevant to different traits (Muller's "features as would be most likely to influence the trait"). The term *trait-relevant environments* has been introduced to deal with this problem. Cases from the extremes of one trait-relevant environment will not necessarily be at the extreme of another trait-relevant environment. One can easily imagine a pair of twins who, while reared apart, both live in affluent homes with unlimited access to books, good education, and so on, but where one twin is loved and showered with affection while the other twin is abused and treated with scorn. We can only hope to capture these types of differences in a sample of MZA twins.

The authors of studies of MZA twins have repeatedly been accused of not adequately studying environments. They are accused, for example, of only examining crude indicators such as education of parent, socioeconomic status of family, family size, physical features of the environment, and fallible self-reports of child-rearing practices by parents. There is no question that this accusation is in part correct. Our reply, however, is that these features are measurable, and because they have often been put forward as explanations of individual differences in ordinary families, the validity of these claims must be tested in the context of an adoption design. Consider the recent discussions of SES and health (Adler, Boyce, Chesney, Cohen, Folkman, Kahn, & Syme, 1994), SES and achievement (White, 1982), and family size and IQ (Blake, 1989). All these authors fail to realize that the correlations they are discussing are confounded, or they hand-wave the possibility of genetic effects away. Adler et al. dismiss the genetic argument and fail to cite the most relevant competing paper that asserts a genetic explanation (West, 1991). It is our contention that in spite of years of concerted effort by psychologists, there is very little knowledge about the trait-relevant environments that influence IQ (Bouchard, 1993b; Jensen, chapter 2, this volume; Locurto, 1988, 1990, 1991) and ordinary personality traits (Bouchard, 1993a). This is not to assert that there are no findings in the environmental domain; rather, the findings are so inconsistent that it is necessary to appeal constantly to higher-order interactions (Wachs, 1992), which are notoriously difficult to replicate. Brand (1993) provides a trenchant critique of this position.

Finally, it is worth mentioning that many so-called environmental variables are not entirely environmental at all. Many of them often have a genetic component. As pointed out previously, SES differences in IQ are now known to have a significant genetic component (Scarr & Weinberg, 1978). Partialling out parental SES from a relationship that involves IQ and some other variable (e.g., occupational success) results in the removal of more genetic variance than environmental variance. In MISTRA, the following correlations between adoptive parental measures and participant's IQ were found: Father's education .10, Mother's education −.001, Father's SES .174. These correlations must be squared in order to estimate variance accounted for. When fit to an appropriate path model, these correlations, taken in conjunction with the degree of placement bias, accounted for only a trivial portion of the MZA similarity in IQ (Bouchard et al., 1990a, Table 3). The best estimate of the correlation between biological parent's SES and offspring IQ, based on a meta-analysis, is .33 (White, 1982). Clearly, a large part of this correlation is genetic in origin. The partialling out of genetic variance in the guise of equating for environmental differences is called the *partialling fallacy*, and it permeates the social science literature (Jensen, 1973; Meehl, 1970, 1971, 1978). It is embarrassing to point it out, but this problem has been well known for over 50 years (Burks, 1938). Recent embarrassing examples can be found in Hoffman (1991) and Tomlinson-Keasy (1990), and detailed criticisms can be found in Bouchard (1993a). Plomin (1994) provides an excellent review of the relevant literature showing that environmental variables are not always what they seem to be.

Constructive replication

Muller was also alert to the problems of method specificity and reliability. He found that Bessie and Jessie were very much alike on two intelligence tests: the Army Alpha Test (Form 8, July 1918) and the Otis Advanced Intelligence Test (Form A, 1922). Their scores were 156 and 153 on the

Alpha and 64 and 62 on the Otis. Since both scores were high (very superior intelligence), it was less likely that the findings were due to chance than if their scores were in the middle range. The replication across tests (a version of constructive replication) also suggested that the findings were reliable. Nevertheless, the twins did not differ very much across their social backgrounds, so Muller felt that he could not draw sweeping conclusions from the IQ findings.

The findings for personality were quite different. The twins differed considerably on the measures used, the Pressey X-O tests and the Downey Individual Will-Temperament tests. The differences were in fact larger than the expected differences between randomly chosen individuals in the norm group for each test. None of these early personality tests has survived in the face of scientific advance in measurement. They simply did not prove to be reliable or valid enough in ordinary usage to become a part of the psychological armamentarium. Very different and much more reliable and valid methods of measuring personality have taken their place (Goldberg, 1971).

One of the striking findings from the IQ data in the MZA studies is the replicability of findings across studies, measures, countries, and cohorts. The studies span over 50 years, involve many different measures of IQ, took place in five different countries, and were conducted in three languages. All the settings were, however, modern industrialized societies. The findings from all kinship studies show similar robustness but are also limited to modern industrialized societies (Bouchard & McGue, 1981).

A closer look at some previous criticisms

The Farber Analysis

Susan Farber in the introductory chapter of her book *Identical Twins Reared Apart: A Reanalysis* (1981) argues, regarding the previous analyses of the MZA data, that

my own evaluation, particularly of the allegedly scientific analysis made of the IQ data, is more caustic. Suffice it to say that it seems that there has been a great deal of action with numbers but not much progress – or sometimes not even much common sense [p. 22].

Farber is echoing the complaints of Leon Kamin (1974), but she and Howard Taylor, who is discussed in the next subsection, go about their debunking in a much more systematic manner, so I will deal with their analyses while recognizing that the results also apply to Kamin. The interested reader may also wish to examine the following reviews of Kamin's work (Bouchard, 1982b; Fulker, 1975; Jackson, 1975). Farber, following

Kamin, spends much of her time trying to demonstrate that various forms of contact and degree of separation account for the similarities. Bouchard reviewed her book in detail elsewhere (1982a), and only a few summary points from that review can be presented here. Regarding the statistical analyses, Bouchard concluded that

the results seriously abuse statistical theory and reinforce the widespread belief that scientists can prove anything with statistics. In sum, the treatment of the IQ data is an exercise in obfuscation. Perhaps this new approach needs a name. I suggest the term "pseudoanalysis" [p. 190].

The most important reanalysis of Farber's data involves the cases that *she herself classifies as highly separated.* Bouchard's comments on this analysis, part of which, it should be noted, Farber did not carry out, are as follows:

By this point I was persuaded that separation had little or no effect on similarity between twins. I decided to calculate intraclass r's for the Highly Separated group for whom I had expected to find an analysis but had not. The results were surprising! For the entire group: $n = 39$, $r_i = .76$, mean = 97.42, SD = 14.28. For the females: $n = 26$, $r_i = .76$, mean = 97.96, SD = 14.29. For the males: $n = 13$, $r_i = .76$, mean = 96.35, SD = 14.20. The three arrays show the slight depression in IQ characteristic of most twin samples, a standard deviation comparable to the normative population, identical intraclass r's that are indistinguishable from the full sample where separation is ignored [p. 191].

These twins admittedly constitute a modest subgroup, but it is large enough to address Taylor's argument regarding the myth of separated twins discussed in the next subsection. As a historical note, it is interesting to find that a full analysis of all the cases reported in Farber's book yield a correlation of .771 (Bouchard, 1982a), precisely the correlation that Kamin accused Burt of fabricating (Kamin, 1974). Farber's data set did not, of course, include Burt's twins. The correlations of .78 and .78 reported in the two recent replications (see Table 5.1) are surprisingly close to this figure as well.

To make it clear that Bouchard's very negative review of this book is not idiosyncratic, I cite a review by Loehlin (1981). After pointing out numerous errors in the reporting of birthweights, he states:

A second aspect of the book is an elaborate statistical treatment of the IQ data from the separated MZ twin studies. Some interesting analyses are provided, but readers are hereby cautioned to watch out for the graphs and summaries in Chapter 7. These suggest that the amount of contact between separated MZ twins accounts for some 20–30% of the IQ variance. Perhaps, but only if one assumes that the mechanisms involved work in the opposite directions in males and females (see Appendix E, p. 350). For the sexes combined, the amount of contact between the twins does not predict their resemblance [p. 297].

Locurto (1991) and Brody (1992) have provided similar critical examinations of the Farber analysis.

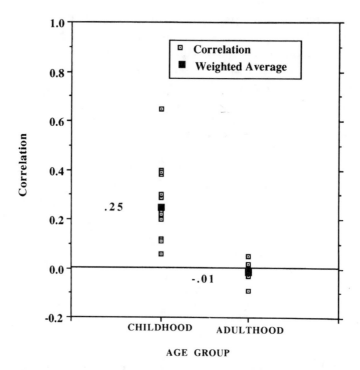

Figure 5.4. IQ correlations among nonbiologically related, but reared-together relatives (both adopted-adopted and adopted-biological pairs). Weighted average correlations were derived using the Fisher Z transformation method. (From McGue, Bouchard, Iacono, & Lykken, 1993, p. 67).

The Taylor Analysis

Howard Taylor, again following Kamin, has also carried out a detailed analysis of the MZA data in his book *The IQ Game* (1980). The chapter dealing with this topic is entitled "The myth of separated identical twins." According to Taylor:

The similarity in educational, socioeconomic, and interpersonal environments, referred to here as social environment, is a central reason why monozygotic twins regarded in the professional literature as separately raised reveal similar IQ scores. MZ twin pairs who have had similar social environment (such as similar schooling) have similar IQs, and twin pairs who have relatively different social environments (especially different schooling) have different IQs [p. 92].

None of these claims is true. Using Taylor's own classification of the twins, Bouchard showed that his findings simply did not replicate when they were tested with a different IQ measure independently obtained from the same sample (Bouchard, 1983). As indicated above, the analysis of Farber's highly separated sample also refutes this claim. Bouchard concluded:

Taylor's conclusion that "it seems reasonable to suggest that the IQ correlations characterizing pairs of individuals with absolutely identical genes and absolutely uncorrelated environments would be extremely low" cannot be substantiated from the evidence at hand [p. 175].

These findings were provided to Taylor prior to submission of the paper for publication, and he did not comment on them. They have yet to be refuted.

The failure to find any influence from the kinds of biasing variables explored by Farber and Taylor extends to both SATSA (Pedersen et al., 1992) and MISTRA (Bouchard et al., 1990a). Neither study has been able to find significant effects on IQ due to placement on these types of variables. Interestingly enough, had Farber's and Taylor's findings been replicable, they would have been an anomaly. We now know that unrelated individuals reared in the same home show no similarity whatsoever if their IQs are measured in adulthood. The results of those studies are shown in Figure 5.4.

Data gathered on unrelated individuals reared together (URTs) and measured as children does show an effect. This influence apparently fades with time and disappears in adulthood. Only one study in Figure 5.4 is longitudinal. We clearly need more studies of this sort. In any event, it is clear that if URT individuals show no similarity in IQ in adulthood after having lived together for years, it is not a surprise that the various measures of environmental similarity used in the MZA studies have failed to explain the similarity in MZA twins.

A more detailed analysis of a commonly hypothesized mediating mechanism: Physical attractiveness

Ford (1993) has argued that the "reported striking physical similarity of MZAs for facial features, height, weight, gait, posture and voice . . ." might explain the personality concordance in MZAs. Ford goes on to assert that

there is a plethora of research indicating that physical features and attractiveness strongly influence how others react to us (Alley, 1988b; Bull & Rumsey, 1988; Feingold, 1992). The literature that examines the related question of the association of personality traits with physical features and attractiveness is less extensive; however, there is evidence of a connection (Borkenau, 1991; Melamed, 1992) [p. 1294].

This criticism has been put forward numerous times at public presentations of the MISTRA findings. It tends to be applied indiscriminately to our findings for abilities, personality, vocational interests, and social attitudes. It is worth pointing out that I have no a priori objection to this or any other explanation that involves mediating mechanisms. Such a process may well be involved in the development of various traits, and it is the goal of psychological research to explicate them. They cannot, however, simply be proposed with a wave of the hand without a close examination of their quantitative implications. Indeed, the adherents of this alternative explanation (physical attractiveness) and others like it do not appear to appreciate its complexity and implausibility. In words, a simple version of the causal model is as follows:

Parent makes judgments about their child's physical attractiveness → they consistently treat the child in some manner → this treatment influences the child's traits (regardless of other attributes which we know are only weakly correlated with physical attractiveness) → physical attractiveness comes to be correlated with a whole host of traits.

I have used a parent–child model because we know that IQ stabilizes early in life and that spouses have a trivial influence over each other's personality over the course of a marriage (Caspi & Herbener, 1993). In order for MZA twins to become correlated, a second set of parents must carry out the same process in the same manner. We also must assume that some of these parents have the capacity to push a trait like IQ as low as 79 and as high as 130 (the range of IQs in MISTRA). I leave it to the reader to decide if such an argument is plausible. If this particular model is unacceptable, simply state another one; nevertheless, the same constraints will apply. Note that we need not assume that the model explains all of the similarity between MZA twins. Even a modest amount of explained variance would make physical attractiveness an interesting variable. In virtually any model of this process, however, physical attractiveness, or whatever trait is considered important, is expected to correlate eventually with the dependent trait (IQ, personality, etc.).

What is the actual evidence regarding physical attractiveness? Let us begin with the meta-analysis of the correlations between physical attractiveness and various traits cited by Ford. Feingold (1992) reports that the experimental literature demonstrates that physically attractive people, as opposed to physically unattractive people, are *perceived* as more dominant, intelligent, mentally healthy, sexually warm, and sociable. In addition, *self-ratings* of physical attractiveness are positively correlated with a wider range of attributes than objectively rated physical attractiveness. The correlational literature, where physical attractiveness is determined *objectively* (not self-ratings), indicates generally trivial relationships. The findings (sample size, median and mean correlations for each trait) taken from Feingold's (1992) table 6, are as follows: Sociability ($N = 1,710$, .00, .04), Dominance ($N = 2,858$, .04, .07), General mental health ($N = 2,597$, .02, .05), Self-esteem ($N = 4,942$, .04, .06), Internal locus of control ($N = 3,683$, .00, .02), Freedom from loneliness ($N = 430$, .04, .15), Freedom from general social anxiety ($N = 1,155$, .06, .09), Freedom from heterosexual anxiety ($N = 1,539$, .19, .22), Freedom from public self-consciousness ($N = 578$, −.20, −.18), Freedom from self-absorption ($N = 746$, .00, −.08), Freedom from manipulativeness ($N = 252$, .03, −.01), Social skills ($N = 1,050$, .25, .23), Popularity ($N = 982$, .04, .08), Intelligence ($N = 3,497$, .00, −.04), and Grades ($N = 3,445$, .07, .02). Feingold concludes, as one might expect, that "good-looking people are not what we think." I would add, given these results, that "the influence of attractiveness on stable personality traits is unlikely to be what some people thought." It is well worth repeating that there is a powerful physical-attractiveness stereotype. People believe that attractive people are at the high end of most of these traits. These data are, of course, not dispositive regarding the influence of differential treatment due to physical attractiveness in childhood and adolescence, the periods during which most traits coalesce. They do, however, throw considerable doubt on the idea that physical attractiveness is an important determinant of any of the traits studied. Note that I am not asserting that physical attractiveness does not influence any social behavior and attitudes. The correlations for Social skills, Freedom from heterosexual anxiety, and Freedom from public self-consciousness are higher than for other traits. In addition, the comparable figures for Noncoital sexual experience are ($N = 1,167$, .16, .13) and for Global sexual experience are ($N = 1,896$, .18, .18). These findings suggest that physical attractiveness has a very modest and a very narrow and specific influence on sexual behavior and attitudes. Other evidence supports this conjecture (Mazur, Halpern, & Udry, 1994).

How about the other references cited by Ford? The Bull and Rumsey (1988) book is in our opinion a thorough, critical, and scholarly work. Nevertheless, it does not deal with the topic under discussion in any straightforward way; moreover, it fails to support the view that Ford would like to foster – namely, that facial appearance is an important determinant of personality and thus significantly influences twin similarity. Each chapter in the Bull and Rumsey book has an excellent summary paragraph, and the skeptical reader is urged to look at each of them. The conclusions reported in the book were such a surprise that I report selected results in the following numbered paragraphs, where I simply cite enough to convey the flavor of the findings.

1. Regarding facial appearance in liking and dating (chapter 2), the authors argue that in the 1960s and 1970s, attractiveness was seen to play a major role. After a review of that literature, they argue that the experimental conditions were very artificial, and while subsequent studies often found a statistically significant effect, "the power or strength of effect was rarely mentioned" (p. 39). They go on to argue that while facial appearance does play a role in marital selection, "no consistent relationship between facial appearance and marital adjustment has been found, although there is the suggestion that facial appearance may play a contributory role" (p. 39).

2. Regarding facial appearance in Persuasion, Politics, Employment,

and Advertising (chapter 3), we find that vis-à-vis advertising, "there is little evidence that actual behavior is affected" (p. 79).

3. Regarding facial appearance in the criminal justice system (chapter 4), the authors conclude that with respect to attributions of responsibility for crimes, the literature is "replete with contradictory findings" (p. 120).

4. Regarding facial appearance and education (chapter 5), the authors conclude, like Feingold, that it is easy to find effects, in this case expectation effects, by using studies that manipulate photographs. "However, the evidence that such expectations have any meaningful resulting effects is much weaker, several studies having found rather limited or no such effects" (p. 150).

The Alley (1988b) book contains little not covered by Feingold (1992). The most relevant chapter is chapter 8, written by the editor (Alley, 1988a). He can speak for himself:

Is facial structure related to personality? Folklore, literary characterizations, and requests for photographs from job applicants, not to mention the practice and promulgation of physiognomy for centuries, indicate that such relationships exist. Scientific research, however, has generally found little or no validity in physiognomy. For instance, in a relatively recent and thorough study, Cohen (1973) "Found it impossible to discover any meaningful relations – even through use of multiple correlations – between physiognomic and psychological characteristics, which could maintain their statistical significance in cross validation on other data: (p. 107)" [p. 172].

Now let's look at the two empirical papers that Ford cites, Borkenau (1991) and Melamed (1992). I quote Borkenau's abstract:

Self-report personality correlates of wearing glasses were investigated. To control for possible effects of social stereotypes on self-reports of personality, judgments by strangers were also collected. The trait that perceivers inferred from spectacles differed from the self-reported traits that actually co-occurred with the presence of glasses. Thus a substantial influence of social stereotypes on self-reports of personality was not reasonable [p. 1125].

The abstract from Melamed (1992) reads as follows:

Physical height was correlated with the 16 PF. Height was significantly related to suspiciousness for both sexes, and to dominance and independence for males [p. 1349].

The correlations with suspiciousness were .29 and .27. As the authors point out, "The correlation found with assertiveness and independence followed expectation. Yet, it is hard to explain the correlation with suspicion and the lack of significant correlation with self-assurance (factor O)" (p. 1349). With results like these, I believe that it might be wise to wait for a few replications.

Rowe (1994, p. 47) presents a path model to represent how weak the influence of physical attractiveness is on twin similarity for personality under unlikely assumptions (correlations between attractiveness and personality are much higher than reported by Feingold). More conclusive, however, is the empirical evidence. Rowe et al. (1987) showed that controlling for attractiveness does not remove the similarity of MZ twins reared together. In addition, two other studies (Matheny, Wilson, & Dolan, 1976; Plomin, Willerman, & Loehlin, 1976) showed that twins who were more alike in personality were not rated more alike in appearance. Burks and Tolman (1932) long ago failed to find such an effect with siblings.

I have devoted a great deal of space to this topic because it is prototypic of the kinds of explanations that are repeatedly bandied about in attempts to explain the similarity of MZA twins. Evidence is needed to support these explanations, and it seldom supports the claims that are made. On numerous occasions, I have found that these explanations are assented to (by well-trained psychologists) within seconds of their being proposed, and the parties wander away with a self-satisfied look that indicates they believe that psychology has again provided a simple and powerful explanation of striking phenomena when it has done no such thing. I note that this is a very general problem in psychology (Dawes, 1994). The striking similarity of IQ and other traits in MZA twins may be mediated by environmental processes of this sort, and I have suggested what such processes might look like (Bouchard et al., 1990a; Bouchard et al., in press). Such theories might, however, also be wrong. Potential IQ, personality, and other psychological traits may simply reside in the brain with a wide range of nonspecific environmental influences being sufficient to mediate their development. Such theories must be tested and shown to be in accord with the best evidence available, not ordinary intuition.

Conclusions

As far as I am aware, no plausible alternative to genetic influence exists to explain the IQ similarity in monozygotic twins reared apart. Since these findings are highly consistent with heritability estimates from other adult kinships, also collected in similar settings (Bouchard et al., in press; McGue et al., 1993), I conclude that genetic factors are the predominant source of variation in adult measured intelligence in modern Western societies.

Acknowledgments

The Minnesota Study of Twins Reared Apart has been supported by grants from the Pioneer Fund, the David H. Koch Charitable Trust, the Seaver Institute, the Spencer Foundation, the National Science Foundation (BNS-7926654), the University of Minnesota Graduate School, and the Harcourt Brace Jovanovich Publishing Co. I would like to thank my colleagues

Margaret Keyes, Auke Tellegen, and David Lykken for numerous suggestions that greatly improved the manuscript. The errors that remain are
mine.

References

Adler, N. E., Boyce, T., Chesney, M. A., Cohen, S., Folkman, S., Kahn, R. L., & Syme, S. L.
    (1994). Socioeconomic status and health: The challenge of the gradient. *American Psy-
    chologist, 49*, 15–24.
Aldhous, P. (1992). Psychologists rethink Burt. *Nature, 356*, 5.
Alley, T. R. (1988a). Physiognomy and social perception. In T. R. Alley (Ed.), *Social and
    applied aspects of perceiving faces*. Hillsdale: Erlbaum.
Alley, T. R. (1988b). *Social and applied aspects of perceiving faces*. Hillsdale, NJ: Erlbaum.
Barrett, G. V., & Depinet, R. L. (1991). A reconsideration of testing for competence rather
    than for intelligence. *American Psychologist, 46*, 1012–24.
Blake, J. (1989). Number of siblings and educational attainment. *Science, 245*, 32–6.
Bollen, K. A. (1989). *Structural equations with latent variables*. New York: Wiley.
Borkenau, P. (1991). Evidence of a correlation between wearing glasses and personality.
    Personality and Individual Differences, 12, 1125–8.
Bouchard, T. J., Jr. (1976). Genetic factors in intelligence. In A. R. Kaplan (Ed.), *Human
    behavior genetics*. Springfield, IL: Charles C. Thomas.
Bouchard, T. J., Jr. (1982a). [Review of "Identical twins reared apart: A reanalysis."] *Contem-
    porary Psychology, 27*, 190–1.
Bouchard, T. J., Jr. (1982b). [Review of "The intelligence controversy."] *American Journal of
    Psychology, 95*, 346–9.
Bouchard, T. J., Jr. (1983). Do environmental similarities explain the similarity in intelligence
    of identical twins reared apart? *Intelligence, 7*, 175–84.
Bouchard, T. J., Jr. (1993a). Genetic and environmental influences on adult personality:
    Evaluating the evidence. In I. Deary & J. Hettema (Eds.), *Basic issues in personality*.
    Dordrecht: Kluwer Academic Publishers.
Bouchard, T. J., Jr. (1993b). The genetic architecture of human intelligence. In P. A. Vernon
    (Ed.), *Biological approaches to the study of human intelligence*. Norwood, NJ: Ablex.
Bouchard, T. J., Jr. (1993c). Twins: Nature's twins told tale. In T. J. Bouchard, Jr., & P. Propping
    (Eds.), *Twins as a tool of behavior genetics*. Chichester, England: Wiley & Sons Ltd.
Bouchard, T. J., Jr., Lykken, D. T., McGue, M., Segal, N. L., & Tellegen, A. (1990a). Sources
    of human psychological differences: The Minnesota study of twins reared apart. *Science,
    250*, 223–8.
Bouchard, T. J., Jr., Lykken, D. T., McGue, M., Segal, N. L., & Tellegen, A. (1990b). When kin
    correlations are not squared. *Science, 250*, 1498.
Bouchard, T. J., Jr., Lykken, D. T., McGue, M., Segal, N. L., & Tellegen, A. (1991). IQ and
    heredity. *Science, 252*, 191–2.
Bouchard, T. J., Jr., Lykken, D. T., Tellegen, A. T., & McGue, M. (in press). Genes, drives,
    environment and experience: EPD theory – revised. In C. P. Benbow & D. Lubinski
    (Eds.), *Psychometrics and social issues concerning intellectual talent*. Baltimore: Johns
    Hopkins University Press.
Bouchard, T. J., Jr., & McGue, M. (1981). Familial studies of intelligence: A review. *Science,
    212*, 1055–9.
Bouchard, T. J., Jr., & McGue, M. (1990). Genetic and environmental influences on adult
    personality: An analysis of adopted twins reared apart. *Journal of Personality, 58*, 263–92.
Bouchard, T. J., Jr., & Segal, N. L. (1985). Environment and IQ. In B. J. Wolman (Ed.),
    Handbook of intelligence: Theories, measurements, and applications. New York: Wiley.

Brand, C. R. (1993). Cognitive abilities: Current theoretical issues. In T. J. Bouchard, Jr., & P. Propping (Eds.), *Twins as a tool of behavioral genetics*. Chichester, England: Wiley.

Brody, N. (1992). *Intelligence* (2nd ed.). San Diego: Academic Press.

Bull, R., & Rumsey, N. (1988). *The social psychology of facial appearance*. New York: Springer-Verlag.

Burks, B., & Tolman, R. (1932). Is mental resemblance related to physical resemblance in sibling pairs? *Journal of Genetic Psychology, 40,* 3–15.

Burks, B. S. (1938). On the relative contributions of nature and nurture to average group differences in intelligence. *Proceedings of the National Academy of Sciences, 24,* 276–82.

Byne, W. (1994). The biological evidence challenged. *Scientific American, May,* 50–5.

Byrne, R. W., & Whiten, A. (Eds.). (1988). *Machiavellian intelligence*. Oxford: Clarendon.

Capron, C., & Duyme, M. (1989). Assessment of effects of socio-economic status on IQ in a full cross-fostering study. *Nature, 340,* 552–4.

Carroll, J. B. (1993). *Human cognitive abilities: A survey of factor-analytic studies*. New York: Cambridge University Press.

Caspi, A., & Herbener, E. S. (1993). Marital assortment and phenotypic convergence: Longitudinal evidence. *Social Biology, 40,* 48–60.

Clausen, J., Keck, D. D., & Hiesey, W. M. (1940). Experimental studies on the nature of species. I. Effects of varied environments on western North American plants. *Carnegie Institution Washington Publication No. 520,* 1–452.

Cohen, J. (1994). The earth is round (*p* < .05). *American Psychologist, 49,* 997–1003.

Cohen, R. (1973). *Patterns of personality judgment*. New York: Academic Press.

Cooper, R., & Zubeck, J. (1958). Effects of enriched and restricted early environments on the learning ability of bright and dull rats. *Canadian Journal of Psychology, 12,* 159–64.

Crow, J. F. (1990). How important is detecting interactions? *Behavior and Brain Sciences, 13,* 126–7.

Darwin, C. (1871). *The descent of man and selection in relation to sex*. London: John Murray (New York: Modern Library, 1967).

Dawes, R. M. (1994). *House of cards: Psychology and psychotherapy built on myth*. New York: Free Press.

DeFries, J. C., Gervais, M. C., & Thomas, E. A. (1978). Response to 30 generations of selection for open-field activity in laboratory mice. *Behavior Genetics, 8,* 3–13.

DeFries, J. C., & Plomin, R. (1978). Behavior genetics. *Annual Review of Psychology, 29,* 473–515.

Degler, C. N. (1991). *In search of human nature: The decline and revival of Darwinism in American social thought*. New York: Oxford University Press.

Detterman, D. K. (1993). Giftedness and intelligence: One and the same? In G. R. Bock & K. Ackrill (Eds.), *The origins and development of high ability*. Chichester, England: Wiley.

Eaves, L. J. (1970). The genetic analysis of continuous variation: A comparison of experimental designs applicable to human data. II. Estimation of heritability and comparison of environmental components. *British Journal of Mathematical and Statistical Psychology, 23,* 189–98.

Eaves, L. J., Last, K., Martin, N. G., & Jinks, J. L. (1977a). A progressive approach to nonadditivity and genotype–environment covariance in the analysis of human differences. *British Journal of Mathematical and Statistical Psychology, 30,* 1–42.

Eaves, L. J., Last, K. A., Young, P. A., & Martin, N. G. (1977b). Model-fitting approaches to the analysis of human behavior. *Heredity, 41,* 249–320.

Ericsson, K. A., & Charness, N. (1994). Expert performance: Its structure and acquisition. *American Psychologists, 49,* 725–47.

Ericsson, K. A., Krampe, T. T., & Heizman, S. (1993). Can we create gifted people? In G. R. Bock & K. Ackrill (Eds.), *The origins and development of high ability*. Chichester, England: Wiley.

Falconer, D. S. (1990). *Introduction to quantitative genetics* (3rd ed.). New York: Longman Group Ltd.

Fancher, R. E. (1985). *The intelligence men: Makers of the IQ controversy.* New York: W. W. Norton & Co.

Farber, S. L. (1981). *Identical twins reared apart: A reanalysis.* New York: Basic books.

Feingold, A. (1992). Good-looking people are not what we think. *Psychological Bulletin, 111,* 304–41.

Feldman, M. W., & Lewontin, R. C. (1975). The heritability hand-up. *Science, 190,* 1163–8.

Fletcher, R. (1991). *Science, ideology, and the media: The Cyril Burt scandal.* New Brunswick, NJ: Transaction Books.

Fletcher, R. (1993). The "Miss Conway" story. *The Psychologist: Bulletin of the British psychological Society, 6,* 214–15.

Ford, B. D. (1993). Emergenesis: An alternative and a confound. *American Psychologist, 48,* 1294.

Fulker, D. W. (1975). [Review of "The science and politics of IQ."] *American Journal of Psychology, 88,* 505–37.

Galton, F. (1869). *Hereditary genius: An inquiry into its laws and consequences.* London: Macmillan.

Galton, F. (1876). The history of twins, as a criterion of the relative powers of nature and nurture. *Journal of the Anthropological Institute of Great Britain and Ireland, V,* 391–406.

Ghiselli, E. E., Campbell, J. P., & Zedeck, S. (1981). *Measurement theory for the behavioral sciences.* San Francisco: Freeman.

Goldberg, L. R. (1971). A historical survey of personality scales and inventories. In P. McReynolds (Ed.), *Advances in psychological assessment.* Palo Alto: Science and Behavior Books.

Gottesman, I. I. (1963). Genetic aspects of intelligent behavior. In N. R. Ellis (Ed.), *Handbook of mental deficiency* (pp. 253–96). New York: McGraw-Hill.

Gottesman, I. I. (1968). Biogenetics of race and class. In M. Deutsch, I. Katz, & A. Jensen (Eds.), *Social class, race and psychological development* (pp. 11–51). New York: Holt, Rinehart & Winston.

Gottfredson, L. S. (1986). The g factor in employment: A special issue of the *Journal of Vocational Behavior. Journal of Vocational Behavior, 29,* 3293–450.

Gottfredson, L. S., & Sharf, J. C. (1988). Fairness in employment testing: A special issue of the *Journal of Vocational Behavior. Journal of Vocational Behavior, 33,* 225–463.

Gould, S. J. (1981). *The mismeasure of man.* New York: W. W. Norton.

Haldane, J. B. S. (1946). The interaction of nature and nurture. *Annals of Eugenics, 13,* 197–205.

Hartigan, J. A., & Wigdor, A. K. (Eds.). (1989). *Fairness in employment testing: Validity generalization, minority issues, and the General Aptitude Test Battery.* Washington, DC: National Academy Press.

Hayes, W. L. (1973). *Statistics for the social sciences* (2nd ed.). New York: Holt, Rinehart & Winston.

Hedges, L. V. (1987). How hard is hard science, how soft is soft science? The empirical cumulativeness of research. *American Psychologist, 42,* 443–55.

Henderson, N. (1990). Why do gene–environment interactions appear more often in laboratory animal studies than in human behavioral genetics? *Behavior and Brain Sciences, 13,* 136–7.

Herrnstein, R. J., & Murray, C. (1994). *The bell curve: Intelligence and class structure in American life.* New York: Free Press.

Hirsch, J. (1975). Jensenism: The bankruptcy of "Science" without scholarship. *Educational Theory, 25,* 3–28.

Hoffman, L. W. (1991). The influence of the family environment on personality: Evidence for sibling differences. *Psychological Bulletin, 110,* 187–203.

Horgan, J. (1993). Eugenics revisited. *Scientific American* (June), 122–32.

Howe, M. J. A. (1990). *Sense and nonsense about hothouse children*. Leicester: British Psychological Society.

Humphreys, L. G. (1992). Commentary: What both critics and users of ability tests need to know. *Psychological Science, 3*, 271–4.

Hyde, J. (1973). Genetic homeostasis and behavior: Analysis, data and theory. *Behavior Genetics, 3*, 233–45.

Hyde, J. (1974). Inheritance of learning ability in mice: A diallel-environment analysis. *Journal of Comparative and Physiological Psychology, 86*, 116–23.

Jackson, D. N. (1975). Intelligence and ideology. *Science, 189*, 1078–80.

Jensen, A. R. (1969). How much can we boost IQ and scholastic achievement? *Harvard Educational Review, 39*, 1–123.

Jensen, A. R. (1971). Note on why genetic correlations are not squared. *Psychological Bulletin, 75*, 223–4.

Jensen, A. R. (1973). Equating for socioeconomic status. In A. R. Jensen (Ed.), *Educability and group differences*. New York: Harper and Row.

Jensen, A. R. (1977). Cumulative deficit in IQ of blacks in the rural south. *Developmental Psychology, 13*, 184–91.

Jensen, A. R. (1980). *Bias in mental testing*. New York: Free Press.

Jensen, A. R. (1994). What is a good g? *Intelligence, 18*, 231–58.

Jinks, J. L., & Fulker, D. W. (1970). Comparison of the biometrical genetical, MAVA, and classical approaches to the analysis of human behavior. *Psychological Bulletin, 73*, 311–49.

Joynson, R. B. (1990). *The Burt affair*. New York: Academic Press.

Juel-Nielsen, N. (1980). *Individual and environment: Monozygotic twins reared apart (revised edition of 1965 monograph)*. New York: International Universities Press.

Kamin, L. J. (1974). *The science and politics of IQ*. Potomac, MD: Erlbaum.

Kendler, K. S., Neale, M. C., Kessler, R. C., Heath, A. C., & Eaves, L. J. (1993). A test of the equal-environment assumption in twin studies of psychiatric illness. *Behavior Genetics, 23*, 21–7.

Layzer, D. (1974). Heritability analysis of IQ scores: Science or numerology. *Science, 183*, 1259–66.

Lewontin, R. C. (1982). *Human diversity*. New York: Scientific American Books.

Lewontin, R. C. (1974). The analysis of variance and the analysis of causes. *American Journal of Human Genetics, 26*, 400–11.

Lewontin, R. C. (1975). Genetic aspects of intelligence. *Annual Review of Genetics, 9*, 387–405.

Lewontin, R. C., Rose, S., & Kamin, L. J. (1984). *Not in our genes: Biology, ideology, and human nature*. Pantheon: New York.

Locurto, C. (1988). On the malleability of IQ. *The Psychologist: Bulletin of the British Psychological Society, 11*, 431–5.

Locurto, C. (1990). The malleability of IQ as judged from adoption studies. *Intelligence, 14*, 275–92.

Locurto, C. (1991). *Sense and nonsense about IQ: The case for uniqueness*. New York: Praeger.

Loehlin, J. C. (1981). [Review of Farber, S. L., "Identical twins reared apart: A reanalysis."] *Acta Geneticae Medicae et Gemellologiae, 30*, 297–8.

Loehlin, J. C., & Nichols, R. C. (1976). *Heredity, environment, & personality: A study of 850 sets of twins*. Austin: University of Texas Press.

Lykken, D. T. (1968). Statistical significance in psychological research. *Psychological Bulletin, 70*, 151–9.

Lykken, D. T., Geisser, S., & Tellegen, A. (1978). *Heritability estimates from twins studies: The efficiency of the MZA design*. Unpublished manuscript, Department of Psychology, University of Minnesota.

Lykken, D. T., McGue, M., Tellegen, A., & Bouchard, T. J., Jr. (1992). Emergenesis: Genetic traits that may not run in families. *American Psychologist, 47*, 1565–77.

Matarazzo, J. D. (1972). *Wechsler's measurement and appraisal of adult intelligence* (5th ed.). Baltimore: Williams & Wilkins.

Matarazzo, J. D. (1992). Biological and physiological correlates of intelligence. *Intelligence, 16*, 257–8.

Matheny, A. P., Wilson, R. S., & Dolan, A. B. (1976). Relations between twins' similarity of appearance and behavioral similarity: Testing an assumption. *Behavior Genetics, 6*, 343–51.

Mazur, A., Halpern, C., & Udry, J. R. (1994). Dominant-looking male teenagers copulate earlier. *Ethology and Sociobiology, 15*, 87–94.

McCartney, K., Harris, M. J., & Bernieri, F. (1990). Growing up and growing apart: A developmental meta-analysis of twin studies. *Psychological Bulletin, 107*, 226–37.

McGue, M. (1989). Nature–nurture and intelligence. *Nature, 340*, 507–8.

McGue, M., Bouchard, T. J., Jr., Iacono, W. G., & Lykken, D. T. (1993). Behavior genetics of cognitive ability: A life-span perspective. In R. Plomin & G. E. McClearn (Eds.), *Nature, nurture and psychology* (pp. 59–76). Washington, DC: American Psychological Association.

Meehl, P. E. (1970). Nuisance variables and the ex post facto design. In M. Radner & S. Winokur (Eds.), *Minnesota studies in the philosophy of science IV*. Minneapolis: University of Minnesota Press.

Meehl, P. E. (1971). High school yearbooks: A reply to Schwarz. *Journal of Abnormal Psychology, 77*, 143–8.

Meehl, P. E. (1978). Theoretical risks and tabular asterisks: Sir Karl, Sir Ronald, and the slow progress of soft psychology. *Journal of Consulting and Clinical Psychology, 46*, 806–34.

Meehl, P. E. (1990). Why summaries of research on psychological theories are often uninterpretable. *Psychological Reports, 80*, 195–244.

Melamed, T. (1992). Personality correlates of physical height. *Personality and Individual Differences, 13*, 1349–50.

Miller, G. A. (1984). The test. *Science, 84, 5*, 55–7.

Miller, J. K., & Levine, D. (1973). Correlation between genetically, matched groups versus reliability theory: A reply to Jensen. *Psychological Bulletin, 79*, 142–4.

Muller, H. J. (1925). Mental traits and heredity. *Journal of Heredity, 16*, 433–48.

Neale, M. C., & Cardon, L. R. (Eds.) (1992). *Methodology for genetic studies of twins and families*. Dordrecht: Kluwer Academic Publishers.

Newman, H. H., Freeman, F. N., & Holzinger, K. J. (1937). *Twins: A study of heredity and environment*. Chicago: University of Chicago Press.

Parker, K. C., Hanson, R. K., & Hunsley, J. (1988). MMPI, Rorschach, and WAIS: A meta-analytic comparison of reliability, stability and validity. *Psychological Bulletin, 103*, 367–73.

Pedersen, N. L., Plomin, R., Nesselroade, J. R., & McClearn, G. E. (1992). A quantitative genetic analysis of cognitive abilities during the second half of the life span. *Psychological Science, 3*, 346–53.

Plomin, R. (1994). The nature of nurture: Family environment. In R. Plomin (Ed.), *Genetics and experience: The interplay between nature and nurture* (pp. 104–48). Beverly Hills: Sage.

Plomin, R., DeFries, J. C., & McClearn, G. E. (1990). *Behavioral genetics: A primer* (3rd ed.). New York: W. H. Freeman.

Plomin, R., Willerman, L., & Loehlin, J. C. (1976). Resemblance in appearance and the equal environments assumption in twin studies of personality traits. *Behavior Genetics, 6*, 43–52.

Popenoe, P. (1922). Twins reared apart. *Journal of Heredity, 5*, 142–4.

Rao, D. C., Morton, N. E., & Yee, S. (1974). Analysis of family resemblance. II. Linear model for familial correlation. *American Journal of Human Genetics, 26*, 331–59.

Reed, T. E. (1984). Mechanisms for heritability of intelligence. *Nature, 311*, 417.

Reed, T. E. (1990). Evolutionary and neurophysiological arguments for the heritability of intelligence. *European Bulletin of Cognitive Psychology, 10,* 659–67.

Richards, R. J. (1987). *Darwin and the emergence of evolutionary theories of mind and behavior.* Chicago: University of Chicago Press.

Robinson, J. P., Shaver, P. R., & Wrightsman, L. S. (Eds.). (1991). *Measures of personality and social psychological attitudes.* San Diego: Academic Press.

Rose, R. J. (1981). [Review of Farber, S. L., "Identical twins reared apart: A reanalysis."] *Science, 215,* 959–60.

Rose, R. J., Kaprio, J., Williams, C. J., Viken, R., & Obremski, K. (1990). Social contact and sibling similarity: Facts, issues, and red herrings. *Behavior Genetics, 20,* 763–78.

Rowe, D. C. (1994). *The limits of family influence: Genes, experience, and behavior.* New York: Guilford Press.

Rowe, D. C., Clapp, M., & Wallis, J. (1987). Physical attractiveness and the personality resemblance of identical twins. *Behavior Genetics, 17,* 191–201.

Rushton, J. P. (1989). The generalizability of genetic estimates. *Personality and Individual Differences, 10,* 985–89.

Scarr, S. (1968). Environmental bias in twin studies. *Eugenics Quarterly, 15,* 34–40.

Scarr, S. (1992). Developmental theories for the 1990's: Development and individual differences. *Child Development, 63,* 1–19.

Scarr, S. (1993). Biological and cultural diversity: The legacy of Darwin for development. *Child Development, 64,* 1333–53.

Scarr, S., & Carter-Saltzman, L. (1979). Twin method: Defense of critical assumption. *Behavior Genetics, 9,* 527–42.

Scarr, S., Scarf, E., & Weinberg, R. A. (1980). Perceived and actual similarities in biological and adoptive families: Does perceived similarity bias genetic inferences? *Behavior Genetics, 10,* 445–58.

Scarr, S., & Weinberg, R. A. (1978). The influence of family background on intellectual attainment. *American Sociological Review, 43,* 674–92.

Scarr, S., & Weinberg, R. A. (1994). Educational and occupational achievement of brothers and sisters in adoptive and biological related families. *Behavior Genetics, 24,* 301–25.

Schiff, M., & Lewontin, R. (1986). *Education and class: The irrelevance of IQ genetic studies.* Oxford: Clarendon Press.

Schmidt, F. L. (1993). Data, theory, and meta-analysis: Response to Hoyle. *American Psychologist, 48,* 1096.

Schmidt, F. L. (1994). Quantitative methods and cumulative knowledge in psychology: Implications for the training of researchers. *Presidential Address to APA Division 5, August 13, 1994.*

Shields, J. (1962). *Monozygotic twins: Brought up apart and brought up together.* London: Oxford University Press.

Snyderman, M., & Rothman, S. (1987). Survey of expert opinion on intelligence and aptitude testing. *American Psychologist, 42,* 137–44.

Snyderman, M., & Rothman, S. (1988). *The IQ controversy: The media and public policy.* New Brunswick, NJ: Transaction Books.

Spearman, C. (1927). *The abilities of man: Their nature and measurement.* New York: Macmillan.

Stanley, J. C. (1973). Accelerating the educational progress of intellectually gifted youths. *Educational Psychology, 10,* 133–46.

Taylor, H. F. (1980). *The IQ game: A methodological inquiry into the heredity environment controversy.* New Brunswick, NJ: Rutgers University Press.

Tomlinson-Keasey, C., & Little, T. D. (1990). Predicting educational attainment, occupational achievement, intellectual skill and personal adjustment among gifted men and women. *Journal of Educational Psychology, 82,* 442–55.

Townsend, M. (1874–75). Letter to Francis Galton. In Galton Archives in the Library of University College, London, List # 112/3.

Turkheimer, E. (1991). Individual and group differences in adoption studies of IQ. *Psychological Bulletin, 110,* 392–405.

Turkheimer, E., & Gottesman, I. I. (1991). Individual differences and the canalization of human behavior. *Developmental Psychology, 27,* 18–22.

Tyler, R. W., & White, S. H. (Chairmen) (1979). *Testing, teaching, and learning: Report of a conference on research on testing.* August 17–26, 1978. Washington, DC: National Institute of Education.

Urbach, P. (1974). Progress and degeneration in the "IQ debate" (II). *British Journal of the Philosophy of Science, 25,* 235–59.

Vernon, P. A. (Ed.). (1993). *Biological approaches to the study of human intelligence.* Norwood, NJ: Ablex.

Wachs, T. H. (1992). *The nature of nurture.* Newbury Park, CA: Sage.

Wahlsten, D. (1990). Insensitivity of the analysis of variance to heredity–environment interaction. *Behavior and Brain Sciences, 13,* 109–61.

West, P. (1991). Rethinking the health selection explanation for health inequalities. *Social Science and Medicine, 32,* 373–84.

White, R. K. (1982). The relation between socioeconomic status and academic achievement. *Psychological Bulletin, 91,* 461–81.

Part II

**Novel theoretical perspectives
on the genes and culture controversy**

6 The invalid separation of effects of nature and nurture: Lessons from animal experimentation

Douglas Wahlsten and Gilbert Gottlieb

Introduction

Several decades of research involving nonhuman animals have led many scientists to a developmental systems explanation of the origins of brain and behavior. Developmentalists recognize the importance of genetic variation for individual differences in behavior but also appreciate that the complex sequence of bidirectional, interacting causes makes it almost impossible to assign a definite role to the genotype unless a major gene can be identified. The prevalent model in human behavior genetics, on the other hand, presumes that heredity and environment are additive, separately acting causes whose contributions to any characteristic can be neatly separated statistically. This presumption is biologically unrealistic in view of all that is known today about the control of gene action and the interdependence of genetic and environmental effects.

In this essay, we focus on lessons learned from animal experimentation that have special relevance for understanding individual differences among humans. We argue that a general theory of proximate causes of behavioral diversity is broadly applicable to all mammals and, in most respects, all vertebrates, but the theory is a developmental one, not one based on the invalid dualism of nature–nurture inherent in quantitative behavior genetics.

Relevance of animal research

Experimentation with animals is relevant to questions about genetic influences on human intelligence because remarkable plasticity of behavior is widespread among animal species, because quantitative genetic models of intelligence do not invoke any process that is uniquely human, and because there is a high degree of homology between all mammals at the genetic level and even with regard to the structure of the brain. The fine details of how this or that behavior first appears and then increases in frequency or complexity may differ greatly between species, but the more abstract, theoretical statements about development are expected to be broadly applicable across most, if not all, species. A principle clearly established for the role of heredity in development of nonhuman animals should constrain our theorizing about human intelligence.

Intelligence and the plasticity of development

Many psychologists contend that intelligence involves the ability to learn and to change adaptively in new situations, although they are still searching for some way to separate capacity for learning from amount that has already been learned. Tests are sought to transcend unique national languages and cultures. Some psychologists contend that information-processing speed is central to intelligence and therefore measures of reaction time can account for much of the individual variation in IQ scores. As conceptions of intelligence move away from higher-level functions involving language and toward simpler and more universal behaviors, they edge closer to things that are done by other species. Because nonhuman animals share some cognitive processes in common with human intelligence (Griffin, 1982) and at the molecular level the mechanisms of memory formation are notably similar (Kandel & Hawkins, 1992), the relevance of knowledge gained from other species is well established.

Many animals show plasticity of behavior that is crucial for reproduction and survival. Sometimes the changes are more dramatic and rapid than anything seen in humans. For instance, the female coral reef fish quickly turns into a physiological male and behaves like a male after the dominant male is captured and removed from the territory (Shapiro, 1981). Specific compounds in the diet of caterpillars of the geometrid moth induce morphological mimicry of very different parts of the host plant (Greene, 1989). Acorn woodpeckers breed communally one year and then form territorial pairs the next year when there is a shortage of food (Stacey & Bock, 1978). As marvelous as these changes may be, it is not at all apparent that they are learned modifications. Indeed, it has always been devilishly difficult to draw a clear line of demarcation between learning and other forms of experience that involve organism–environment interaction, especially when behavior outside the laboratory is at issue. Studies of lowly insects, which some of our contemporaries contend are governed by blind instinct, provide many examples of the participation of experience in the development of behavior. Defense of a colony of ants against intruders depends on early experience with chemical cues from the colony that define the properties of a stranger (Isingrini, Lenoir, & Jaisson, 1985). Fruit flies can learn to avoid chemical cues paired with aversive stimuli (Heisenberg, 1989) and to respond to tastes associated with positive reinforcement (Lofdahl, Holliday, & Hirsch, 1992). Evolution proceeds as a branching, bushlike process rather than a linear scale of nature with humans at the apex, and each species is relatively well adapted to its own ecological niche. How they adapt so well cannot be determined by description alone. Animal psychologists have concluded that the often nonobvious contribution of experience to adaptive behavior can be determined only with properly controlled experiments that vary potentially influential factors in a systematic way. Such experiments are usually impractical or unethical with humans, and knowledge gained from animal studies is our best source of answers to many vexatious questions about development. Our conclusion from years of animal research is that some of the foundational assumptions frequently made by behavior geneticists studying human intelligence are invalid.

Quantitative genetic theory is general

The fundamental principles of Mendelian heredity with its classic 3:1 ratio of phenotypes in a dihybrid cross were first elucidated with certain varieties of garden pea and later found to apply to many characteristics of other sexually reproducing organisms, although non-Mendelian heredity is also well established in biology (e.g., Avise, Quattro, & Vrijenhoek, 1992; Preer, 1993). Quantitative genetic analysis based on a strictly Mendelian model is employed when the investigator has no information about precisely how many or which genes may be relevant and, consequently, no argument can be made for the uniqueness of quantitative genetic models of human functions. The laws of resemblance of relatives as employed by quantitative geneticists are said to apply to any phenotype we might care to measure. For psychologists who believe that quantitative genetics can be applied fruitfully to human intelligence, lessons from other species should be relevant and timely. The fact that so many colleagues specializing in the genetic analysis of animal brain and behavior have rejected heritability analysis should occasion a reexamination of the conceptual foundations of human behavior genetics.

Homology at the genetic level

Not only are the quantitative models of genetic effects the same for human and mouse behavior, but the genes themselves are remarkably similar. Indeed, large expanses of the chromosomes of the two species contain virtually the same genes in the same linear order, and numerous diseases caused by a defect in a specific human gene are also known in mice

(Copeland et al., 1993). Most of these similarities are the daughters of homology, descent from a common ancestor, rather than convergent analogy. Even the humble fruit fly shares many genes in common with exalted humanity (McGinnis & Kuziora, 1994; Merriam, Ashburner, Hartl, & Kafatos, 1991). Although there are genes in humans that are not found in mice or flies, the vast majority of genes of documented importance for healthy brain activity exists in many vertebrate species. Likewise, the principles of the control of gene action at the molecular level obtain across a wide range of species. At the molecular level, there is no evidence of human uniqueness that makes us exempt from the generalizations induced from elegant experiments with simpler and smaller animals.

The concept of levels of analysis (Capitanio, 1991; Feibleman, 1954) is critically important for any discussion of genetic aspects of human thinking, because thought is the product of an entire organism existing in a societal context, whereas genes are large molecules functioning at the subcellular level. Generally speaking, principles enunciated at the more molecular, microscopic level tend to be valid across a wider range of species, whereas phenomena at the more macroscopic, societal level are likely to be species specific. Even chimpanzees, so closely related to us genetically, are bereft of spoken language (Locke, 1993), whereas all placental mammals share remarkable similarities of early embryonic development. With the greatest of hubris, quantitative behavior genetics strives to traverse the molecular and psychological levels in one grand inferential leap. By so doing, this discipline forfeits any claim to human uniqueness in its theorizing. The plea that gene–environment interaction, for example, is not important for models of human development (Detterman, 1990), despite being well documented with laboratory animals, must be rejected by virtue of continuity across species.

We argue that the similarities of humans and nonhumans with respect to genetic principles as well as processes of behavioral plasticity establish the relevance of a large body of research with animals. Those who discuss the genetic aspects of human intelligence should not ignore the findings from animal research because results from developmental-genetic studies of animals condemn the quantitative genetic models to eventual extinction.

Implications for design and interpretation of experiments

A number of principles well established with laboratory animals show the invalidity of attempts to separate the contributions of heredity and environment to human intelligence. Certain of these principles pertain more to research methodology than theory. For practical reasons, it is usually not possible to isolate the effects of multiple genes on human development.

The adoption method and its limitations

In species with a long period of parental care of the young, there is obviously a confounding of genetic and environmental causes of parent–offspring resemblance. Comparison of genetically identical, monozygotic (MZ) twins with genetically different, dizygotic (DZ) twins cannot circumvent this problem when the twins are reared together, because the experiences of MZ twins tend to be more similar than those of DZ twins. Many behavior geneticists therefore rely on adoptions to separate the contributions of nature and nurture. Unfortunately, *the adoption method cannot separate the effects of heredity and environment* because of the long-term importance of a shared prenatal environment.

Birth or hatching is a momentous event in the life of an animal, yet it occurs at radically different stages of development in different species, ranging from the bare embryo of the kangaroo to the remarkably competent wood duck and mountain sheep. In no bird or mammal can we fairly state that environment begins to influence development only postnatally. Environment, consisting of those aspects of an organism's surroundings that impinge on it, comes into being at conception; there can be no organism without an environment. Experimental studies of early embryonic development in an artificial medium have confirmed the great importance of many features such as temperature, gravity, and chemical composition (e.g., Yoshinaga & Mori, 1989). By grafting the ovarian follicle cells of one inbred strain of mouse into a hybrid female, it is possible to assess the role of uterine environment, and several studies have demonstrated the importance of prenatal maternal environment for rate of embryonic development, brain growth, and even later adult behavior (Bulman-Fleming & Wahlsten, 1988; Carlier, Nosten-Bertrand, & Michard-Vanhée, 1992). The uterine location of male and female fetuses can also influence adult reproductive behavior (Crews, 1994; vom Saal, 1984). Stress as well as drugs and poor nutrition during pregnancy affect the fetus and have lasting effects (Morgane et al., 1993; Ward, 1983).

In addition to the physical and chemical aspects of the prenatal environment, the sensory environment also begins to sculpt the nervous system prenatally, especially the gustatory, tactile, and auditory senses. Sound travels quite well through a fluid; consequently, a complex repertoire of sounds, especially maternal speech, is presented to the fetus, sometimes with lasting effects (Busnel, Granier-Deferre, & Lecaunet, 1992; DeCasper & Spence, 1986). Mammals in the womb begin to form associations with tastes transmitted via the mother's blood as well as the amniotic fluid, and these can affect later feeding preferences (Petersen & Blass, 1982).

The human infant is born with all of its senses functioning to some extent. The fetus may have had no prior experience with patterned visual stimuli, but other modalities are far from naive at parturition (Gottlieb, 1971). What may be regarded as uniquely human neuroanatomical features and the cortical substrates for learning and memory emerge prior to birth. Thus, a claim that prenatal ontogeny is governed purely by genes, whereas environment is merely the essential fuel of nutrition, cannot be justified. Prior to birth, the fetus lives in a milieu provided by its genetic mother, and adoption occurs too late in life to separate hereditary and environmental effects in the way they can be cleaved by ovarian grafting.

Leading advocates of quantitative genetic analysis nevertheless assert that adoption can separate nature from nurture. Although the authors of the French adoption studies of IQ and the director of this research effort have made no claims that their research can answer the nature–nurture question (Duyme & Capron, 1992; Roubertoux & Capron, 1990), some behavior geneticists have made this unwarranted interpretation of their work (e.g., McGue, 1989). We concur with the views of our French colleagues; abundant evidence from laboratory animals persuades us that the adoption method is a good design for studying the importance of different postnatal environments but cannot teach us about the role of genes. Much of the contemporary writing on human behavior genetics contains a flawed presumption about the powers of the adoption method.

Critical and sensitive periods for modifying intelligence

The case of MZ twins reared apart is often touted as the ultimate experimental design to separate nature and nurture with humans. However, several kinds of correlated environments can occur for supposedly separated MZ twins. As discussed in the previous section, there is certain to be a correlation of their prenatal and early postnatal environments because they are wombmates. There is also likely to be a low to moderate correlation of childhood environments because of selective placement that is commonly practiced by adoption agencies, and there may be similarity of adult environments if and when the twins are reunited prior to psychological testing, such as occurred in the Minnesota Twin Study. Over 40 MZ twin pairs were separated at an average age of 5 months, reunited at an average age of 30 years, and then tested by psychologists an average of more than 10 years after being reunited (Bouchard, Lykken, McGue, Segal, & Tellegen, 1990). We regard this as an unacceptable confounding of two factors. Bouchard et al. (1990) argued that the similar experiences of adult MZ twins were not important because they had already passed through the early "formative years" for intelligence; therefore, they argued, correlated adult experiences would have no effects on test scores. Several facts negate this argument, however.

Some characteristics do exhibit a distinct critical period with abrupt beginning and end, such as determination of sex by incubation temperature in reptiles (Deeming & Ferguson, 1992), whereas others show a gradual waning of sensitivity to environmental modification. In particular, many aspects of brain and behavior retain significant plasticity in adults. The number of synapses and the arborization of neuronal dendrites are responsive to experience in adult animals, even those considered quite old. Part of the apparent decline in nervous system complexity and competence seen in aging laboratory animals is attributable to months and years of tedium in an impoverished sensorimotor environment (Greenough, McDonald, Parnisari, & Camel, 1986) rather than an inevitable, biologically programmed degeneration. We now know that new neurons can be generated each season in the brain of an adult songbird (Nottebohm, 1985). Neuroscience is moving toward a revised view of the brain in which multifaceted sensory experience is essential for maintaining healthy neurons and viable connections among neurons (Barinaga, 1992) and in which learning actually modifies the physical structure of cells, not just their chemical contents (Purves, 1988). The distinction between hardware and software that is so obvious in a computer is not present in the living brain, where experience continues to alter the connections throughout life. If brain structure can be altered significantly in adult animals, it seems likely that the intelligence of adult humans is also modifiable.

Direct evidence is also available on this question of an early critical period for changing human intelligence. Children can sometimes recover dramatically from several years of severe deprivation early in life (Clarke & Clarke, 1976), and several studies have indicated that IQ can be altered throughout childhood. It is well established that IQ score of an adult can fluctuate by 20 points or more over a period of years (Tyler, 1972), although the causes of such large changes are not well understood. A 20-point fluctuation is similar to the 15-point change effected by adoption of an infant into a substantially different environment (Duyme & Capron, 1992). This evidence challenges the credibility of a critical period for determination of human IQ by early experience.

If there is no critical period for changing human intelligence, many of the reunited twin pairs in the behavior-genetics literature yield inconclusive results. The practice of bringing these twins to the same laboratory for psychological testing at the same time may inflate the apparent heritability of IQ by increasing the correlation of adult twin experiences.

Continuity versus thresholds for environmental effects

One behavior geneticist maintains that confounding of heredity and environment in twin and adoption studies is not a serious shortcoming because variations across a wide range of environments have little impact on IQ, which will be altered only by extreme environments (Scarr, 1992). This argument is also invoked to deny the relevance of a large body of research on animal development because the treatments are supposedly too extreme (Scarr, 1993). These arguments in effect claim that there is a *threshold* for treatment effects on IQ. Three considerations lead us to reject this claim.

First, the quantitative genetic model of individual differences asserts that an individual's phenotype is an additive sum of genetic and environmental components $(P = G + E)$. In this expression, genotype is polygenic; many loci, each with small effect, are said to combine by addition. There is said to be continuity and linearity of genetic effects on IQ, so there is no reason at all why environmental effects on IQ should be suppressed by a threshold. According to the additive $(G + E)$ model, what is good for the Mendelian goose must be just as good for the environmentalist gander. Internal consistency of the quantitative genetic model requires this. (As will be discussed in greater detail later – see "The developmental systems approach" – the bedrock notion of the developmental systems concept is that interaction or coaction obtains everywhere and anywhere, so that, in any event, a simple summation of influences is psychologically and biologically invalid.)

Second, although thresholds are sometimes observed, as for toxins in the diet that cause gross malformations, they are not apparent for many variations that are commonly encountered by the species. For example, the number of mice in a litter has linear effects on brain growth over a wide range of litter size (Wainwright, Pelkman, & Wahlsten, 1989). Many nonlinear but continuous effects have also been observed, especially for drugs where there is often an intermediate dose with optimal effects. The concept of canalization (Gottlieb, 1991), involving constancy of form despite wide fluctuations in environment, may aid in the understanding of why almost all members of a species usually have the same number of eyes, limbs, and digits, but it does not apply to phenotypes which themselves show continuous variation in a population, as does IQ. Available evidence supports the idea that the larger the difference between two environments, the larger will be the difference in average test score. The cohort effect on IQ, manifest as higher scores for people born longer after the test was standardized, is continuous and nearly linear with a slope of about .7 point per year in the Netherlands (Flynn, 1987). This effect represents the joint functioning of many features of the environment that are commonly encountered in industrialized societies. Birth order also has a continuous, although somewhat nonlinear, effect on IQ, as does family size (Zajonc, 1983). The birth-order effect is rather small, partly because it is counteracted by the cohort effect, but the cohort effect itself is large, accomplishing in one generation what adoption from an impoverished to an enriched environment can achieve in much shorter time. The phenotypic resemblance of separated twins also declines monotonically as the duration of separation increases (Hayakawa, Shimizu, Ohba, & Tomioka, 1992).

Third, and finally, statistically significant effects are sometimes observed for especially potent treatments but not for milder interventions simply because the sample size is inadequate to reveal smaller effects. By accepting a convention that the null hypothesis of no treatment effect is to be rejected only when the probability of wrongly rejecting a true null is less than $\alpha = .05$, the statistical methodology imposes an apparent discontinuity that need not exist in reality. As the magnitude of a treatment increases, the observed effect will rather abruptly leap from nil to significance, even if the true effect size is a linear function of treatment strength over a wide range.

Researchers typically employ extreme environments in laboratory experiments with animals because this enhances the clarity of results and sensitivity of tests, which in turn economizes on the numbers of animals. Few studies have rigorously tested the linearity of graded effect sizes, in part because the requisite sample size increases in inverse proportion to the square of the effect size, which calls for four times as many animals to detect an effect half as large (Wahlsten, 1991). When continuity or linearity has been tested formally, it has often been confirmed, and this result casts doubt on the argument that experience is important for the development of intelligence only at the extremes. Available evidence implies that there is a continuum for most environmental effects, and theories of human intelligence should reflect this knowledge.

These three methodological points create grave doubts about the capability of quantitative genetic analysis to isolate the effects of a multifaceted genotype on development of any human characteristic.

Implications for theories of development

Even if an ideal experiment could be designed and executed with humans, the persistent difficulties in separating genetic and environmental effects would not evaporate. Elegantly controlled experiments on animal development have taught us that these factors simply do not act separately. Instead, they are interdependent causes that render statistical separation invalid. Hopefully, the results of these animal experiments will provide a corrective for ideas about human development that are endemic to the quantitative genetic theory of intelligence.

Reaction range versus reaction norm

As discussed previously, some advocates of quantitative genetic analysis of human behavior acknowledge that environment can be important for mental development but argue that environmental effects over a wide range are minor and similar for all genotypes, maintaining their rank orders, and that the consequences of genetic differences are markedly altered only in extreme circumstances (Scarr, 1992, 1993). Waddington's (1957) concept of genetic canalization of development leads quite logically to the *reaction-range* idea, which has wide currency in developmental psychology texts (e.g., Berk, 1991, pp. 110–11; Newman & Newman, 1991, pp. 141–2; Shaffer, 1992, pp. 109–10; Vasta, Haith, & Miller, 1992, pp. 109–10; Zigler & Stevenson, 1993, pp. 78–9). All of these texts reproduce a figure from Gottesman (1963), in which he shows a schematic of the reaction-range concept for mean IQ for four hypothetical genotypes reared in different environments (reproduced here in Figure 6.1). According to Gottesman (1963), "each genotype has its own more or less natural habitat . . . [which] would include an adequate diet, freedom from gross organic defects, exposure to our system of compulsory education, and being reared in a home by one's own parents" (p. 256). His notion sets strict and predictable upper and lower limits for IQ of a given genotype, and it maintains relative rank orders when environment changes over a wide range. The predicted, nearly parallel lines are a characteristic of the reaction-range concept.

Although Scarr (1993) confuses the reaction-range with the reaction-norm concept, the two are quite distinct (Platt & Sanislow, 1988). The reaction norm asserts that each genotype is associated with a characteristic pattern of phenotypic changes in response to alteration of the environment, but it does not require parallel response profiles for all individuals over a wide range. The reaction norm is something to be discovered through experimentation, and it allows no prediction of what will happen when an individual is placed beyond a rearing environment that has already been observed in scientific studies. Rank orders of individuals can change appreciably and unpredictably under novel conditions. This is the essence of the conceptual difference between a *norm* of reaction and a *range* of reaction.

The norm of reaction was first proposed by R. Woltereck (1909) to help to define operationally and experimentally Johannsen's (1909) newly coined notions of gene, genotype, and phenotype. As pointed out by Dunn (1965), Johannsen's synthesis was magnificent and stimulated great

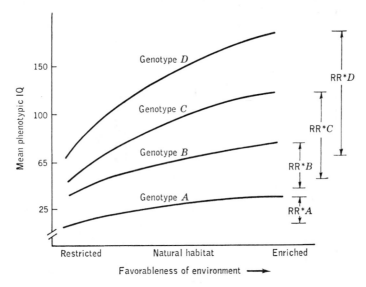

Figure 6.1. Gottesman's schematic illustration of the reaction-range concept for IQ in humans. (From Gottesman, 1963, p. 255.)

progress in experimental genetics because of the analytic clarity of his concepts. However, Woltereck, while acknowledging the general utility of the constructs, believed that Johannsen's notion of genotypic influence on phenotypic outcomes under different rearing circumstances was incorrect. Woltereck portrayed Johannsen's understanding of phenotypic development as what is here termed the reaction range involving parallel lines (see Figure 6.2). When Woltereck himself experimented with different quantities of nutrition affecting head development of three strains of the water flea (*Hyalodaphnia cucullata*), he observed three very different curves (Figure 6.2). Woltereck regarded the complete profile of responses to different environments as the genotype, and he found that the profile was unpredictable in new conditions.

A similar pattern was observed for the McGill rat lines selectively bred for fast or slow learning of the Hebb-Williams mazes. The selection experiment conducted with animals reared in the normal laboratory environment yielded a large line difference in rate of learning. When Cooper and Zubek (1958) later reared the two lines of rats in either an enriched or a restricted environment and then tested learning on the same set of mazes, they expected a result in accord with the reaction-range concept, whereby the lines would both improve or decline (see Figure 6.3) but instead obtained clear evidence of a reaction norm. Even though the line difference com-

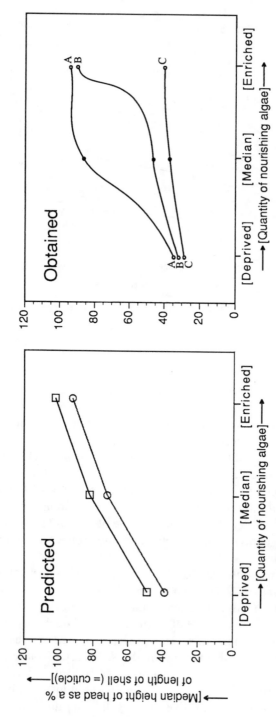

Figure 6.2. Woltereck's interpretation of Johannsen's notion of the genotype's influence on phenotypic expression (predicted) and the actual results (obtained) of rearing three geographic varieties of *Hyalodaphnia cucullata* on different levels of nourishment: reaction range (predicted) versus reaction norm (obtained). (Translated and redrawn from Woltereck [1909], Figures 11 and 12, pp. 138–9.)

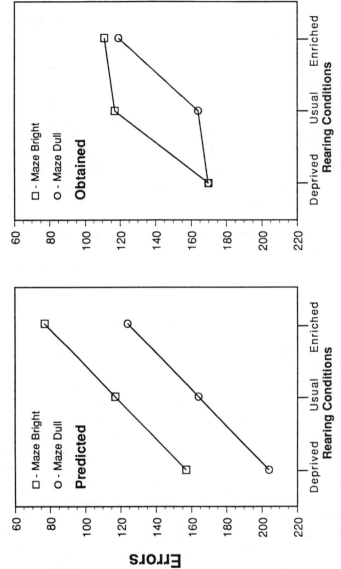

Figure 6.3. Mean errors on the Hebb-Williams mazes for the McGill maze bright and dull lines of rats reared in three different laboratory environments. Results expected according to the reaction-range concept would be two parallel lines (predicted), whereas the actual data (obtained) support the reaction-norm idea. Obtained results are based on information in Cooper and Zubek (1958).

pletely disappeared after enriched or restricted rearing, Gottesman (1963, p. 273) cited this study as support for the reaction-range concept. The reaction-range concept is compatible with an additive model of heredity and environment, whereas the Cooper and Zubek (1958) study instead demonstrated how the expression of hereditary differences is strongly modified by rearing environment. Their surprising findings illustrate organism–environment interaction, the bane of heritability analysis. These data do not in the least substantiate Scarr's (1992) claims that only extreme environments matter for learning ability, because only two deviations from the normal animal housing were studied. Even the enriched condition in the Cooper and Zubek (1958) study is relatively impoverished in comparison with the variety of experiences encountered by wild rats. Perhaps in a much more varied environment, the McGill maze "dull" line would even have exceeded the "brights." Without trying this experiment, there is no way of knowing the results because the reaction norm precludes prediction. (Because the McGill lines are now extinct, the necessary experiments can never be done with them, and the original findings cannot even be replicated.)

Interdependence of genetic and environmental causes

Attempts to weigh the relative importance of genetic and environmental variations rely on their presumed additivity, which is one feature of the reaction-range concept, and additivity demands separation and independence of causes, such that the contribution of genotype (G) is not influenced in any way by the rearing environment (E) and vice versa. This stipulation can be evaluated easily with a two-way factorial experiment and an analysis of variance (ANOVA) to test for the presence of $G \times E$ interaction. Such an experiment requires replicated genotypes, something easily achieved with highly inbred mice that are genetically identical at all loci, but effectively impossible with humans unless the test involves a single, identifiable genetic locus. Although MZ twins have the same G, a difference between intrapair phenotypic differences cannot be ascribed entirely to $G \times E$ interaction and could instead reflect ordinary variation in co-twin environment. MZ twins could provide good material for the ANOVA only if several pairs were available with the same genotype. This is not feasible when the specific loci pertinent to a trait are unknown.

A further difficulty is posed by the extraordinary sample sizes required to confer adequate statistical power on the test of $G \times E$ interaction (Wahlsten, 1990). Even if a very rare event such as two pairs of MZ twins with the same parents occurred, the sample size would still be woefully insufficient. With laboratory strains, on the other hand, the only limiting factor is the amount of research funding. We conclude that for all practical purposes, the existence of G × E interaction cannot be evaluated in species where breeding cannot be controlled and consequently with them the hypothesis of additivity is unfalsifiable.

Woltereck's idea of the unpredictability of development in novel environments has been confirmed repeatedly in biological research with animals. Numerous observations support the notion that epigenetic outcomes are probabilistic rather than predetermined (Gottlieb, 1970, 1971). For example, Gupta and Lewontin (1982) examined several characteristics of fruit flies (*Drosophila pseudoobscura*) under two egg densities and three temperatures, and they found 30–45% reversals in relative rank orders among strains. Rodent strain differences also are known to disappear under novel rearing conditions (Cooper & Zubek, 1958) or even reverse when testing conditions are altered (see Wahlsten, 1978). Many factorial experiments using well-defined laboratory strains have been reported, and those using adequate samples generally support the reaction-norm concept. Inbred strains or selectively bred lines differing at a large number of genetic loci are informative even when the specific loci are unidentified. Many instances of significant strain-dependent norms of reaction, including disordinal interaction where strain rank orders change, have been documented (Erlenmeyer-Kimling, 1972; Wahlsten, 1990). Genuine additivity, apparent in parallel reaction norms, is an exceptional result.

Perhaps the most impressive evidence comes from coisogenic mice differing at only one genetic locus. In mouse research, there is a tradition of naming a newly discovered gene for its most obvious phenotypic effect. Thus, we find colorful entries in catalogues of mouse genes such as *fatty*, *obese*, *reeler*, *staggerer*, and *varitint waddler*. However, the typical phenotype sometimes depends strongly on the conditions of rearing. In the usual mouse colony, mice with two copies of the *diabetes* gene become obese and exhibit high levels of blood glucose and insulin. Preventing their overeating by restricting food to the amount eaten by normal siblings, however, prevents obesity and forestalls the symptoms of diabetes (Lee & Bressler, 1981). Likewise, the abysmal breeding and parental behavior of *staggerer* mice can be ameliorated by a regimen of mouse-style sex therapy and a special cage (Guastavino, Larsson, & Jaisson, 1992). For the diabetic and staggerer mice as well as many other mutations in numerous species, it commonly happens that a mutation *increases the sensitivity of an animal to alterations of environment* by impairing physiological and behavioral regulatory mechanisms.

Perhaps the decisive evidence against the validity of separating genetic and environmental causes is provided by modern molecular biology. Under appropriate conditions, the gene, a segment of DNA molecule, is transcribed into messenger RNA and then translated into a polypeptide molecule that may function as an enzyme, hormone, or structural element of a cell. The gene codes for the sequence of amino acids in the polypeptide, and the place and time when a gene is metabolically active, can be documented by antibodies specific for the protein in question or even by complementary DNA probes that bind to a specific region of the mRNA molecule. Abundant results prove that most genes are active for restricted periods of time and in a limited number of places in the organism. The gene itself is subject to control by its surroundings and cannot be self-governing. The stimuli switching a gene on or off are transmitted by the cytoplasm of the cell (Blau et al., 1985) and can originate in the external environment of the animal (e.g., Zawilska & Wawrocka, 1993). A variety of environmental stresses provoke a cascade of events that turn on a class of "immediate-early" genes, including *c-fos* and *c-jun*, which, in turn, unleash a diversity of further molecular events (Sagar & Sharp, 1993). Learning and memory involve the controlled actions of numerous genes (Kaczmarek, 1993). There is no longer any refuge for doubt on this question. Environment regulates the actions of genes, and genes, via changes in the nervous system, influence the sensitivity of an organism to changes in the environment. The two causes are not separable developmentally. Statistical procedures that *appear* to separate variance according to genetic and environmental causes do not provide a valid representation of physiological reality.

Different genes do not act in isolation

Genes and environment do not act separately, and neither do the genes at various chromosomal loci act in isolation from one another. The linear $P = G + E$ model requires that the genetic effects at many loci be physiologically independent so that their effects will be algebraically additive. This latter assumption can never be evaluated when the number and identity of relevant genes are unknown, which is always the case in human heritability analysis. Nevertheless, a considerable body of evidence from the laboratory points to the profound interconnectedness of physiological activities.

One vivid illustration of the manifold interactions among parts of the cellular system is provided by genetic engineering of bacteria to synthesize large amounts of commercially valuable products. Understanding the local effects of a new gene inserted into a cell often proves insufficient for understanding metabolic functioning of the whole organism in the company of many other genes. As Bailey (1991) summarized this problem, "complex cellular responses to genetic perturbation can complicate predictive design." A large fraction of scientific activity has been devoted to analysis of things into their parts. The reductionistic-analytical approach has had great success in detecting and understanding genes at the molecular level but is far from adequate for understanding how a gene functions in the larger context, where there is a web of feedback relations, nonlinear interactions, and multifactorial contingencies.

Formal tests of independence of genes are possible in animals where several loci have already been identified. With complicated cross-breeding schemes, it is possible to produce "double mutants" afflicted by two different genetic disorders in order to test whether their effects are additive or whether, as the developmental systems view predicts, there is instead some kind of nonlinear, nonadditive synergism, exacerbation, or even sparing. Neuroscientists have examined several well-known neurological mutations and detected epistatic (gene–gene) interactions between effects of different loci. For example, the *Lurcher* mutation in mice results in a total loss of Purkinje cells in the cerebellum, whereas the *staggerer* defect causes about half of the Purkinje cells to degenerate. Mice with both the *Lurcher* and *staggerer* defects do not suffer 100% loss of Purkinje cells, which means the devastating consequences of the *Lurcher* mutation are prevented by combining it with the *staggerer* trait (Messer, Eisenberg, & Plummer, 1991). Two mutations (*jimpy* and *shiverer*) that impair formation of the myelin sheath around axons also show this kind of interaction (Billings-Gagliardi & Wolf, 1988). Another form of contextual dependency is apparent when the same mutation is placed onto two different genetic backgrounds. The *obese* and *diabetes* mutations in mice lead to very different forms of diabetes on the inbred C57BL/6J and C57BL/KsJ backgrounds, even though neither inbred strain has anything resembling diabetes under normal circumstances (Coleman, 1981). Likewise, the *reeler* mutation on an inbred C57BL/6 background causes serious deficits in motor coordination, reproduction, and viability, whereas the same mutation on an F_1 hybrid background leads to a more viable and better coordinated animal that can breed, although the gross neuroanatomical defect caused by the mutation on both backgrounds is the same (Caviness, So, & Sidman, 1972).

The physiological pathways responsible for these epistatic interactions in mice are not yet understood, but enough is known about the structure and function of the nervous system to make them appear plausible. Interconnectedness is the very essence of the nervous system. A single neuron in the cerebral cortex or cerebellum may have more than 10,000 synaptic inputs from other neurons plus contacts with hundreds of glial cells that in turn communicate with the circulatory system. During development, a loss of function of one part of the system can provoke a cascade of effects with widespread ramifications for different functions (Herrup & Vogel,

1988). Interdependency of parts of the system teaches us that the reductionistic approach, while being successful in analysis, is incapable of explaining development of the intact animal. The $P = G + E$ model is thoroughly reductionistic and is invalidated by recent advances in molecular biology and developmental neuroscience.

Three sources of individual differences

For many theorists, it is gospel that if some characteristic is not specified by environment, it must therefore arise from genetic information. This is the rationale for the Lorenzian interpretation of the deprivation experiment with bird song (e.g., Eibl-Eibesfeldt, 1979); if a bird sings the species-typical song without having heard it before, the song must be coded in the genes. The logic of this interpretation has been criticized on the grounds that neither genes nor environment can be the exclusive source of information specifying a behavioral phenotype (Oyama, 1985) and that the deprivation experiment can tell us nothing about the role of genes because it varies only experience (Johnston, 1988; Lehrman, 1970). There is yet another fundamental flaw with the argument because of the inherent variability in organismic development. Not all noteworthy individual differences in brain and behavior arise from either hereditary or environmental variations; instead of two, there are three basic sources of developmental variation, the third being a result of processes occurring internally to the organism that are sometimes termed *developmental noise* (Lewontin, 1991) or *randomness* (Gärtner, 1990; Layton, 1976). If there are three potential sources of individual variation, it is utterly impossible and illogical to prove that something is genetic by holding environment constant and vice versa. The correlational methods of behavior genetics are also incapable of distinguishing between environmental effects unique to the individual and those arising spontaneously from within the individual (Molenaar, Boomsma, & Dolan, 1993).

Some theorists have recognized a third source of individual differences but retained an untested faith in its environmental origins. For example, Wright (1920) statistically partitioned variation in guinea pig coat color into components arising from heredity (H), environment common to littermates (E), and "irregularity in development" (D). While arguing that "the peculiarities of an individual are entirely determined by heredity and environment," he referred to the irregular D as something "intangible . . . to which the word chance is applied." Analyzing the causes of severe embryonic malformations, he asserted that "randomness of occurrence within litters . . . indicates that each monster is due to a highly localized chance event" (Wright, 1934), but he continued to advocate two "ultimate factors," heredity and "the succession of external influences" (Wright, 1968).

More recently, several authors have entertained the possibility that nongenetic developmental irregularity may not be environmental. The idea of randomness has been invoked to explain differences in hand preference, for example. Collins (1985) posits an "asymmetry lottery" to account for the frequent discordance in monozygotic twins and a bimodal distribution in an inbred strain, and a similar idea is incorporated in the model of McManus and Bryden (1992). Randomness is invoked to account for nonhereditary asymmetries of the internal viscera (Layton, 1976) and to explain congenital heart anomalies (Kurnit, Layton, & Matthyse, 1987).

These examples establish the plausibility of a third source of individual differences but do not prove it rigorously. Gärtner (1990) made a serious effort to eliminate individual variations in organ development by rearing genetically uniform inbred mice in strictly controlled laboratory conditions for many generations, yet differences persisted and were attributed to random processes. However, controlling the temperature, food, water, and other global aspects of the environment does not render environment homogeneous in birds or mammals. In rodents, the number of offspring in a litter will continue to fluctuate, and this variable can strongly influence growth (Wahlsten & Bulman-Fleming, 1987). Likewise, the sex composition of the fetuses in the uterus, which is truly random for genetic reasons, engenders significant variations in the hormonal environment that can influence adult physiology and behavior; genetic females can be masculinized and genetic males feminized in their adult sexual behavior if they were located between two fetuses of the opposite sex prenatally (vom Saal, 1984). This intrauterine effect may also be important for human dizygotic twins (McFadden, 1993). Thus, a randomly distributed feature of the environment can easily mimic a pattern that is strictly nonenvironmental, which renders proof of the third source of variation exceedingly difficult.

Reasonably strong proof has been provided for one characteristic, absence of the corpus callosum (CC) that usually unites the cerebral hemispheres in inbred mice. In the strain BALB/c, less than half of the animals lack the CC, yet the defect is not hereditary *within* the strain, it cannot be attributed to an array of variations in the uterine environment, and it is randomly distributed among littermates (Bulman-Fleming & Wahlsten, 1991). Embryological studies reveal a distinct threshold for formation of a normal CC, whereby very small differences in timing of the growth of axons toward the middle of the brain in relation to formation of a bridge where they can cross a widening gap between the hemispheres leads to a bimodal distribution of CC size in the adult (Ozaki & Wahlsten, 1993).

Careful measurements of the developing brain reveal that almost everything varies among individuals to some extent but that the variations usually do not alter adult brain structure qualitatively (canalization). If the
microscopic variations become a little too extreme, however, they can
exceed a threshold and create macroscopic anomalies in the adult. The
location of the threshold, which reflects the relative averages of distributions of interacting growth processes, can itself depend on hereditary and
environmental causes. Thus, the frequency of absent CC in BALB/c mice
(the degree of "penetrance" in genetic parlance) can be influenced strongly
by rearing conditions (Wahlsten, 1982) as well as a change in heredity that
creates a new substrain (Wahlsten, 1989). Just as heredity–environment
interaction negates attempts to partition phenotypic variance into two mutually exclusive components, the interdependence of all three sources of
variation renders a nondevelopmental approach to individual differences
invalid.

In arguing for a third source of individual differences, one need not
invoke the concept of randomness, which is often regarded as an analytically empty idea with mystical connotations of indeterminacy (e.g., Scarr,
1993). Randomness is important in probability theory when an event is
inherently unpredictable. If, during development, a complex, nonlinear
process is exceedingly sensitive to small variations in its initial conditions,
then for all practical purposes, the macroscopic outcome will be unpredictable, even though the precise trajectory of ontogeny is entirely lawful and
deterministic. By analogy, it may appear to be random. Nevertheless, as the
theory of chaos has taught us (Prigogine & Stengers, 1984), deterministic
physical laws can give rise to unpredictable results (e.g., Sussman & Wisdom, 1988). The key point for our argument is that individual differences
can emerge from processes internal to the developing organism, but the
statistical methods of quantitative behavioral genetics cannot separate this
source from genetic or environmental sources of variation.

Thus, biological experimentation with laboratory animals has revealed
that the genes and environment are distinct entities but do not act in
isolation from one another, that different genes do not act separately, and
that development itself is not completely specified by an individual's heredity and environment. Consequently, it is misleading to claim that influences
of genes and environment can be separated statistically in a psychologically
meaningful way.

The developmental systems approach

The above paragraphs outline what we regard as decisive evidence against
the statistical models that form the conceptual core of quantitative genetic

BIDIRECTIONAL INFLUENCES

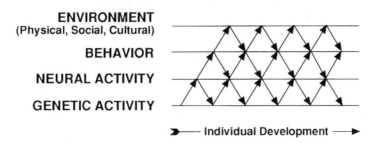

ENVIRONMENT
(Physical, Social, Cultural)

BEHAVIOR

NEURAL ACTIVITY

GENETIC ACTIVITY

Individual Development

Figure 6.4. Depiction of the completely bidirectional nature of genetic, neural, behavioral, and environmental influences over the course of individual development. (From Gottlieb, 1992, p. 186.)

analysis of human behavior. Perhaps at the time when Fisher (1918) first elucidated the theory of individual differences from a Mendelian perspective, there was wide latitude for debate on this question. Seventy-seven years later, however, sufficient evidence has accumulated to warrant rejection of the supposition that heredity acts independently of other causes during development. Instead of pursuing sterile disputes about the degree of heritability of a trait, many scientists have opted for a more illuminating study of the dynamics of development (see Wahlsten, 1994). From several decades of this kind of work has emerged a positive alternative – a developmental systems approach.

Insight into what would constitute a truly developmental behavior-genetic analysis is provided by a diagram of the four major levels that are involved in moving from genetic activity to behavior and back again (Figure 6.4). This is a fully bidirectional coactional system, where developmental psychologists work at the Behavior ↔ Environment level, developmental neuropsychologists and neuroscientists work at the Brain ↔ Behavior ↔ Environment levels, and biologically trained persons work at the Gene Activity ↔ Neural Activity levels. Viewed in this way, developmental understanding or explanation is a multilevel affair involving at least culture, society, immediate social and physical environments, anatomy, physiology, hormones, cytoplasm, and genes. Although it is hierarchical, the multilevel or developmental systems analysis is not theoretically reductionistic. Psychological explanation does not come from or reside only in the lower levels; both the higher and lower levels of analysis are necessary to explain developmental outcomes.

Several means whereby genetic activity can be influenced by supragenetic factors are diagrammed in Figure 6.5. Gene expression is

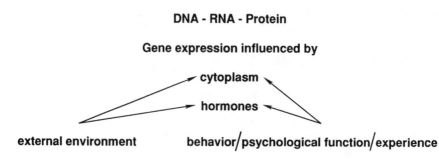

**Different proteins formed depending on particular factors
influencing gene expression during course of development**

Figure 6.5. Some of the supragenetic influences on gene expression and protein
formation during the course of individual development.

influenced by cytoplasmic factors that provide the immediate surroundings
of the cell nucleus where most of the structural DNA occurs. Cytoplasm in
turn can be influenced by the external environment of the organism
(reviewed in Ho, 1984) via the behavior of the organism, which moves it
into and out of different environmental influences (physical, social, cultural). Likewise, hormones make their way into the nucleus of the cell and
trigger gene expression, and hormones themselves are responsive to the
organism's experiences in the external world. Finally, different proteins are
formed depending on the particular factors influencing gene expression
(reviewed in Davidson, 1986). A clear exposition of the place of genetic
activity in the larger system of development is provided by Johnston and
Hyatt (1993).

Although probably not exhaustive, four kinds of coactional processes
have been thus far identified by the way they contribute to developmental
outcomes: canalization, induction, facilitation, and maintenance. Canalizing
coactions channel development, making some outcomes more probable
than others and completely ruling out other outcomes. Induction is the
main way in which canalization is achieved. Inductive coactions make
things happen that would not otherwise occur, whereas facilitating
coactions accelerate the rate and other quantitative aspects of development, and maintaining coactions keep induced outcomes functional. An
extensive review of the operation of these various coactions at the neural
and behavioral levels is presented in Gottlieb (1976).

In summary, developmental analysis is a multilevel, coactional affair,
being neither exclusively top–down nor bottom–up. Causes do not reside
in any single level because no level is considered more basic or primary than

any other level, and there are at least four ways in which coactions influence developmental outcomes between and among levels. In this way of looking at development, coactions (specifiable relationships between and among levels) serve as causes and, thus, explanations of developmental outcomes. Full causal explanation in developmental behavior-genetic analysis will eventually involve all four levels of analysis portrayed in Figure 6.4.

Many of the arguments we have made here are accepted by some developmental psychologists as relevant to human behavior (e.g., Ford & Lerner, 1992; Wachs, 1992), and there are examples of a nonreductionistic approach to understanding human intelligence (e.g., Keating, 1996). Our criticisms in this chapter are directed at a model of human development that still has a pervasive influence on many other psychologists. It is hoped that those who begin to doubt the validity of heritability analysis of human IQ will find the alternative – developmental systems theory – sufficiently well elaborated to serve as a guide to future research.

A multilevel research program will undoubtedly involve a large amount of statistical analysis of data, but in the context of a developmental systems approach, the investigator will utilize a wide variety of generic statistical tests appropriate for the data at hand. There is no need to confine discussion to a narrow range of algebraic models that assume additivity of genetic effects and normality of data. Alternatives exist (e.g., Efron & Tibshirani, 1991) and should be utilized more often. This is especially important when many readers may be misled by intricate and sometimes baffling statistical procedures that are based on unstated assumptions of additivity and independence of causes.

Summary and conclusions

Quantitative behavior genetics derives from the population-genetic tradition in evolutionary biology, which is a quite separate tradition from developmental genetics or embryology (Gottlieb, 1992). The analysis of variance comes from the population-genetic tradition, which sees two sources of variation – one heritable (genetic), and the other nonheritable (environmental) – and regards only the heritable source as important for evolution. The population-genetic approach does not concern itself with individual development, and instead it simply presumes that heritable phenotypic differences within or between populations of organisms (species, subspecies, varieties) are somehow specified directly by genetic information.

Some behavior geneticists have attempted to apply the population-genetic approach to human development (e.g., Plomin & McClearn, in press; Scarr, 1992, 1993). On the other hand, it is widely recognized in evolutionary biology today that an important deficiency in the population-genetic approach is its failure to deal with individual development (see various critiques summarized in Gottlieb, 1992). Consequently, the adoption of the statistical methods of population genetics by behavior geneticists and the application of those methods to the study of individual differences is bound to result in a nondevelopmental understanding of developmental outcomes.

In his 1951 review of the field that he anointed *psychogenetics*, Calvin Hall wrote:

The main objectives of psychogenetics are four in number: (1) to discover whether a given behavior pattern is transmitted from generation to generation, (2) to determine the number and nature of the genetic factors involved in the trait, (3) to locate the gene or genes on the chromosomes, and (4) to determine the manner in which the genes act to produce the trait [Hall, 1951, p. 304].

Of the four objectives set forth by Hall, one (4) deals with development, and none is especially concerned with estimating components of variation. Of the developmental explanation, what was said by Snyder in 1940 in a chapter on "How genes act" is still true today: "Between the presence of the genes in the chromosomes and the appearance of the developed characters in the individual organism, there exists a gap which is not yet thoroughly understood" (p. 356). It is probably fair to say by virtue of considerable research in the ensuing 50 years that we now realize the gap is more complex and even greater than Snyder supposed. In Hall's time, psychologists could imagine genes coding for neural structures that underlie behavior, whereas today we realize that genes do not make finished neural structures but rather code for protein structure (DNA → RNA → protein), which is part of an amazingly complex metabolic system comprising a cell that becomes a neuroblast, migrates through a maze of other cells, and eventually serves as a differentiated neuron connected in a network of cells in a functionally active brain. The developmental fate of the protein is not determined upon its production; influences from higher, supragenetic levels in the developmental system play a role in specifying the incorporation of the protein into a mature, functional unit that is integrated with other parts of the nervous system (Gottlieb, 1991). A specific protein can influence the progress of development, but the same protein can usually participate in many kinds of developmental outcomes, depending on the context. A deleterious mutation can initiate cellular changes which then grossly impair development of a part of the brain, even though the gene in question does not code for that brain structure, and events at higher levels can sometimes compensate for a serious deficiency at the molecular level.

If one accepts that the processes of neural development and the substrates of learning and memory involve the metabolic actions of thousands of genes, then a kind of genetics of intelligence is possible. Thinking engages the activities of many different genes in the nerve cells, and different patterns of thinking invoke different patterns of gene activation in the cerebral cortex. Memory storage requires numerous changes in gene expression. However, genetic information does not specify what we think or remember, any more than the colors of the artist's palette dictate the artistic quality of the painting. Nevertheless, at the molecular level, genetic analysis has led to considerable advance in our understanding of what happens in the cell when learning and memory occur. Molecular genetics can elucidate the roles of genes that are uniform in the population and do not produce marked differences between individuals (e.g., Capecchi, 1994). The same cannot be said for quantitative genetic analysis with twins or adoptees, which has yet to uncover even one new gene relevant to human intelligence. Genetic knowledge can help us to understand learning, memory, and intelligence only when a specific gene is identified and its relations with its surroundings are characterized. There can be no meaningful and valid developmental genetics of the human brain and behavior without positive knowledge of real genes. Quantitative genetic analysis and the overweening focus on heritability of intelligence has retarded progress toward a deeper comprehension of the role of heredity in human mental development.

References

Avise, J. C., Quattro, J. M., & Vrijenhoek, R. C. (1992). Molecular clones within organismal clones. Mitochondrial DNA phylogenies and the evolutionary history of unisexual vertebrates. In M. K. Hecht, B. Wallace, & R. J. MacIntyre (Eds.), *Evolutionary biology* (Vol. 26) (pp. 225–46). New York: Plenum.

Bailey, J. E. (1991). Toward a science of metabolic engineering. *Science, 252*, 1668–74.

Barinaga, M. (1992). The brain remaps its own contours. *Science, 258*, 216–18.

Berk, L. E. (1991). *Child development* (2nd ed.). Boston: Allyn & Bacon.

Billings-Gagliardi, S., & Wolf, M. K. (1988). Shiverer*jimpy double mutant mice. IV. Five combinations of allelic mutations produce three morphological phenotypes. *Brain Research, 455*, 271–82.

Blau, H. M., Pavlath, G. K., Hardeman, E. C., Choy-Pik, C., Silberstein, L., Webster, S. G., Miller, S. C., & Webster, C. (1985). Plasticity of the differentiated state. *Science, 230*, 758–66.

Bouchard, T. J., Lykken, D. T., McGue, M., Segal, N. L., & Tellegen, A. (1990). Sources of human psychological differences: The Minnesota study of twins reared apart. *Science, 250*, 223–8.

Bulman-Fleming, B., & Wahlsten, D. (1988). Effects of a hybrid maternal environment on brain growth and corpus callosum defects of inbred BALB/c mice: A study using ovarian grafting. *Experimental Neurology, 99*, 636–46.

Bulman-Fleming, B., & Wahlsten, D. (1991). The effects of intrauterine position on the degree of corpus callosum deficiency in two substrains of BALB/c mice. *Developmental Psychobiology, 24*, 395–412.

Busnel, M. C., Granier-Deferre, C., & Lecanuet, J. P. (1992). Fetal audition. *Annals of the New York Academy of Sciences, 662*, 118–34.

Capecchi, M. R. (1994). Targeted gene replacement. *Scientific American, 270*(3), 52–9.

Capitanio, J. P. (1991). Levels of integration and the "inheritance of dominance." *Animal Behaviour, 42,* 495–6.

Carlier, M., Nosten-Bertrand, M., & Michard-Vanhée, C. (1992). Separating genetic effects from maternal environmental effects. In D. Goldowitz, D. Wahlsten, & R. E. Wimer (Eds.), *Techniques for the genetic analysis of brain and behavior* (pp. 111–26). Amsterdam: Elsevier.

Caviness, V. S., Jr., So, D. K., & Sidman, R. L. (1972). The hybrid reeler mouse. *Journal of Heredity, 63,* 341–6.

Clarke, A. M., & Clarke, A. D. B. (Eds.) (1976). *Early experience: Myth and evidence.* New York: Free Press.

Coleman, D. L. (1981). Inherited obesity-diabetes syndromes in the mouse. In E. S. Russell (Ed.), *Mammalian genetics and cancer* (pp. 145–58). New York: Liss.

Collins, R. L. (1985). On the inheritance of direction and degree of asymmetry. In S. D. Glick (Ed.), *Cerebral lateralization in nonhuman species* (pp. 41–71). New York: Academic Press.

Cooper, R. M., & Zubek, J. P. (1958). Effects of enriched and restricted early environments on the learning ability of bright and dull rats. *Canadian Journal of Psychology, 12,* 159–64.

Copeland, N. G., Jenkins, N. A., Gilbert, D. J., Eppig, J. T., Maltais, L. J., Miller, J. C., Dietrich, W. F., Weaver, A., Lincoln, S. E., Steen, R. G., Stein, L. D., Nadeau, J. H., & Lander, E. S. (1993). A genetic linkage map of the mouse: Current applications and future prospects. *Science, 262,* 57–66.

Crews, D. (1994). Animal sexuality. *Scientific American, 270,* 108–14.

Davidson, E. (1986). *Gene activity in early development* (2nd ed.). San Diego, CA: Academic Press.

DeCasper, A. J., & Spence, M. J. (1986). Prenatal maternal speech influences newborns' perception of speech sounds. *Infant Behavior and Development, 9,* 133–50.

Deeming, D. C., & Ferguson, M. W. J. (Eds.) (1992). *Egg incubation. Its effects on embryonic development in birds and reptiles.* New York: Cambridge University Press.

Detterman, D. K. (1990). Don't kill the ANOVA messenger for bearing bad interaction news. *Behavioral and Brain Sciences, 13,* 131–2.

Dunn, L. C. (1965). *A short history of genetics.* New York: McGraw-Hill.

Duyme, M., & Capron, C. (1992). Socioeconomic status and IQ: What is the meaning of the French adoption studies? *European Bulletin of Cognitive Psychology, 12,* 585–604.

Efron, B., & Tibshirani, R. (1991). Statistical data analysis in the computer age. *Science, 253,* 390–5.

Eibl-Eibesfeldt, I. (1979). Human ethology: Concepts and implications for the sciences of man. *Behavioral and Brain Sciences, 2,* 1–58.

Erlenmeyer-Kimling, L. (1972). Genotype–environment interactions and the variability of behavior. In L. Ehrman, G. S. Omenn, & E. Caspari (Eds.), *Genetics, environment and behavior* (pp. 181–208). New York: Academic Press.

Feibleman, J. K. (1954). Theory of integrative levels. *British Journal of the Philosophy of Science, 5,* 59–66.

Fisher, R. A. (1918). The correlation between relatives on the supposition of Mendelian inheritance. *Transactions of the Royal Society of Edinburgh, 52,* 399–433.

Flynn, J. R. (1987). Massive IQ gains in 14 nations: What IQ tests really measure. *Psychological Bulletin, 101,* 171–91.

Ford, D. H., & Lerner, R. M. (1992). *Developmental systems theory.* Newbury Park, CA: Sage.

Gärtner, K. (1990). A third component causing random variability beside environment and genotype. A reason for the limited success of a 30 year long effort to standardize laboratory animals? *Laboratory Animals, 24,* 71–7.

Gottesman, I. I. (1963). Genetic aspects of intelligent behavior. In N. R. Ellis (Ed.), *Handbook of mental deficiency: Psychological theory and research* (pp. 253–96). New York: McGraw-Hill.

Gottlieb, G. (1970). Conceptions of prenatal behavior. In L. R. Aronson, E. Tobach, D. S. Lehrman, & J. S. Rosenblatt (Eds.), *Development and evolution of behavior* (pp. 111–37). San Francisco: Freeman.

Gottlieb, G. (1971). Ontogenesis of sensory function in birds and mammals. In E. Tobach, L. R. Aronson, & E. Shaw (Eds.), *The biopsychology of development* (pp. 67–128). New York: Academic Press.

Gottlieb, G. (1976). The roles of experience in the development of behavior and the nervous system. In G. Gottlieb (Ed.), *Neural and behavioral specificity* (pp. 25–54). New York: Academic Press.

Gottlieb, G. (1991). Experiential canalization of behavioral development: Theory. *Developmental Psychology, 27*, 4–13.

Gottlieb, G. (1992). *Individual development and evolution: The genesis of novel behavior.* New York: Oxford University Press.

Greene, E. (1989). A diet-induced developmental polymorphism in a caterpillar. *Science, 243*, 643–6.

Greenough, W. T., McDonald, J. W., Parnisari, R. M., & Camel, J. E. (1986). Environmental conditions modulate degeneration and new dendrite growth in cerebellum of senescent rats. *Brain Research, 380*, 136–43.

Griffin, D. R. (Ed.) (1982). *Animal mind – human mind.* New York: Springer-Verlag.

Guastavino, J. -M., Larsson, K., & Jaisson, P. (1992). Neurological murine mutants as models for single-gene effects on behavior. In D. Goldowitz, D. Wahlsten, & R. E. Wimer (Eds.), *Techniques for the genetic analysis of brain and behavior* (pp. 375–90). Amsterdam: Elsevier.

Gupta, A. P., & Lewontin, R. C. (1982). A study of reaction norms in natural populations of *Drosophila pseudoobscura. Evolution, 36*, 934–48.

Hall, C. S. (1951). The genetics of behavior. In S. S. Stevens (Ed.), *Handbook of experimental psychology* (pp. 304–29). New York: Wiley.

Hayakawa, K., Shimizu, T., Ohba, Y., & Tomioka, S. (1992). Risk factors for cognitive aging in adult twins. *Acta Genetica Medicae et Gemelloligae, 41*, 187–95.

Heisenberg, M. (1989). Genetic approach to learning and memory (mnemogenetics) in *Drosophila melanogaster.* In H. Rahmann (Ed.), *Fundamentals of memory formation: Neuronal plasticity and brain function* (pp. 3–45). Stuttgart: Gustav Fischer Verlag.

Herrup, K., & Vogel, M. W. (1988). Genetic mosaics as a means to explore mammalian CNS development. In S. S. Easter, Jr., K. F. Barald, & B. M. Carlson (Eds.), *From message to mind* (pp. 238–51). Sunderland, MA: Sinauer.

Ho, M.-W. (1984). Environment and heredity in development and evolution. In M.-W. Ho & P. T. Saunders (Eds.), *Beyond neo-Darwinism: An introduction to the new evolutionary paradigm* (pp. 267–89). San Diego, CA: Academic Press.

Isingrini, M., Lenoir, A., & Jaisson, P. (1985). Preimaginal learning as a basis of colony-brood recognition in the ant *Cataglyphis cursor. Proceedings of the National Academy of Sciences U.S.A., 82*, 8545–7.

Johannsen, W. (1909). *Elemente der Exacten Erblichkeitslehre,* Jena: G. Fischer.

Johnston, T. D. (1988). Developmental explanation and the ontogeny of birdsong: Nature/nurture redux. *Behavioral and Brain Sciences, 11*, 617–63.

Johnston, T. D., & Hyatt, L. E. (1993). An interactionist model of behavioral development that incorporates genetic influences. Paper presented at International Scientific Developmental Psychobiology, Alexandria, Virginia.

Kaczmarek, L. (1993). Molecular biology of vertebrate learning: Is *c-fos* a new beginning? *Journal of Neuroscience Research, 34*, 377–81.

Kandel, E. R., & Hawkins, R. D. (1992). The biological basis of learning and individuality.

Scientific American, 267(3), 79–86.

Keating, D. P. (1996). Habits of mind: Developmental diversity in competence and coping. In D. K. Detterman (Ed.), *Current topics in human intelligence* (Vol. 5) (pp. 31–44). Norwood, NJ: Ablex.

Kurnit, D. M., Layton, W. M., & Matthyse, S. (1987). Genetics, chance and morphogenesis. *American Journal of Human Genetics, 41*, 979–95.

Layton, W. M., Jr. (1976). Random determination of a developmental process. Reversal of normal visceral asymmetry in the mouse. *Journal of Heredity, 67*, 336–8.

Lee, S. M., & Bressler, R. (1981). Prevention of diabetic nephropathy by diet control in the *db/db* mouse. *Diabetes, 30*, 106–11.

Lehrman, D. S. (1970). Semantic and conceptual issues in the nature–nurture problem. In L. R. Aronson, E. Tobach, D. S. Lehrman, & J. S. Rosenblatt (Eds.), *Development and evolution of behavior* (pp. 17–52). San Francisco: Freeman.

Lewontin, R. C. (1991). *Biology as ideology.* Concord, Ontario: House of Anansi Press.

Locke, J. L. (1993). *The child's path to spoken language.* Cambridge, MA: Harvard University Press.

Lofdahl, K. L., Holliday, M., & Hirsch, J. (1992). Selection for conditionability in *Drosophila melanogaster. Journal of Comparative Psychology, 106*, 172–83.

McFadden, D. (1993). A masculinizing effect on the auditory systems of human females having male co-twins. *Proceedings of the National Academy of Science U.S.A., 90*, 11900–4.

McGinnis, W., & Kuziora, M. (1994). The molecular architects of body design. *Scientific American, 270*, 58–66.

McGue, M. (1989). Nature–nurture and intelligence. *Nature, 340*, 507–8.

McManus, I. C., & Bryden, M. P. (1992). The genetics of handedness, cerebral dominance and lateralization. In I. Rapin & S. J. Segalowitz (Eds.), *Handbook of neuropsychology: Vol. 6. Child neuropsychology* (pp. 115–44). Amsterdam: Elsevier.

Merriam, J., Ashburner, M., Hartl, D. L., & Kafatos, F. C. (1991). Toward cloning and mapping the genome of *Drosophila. Science, 254*, 221–5.

Messer, A., Eisenberg, B., & Plummer, J. (1991). The *Lurcher* cerebellar mutant phenotype is not expressed on a *staggerer* mutant background. *Journal of Neuroscience, 11*, 2295–302.

Molenaar, P. C. M., Boomsma, D. I., & Dolan, C. V. (1993). A third source of developmental differences. *Behavior Genetics, 23*, 519–24.

Morgane, P. J., Austin-LaFrance, R., Bronzino, J., Tonkiss, J., Diaz-Cintra, S., Cintra, L., Kemper, T., & Galler, J. R. (1993). Prenatal malnutrition and development of the brain. *Neuroscience and Biobehavioral Reviews, 17*, 91–128.

Newman, B. M., & Newman, P. R. (1991). *Development through life* (5th ed.). Pacific Grove, CA: Brooks/Cole.

Nottebohm, F. (1985). Neuronal replacement in adulthood. *Annals of the New York Academy of Sciences, 457*, 143–61.

Oyama, S. (1985). *The ontogeny of information. Developmental systems and evolution.* Cambridge: Cambridge University Press.

Ozaki, H. S., & Wahlsten, D. (1993). Cortical axon trajectories and growth cone morphologies in fetuses of acallosal mouse strains. *Journal of Comparative Neurology, 336*, 595–604.

Petersen, P. E., & Blass, E. M. (1982). Prenatal and postnatal determinants of 1st suckling episode in albino rats. *Developmental Psychobiology, 15*, 349–55.

Platt, S. A., & Sanislow, C. A., III. (1988). Norm-of-reaction: Definition and misinterpretation of animal research. *Journal of Comparative Psychology, 102*, 254–61.

Plomin, R., & McClearn, G. E. (Eds.) (in press) *Nature, nurture, and psychology.* Washington, DC: APA Books.

Preer, J. R., Jr. (1993). Nonconventional genetic systems. *Perspectives in Biology and Medicine, 36*, 395–419.

Prigogine, I., & Stengers, I. (1984). *Order out of chaos. Man's new dialogue with nature.*

Toronto: Bantam Books.

Purves, D. (1988). *Body and brain. A trophic theory of neural connections.* Cambridge, MA: Harvard University Press.

Roubertoux, P. L., & Capron, C. (1990). Are intelligence differences hereditarily transmitted? *European Bulletin of Cognitive Psychology, 10*, 555–94.

Sagar, S. M., & Sharp, F. R. (1993). Early response genes as markers of neuronal activity and growth factor action. *Advances in Neurology, 59*, 273–84.

Scarr, S. (1992). Developmental theories for the 1990s: Development and individual differences. *Child Development, 63*, 1–19.

Scarr, S. (1993). Biological and cultural diversity: The legacy of Darwin for development. *Child Development, 64*, 1333–53.

Shaffer, D. R. (1992). *Developmental psychology* (3rd ed.). Pacific Grove, CA: Brooks/Cole.

Shapiro, D. Y. (1981). Serial female sex changes after simultaneous removal of males from social groups of coral reef fish. *Science, 209*, 1136–7.

Snyder, L. H. (1940). *The principles of heredity* (2nd ed.). New York: Heath.

Stacey, P. B., & Bock, C. E. (1978). Social plasticity in the acorn woodpecker. *Science, 202*, 1298–1300.

Sussman, G. J., & Wisdom, J. (1988). Numerical evidence that the motion of Pluto is chaotic. *Science, 241*, 433–7.

Tyler, L. E. (1972). Human abilities. *Annual Review of Psychology, 23*, 177–206.

Vasta, R., Haith, M. M., & Miller, S. A. (1992). *Child psychology.* New York: Wiley.

vom Saal, F. S. (1984). The intrauterine position phenomenon: Effects on physiology, aggressive behavior, and population dynamics in house mice. In K. J. Flennelly, R. J. Blanchard, & D. C. Blanchard (Eds.), *Progress in clinical and biological research: Vol. 169. Biological perspectives on aggression* (pp. 135–79). New York: Liss.

Wachs, T. D. (1992). *The nature of nurture.* Newbury Park, CA: Sage.

Waddington, C. H. (1957). *The strategy of the genes.* London: Allen & Unwin.

Wahlsten, D. (1978). Behavioral genetics and animal learning. In H. Anisman & G. Bignami (Eds.), *Psychopharmacology of aversively motivated behavior* (pp. 63–118). New York: Plenum.

Wahlsten, D. (1982). Deficiency of the corpus callosum varies with strain and supplier of the mice. *Brain Research, 239*, 329–47.

Wahlsten, D. (1989). Deficiency of the corpus callosum: Incomplete penetrance and substrain differentiation in BALB/c mice. *Journal of Neurogenetics, 5*, 61–76.

Wahlsten, D. (1990). Insensitivity of the analysis of variance to heredity–environment interaction. *Behavioral and Brain Sciences, 13*, 109–61.

Wahlsten, D. (1991). Sample size to detect a planned contrast and a one degree-of-freedom interaction effect. *Psychological Bulletin, 110*, 587–95.

Wahlsten, D. (1994). The intelligence of heritability. *Canadian Psychology, 35*, 244–60.

Wahlsten, D., & Bulman-Fleming, B. (1987). The magnitudes of litter size and sex effects on brain growth of BALB/c mice. *Growth, 51*, 240–8.

Wainwright, P., Pelkman, C., & Wahlsten, D. (1989). The quantitative relationship between nutritional effects on preweaning growth and behavioral development in mice. *Developmental Psychobiology, 22*, 183–95.

Ward, I. L. (1983). Effects of maternal stress on the sexual behavior of male offspring. *Monographs in Neural Science, 9*, 169–75.

Woltereck, R. (1909). Weitere experimentelle Untersuchungen über das Wesen quantitatitiver Artunterschieder bei Daphniden. *Verhandlungen der Deutschen Zoologischen Gesellschaft, 19*, 110–73.

Wright, S. (1920). The relative importance of heredity and environment in determining the piebald pattern of guinea pigs. *Proceedings of the National Academy of Sciences U.S.A., 6*, 320–32.

Wright, S. (1934). On the genetics of subnormal development of the head (otocephaly) in the

guinea pig. *Genetics, 19,* 471–505.
Wright, S. (1968). *Evolution and the genetics of populations: Vol. 1. Genetic and biometric foundations.* Chicago: University of Chicago Press.
Yoshinaga, K., & Mori, T. (Eds.) (1989). *Development of preimplantation embryos and their environment.* New York: Wiley-Liss.
Zajonc, R. B. (1983). Validating the confluence model. *Psychological Bulletin, 93,* 457–80.
Zawilska, J. B., & Wawrocka, M. (1993). Chick retina and pineal gland differentially respond to constant light and darkness: In vivo studies on serotonin *N*-acetyltransferase (NAT) activity and melatonin content. *Neuroscience Letters, 153,* 21–4.
Zigler, E. F., & Stevenson, M. F. (1993). *Children in a changing world* (2nd ed.). Pacific Grove, CA: Brooks/Cole.

7   Between nature and nurture: The role of human agency in the epigenesis of intelligence

Thomas R. Bidell and Kurt W. Fischer

On both personal and collective levels, human beings are active and creative participants in their own intellectual development. In the course of our everyday lives, we actively solve problems, make decisions, and make sense out of world events and our own roles in them. We are the creative agents in the formation of our own opinions and in the solutions to our life crises, even while we acknowledge the contributions of others to our lives. At a collective level, human beings create culture and civilization, and they participate in the collective problem solving needed to advance social progress in areas such as medicine, education, and civil rights. Indeed, the denial of this capacity for self-determination by traumatic events or repressive social conditions is profoundly damaging to our sense of self (Herman, 1992). Whether on a personal or a collective level, human intellectual development is unmistakably a process of creative self-determination.

This experience of creativity and self-determination is reflected in contemporary scientific accounts of intellectual development as a constructive, self-regulated process. Since the late 1950s, cognitive developmental researchers from virtually every theoretical tradition (Bandura, 1977; Piaget, 1977; Resnick, 1985; White, 1976) have contributed to the now nearly universal consensus that our intellectual skills and abilities are in one way or another the products of our own self-governed activity in relation to the world (Bruner, 1990; Gardner, 1985).

Considering the breadth of consensus on the constructive nature of intellectual (as well as social and emotional) development, it is paradoxical that scientific discussions about the relative contributions of genetics and environment in human intelligence have rarely considered seriously the role of the human agent in his or her own intellectual development. The oscillating nature-versus-nurture debate regularly swings from claims of genetically determined linguistic and cognitive capacities to affirmations of environmentally induced behaviors and ideas, as if these two categories

193

exhaust all sources of explanation in accounting for the remarkable range of human intellectual achievements. Periodic calls for interactive approaches in which genes and environment jointly determine our intellectual capacities only confirm the exclusivity of these two explanatory categories. Strangely absent from genetic, environmental, and interactionist approaches alike is the constructive activity that is so widely recognized as the core of the developmental process. The human experience of self-determined insights, ideas, interpretations, decisions, and solutions seems to vanish into the narrow gap between nature and nurture.

Why does the role of constructive activity get lost in the nature–nurture debate? Is our sense of self-efficacy and self-determination in our own thinking an illusion that dissipates in the cold light of empirical analysis? Is constructive, goal-oriented activity merely derivative – an outgrowth of more fundamental genetic and environmental processes?

We will argue that the role of constructive activity in intellectual development is neither illusory nor derivative. Instead, it is the central motivating force in the development of human intellectual abilities and the key to the interrelations of genome and environment. The failure of contemporary psychology to recognize adequately the central role of human agency in linking genome and environment is a direct outgrowth of the conceptual fallacy of reductionism (Bidell, 1988; Levins & Lewontin, 1985; Overton, 1973, 1994; Oyama, 1985; Weiss, 1970), which artificially reduces multileveled, intrinsically related, self-organizing processes to dualistic abstractions like heredity and environment. These abstractions isolate real genetic and environmental activity from the context of their joint participation with other levels of biological, social, and mental activity in developing person–context systems. In so doing, they obscure the role of the person as agent in development. The role of the human agent in development is to continually create new *relationships* between multiple levels of biological and environmental systems through our unique capacity to integrate skills and ideas whose roots extend down through organismic systems to the genetic level and whose branches extend out into systems of socially patterned activity. The reductionistic separation of heredity and environment into just two independent factors systematically eliminates the very category that most fundamentally accounts for change in cognitive development: the constructive activity of the person. The result is an illusion that just two factors are responsible for the development of intelligence.

This illusion – that genetics and environment exclusively determine intellectual development – is reinforced by research methodologies based on that assumption. Because these two factors are abstracted from their natural relations as parts of a larger person–context system, they are seen as mutually exclusive and are treated as the source of all meaningful variability in developmental processes. The partitioning of variability into two mutually exclusive categories diverts the attention of researchers away from the real sources of developmental variation – the dynamic activity of the person in context.

Alternative methods are available that support treating behavior as a product of self-regulating dynamic systems. When intellectual development is reconceptualized in terms of dynamic, constructive, person-in-context systems, multiple sources of developmental variability become apparent. Fischer's (1980a; Bidell & Fischer, 1992a; Fischer & Rose, 1994) *dynamic skills theory* provides both a theoretical framework and a set of methodological tools for the study of developmental variability in person-in-context systems. Using these tools, researchers can begin to sort out the developmental relations among the multiple systems, including genetic regulation, sociocultural activity, and personal construction, all participating in the epigenesis of human intelligence.

Intellectual development as self-organization of integrative systems

The intractability of the nature–nurture controversy may be traced to a fundamental fallacy in the framework in which the debate is conceptualized. Within this framework, nature and nurture are conceived as separate, independent forces, each exerting its own influence toward the determination of cognitive outcomes (Overton, 1973). This conceptualization, however, is contrary to the view that has emerged from the fields of molecular genetics and developmental biology and among a growing number of cognitive developmental researchers (Fischer & Silvern, 1985; Gottlieb, 1992; Lerner, 1991; Oyama, 1985). This newer epigenetic view treats genetics and environment not as factors that stand apart from development, determining its course, but as active, self-organizing systems that co-participate with other self-organizing systems to determine intellectual development jointly. In this integrative-systems perspective, intellectual development is a self-organizing process. The integrative-systems perspective provides an alternative to the sterile nature–nurture debate in the quest to understand the role of genetics and environment in the origins of intelligence. To appreciate this alternative, it is necessary to begin with a discussion of the integrative-systems perspective in general, outlining the ways in which it differs from the nature–nurture framework in its treatment of the relations between genetics and environment.

Nature of living systems

Twentieth-century science has taught us that living processes are multileveled, self-organizing, integrative systems (von Bertalanfy, 1968; Weiss, 1970). By *integrative systems*, we mean self-organizing systems in which the parts do not and cannot exist prior to or separately from one another or from the whole in which they participate. The term *integrative system* is used in this chapter to emphasize the *interparticipatory* character of living systems and their constituent subsystems. Figure 7.1 presents a diagram in which developmental biologist Paul Weiss (1970) has schematically represented some of the major forms of interparticipation among different levels of integrative systems. The arrows in this figure may be understood to represent forms of interparticipation and mutual regulation of one system upon another.

The human body is an obvious example of the integrative character of living systems. Living human beings are composed of multiple interdependent systems including the nervous system, the circulatory system, the respiratory system, the endocrine system, and many others including innumerable subcellular systems. However, the recognition of simple interdependence in human biological systems falls far short of the point. The stones in a well-crafted archway are interdependent, each depending on the weight and shape of the others to maintain its position, but they are not organized into a living system. The subsystems of living organisms not only depend on one another but actively participate in one another. They take an active part in one another's functioning as well as in the life of the individual as a whole. One can draw no absolute division between the cardiovascular system, which participates in the life of every bodily organ, and those organs in which it participates, or between the cardiovascular system and other systems like the immune system, which carry out their functions in and through the agency of cardiovascular activity. Biological systems are thus composed not of atomistic building blocks but of inextricably interpenetrating living systems that participate to greater or lesser degrees in the activity and functioning of other systems.

Because they are integrative, living systems *cannot be subtractively decomposed*. That is, none of the systems can exist or function separately from the whole (the living person) in whom they take part. Indeed, it doesn't even make sense to imagine a cardiovascular system, pulmonary system, endocrine system, or nervous system living or functioning outside of the context of the other systems in a living organism. Even single organ transplants must be accepted and integrated into a functioning system to survive beyond a short time.

Moreover, because no living system can exist outside the context of the

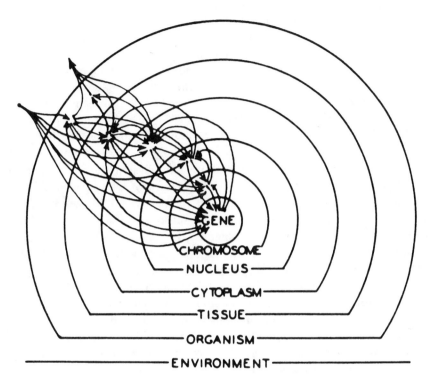

Figure 7.1. A schematic diagram indicating the interparticipation and mutual regulation of multiple levels of biological, cognitive, and sociocultural systems in the developing person. (From Weiss, 1959).

living whole, none of the systems, including the genetic system, can have existed prior to the existence of the individual. In the ontogenesis of a given individual, each system has evolved into its present form and function in coordination with the ontogenesis of all the other systems. The nervous system evolved along with the circulatory system, which it regulates and which supplies it with oxygen and nutrients. There is no prime mover or privileged system, standing outside of and directing the activity and development of the others, and the whole cannot be built atomistically from preexisting parts. The system, whole and part, must evolve together through the agency of its own activity. This point will be taken up in more detail with respect to the role of the genetic system shortly.

Another consequence of living systems' interparticipation in one another is that they *mutually regulate* one another's functioning and activity. As an individual climbs a mountain, the increase in muscular activity and decrease in oxygen intake are compensated by increased cardiopulmonary and metabolic activity. While systems are partly independent, each carrying out its own functions toward more or less separate ends, the separate ends are all subordinated to the same overall function of maintaining the life and activity of the individual. To the extent that the systems are thus integrated, they adjust their functioning to compensate for and to cooperate with the functioning of related systems. Those systems more intimately involved in a particular function may respond more directly than those more distally involved, but in general, all systems participating in a given function adjust their own functioning to compensate for changes in other systems involved in the joint activity.

The mutual regulation of integrally related systems leads to a third fundamental consequence of the integral nature of living systems: Their development is *profoundly nonlinear*. The joint activity of living subsystems is simultaneous and dynamic, with many self-organizing subsystems both competing and cooperating simultaneously to produce qualitative changes jointly in a person's overall activity in a given context (Fischer & Rose, 1994). In contrast, mechanical processes operate in a more linear fashion, with the outcome of one action forming the input of another with little or no self-regulation. For instance, in an engine, a spark leads to fuel combustion, which leads to gaseous expansion, which impels the cylinder outward. The simultaneous action of these different functions in this mechanical system would lead to engine seizure. In living systems, however, the simultaneous action of systems with partially opposing tendencies or functions contributes to the general development of the whole. Changes in one system lead to adjustments and reorganizations in other systems, creating an altered systemic context for further adjustments and reorganizations.

This dynamic interparticipation of partially opposing systems leads to one of the most characteristic features of the developmental process: its *unevenness and variability*. The measurement of this variability provides a key to the empirical study of the self-organizing processes of complex dynamic systems. Methods for the empirical study of such developmental variability are discussed later in this chapter.

Function of the genome in living systems

Contemporary research in molecular genetics leaves no doubt that the genome is one of the living systems that integrally participates in and collaborates with other biological systems in the production and regulation of organismic activity (Hirsch, 1990; Ho, 1983; Lerner, 1993). The genes in every cell are virtually identical. Yet this identical material must participate in a multitude of different local functions throughout the organism, and the nature of these functions frequently changes as the organism develops.

How can this be possible if each gene contains the same sequences of DNA?

The answer provided by contemporary genetics is that the function of genetic systems is *context dependent*. Contrary to popular misconceptions of genes as self-contained messengers, standing outside of and directing the activity of other organismic systems, the genetic system participates in the functioning of local nucleic and subcellular systems, affecting those systems in ways consistent with the genes' organization and, in turn, being affected in their own functioning by the activity of multiple local and distal systems, which interparticipate with subcellular and nucleic systems (Lerner, 1991, 1993; Weiss, 1970). One of the most dramatic demonstrations of this interparticipatory character of the genetic system comes from situations, either natural or experimental, in which genetic material is relocated to different organismic contexts, resulting in markedly different developmental outcomes from the same genetic material. One fascinating instance cited by Gottlieb (1992) is the case of a parasitic wasp that lays its eggs in two different hosts – a fly and a butterfly. Although the genetic "code" for wasp morphology is identical in both cases, offspring developing in the butterfly host develop wings and other morphological features that are absent in the individuals developing in the fly hosts. Thus, the same genetic material, operating in different developmental contexts, can lead to strikingly different developmental outcomes.

An even more dramatic demonstration of the context dependence of genetic action comes from experiments in which induced changes in the subcellular cytoplasmic context of genetic activity result not only in new developmental outcomes but in the transmission of these outcomes across generations. In one such experiment, Ho and Saunders (1984) exposed fruit fly embryos to ether, a treatment that alters the cytoplasmic context in which the genes operate, leading to adult specimens that developed an extra set of wings. The effect of the altered cytoplasmic environment was not restricted to treated individuals; it extended to the offspring of treated mothers, who reproduced the altered context of genetic function by passing the altered cytoplasm to the next generation through their eggs. Thus, without a structural alteration in genes, a change in the cytoplasmic context of gene activity led to a new morphological trait that was transmitted across generations.

Such contextual effects arise because developmental outcomes are not simple readouts of stored templates but the products of the joint activities or "coactions" (Gottlieb, 1992, p. 161) of multiple levels of organismic (and environmental) systems working together with the genetic system. As Gottlieb has written,

The genes are part of the developmental system in the same sense as other components (cell, tissue, organism), so genes must be susceptible to influence from other levels during the process of individual development. (p. 167)

In other words, just as the cardiovascular system, nervous system, respiratory system, and sensorimotor system work together to produce changes in breathing and pulse rates as we climb a mountain, so the genetic system works together with systems of subcellular tissues, organs, and sensorimotor actions and perceptions in the ontogenesis of the individual. The genes do not direct the activities of these other systems any more than the cardiovascular system directs the nervous system. But like the cardiovascular and nervous systems, the genes participate together with the other systems through mutual regulation and joint activity, as suggested in Figure 7.1. If the genes were simply an encapsulated code spelling out when and how every organ should take shape and every system should commence functioning, their function would not depend on the organismic context in which they operate. Indeed, even the basic mechanisms of production of RNA and DNA involve co-participation of the cytoplasmic system (Gottlieb, 1992; Weiss, 1970).

This contemporary understanding of the function of the genome in ontogenesis contradicts older reductionist accounts of genes as lone determinants of developmental outcomes. In standard versions of this view, depicted in Figure 7.2, DNA provides a blueprint or a "master tape" that dictates precise instructions to "slave" cellular processes, which then pass the instructions on, domino style, to obedient tissue, organismic, and behavioral-level systems. In one of the most extreme versions, human bodies are merely "robot" carriers of "selfish genes" that dictate behavior such as criminality, aggression, or mental illness (Dawkins, 1976).

Such views continue to receive popular acceptance partly because the newer findings of molecular and developmental genetics have not reached a wide audience outside of their fields (Gottlieb, 1992), and partly because of a reductionist current within Western culture and science that privileges linear-causal explanations (Levins & Lewontin, 1985). However, advances in our understanding of the role of genetics in the development of human abilities (including intelligence) demand that we begin to integrate these findings into our models of developmental processes. As Ho (1983, p. 285, quoted in Gottlieb, 1992) has written,

Forever exorcised from our collective consciousness is any remaining illusion of development as a genetic programme involving the readout of the DNA "master" tape by the cellular "slave" machinery. On the contrary, it is the cellular machinery that imposes control over the genes. . . . The classical view of an ultraconservative genome [as] the unmoved mover of development is completely turned around.

Consistent with the principles of integral living systems, the genetic system both influences and is influenced by other systems at every level in

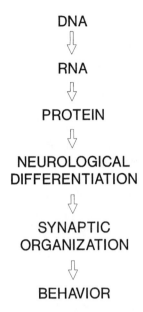

DNA
⇩
RNA
⇩
PROTEIN
⇩
NEUROLOGICAL
DIFFERENTIATION
⇩
SYNAPTIC
ORGANIZATION
⇩
BEHAVIOR

Figure 7.2. The popular but empirically inaccurate domino model of genetic influence on cognition and behavior.

the process of development (Figure 7.1). The genes are not unmoved movers acting to determine linearly the action of other systems; instead, they are integrated co-participants with other systems in the activity and development of a person.

Cognitive-behavioral systems

As with organismic systems, cognitive-behavioral systems are self-organizing, multileveled, and integral. It is therefore not surprising that the cognitive sciences, like the biological sciences, have evolved away from atomistic models toward models of hierarchically integrated systems involving active control over mental and behavioral activities. However, researchers have been relatively slow to recognize the interpenetration of cognitive systems with both biological and social systems. The same reductionist tendencies that have supported atomistic conceptions of behavior and consciousness also support dualistic conceptual separations between mind and body and between mind and social context.

Developmental research has begun to show how the cognitive system is integrally united with both biological and social systems. On the biological side, there is a growing literature challenging linear, reductionist views of the biological role of intelligence. In the past, research on biological bases of intelligence has too often fallen victim to familiar assumptions about linear causality. The assumption of a linear-causal chain running from gene to cell to neural organization and then behavior leads to hypotheses that certain psychological characteristics are hardwired, such as linguistic abilities or personality traits. Framing questions within this reductionist framework eliminates from the outset conceptions of multilevel systemic co-participation and results in erroneous conclusions of genetically determined characteristics.

A recent study of the emergence of height wariness during sensorimotor development (Campos, Bertenthal, & Kermoian, 1992) illustrates the developmental relations among biological, cognitive, affective, and environmental systems in development. The relatively sudden onset of fear of heights in infants during the 6–10-month age range has frequently been interpreted in terms of one-way, linear-causal assumptions about the actions of genes in determining an organ structure and, in turn, a behavioral function. As Campos and his colleagues put it,

These changes in fearfulness occur so abruptly, involve so many different elicitors, and have such biologically adaptive value [e.g., preventing falls] that many investigators propose maturational explanations for this developmental shift (Emde, Gaensbauer, & Harmon, 1976; Kagan, Kearsley, & Zelazo, 1978). For such theorists, the development of neurophysiological structures (e.g., the frontal lobes) precedes and accounts for changes in affect [p. 61].

Yet, in a series of carefully controlled studies, Campos and his colleagues uncovered quite a different picture of the developmental mechanisms underlying the emergence of fear of heights. These authors found that when age (and therefore maturation) was controlled, infants' own activity, in the form of locomotion, emerged as the primary factor associated with fear of heights. Both experimental studies with the visual cliff and a single-subject longitudinal study of an orthopedically handicapped child showed that fear of heights commenced only after children had ample opportunities for crawling. Moreover, fear of falling was found to increase as a function of the duration of locomotor activity, regardless of the age of the subject. Of course, crawling brings with it a host of concomitant activities, including new experiences of falling and new ways of exploration.

These findings fail to support a one-way linear view of the relation between biology and behavior. The primacy of locomotion in the process means that the relations between neurological maturation, sensorimotor activity, and affective development are necessarily complex and nonlinear. It is worth quoting Campos et al.'s interpretation of this complex interrelationship.

We believe that locomotion is initially a goal in itself, with affect solely dependent on the success or failure of implementing the act of moving. . . . However, as a result of locomotor experience, infants acquire a sense of both the efficacy and the limitations of their own actions. Locomotion stops being an end in itself, and begins to be goal-corrected and coordinated with the environmental surround [p. 63].

Thus, through its own self-organized activity in the environment, an infant constructs a coordinated understanding of the possibilities of action in the environment, including the possibilities of falling, creating a new context-dependent emotional structure. This active agency on the part of the infant creates a link between biological and environmental systems because it creates a new intersystemic context involving new forms of mutual regulation. The infant's construction of locomotion skills leads to new understanding of possible environmental threats, new protective responses from parents, new possibilities for expanding emotional attachment, and so forth. As Campos et al. concluded,

> It seems clear from these findings obtained in the line of research that new levels of functioning in one behavioral domain can generate experiences that profoundly affect other developmental domains, including affective, social, cognitive, and sensorimotor ones [p. 64].

Once the relative contributions of multiple biological, behavioral, and environmental systems begin to be empirically unpacked, neither nature nor nurture appears as the prime mover in the developmental process. Instead, it is the self-organizing activity of the human agent that drives the system, uniting environmental and biological systems at many levels.

Indeed, recent research findings have begun to reveal the depth of influence that the behavioral system has on the biological system through the agency of mutual regulation. Studies in developmental neuropsychology, for instance, show that excess synaptic connections generated early in neurological development are selectively pruned as a function of differential behavioral activity (Edelman, 1987). The well-known experiments by Hubel and Wiesel (1977) demonstrated the need for perceptual activity in sustaining and integrating the neonatal neurological organization of perception. Singer (1979) and others have shown that to have an effect, visual activity must be controlled by the animal; passive experience does not affect neurological organization the way that active experience does. As Gottlieb (1992) notes, variations in sensorimotor activity, social activity, and nutrition in rats and chicks can each result in changes in brain cell size and numbers, as well as in the relative amount of RNA produced in the brain cells.

Empirical findings of these kinds provide growing evidence that in behavioral development, as in biological development, the genetic system is not a master code that simply kicks off the first domino in the chain from gene to behavior; rather, it is one among many co-participating systems whose joint activities produce not only developmental change in behavior but also changes in the organization of and interrelations among all the developing systems.

Systems of social and cultural activity

It is nothing new to suggest that social and cultural systems are also integral, interparticipatory, and self-organizing (Bronfenbrenner, 1979; Lerner 1991; Sameroff, 1975; Rogoff, 1990). Family systems provide a well-studied example of the mutual regulation and co-participation involved in the organization and development of human interpersonal relations. Moreover, there exists extensive evidence of the ways in which sociocultural subsystems, such as families, mother–child dyads, or classroom organizations, vary in structure and function across forms of cultural organization and social stratification. In the context of the present argument, however, it is important to emphasize the specific nature of the interparticipation between cognitive-behavioral and sociocultural systems: The tenacious forms of dualistic thinking that have divided mind from body also lead to artificial divisions between cognitive-behavioral systems and the environmental systems in which they participate.

Cultural systems are not simply containers for a collection of individuals independently constructing their personal intellectual systems. Instead, cultural systems are *constitutive* (Bruner, 1990; Geertz, 1973) of the personal intellectual abilities being constructed, and vice versa. The process of human intellectual development is a simultaneous construction of personal and cultural meaning systems. At the level of the person, intellectual development is a process of acquiring culture while making sense of oneself and the world. At the level of the cultural system, this bipolar process of mental development is a reproductive mechanism that maintains, renews, and sometimes challenges the cultural system itself.

Central to these mechanisms is the reconstruction of sociocultural activity at the personal level. Vygotsky first described this process with the somewhat vague term *internalization*, which Wertsch (1985) has defined more specifically as a shift from interpersonal regulation of activity to intrapersonal regulation. Interpersonal activities such as joint problem solving (Wertsch, McNamee, McLane, & Budwig, 1980; Westerman, 1990), group planning (Rogoff, Baker-Sennett, & Matusov, 1994), communication (Cook-Gumperz, 1981; Diaz, Neal, & Amaya-Williams, 1990), and classroom interactions (Mehan, 1979) are all organized in very specific ways as functions of their roles in the larger sociocultural system. As children participate in such culturally patterned interpersonal activities, they construct cognitive-behavioral control systems that are organized and adjusted to take part in these specialized activities. By gradually constructing increasingly more adequate control systems for participation in culturally patterned activity, children take increasing personal control of what was at first culturally patterned joint activity. The specific organization of cognitive-behavioral systems are thus guided (Rogoff, 1990) or scaffolded (Bruner, 1982; Wood, 1980) along culturally specific pathways.

For present purposes, two crucial consequences of this mechanism should be noted. First, because cognitive control systems are constructed through and for participation in culturally patterned activities, they are just as much a part of the cultural system as they are a part of the person. Thus, it makes little sense to speak of such skills as either genetically inherited or environmentally induced. Instead, they are actively constructed integral parts of developing person–context systems (Fischer, Bullock, Rothenberg, & Raya, 1993; Lerner, 1991).

Second, self-organizing activity is a key process in the transition from inter- to intrapersonal regulation of activity. A person's construction of new context-embedded cognitive skills drives the internalization of culturally patterned activity. No amount of social support or shared meaning is sufficient in itself to explain developmental advance in the skills for participating in culturally patterned activity. For cognition, as for biology, a person's own self-organizing activity must be included if we are to understand the mechanisms by which the cultural patterning of social activity guides intellectual development into context-specific forms.

Summary: Dynamics of integrative systems

We have reviewed some of the theory and evidence deriving from contemporary developmental sciences on the nature of biological, cognitive-behavioral, and sociocultural systems. Three major conclusions may be drawn from our review. First, given what is known about living systems, it makes little sense to treat any system or its components as isolable elements whose separate contributions to a developmental outcome can be expressed as a constant number or percentage. Any such attempt ignores the context-dependent nature of each component's activity, which is regulated by, and in turn regulates, the many other systemic components jointly participating in the developmental process.

For example, in the development of butterflylike wings in the parasitic wasps described earlier, the proportional contribution of genes to the eye's structure *depends* on the systemic context of the host in which the genes operate. If the same genetic material that produces wings in one context produces no wings in another context, then in the second case, its contribution to wing making is reduced to zero. Since the contribution of genetic structure to eye morphology varies radically with the systemic context of joint activity, attempts to determine the proportional contribution of genes to morphological features are ill founded.

Likewise, if researchers ask what part of intelligence is environmental, they are confronted with the fact that the intelligence systems being assessed are constitutive of and participate in the environmental (socially patterned) activities that are being used to explain intelligence. Moreover, the culturally specific ways of thinking, recalling, and perceiving that we construct as participants in the culture are simultaneously the vehicles by which the culture participates in us.

From an integral dynamic-systems viewpoint, research questions aimed at determining the relative contribution of a given system in isolation cannot yield useful information. At worst, such questions can mislead researchers into futile debates over the real source of developmental changes and obscure the role of self-organizing activity in producing developmental variability.

A potentially more productive set of research questions centers on the developmental mechanisms by which the interparticipating biological, cognitive-behavioral, and sociocultural systems jointly construct specific cognitive outcomes. Researchers need to specify not isolated components but developmental relations among integrally related systems as well as processes of change, including the effects of contexts and domains. As Anastasi (1958) argued at an earlier point in this debate, the important question is *how* genes and environment contribute to epigenetic outcomes, not *how much* they contribute.

The second conclusion of our review bears directly on Anastasi's how question. The primary mechanism of epigenetic change is the self-organizing activity of integral systems themselves. The sequence of epigenetic changes observed either in physical or psychological development cannot be explained by the linear-causal chains advanced in reductionist conceptions, regardless of whether these chains consist of genetic events, environmental events, or both. Because genes and environment interparticipate in multiple mutually regulating systems from the subcellular to the cultural level, the changes leading to epigenetic sequences must be understood in terms of complex adjustments and reorganizations among the many mutually regulated subsystems.

Third, the cognitive-behavioral system plays a unique leading role with respect to the epigenesis of intelligence. The active construction of new cognitive skills by the developing person weaves biological and sociocultural systems into new patterns of relations, creating new context-specific forms of intellectual abilities. Since epigenesis proceeds through changing relations among systems, not through one-way genetic instructions or environmental reinforcements, the construction of new forms of personal/cultural meaning creates new contexts for the interparticipation of biological and environmental systems in intelligence. The construction of meaning in social context thus effectively may be seen as driving the process of cognitive epigenesis. In the section that follows, we explicate this claim, detailing some of the mechanisms by which constructive cognitive activity leads the epigenesis of mind. In subsequent sections, we examine methodological implications of this approach, contrasting it with reductionist methodologies in the study of nature–nurture relations.

Constructive epigenesis

The epigenetic conception of development originated as an alternative to the 18th-century preformationist view in which the embryo was viewed as a tiny, fully formed – if hard to see – version of the adult, which had only to grow in size (Gottlieb, 1992). This strictly quantitative version of embryonic development was challenged by the theory of epigenesis, which held that changes in the developing embryo were qualitative, with new structures and features emerging in a predictable sequence during development. With the advent of improved observational techniques, the debate was settled on the side of epigenesis, and the epigenetic conception was gradually extended beyond embryogenesis to encompass physical, cognitive-behavioral, and emotional development at least into adulthood (e.g., Erikson, 1964; Gesell, 1946; Hall, 1904; Piaget, 1947/1960; Werner, 1948).

Although emergence of qualitatively new features, structures, and behaviors is generally accepted in contemporary theories of development, the mechanisms that bring about these qualitative changes remain a subject of debate. In one view, called *predetermined epigenesis* (Gottlieb, 1970, 1992; Lerner, 1980), the emergence of new behaviors and mental capacities remains preformed in a genetically programmed sequence. In this view, the genetic code is the prime mover in the developmental process. Conditions in other systems or in the environment may switch on the genetic instructions, but it is these instructions that drive the emergence of new abilities. The genetic code is seen as determining neurological structures independently of social or behavioral systems. Only when new neurological structures or hardwiring have been set in place by this genetically driven process are new forms of ideas and behaviors possible. The emergence of new cognitive abilities is reduced to the assumption that new neurological structures can receive the new types of environmental input that they are preadapted to interpret. Although lip service may be paid to constructivism, the role of the human agent in this model is reduced to that of a passive bystander in a predetermined genetic unfolding.

A major flaw in this viewpoint is that it depends on the linear model of genetic regulation (Figure 7.2), which is inconsistent with contemporary genetic research. In the linear model, genes switch on at their appointed times and set off a chain reaction that produces a new structure or behavior. This focus on the genes as the causal agents ignores the contributions of self-organizing activity in development.

An alternative view of the epigenetic process is that emergence of each structure or behavior is the product of the self-organizing activity of previously developed systems (Gottlieb, 1992; Lerner, 1991; Oyama, 1985; Thelen & Smith, 1994; van Geert, 1994). The genetic system is one among many systems participating in and helping to regulate the emergence of a new structure, but its actions are dependent on the context of intersystemic activity, as represented in Figure 7.1. Gottlieb calls this concept of epigenesis *probabilistic* in order to emphasize the fact that a given developmental outcome is not genetically predetermined but free to vary depending on systemic context. However, we prefer to call this type *constructive epigenesis* in order to emphasize the leading role of self-organizing activity in the emergence of qualitatively new structures and functions during development.

A central mechanism in the process of constructive epigenesis is the coordination of component processes or subsystems. As Weiss (1970) has demonstrated, integral systems are always hierarchically organized, with the component systems not only pursuing more or less separate functions but also working together under the regulation of the larger system in which they take part. Weiss gives an example of the tissue structure of taste buds, which, when microscopically examined, all exhibit the same shape and size, which they maintain over time. Underlying this similarity, the cells of each taste bud structure are constantly moving, dying, and dividing, presenting different patterns of spatial arrangement within each taste bud structure. Thus, despite the great divergence of cellular activity and the numerous independent functions of the component cellular systems, "the patterned structure of the dynamics of the system as a whole 'co-ordinates' the activities of the constituents" (Weiss, 1970, p. 29).

Because an integral system, by definition, involves the coordinated action of the components within the whole system, the emergence of a new integral system requires that components be actively brought into coordinated action to form an integrated whole. In other words, the formation of a new integral system requires an *act of integration*. Existing systems must be actively intercoordinated to form a larger organization whose structures, functions, and goals go beyond those of the component parts by themselves.

This process of hierarchical integration through the active coordination of subsystems constitutes the primary mechanism driving constructive epigenesis. Active coordination thus replaces the one-way linear-causal mechanisms of predetermined epigenesis. Instead of genetic determination of new structures through a domino sequence of causal reactions, the emergence of new forms at every level is self-determined by the needs and activity of the developing system.

In constructive epigenesis, the need for a new emergent form of organization at any given point in development is not determined by a preset program but by the changing contexts produced by ongoing constructive activity. Changes in one system create new conditions affecting the activity of interparticipating systems. Because the subsystems composing a larger system mutually regulate one another's activities, the construction of a new hierarchically integrated system changes in the constellation of mutual regulation among its component systems, requiring the readjustment or reorganization of the components' activities.

In embryogenesis, for example, growth in the number and density of the ball of cells constituting the blastula eventually necessitates a new general organization, the gastrula, which in turn requires new types of cellular and subcellular structures. These new structures are neither called forth ex nihilo nor dictated by a predetermined blueprint. Instead, the general structure and the cellular subsystems, including the genetic subsystem, participate together in the production of the new cellular structures called for in the developmental process. This organized intersystemic process would not be possible if the genetic system were, for instance, that of a paramecium. The structure of the human genetic system is essential to the success of the cooperative effort. But the fact that the genes play one part by no means implies genetic predetermination of the vast set of cellular and subcellular processes that cooperate in the formation of new cell types and tissue structures. On the contrary, the integrative process of coordination creates new forms of both intra- and intercellular organization and thus sets the context for the next round of integrative activity, in which the genetic system will again participate, with a revised systemic organization and context.

In constructive epigenesis, the role of the genetic system is thus subordinate to the primary mechanism of hierarchical integration by active coordination of subsystems into new superordinate wholes. Coordination drives epigenesis by constantly setting the agenda for intersystemic regulation and reorganization at the subordinate levels. Since genetic action, as with other forms of living activity, is dependent on the context of the intersystemic activity in which it participates, the constructive process of coordination ultimately helps to channel the activity of the genes through the creation of new intersystemic relations.

Epigenesis of cognition: Constructor's role in constructivism

The framework of constructive epigenesis provides an alternative to the reductionist nature–nurture dichotomy in conceiving the role of genetics and environment in the development of intelligence. Within this framework, the epigenesis of intelligence is a special case of constructive epigenesis. Just as coordination of biological systems is central to epigenesis of the body, coordination of cognitive systems is the primary mechanism propelling epigenesis of the mind.

This constructivist view stands the reductionist model on its head. Instead of genetics and environment determining cognitive outcomes, the process of cognitive development is seen as self-determining through constructive activity. This constructive activity sets the agenda for the participation of genetics and environment in intellectual epigenesis. Since cognitive-behavioral systems participate with both sociocultural systems and biological-genetic systems, the coordination of component systems into new hierarchically integrated cognitive systems creates a new relational context in which biological-genetic systems co-participate with internalized sociocultural patterns of activity. And since genetic activity, as we have shown, is context dependent, the process of constructing new cognitive systems effectively *leads* the process of cognitive epigenesis.

For researchers, the constructive epigenesis model implies a shift of focus away from the unproductive nature–nurture debate to questions about the processes by which hierarchical integration functions to weave nature and nurture together. Unfortunately, research on hierarchical integration has been remarkably vague and unsatisfying, especially in the behavioral sciences (see critiques by Fischer, 1995; Thelen & Smith, 1994; van Geert, 1994). This has been due in part to the vagueness of classic theories of cognitive development in describing hierarchical integration (e.g., Piaget, 1936/1952; Vygotsky, 1934/1987; Werner, 1948). Recently, however, neo-Piagetian theories have built upon these classic theories to provide conceptual and methodological tools for more precise and powerful description of epigenetic processes (Case, 1985; Fischer, 1980a; Siegler, 1981). With these new tools, researchers are in an unprecedented position for studying developmental processes in the epigenesis of intelligence. The tools of skill theory are especially suitable for application to the problem of cognitive epigenesis because they include specification of mechanisms of hierarchical integration through coordination (Bidell & Fischer, 1994; Fischer, 1980a; Fischer, in press; Fischer & Rose, 1994).

In the remainder of this section, we explicate the constructivist approach to epigenesis of mind, outlining in more detail the role of hierarchical integration and then presenting a model of coordination. Then, in the final section of the chapter, we outline methodological principles that support the application of constructivist models to research on cognitive epigenesis (including but not limited to skill theory).

Processes of coordination in epigenesis of mind

The fundamental problem in accounting for the emergence of new forms of living systems during biological epigenesis is to show how systems functioning relatively independently can become unified, operating in concert as one system that controls and directs relationships and activities among subsystems. Just as the mechanism of coordination provides a foundation for understanding biological epigenesis as a process of self-organization among living systems, in cognitive epigenesis the same principle of coordination operates, but the units coordinated are forms of cognitive-behavioral activity, not just forms of biological activity. This process of cognitive coordination is the key to understanding developmental relations between biological, cognitive, and sociocultural systems.

By now, it should be clear that the reductionist chain-reaction model of genetic instructions dictating the maturation of neurological structure with consequent new cognitive forms is untenable. The reductionist vision of genetic, neurological, cognitive, and social systems as fundamentally isolated, influencing each other only through the cumbersome mechanism of myriad chains of instructions somehow directed by a mysterious master code, is inconsistent with contemporary knowledge of molecular genetics, developmental biology, and psychosocial systems. But in rejecting this traditional view, what alternative model can be advanced to explain the role of genetics and environment in the emergence of new cognitive skills during ontogenesis?

A simpler, more direct, and much more elegant solution presents itself in the form of self-organized hierarchical integration. In this view, the vehicle for participation of both genome and environment in cognitive development is the creation of new developmental contexts through the coordination of multiple systems into new forms of collaboration and mutual regulation. The participation of genetic and environmental systems in cognitive development is organized by the cognitive system itself through its activities of interpreting and operating on the world. The creation of new hierarchically integrated cognitive systems leads to new relations of mutual regulation among multiple participating systems, including environmental and genetic.

As an illustration, let's take the development of the concept of object permanence in infancy, which Piaget (1937/1954) first described. At any given level of object concept, a child has constructed stable skills for interpreting and acting upon the world in certain ways. For instance, skills that are sensorimotor mappings (similar to what Piaget [1936/1952] called *secondary circular reactions*) involve controlling one action in relation to another one, such as using one as a means to produce the other. An infant who hits a rattle (first action is hitting the rattle) and likes the sound (second action is listening to the rattle) can repeat the action when the rattle is accessible. Or an infant whose babbling has just elicited a desirable vocalization from a parent can repeat the babbling activity to hear more of the parent's vocalization. In both these cases, the organization of the infant's activities provides a specific type of knowledge about objects and their permanence. Through this type of sensorimotor activity, an infant begins to understand how objects behave independently of his or her body and how they can be influenced by his or her actions. At the same time, this organization of activity also limits knowledge, so that, for example, an infant still does not understand that objects can continue to exist when they are no longer directly observable, as when they are hidden by a barrier. This piece of knowledge will be acquired as infants gradually construct new forms of organized activity with objects in the world.

Any such skill, like the skill for repeating babbling to elicit a desired parental response, depends on the simultaneous functioning of multiple systems that can be self-sustaining. For instance, in the babbling example, a parental response must be available (social system), a vocal-neurological structure must be in place (vocalization system), a cognitive coordination must occur (cognitive-neural system), and the entire bodily system, including cellular structure, neural patterns, and much more, must function effectively. The genetic system participates in this hierarchy of systems of helping to sustain the cellular structure and functioning needed to organize this form of activity. The environment participates by providing a specific form of support to the activity. The skill is not determined by a causal chain originating in either genes or environment, but by the interparticipation of multilevel integrative systems, with diverse relevant components mutually regulating one another to maintain the integrity of the whole.

A key to the creation and maintenance of this integrative whole is the functional activity of the skill itself – that is, its use in interpreting and acting upon the world. The very neurological structures that support the skill are dependent on cognitive activity for their development and maintenance. The construction of a new form of cognitive control, governing components in a new way, creates a different context for participation of neural activity, which redirects the growth and pruning of neural connections to support the current form of activity. We noted earlier that basic neurological connections are stimulated, suppressed, or pruned based on the cognitive-behavioral activity in which they participate (Edelman, 1987).

Contrary to the widely held assumption that new hardwiring must precede new cognitive functioning, new biological structures emerge and function in a way that depends on functional activity. Both individual neurons and synaptic connections must compete to survive and grow (Changeux & Danchin, 1976). Neurons that are not active because they are not participating in functional activity die away, and neurons that are active are sustained. Similarly, in a process of competition among synapses, those that are highly active thrive, while those that are inactive become weaker or are pruned out. A young kitten or child, for example, that sees actively through only one eye will sustain connections to the active eye only, losing neural connections to the inactive one, even if it receives visual input (unlinked to action) (Hubel & Wiesel, 1977; Singer, 1979).

The functional activity of the cognitive skill fundamentally affects which groupings of neurons and neural connections will develop and be maintained and developed, and which will disappear. Component cognitive systems are already supported by a certain level of neurological development, and the direction of further development is determined to a large extent by the new types of activities in which they are required to participate. Through the active process of coordinating cognitive component systems, new neurological structures are gradually initiated at the same time that the new cognitive-behavioral system is established.

These findings do not imply a one-way causality between cognitive activity and neurological growth. Cognitive skills clearly depend on neurological systems, most obviously with skills such as binocular vision or human speech, which are supported by highly specialized neurological and sensorimotor structures. Our argument is simply that no version of one-way causality is consistent with the ways these systems develop and function. Cognitive epigenesis involves both cognitive and neurological systems participating in one another. However, in important ways, the cognitive-behavioral system *leads* this developmental process, contributing significantly to setting the pattern for neurological development.

At the same time that the coordination mechanism redirects neurological development, it restructures the form of mutual participation between the cognitive and sociocultural systems. The construction of a new cognitive skill implies a qualitatively different means of participation in sociocultural activities and, through further internalization of those patterns, a new form of sociocultural patterning of cognitive activity. The coordination of babbling with the auditory activity of listening to parental vocalizations, for example, not only establishes a cognitive skill but also creates a qualitatively new set of relations between social world and infant. The infant now possesses another strategy besides crying for controlling parents' behavior. Individuals participate in socially and culturally patterned activities that require specific skills. As individuals construct these skills, internalizing culturally patterned forms of activity, they create the means of inter–participation among environmental, neurological, and genetic systems and thus the means of mutual regulation among these systems.

The coordination mechanism that integrates the cognitive, neurological, and social systems serves an essential mediating function between the genetic system and the environment. The coordination of cognitive component systems gives rise to new forms of sociocultural patterning of cognitive activity, which in turn pattern the neurological activity that supports the coordination. In this way, the coordination mechanism provides a vehicle through which forms of sociocultural activity help to pattern genetic activity. By creating new contexts for genetic participation in neurological activity, the socially patterned activity sets new directions for neurological development. As neurological structures are stimulated and pruned differentially through specific forms of socially patterned activity, genetic systems are called upon to support the emergence of new intra- and intercellular structures and then to continue functioning in the ongoing maintenance of these cellular structures and connections.

At the same time, neither neurological nor cognitive systems would develop without the participation of the genetic system. There is no evidence of a one-way determination of either neurological or cognitive structure by the genes, or vice versa. Instead, the genes' activities are highly dependent on the intra- and intercellular context. With every cell in the body equivalent in its genetic potential, the current developmental status of the neurological system determines how the genetic system operates at any given point. Since development of specific neurological structures is dependent on the cognitive activities in which they participate, the nature and timing of the genetic system is partly determined by a human agent's cognitive activities – how he or she acts in the world and interprets it.

From this perspective, genetic and environmental factors do not interact solely with each other in determining cognitive outcomes. On the contrary, although both factors are involved in the epigenesis of intellectual skills, they are woven into a fabric of multiple component systems producing specific cognitive outcomes through the constructive agency of a person. Without this constructive agency producing and directing cognitive development, neither genetic nor environmental factors would have any vehicle by which to produce cognitive outcomes. Genetics and environment play indispensable roles in the epigenesis of cognitive skills, but they do so only in and through the constructive activity of a person, enabling and directing their influence.

Hierarchical integration as coordination of dynamic skills

If hierarchical integration through coordination plays a leading role in cognitive epigenesis, then well-specified models of that mechanism are needed to support research, theory, and practice. Piaget (1936/1952) and Werner (1948) both advanced concepts of hierarchical integration as the primary developmental mechanism, and Vygotsky (1934/1987) proposed a similar process in which intellectual development progressed through the active formation of connections or generalizations constituting new, more inclusive hierarchical relations within and among concepts. However, these theories lacked the specificity needed for rigorous empirical study of the process.

Neo-Piagetian skill theory (Fischer, 1980a) has built upon the insights of these classic theories to offer a contextual framework and task-analytic methodology for analyzing development of cognitive and emotional systems in specific contexts and situations. These conceptual-methodological tools also provide a specific dynamic model of the mechanisms by which people coordinate cognitive systems to form new hierarchically integrated systems. The specificity of skill theory allows researchers to define in considerable detail the form and content of particular cognitive structures, including component and superordinate relations among skills. Among the central concepts in the model of coordination are skill (structures of action and thought), and mechanisms of coordination in transition. (For more detailed treatments of skill theory, see Bidell & Fischer, 1992a; Fischer, 1980; Fischer & Rose, 1994; Fischer, Shaver, & Carnochan, 1990.)

Definition of cognitive skills. In skill theory, cognitive systems of all kinds (i.e., from early reflex and sensorimotor actions to systems of abstract concepts) are characterized as active skills, which are contextualized integrative systems of action, thought, and feeling, not decontextualized logical structures. A skill is defined as a cognitive control structure built to relate and organize mental, emotional, and physical activity in a specific context. Skills are integrative systems in the sense used throughout this chapter, and so any given skill entails all levels of participation from biological to social. Cognitive skills are not strictly within-person structures but integrative systems that interparticipate with social systems (Bidell & Fischer, 1992b).

In English usage, *skill* suggests a combination of person and context (Bruner, 1973; Fischer, 1980a). Every skill is constructed for participation in a specific context. The context may be relatively particular, as when we initially construct a skill for using a particular strategy for interpersonal negotiation with one friend or acquaintance, or it may be relatively general, as when we generalize that skill to many relationships connecting it with participation in a cultural style of interpersonal negotiation (Fischer & Ayoub, 1993; Selman & Schultz, 1990). In either case, the skill consists of a control structure for organizing actions in context, and both the organization of the skill and the actions that it controls are definable and measurable.

We use the terms *level* and *step* to describe, respectively, large and small advances in skill organization. These terms contrast with *stage*, which is usually taken to mean that people have fixed structures for acting and thinking. Dynamic skills are not fixed but vary across several levels (and many steps) depending on the complex relations among interparticipating systems. Individual children and adults function at different skill levels for different tasks (e.g., more or less complex tasks), states (e.g., relative fatigue, glucose level, emotional state), and situations (e.g., highly supportive vs. nonsupportive contexts). Instead of a fixed level, individuals exhibit a *developmental range* (Bidell & Fischer, 1992a; Lamborn & Fischer, 1988), defined as the range of skill levels between a person's optimal level and functional level of performance in a domain. The *optimal level* is the most complex type of skill (the limit) that a person can consistently control under optimal conditions, including an alert state, a familiar context, practice with the task, conditions that support optimal performance (such as priming key components), and an emotional state consonant with the skill. Without optimal conditions, people usually perform at lower steps or levels on a developmental scale. The *functional level* is the limit at which a person can consistently produce within a context where there is no special support – the highest level beyond which a person's skills do not go without environmental support.

Mechanism of skill coordination.   Development from one level to another takes place through the active coordination of previously existing skills. As with other dynamic systems, new forms of hierarchical control – new cognitive skills – result from the active coordination of components that originally operated as mostly independent. Initially, these systems compete with each other, vying for a person's attention and resources in specific contexts. The sources of variation controlled by each skill are dealt with separately, or at best, in succession. For example, in object permanence, a baby at first can grasp either a desirable toy or a piece of cloth. These are independent, competing activities until he or she coordinates them to form a skill for removing the cloth in order to uncover the object and grasp it. When separate skills are thus coordinated with one another, they produce a larger dynamic skill in which they become components functioning together to produce simultaneous control of the previously separate sources of variation. To find a hidden object, an infant must coordinate the component skill of seeing and holding the cloth with that of seeing and holding the object, so as to keep in mind where the object disappeared and how it can be grasped when it is recovered.

As this example suggests, a key to the coordination process is the ability to hold in mind simultaneously the component cognitive skills. Unless the component skills can be somehow brought together simultaneously, there is no chance to coordinate them since they function successively – first one, then the other. Indeed, research on cognitive transitions and task simplification confirms this pattern. Earlier developmental levels are characterized by shifting between components that need to be coordinated. In research with dozens of different tasks across the entire age range from infancy to adulthood, people show a common type of error or simplification before they develop a capacity for coordinating co-occurrence: They keep components separate that need to be coordinated, shifting between them instead of coordinating them into a unit (Fischer, 1980a; Fischer & Elmendorf, 1986; Gottlieb, Taylor, & Ruderman, 1977; Perry, Church, & Goldin-Meadow, 1988; Roberts, 1981). For instance, when 13-year-olds are asked how addition and multiplication relate to each other in general, many of them first explain one operation and then shift to the other, never specifying the general relation between the two (Fischer, Hand, & Russell, 1984). When 5-year-olds explain a story about a doctor helping his sick daughter, many of them likewise split the story in two, telling one story about the father and daughter and a second one about doctor and patient (Watson & Fischer, 1980).

When initially independent skills can be held in mind together in a context that supports their coordination, they can be intercoordinated into a new hierarchically organized skill in which they function together. This intercoordination takes place gradually, in a microdevelopmental process that typically involves gradual steps toward integration (Bidell, 1990; Bidell & Fischer, 1994; Fischer, 1980b; Granott, 1991). Through extended attempts at integration of component skills within a given context, a person forms a new, more extensive skill that simultaneously organizes sources of variation that could only be controlled successively before.

Analyzing structures of component and outcome skills. Using Skill Theory task analysis, the mechanism of coordination can be directly studied by analyzing the structural organization of the component and outcome skills. An example comes from a recent study of cognitive transition mechanisms in children's planning skills in the tower of Hanoi problem-solving task (Bidell, 1990; Bidell & Fischer, 1994). In this task, children must move a stack of three graduated discs from a start post to a goal post with the help of an intermediary rest post. Since children may move only one disc at a time and may not place larger discs on top of smaller ones, the rest post must be used for temporary storage.

The goal of Skill Theory task analysis is to determine what sources of variation in children's problem-solving activity on this task must be cognitively controlled in order to function effectively. This involves identifying not only the succession of moves that a child must anticipate but also the *relations* among these moves that the child must be able to hold in mind simultaneously in order to organize effectively the correct move sequence. The task analysis proceeds by identifying simple component skills and then defining the specific ways in which these components must be related in forming larger units of meaning.

Space will not permit a full exposition of the task analysis (for details, see Bidell & Fischer, 1994), but Figure 7.3 summarizes its results. This figure shows the complex interrelations among the first four moves of the task – the key moves in solving it. At the simplest level, a child must cognitively control (hold in mind, i.e., working memory) a single move of one disc from one post to another (object–action–outcome). In Skill Theory, this corresponds to the developmental level of *single representations*, a type of skill that starts emerging around 2 years of age.

But to know *which* of the possible moves to perform first, children must also be able to organize these simple component skills into more inclusive types of skills. One essential type of skill for the present task is *representational mapping*. In a representational mapping, two single representations are coordinated in a one-way or dependency relationship. With this skill, a child can represent a relation in which the events represented in one single representation depend on or are the result of the events represented with another skill. In the tower of Hanoi, this means that children can represent the dependency of one move on another. For the example in Figure 7.3, two mappings are of importance. In one of these mappings, a child mentally controls the movement of disc A to the goal post *in order to* move disc B to the rest post (thus, the second move depends on the first). In the other mapping, the child mentally controls the movement of disc A to the rest post *in order to* move disc C to the goal post.

Coordinating single representations into mappings is usually trivial for school-aged children. However, in order to solve the first four moves of the present task, a much more complex and hierarchically inclusive type of skill must be constructed, using the mapping-level skills themselves as components. This is because the significance of the events related by either of the

Representational Mapping A

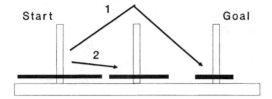

Coordinated with...

Representational Mapping B

Yields...

Representational System

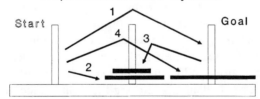

Figure 7.3. Component cognitive skills for planning in the tower of Hanoi task are actively intercoordinated to form a hierarchically inclusive and successful plan.

foregoing mappings cannot be grasped by the individual unless these mappings themselves are brought into relationship. The movement of disc A to the goal post in order to move disc B to rest only makes sense if this mapping is itself understood as a *means* of accomplishing the moves represented in the first mapping. That is, a child must establish relations between the mapping relations.

It is important to note, however, that the relations between the mappings are not merely sequential. Although the moves involved must be carried out sequentially, in order for a child to appreciate why a particular order is required, he or she must be able mentally to represent the relations among all four moves *simultaneously*. This can be seen in Figure 7.3. To anticipate the correct first move of disc A to the goal, the child must appreciate that disc B cannot be placed on the goal since disc C will have to be placed there in several moves; at the same time, the child must grasp that to get disc B out of the way of disc C, disc A must first be placed directly on the goal, only to be removed again to be placed on top of disc B once that disc has been placed on rest. Note that these moves cannot be considered simply in the order of their occurrence in the final action sequence. Moves from each of the two mappings must be combined so that the interrelations among all the moves can be represented simultaneously.

Representational skills involving this type of simultaneous control of multiple relations between two or more mapping-level skills and their components are called *representational systems*. The construction of a representational-systems skill for this task provides a good example of need for a mechanism of coordination. Unless the component mapping skills are actively brought together to form a more inclusive representational system, a child will be forced to alternate between one or another mapping-level interpretation of the problem. Indeed, when they first approach this task, most children do not immediately see the nature of these interrelations. The most common mistake of school-aged children on this version of the tower of Hanoi is to make their first move to the rest post instead of the goal post, a move consistent with less complex planning skills that anticipate only one-way dependency relations. Somehow a child must bring the two dependency relations into co-occurrence in working memory and organize them into the *simultaneous* system of interdependent actions needed to anticipate the correct sequence of moves. By representing a succession of possible mapping-level plans, a child can create a series of potential component representations. The challenge is to bring the right component plans into co-occurrence and then to coordinate them into a new, more complex and inclusive plan, constituting a hierarchically new level of planning skill.

This example illustrates the application of Skill Theory task analysis to the study of the hierarchical integration process. Other cognitive developmental theories also provide task analytic methods that can be applied to the question of hierarchical integration. The primary point we wish to make here is that contemporary neo-Piagetian cognitive developmental theories have moved beyond the vague generalizations about hierarchical integration found in the classical theories of Piaget, Werner, and Vygotsky. Contemporary theories offer both specific constructs and task-analytic methods that can be applied to researching this process. Hierarchical integration should no longer be treated as simply a metaphor but rather as a key mechanism in cognitive development and cognitive epigenesis; as such, it should be made the object of systematic research.

Tracking the constructive agent: Principles for research in cognitive epigenesis

From the perspective of constructive epigenesis that we have outlined, research questions and research methods based on a reductionist separation of nature from nurture may be seen as misdirected because they systematically exclude a key source of developmental variation: the activity of the constructive agent. Research approaches based on partitioning sources of variation into heredity and environment typically obscure the role of constructive activity because any variability arising from this source is assigned a priori to one of the other two. With the origins of cognitive skills apparently explained by genes and environment, questions about the role of the constructive agent appear superfluous, and the only available mechanism of cognitive epigenesis seems to be the empirically untenable predeterministic model.

For these reasons, satisfying answers to Anastasi's (1958) oft-quoted question of how genes and environment operate in development will not be found within the traditional nature–nurture framework. Research capable of uncovering the mechanisms of cognitive epigenesis demands a conceptual shift to a dynamic integrative-systems approach of the sort we have outlined. Only within a framework that is capable of positively representing the active construction of new relations among the many levels of participating systems can the mechanisms of cognitive epigenesis be productively studied.

If, as we have argued, the central mechanism in the reorganization of relations among these systems is active hierarchical integration, then a fundamental goal in the study of cognitive epigenesis must be to pick up the trail of these integrations as they lead from earlier to later forms of intersystemic relations. In other words, the role of any given system can only be understood when it is placed in a developmental context that allows us to understand both its current relations with other systems and the history of changing interrelationships that brought it to the present point.

There are a number of specific theories and models (Bronfenbrenner,

1979; Fischer, 1980a; Gottlieb, 1992; Lerner, 1991, 1993; Molenaar, Raijmakers, & Hartelman, 1994) that offer a range of valuable constructs and methods consistent with this approach. Since a review of all these approaches goes beyond the scope of this chapter, we will limit ourselves in this final section to (1) discussing some of the conceptual fallacies that arise when research methods fail to take into account the constructive context of a cognitive skill, and (2) presenting and illustrating some general methodological principles that we believe are needed to guide the application of integrative-systems theories to research on the epigenesis of intelligence.

Losing the trail

There are two main ways in which the reductionist assumptions entailed in the nature–nurture conceptualization can mislead researchers. First, the assumption of a nature–nurture dichotomy tends to channel research questions into a fruitless debate over relative proportional contribution of genetics as compared with the environment, while discouraging research into the mechanisms of cognitive epigenesis and the interrelations among the multiple participating systems. Second, by spawning and sustaining a mechanistic, predetermined view of cognitive epigenesis, the nature–nurture framework leads to methodological shortcomings when studying the mechanisms of cognitive epigenesis.

The proportional pitfall.   The assumption that all variability in behavior can be partitioned meaningfully into just two categories, genetics and environment, leads to a hopeless quest to determine the "real" proportional contributions of genetics and environment to behavioral factors, such as intelligence or genetics, and directs attention away from the mechanisms by which constructive activity produces new contexts of interrelations among developing systems. We say "hopeless quest" partly because the effects of genetic activity are so profoundly context dependent. The proportion of influence can vary dramatically with changes in the activity of any context from the subnucleic to the cultural.

At a more fundamental level, the interparticipatory nature of the genetic and environmental systems with multiple other systems involved in cognitive development renders the question of proportional contributions somewhat absurd. It is analogous to asking what proportional contribution is made by the blood vessels and the blood to the process of circulation. Each is necessary, neither is sufficient, and the question of scientific interest is not how *much* each contributes but the mechanisms by which they work together in maintaining the life of the organism.

This analogy points up an even greater problem with research questions about fixed relative proportions of genetic and environmental contributions to development: Such questions direct research attention away from the more fundamental questions about the mechanism of constructive activity in cognitive epigenesis. The perpetual debate over just how much of a given behavior is environmentally determined versus how much is genetically determined generates a scientific discourse in which the nature–nurture dichotomy dominates, and questions about the constructive activity of the participating systems seem out of place – as if constructivist mechanisms somehow operate in a different realm, unrelated to the production of intellectual or behavioral abilities.

This situation creates a vicious circle in which the assumption of a na-ture–nurture dichotomy leads to research approaches that produce evidence related only to heredity and environment. With no other sources of variation apparent, intellectual development appears to be explained entirely by these two factors, reinforcing the original assumption. This circle of assumptions and methods is seriously misleading, especially to students and beginning researchers, since the bulk of the research creates the appearance that constructive agency is not a relevant factor or that there are no adequate research methods for studying its role and effects in the development of intelligence.

In twin studies, for instance, where genetic factors are presumed to be held constant, variation due to constructive agency typically goes undetected because, as a source of between-persons differences, it is not distinguished from environmental factors (Molenaar et al., 1994). And attempts to more closely identify the origins of behaviors by further subdividing either nature or nurture into, for instance, shared versus nonshared environment (Dunn & Plomin, 1990) only seems to confirm the exclusivity of genetics and environment as sources of developmental variability.

Because the nature–nurture framework lacks conceptual categories for relating human agency to behavioral outcomes, researchers are often led to meaningless, even absurd conclusions. For instance, one recent study (Plomin, Corley, DeFries, & Fulker, 1990) reported "significant genetic influences" (p. 376) on television viewing in young children. How can we interpret a conclusion of this kind? Has evolution acted so quickly in the short time since the invention of television? Although researchers in this tradition are quick to disclaim causal attribution in heritability estimates, the lack of conceptual categories for describing the role of the person's own agency in such behaviors creates a powerful impression to the contrary – one that influences both scientific and lay communities alike.

Our critique of the assumptions entailed in the nature–nurture framework should not be taken as an indictment of the science of behavioral genetics. Discovering the relations between genetics and forms of human behavior is a valid and vital scientific enterprise (Molenaar et al., 1994). Our concern is that this enterprise not be diverted into a futile quest for the "true" percentage of genetic contribution to behaviors, but instead focus on the more meaningful question of the developmental mechanisms by which that contribution is made.

The preformationist peril. If we turn to the mechanisms of cognitive epigenesis, however, we once again encounter the misleading influence of the reductionist assumptions embedded in the nature–nurture dichotomy. As noted earlier, the reduction of the complex relations of inter-participation among the multiple systems to the static categories of nature and nurture creates the illusion that the only mechanism by which genes can influence development is through a one-way chain of messages in which the cognitive outcome is essentially preformed in the genetic code that initiates the message. With the leading role of constructive activity systematically ruled out of the picture, there appears to be no other means for genes to participate than to predetermine certain cognitive abilities. And, with a predeterministic model of genetic contribution, the role of the environment is reduced to either triggering the onset of a predetermined behavior or filling in genetically determined cognitive structures with information.

Research approaches that are tacitly or explicitly based on a pre-deterministic model employ a circular logic that rules out evidence of constructive activity has leads to the appearance of genetically determined behaviors or ideas. Since the role of constructive activity in integrating new systems and changing the nature of intersystemic relations is not directly represented in the predeterministic model, there appears to be no particular need to introduce methodological techniques for setting cognitive skills in their developmental contexts. Lacking such methods, evidence that might relate to the constructive mechanisms involved in epigenesis goes undetected. With no evidence of a constructive mechanism at work, innate determination of cognitive skills appears to be a logical alternative.

Perhaps the best contemporary example of this pattern is found in recent studies of infant cognition (Baillargeon, 1987; Baillargeon, Spelke, & Wassermann, 1985) and the nativist interpretations that they have typically been given (Spelke, 1988, 1991; Gelman 1991). These studies tend to follow a pattern in which an isolated behavior such as dishabituation to a given stimulus (looking longer at a new event than at a similar event that one is used to seeing) is taken a priori as an indication of a cognitive skill such as object permanence or number concept. Then task and assessment conditions are manipulated so as to find the earliest possible onset of this behavior. This, of course, is a valid methodology for finding the earliest age of onset for specific dishabituation skills. However, typically, there is no attempt made to pin down the meaning of an infant's dishabituation behavior by placing it in *developmental contexts* (Bidell & Fischer, 1992b; Fischer & Bullock, 1981), such as the range of ages at which the skill appears, the sequence of skills leading up to it, or the clusters of similar skills and behaviors that emerge concurrently. With no developmental history or context, there appears to be no constructive mechanism at work, making innate determination seem to be a reasonable alternative.

An example of this approach is the well-known study by Baillargeon (1987) in which $3\frac{1}{2}$–$4\frac{1}{2}$-month-old infants were habituated to seeing a door in front of them swing up away from them through a 180° arc and return with nothing to hamper it. They then witnessed two events involving a block behind the door. In one event, which would seem quite possible to an adult, a block is placed in the path of the door, and the door stops when it gets to the block (the block at this point is hidden by the door). The second event, which adults would consider impossible, was the same in all ways except that the door seemed to swing right through the space where the block had been (really, the block was surreptitiously lowered out of sight while the door hid it). Baillargeon found that the infants looked significantly longer at the second, impossible event.

Baillargeon (1987), Baillargeon, Spelke, and Wasserman (1985), and others (Gelman, 1991; Kellman & Spelke, 1983; Spelke, 1988, 1991) have inferred from this dishabituation behavior that 3–4-month-old infants possess a concept of object permanence similar to that which leads adults to view the second event as impossible. Based on this interpretation, these authors have claimed that the relatively early onset of this and similar behaviors is an indication that object concept is somehow innately determined. Similar claims have been advanced or implied for other concepts including concepts of Euclidean geometry (Landau, Spelke, & Gleitman, 1984) and number (Starkey & Cooper, 1980). Others, perhaps sensitive to the problems inherent in a strictly nativist stance, have advanced innate learning mechanisms in which concepts per se are not predetermined, but instead, the process by which they are learned is predetermined (e.g., Mandler, 1992) – a subtle distinction at best.

Typically, authors advancing these nativist positions have argued that their data reject Piaget's constructivist model, in which concepts like object permanence are constructed gradually through the coordination of sensorimotor systems. The claim is that since the dishabituation behaviors appear earlier than the behaviors Piaget took to represent concepts such as object permanence, Piaget's constructivist model of infant cognitive development is invalidated.

As surprising as it is to hear that infants possess innate conceptions of object permanence or Euclidean geometry, it can be difficult to challenge these interpretations because the methods employed do not include any means of placing the behaviors in the developmental contexts that would allow interpretations of their developmental relations to other skills and behaviors. One very direct way to contextualize these behaviors would be to include additional tasks to detect precursor and subsequent skills. By testing to see whether a behavior fits into a predicted sequence of developmental constructions, one can better determine whether the behavior is arising suddenly of its own accord or as the result of a constructive process.

For instance, Piaget's constructivist model posited a gradual series of reorganizations in the emergence of object concept resulting in a developmental sequence of changing ideas about objects during infancy. The classic skill of searching for hidden objects was a relatively late appearing and quite sophisticated version of object concept, constructed out of precursor skills and forming the basis of even more complex object-concept skills later. How do we know that the dishabituation behavior observed by Baillargeon does not just represent an earlier step in the constructive process Piaget outlined? It is difficult to answer this question because the dishabituation task is the only task used in the study. Taken out of the context of a development sequence, we have no means of gauging how the behavior exhibited on this task relates to developmentally prior or subsequent behaviors in a similar task context, or how the behavior relates to those shown at the same age on other tasks such as those designed to assess Piaget's model (e.g., Uzgiris & Hunt, 1987).

Another type of developmental context is the range of variability in the age at which a skill is typically manifested. It is well known that the age of acquisition of a given cognitive skill varies greatly with changes in task context or task complexity (Bidell & Fischer, 1992a; Gelman & Baillargeon, 1983). For example, the age at which children can perform number conservation tasks can vary by several years depending on how many objects are included in the sets, the nature of the objects used (e.g., counters vs. candy), and the familiarity of the task. Neonativist researchers commonly advance the view that when behaviors like the dishabituation to an apparent violation of object permanence occur at an earlier age than established norm, Piaget's constructivist model is somehow invalidated (e.g., Spelke, 1991). Fischer and Bidell (1991) have termed this the "argument from precocity" because it depends on viewing the behavior in question as a surprisingly precocious instance of a normally later-occurring skill. Yet how do we know that a seemingly early age of acquisition – really only a few months prior to the acquisition of full-blown search for hidden objects – is not within the normal range of variability for object permanence? How do we know that such procedures are not simply tapping into the skill at an earlier point in the constructive process?

Once again, the only way to answer this question is to introduce methods that allow us to place the observed behavior in context – in this case, in the context of the range of variation in the age at which the skill normally appears. To do this requires that relevant task and situational factors be *manipulated systematically*, to determine the upper and lower age limits. Typically, however, studies of this type manipulate the age of acquisition only in one direction – downward. Baillargeon's task, for example, was designed to be maximally supportive. The habituation procedure, by definition, familiarizes infants to the point where the situation no longer seems novel. Other supporting conditions included a mother comforting and holding an infant in just the right position to see the display so that it did not need to engage any activities that might compete for cognitive resources. Such conditions are perfectly appropriate for determining the lower end of the age range for a given skill. But this tells us only half the story. Unless the conditions are further varied so that we learn the full range of ages over which the skill is acquired, there is no valid standard by which to judge whether this behavior is early or not. Schlesinger and Langer (1994), for instance, found that on a similar possible–impossible task in which there was no habituation to the situation, and thus less supportive conditions, even 10-month-old infants did not look longer at the event that violated object permanence.

By manipulating the age of acquisition maximally downward, without the complementary conditions to probe its upper ranges, researchers artificially isolate behaviors at the lower end of their age range, creating the appearance that a cognitive skill has emerged out of nowhere, and inviting a predeterministic interpretation of genetic involvement. To tell the whole story requires a systematic approach to uncovering the developmental continuity of the skill across age. When behaviors are thus contextualized, they no longer appear precocious, and the apparent need for predeterministic interpretations is removed.

Finally, a third type of developmental context involves the relations of a behavior or skill of interest with comparable skills and behavior developing around the same time. Although cognitive skills are not organized into unified stages, children nevertheless tend to construct similar types of skills in a given domain around the same time. Thus, visual search tends to appear around the same time as tactile search. Often, when questions arise as to the meaning of a behavior in one situation, setting it in the context of the range of contemporary behaviors that a person has constructed will reveal patterns of similarity and difference that help to interpret the behavior.

However, in recent nativist interpretations of infant behavior, this form of developmental context has almost always been neglected. Since the tasks involved, like the one described earlier in Baillargeon's experiment, are novel and involve unusual conditions, this type of cross-comparison would seem to be particularly relevant. But instead of systematically comparing behavior across a range of tasks and conditions, most of the research on which nativist claims are being based reproduces key aspects of unusual tasks and conditions found in Baillargeon's study. These include (1) tasks involving highly unusual stimuli which create illusions that must be interpreted by an infant, (2) only a single sensory modality is used to gather information about the unusual stimulus – usually visual, and seldom demanding coordinated modalities such as visual/kinesthetic, (3) only a single response mode – simply looking at the target stimulus, and (4) habituation to the main features of the task situation. How do infants' reactions under these highly specialized and restricted conditions relate to the abilities they display across a range of differently structured situations? Without this contextual information, infant behavior is, once again, isolated from its developmental context, making a nativist interpretation seem more plausible than it might be if we knew the full range of an infant's capabilities at a given point in development.

As Fischer and Bidell (1991) have argued, once the relevant developmental contexts have been considered in relation to the phenomena that give rise to nativist interpretations, there remains no compelling evidence of innately determined concepts. When behaviors are placed in a context in which their developmental relations to other concepts can be seen, a range of constructivist hypotheses present themselves. In the appropriate developmental context, it is possible to see how seemingly precocious and apparently unrelated behaviors can be understood as part of a constructive integrative process of self-organization and constructive epigenesis. All of infant cognition was not explained by Piaget, and developmental psychology clearly needs the innovative type of work now being done in infant cognitive development. But in interpreting new findings from new approaches, we must set them firmly in the context of their developmental relations with known phenomena or risk losing the trail that leads to a fuller understanding of the dimensions of cognitive epigenesis.

Finding the path

How can we design research to answer the questions we have raised about nativist claims, and more generally, how can we study the process of cognitive epigenesis from a constructive instead of predeterministic perspective? It seems to us that the answer lies in rethinking the way developmental variability has been treated in the nature–nurture framework. From the nature–nurture point to view, all variability in behavioral outcomes is ultimately attributable to variations in genetics or environment: Differences across individuals in any concept or skill must be traced back to some source in their genetic code or to some impinging events around them. But in the constructivist view, developmental variability is primarily due to the *dynamics of self-organizing activities.* Genetic and environmental systems participate, but their participation is directed and organized through the activity of hierarchical integration that creates new contexts of participation and new functional relations among cognitive, biological, and social systems.

In this view, the key to the study of developmental variability is the description of the series of changes in interrelations among systems – the "trail" of the constructive agent – in the process of constructing cognitive skills in social and cultural contexts. And the primary method for detecting this trail is to place cognitive skills and related behaviors in their developmental context. To begin to understand the actual role of genetic and environmental activity in the emergence of a new cognitive skill, we must first be able to locate that skill in its constructive relation to other cognitive skills. Only when we can say with confidence that we know the developmental history and trajectory of a given skill, as well as its relations with contemporary skills in other task contexts and domains, can we begin to form objective hypotheses about its relations with genetic or environmental systems.

To form hypotheses about the modes of participation of genetic or environmental systems in the emergence of object permanence, we must have information about the range of developmental variability associated with that skill. As the previous section suggested, we need to be able to determine with confidence the range of ages at which it is acquired, the range of other skills that typically emerge with it, the range of contexts and task conditions across which it appears, and the range of developmental precursors and subsequently emerging object-permanence skills. With this information, variability in object-permanence skills can be related to variability in biological and environmental systems to form hypotheses about the nature of each system's contribution to the emergence of object permanence. In the remainder of this section, we describe some general tools and principles for picking up the trail of constructive integrations by placing cognitive skills in a multidimensional developmental context.

Developmental web. One of the most useful tools we have discovered for placing behaviors in a developmental context is a conceptual one. Typically, development is conceived of as a ladder of some kind, with a fixed sequence

of steps. Real development, however, does not conform to this uni-
dimensional image. Because development takes place in multidimensional
contexts, the process of analyzing the roles of differing context in development is facilitated by thinking of development in terms of a web instead of
a ladder (Bidell & Fischer, 1992a,b). A web supports thinking about development as the product of a constructive agent, actively making new connections. The sequential aspect of these constructions is represented, but so too
is the potential of multidirectionality in developmental pathways that arises
through the interparticipation of other systems including social and biological. Thinking of development as a weblike process of construction instead
of a ladderlike trek along a fixed sequence helps to support thinking about
research as a process of detecting constructive activity within a multidimensional developmental context.

Detecting developmental sequences.   One fundamental dimension of developmental context is developmental sequence. Since the construction of new
skills involves the integration of prior-level component skills, the constructive agent always leaves a trail in the form of a developmental sequence.
When researchers can show that a behavior such as a particular type of
visually guided reaching forms a step in a developmental sequence with
other behaviors, the sequence aids them in interpreting the behavior. Placing the behavior in the sequence circumvents fruitless arguments about
whether a particular behavior really demonstrates a particular capacity.
Each successive step in the sequence is a further realization of the capacity,
which emerges in the gradual manner typical of epigenesis.

Infants show a regular developmental sequence of visually guided
prereaching in the situation of sitting upright facing a ball (Bruner &
Koslowski, 1972; von Hofsten, 1984, 1994). The work of Piaget (1937/1954)
and many others on object permanence documents a sequence for more
advanced, visually guided reaching up to 2 years of age (Diamond, 1991;
Uzgiris & Hunt, 1975). Fischer and Hogan (1989) present a framework for
describing and integrating these sequences into an epigenetic portrait for
visually guided reaching.

This framework moves beyond arguments about whether a single isolated behavior demonstrates the presence of an underlying capacity. Instead, the behavior is interpreted as fitting a point into the pattern of a
developmental context for visually guided reaching. A general outline of
the developmental sequence for visually guided reaching in middle-class
American and European infants runs as follows.

1. At 1 month of age, infants are able to look at an object in front of
them if they are facing it and their posture does not prevent looking.
Likewise, they grasp a ball placed in their hand (Thelen & Fogel, 1989). By

approximately 2 months, infants sometimes show a simple reaching behavior: They look at the ball and produce a coarse reach, in which they move their arm toward it with their fist closed (von Hofsten, 1984). Infants at this age can often grope toward a parent's face, but only if the face is located in a position in which the infant's arm happens to be moving, and only if the infant is in a position that supports the movement. Thus, in a highly constrained situation, a limited type of visually guided reaching is possible, but the behavior occurs only with strong contextual supports.

2. At 3 months, infants are able to open their hands while extending their arms toward objects, but they lack accuracy in their attempts to grasp the object. By 5 months, many infants can reach adeptly for familiar objects but still flail awkwardly at unfamiliar objects. Over the next few months, they gradually generalize and elaborate these skills until by 7 to 8 months, they can flexibly use looking at an object to guide how they reach for it from virtually any position and without support. At this point, they even begin to search for objects hidden under cloths or behind screens (Diamond, 1991; Piaget, 1937/1954).

3. Of course, even this advanced capacity is not the end of the developmental sequence. By 12 to 13 months, babies become much more facile with coordinating vision and prehension. They can skillfully pick up a ball and move it around, using what they see to guide what they do and anticipating many of the consequences of moving the ball a particular way. They even carry out little experiments in action relating how they move the ball with what they see so as to determine how to accomplish special goals, such as dropping the ball through a small hole in a box. This skill continues to grow for years to come.

As these examples demonstrate, placing individual behaviors in the framework of a developmental sequence illuminates their significance. Each behavior becomes a step in a pathway toward a broad capacity. No particular behavior is treated as the one, true demonstration of the capacity. Arguments among researchers about which behavior indeed shows the capacity and when the capacity first develops can be eliminated.

Defining developmental range. Another dimension of developmental variability is in the age of acquisition for a given skill or behavior. The age of acquisition typically varies through a considerable range depending on conditions under which it is assessed. It is useful to recognize this principle of age variability by naming it the *developmental range* (Lamborn & Fischer, 1988). Because the age of acquisition is highly sensitive to assessment conditions, the developmental range is a useful tool for contextualizing emerging cognitive skills. Instead of characterizing a skill as simply present or absent, it is possible to describe the conditions under which the age varies through the developmental range. Moreover, because assessment conditions include relations with both biological and social systems, a systematic description of the way in which age of acquisition behaves under differing conditions provides a useful tool for generating hypotheses about the developmental relations between the emergence of new skills and biological and social systems.

The developmental range for a given individual may be defined in terms of an optimal level and a functional level of cognitive performance. The functional level is the developmental level at which the individual functions independently, without a supportive social context. The optimal level is the highest developmental level that a person is capable of in a given task under optimum social-support conditions. A key to interpreting any behavior is determining its relative location between what is functional and what is optimal for a person at a given moment in development. With reference to the development of visually guided reaching described earlier, it is easy to see how a 2-month-old's early groping toward a face, which succeeds under highly supportive and specialized conditions, could be mistaken for the mature visually guided reaching of the 8-month-old if we fail to recognize that the former skill represents an optimal level and the latter, a functional level. By failing to control systematically for factors contributing to variation in the age at which behaviors are manifested, it is easy to create the mistaken impression that a given skill is emerging much earlier than it was previously believed to appear.

Testing for developmental synchronies. Besides the vertical frameworks of developmental sequence and developmental range, the interpretation of a behavior is also aided by relating it to a horizontal framework – the other behaviors that an infant demonstrates at about the same age. Although children show only a few tight synchronies in development at any given age, there are general correspondences between behaviors that develop on average at about the same time (Fischer & Bullock, 1981; Fischer & Farrar, 1987). Developmental correspondences help the investigator to place individual behaviors in a developmental context of comparable skills. If children consistently acquire skills for performing certain specific tasks all at about the same time, there is good reason to believe that these tasks tap into skills emerging around that age. By examining what the range of tasks have in common, we may formulate hypotheses about the nature of the emerging skill.

Finding developmental clusters. Another tool for detecting developmental relations is finding clusters of stable behaviors that change together under changing contextual conditions. As relevant conditions vary, some behaviors move together, either remaining tied in a specific sequence or remaining in close synchrony even though the age of emergence for each skill in the cluster has changed. Such developmental clusters provide important evidence about the developmental relations among the behaviors.

One of the best examples involves the onset of crawling at about 8 months of age. When children coordinate the means to crawl, it seems to induce development of a cluster of spatial skills, including visually guided search, according to Campos (Campos & Bertenthal, 1987; Campos et al., 1992). In normally developing infants, the appearance of upright crawling appears to induce advances in skills of searching for hidden objects and appreciating the danger of heights, among others. A cluster of these behaviors seems to develop a few weeks after infants can crawl on their hands and knees.

Various studies converge on the conclusion that it is the experience of crawling itself that induces the new behaviors. For example, upright crawling – on the hands and knees – seems to be necessary to bring about the change. Cruder forms of crawling, such as dragging oneself along on the belly, do not seem to have the same effect. Likewise, handicaps that prevent crawling, such as spina bifida and orthopedic problems, prevent the development of these spatial skills in infants who are otherwise cognitively normal. The handicapped infants do not demonstrate the relevant spatial skills until they have developed crawling, even when the delay involves several months.

The development of a cluster of spatial behaviors after the emergence of crawling suggests that the behaviors are all activated by the experiences that arise from crawling. The covariation of the behaviors with crawling thus helps to constrain inferences about how the behaviors develop. This kind of self-activation of skills is a central component in epigenesis and is a crucial part of a number of models of how innate factors affect development (Changeux, 1983/1985; Gottlieb, 1983; Marler, 1991).

Mapping developmental pathways. A final dimension of developmental variability is developmental pathways – what Werner (1957) called *multilinearity in development.* As indicated by the metaphor of the developmental web discussed earlier, different individuals, constructing a given type of cognitive skill in different context, may follow quite different pathways in this constructive activity, depending on the nature of the intersystemic contexts (Bidell & Fischer, 1992a,b). An important tool in describing developmental relations among subsystems is to trace alternative developmental pathways as they vary across individuals and situations.

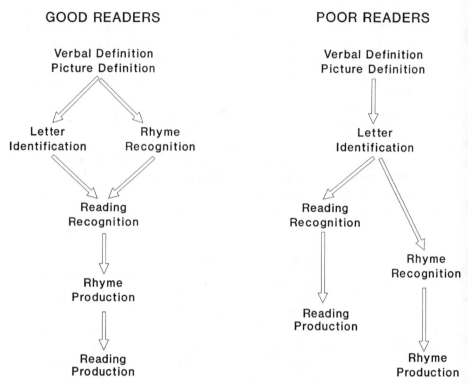

Figure 7.4. Dendrograms showing the alternative developmental pathways constructed by two groups of children en route to the skill of word reading. *Note*: These
diagrams show only order relations: Differences in the distance between skills in the
two groups do not indicate whether the skill was acquired sooner or later in one
group. Also, note that this figure reads from top to bottom, consistent with the
convention for this type of analysis.)

An example comes from a study of the emergence of reading skills
(Fischer & Knight, 1990). Poor readers in first to third grade were found to
follow different developmental pathways from normal readers through a
set of reading-related skills. For each group, the order of acquisition for six
reading tasks was tested using a statistical technique called *partially ordering scaling* that is based on the logic of Guttman scaling (Kuleck & Fischer,
1989; Krus, 1977). The results are presented as dendrograms in Figure 7.4.
Tasks acquired first are shown at the top of the sequence, and later acquisitions are shown below them. A line between two tasks means that the
ordering is statistically reliable, and tasks that are parallel but have no lines
between them are acquired at about the same time. A comparison of the
two developmental pathways shows that the poor readers are not simply

delayed with respect to a universal sequence of acquisitions but actually follow *different* pathways in acquiring these skills. Normal readers all showed one common pathway, but poor readers showed three different pathways (only one of these is shown in Figure 7.4), all different from the standard one.

Detecting this kind of variation in developmental pathways makes it possible to ask informed questions about the factors, both biological and social, that might be affecting the order of acquisition of these tasks. The assumption of a single ladderlike sequence of acquisitions can obscure possible relations between the construction of reading skills and factors, such as the relative rates of development of perceptual analysis skills on one hand, or cultural/linguistic contexts on the other. Lacking this evidence of variability, it would be easier to conclude that there exists only one, maturationally controlled pathway in the development of reading skills. Placing the children's development of reading skills in the context of a web of possible developmental pathways highlights the variability that can arise when skills are constructed by different individuals in different contexts. This approach makes it possible to generate and test hypotheses about developmental relations between constructive processes, perceptual-linguistic processes, cognitive levels, and sociocultural context, instead of simply attributing developmental outcomes to a predetermined maturational timetable.

Conclusion

If the question of how heredity and environment influence intelligence has met with little consensus in the more than three decades since Anastasi (1958) brought it to center stage of the nature–nurture debate, perhaps it is because the very framework of that debate has precluded a satisfying answer. By artificially partitioning all variability in intelligent behavior into just two mutually exclusive sources, the traditional nature–nurture framework defines the problem in a way that excludes the integrative role of constructive activity, lending plausibility to a predeterministic model of epigenetic mechanisms. Despite the inability of the predeterministic model to account for the ever-growing volume of findings in molecular genetics, developmental biology, and developmental psychology, the dominance of the nature–nurture framework has sustained the idea of a preset maturational program with environmental inputs as the default model of the epigenesis of human intelligence.

In this chapter, we have advanced one version of a growing alternative framework (Bronfenbrenner & Ceci, 1993; Gottlieb, 1970, 1992; Lerner, 1980, 1991; Oyama, 1985) for understanding the role of genes and environment in the development of intelligence. In the alternative framework, genetics and environment are viewed not as separate, independent forces influencing intelligence but as intrinsically related integrative systems that take part in and are parts of the developing person. Change in such systems is a product of self-organizing activity; for this reason, the epigenesis of intelligence is a constructive, not predetermined, process. Instead of genes and environment either singly or jointly determining cognitive outcomes, we – as human agents – determine our own cognitive outcomes as we actively make sense out of our world and build skills for participation in it. The creative construction of new concepts and cognitive skills through acts of hierarchical integration produces new contexts for the participation of both biological and environmental systems, providing a concrete vehicle for their operation in development, and setting the direction and conditions for their activities.

From the point of view of constructive epigenesis, the structure, sequence, timing, and pathways of intellectual development are all jointly determined by the constructive activity of many participating systems and are directed by the construction of new cognitive skills in sociocultural contexts. For this reason, cognitive skills vary in specific ways that are consistent with the particular history and context of a given developmental process. This variability due to the constructive dynamics of integrative systems is the key to detecting the mechanisms of cognitive epigenesis by uncovering the changing developmental relations among subsystems in the wake of constructive activity. Instead of arbitrarily partitioning variability into two mutually exclusive categories – nature and nurture – the goal in this approach is to use developmental variability to tease out the developmental relations among the contributing systems at the biological, cognitive-behavioral, and sociocultural levels of analysis. Research questions about how much of a given behavior is due to genetics or environment, or how early we can discover behaviors related to a given cognitive skill, are misleading because they direct attention away from the mechanisms by which cognitive skills are constructed. Instead, contextualizing methods allow researchers to follow the trail of constructive agency in its weaving together of biological and sociocultural systems. Only by setting cognitive skills in their developmental context, and by describing the full range of variability in the emergence of new skills, can we begin to make objective assessments of the mechanisms in which co-participating systems influence the course of cognitive construction.

Between nature and nurture stands the human agent whose unique integrative capacities drive the epigenesis of intelligence and organize biological and environmental contributions to the process. For too long the role of human agency in the origins of human intellectual abilities has been obscured by the dominance of the nature–nurture framework in the scientific discourse on this topic. Progress in understanding mechanisms in the epigenesis of intelligence now depends on our ability and willingness to move beyond the nature–nurture framework and take advantage of the growing repertoire of constructs and methods for the study of cognitive epigenesis as a constructive process.

Acknowledgments

Preparation of this chapter was supported by grants from Mr. and Mrs. Frederick P. Rose, the Spencer Foundation, Boston College, and Harvard University. The authors thank Daniel Bullock, Nira Granott, Samuel P. Rose, Robert Thatcher, and Paul van Geert for their contributions to our development of the arguments and evidence presented in this chapter.

References

Anastasi, A. (1958). Heredity, environment, and the question "how?" *Psychological Review*, *65*, 197–208.

Baillargeon, R. (1987). Object permanence in $3\frac{1}{2}$- and $4\frac{1}{2}$-month-old infants. *Developmental Psychology*, *23*, 655–64.

Baillargeon, R., Spelke, E. S., & Wasserman, S. (1985). Object permanence in five-month-old infants. *Cognition*, *20*, 191–208.

Bandura, A. (1977). Self-efficacy: Towards a unifying theory of behavior. *Psychological Review*, *84*, 118–42.

Bidell, T. R. (1988). Vygotsky, Piaget, and the dialectic of development. *Human Development*, *31*, 329–48.

Bidell, T. R. (1990). *Mechanisms of cognitive development during problem-solving: A structural integration approach*. Unpublished doctoral dissertation, Harvard University, Cambridge, MA.

Bidell, T. R., & Fischer, K. W. (1992a). Beyond the stage debate: Action, structure, and variability in Piagetian theory and research. In C. A. Berg & R. J. Sternberg (Eds.), *Intellectual development*. New York: Cambridge University Press.

Bidell, T. R., & Fischer, K. W. (1992b). Cognitive development in educational contexts: Implications of skill theory. In A. Demetriou, M. Shayer, & A. Efklides (Eds.), *The neo-Piagetian theories of cognitive development: Implications and applications for education* (pp. 11–30). London: Routledge & Kegan Paul.

Bidell, T. R., & Fischer, K. W. (1994). Developmental transitions in children's early on-line planning. In M. M. Haith, J. B. Benson, R. J. Roberts Jr., & B. F. Pennington (Eds.), *The development of future oriented processes* (pp. 141–76). Chicago: University of Chicago Press.

Bronfenbrenner, U. (1979). *The ecology of human development*. Cambridge, MA: Harvard University Press.

Bronfenbrenner, U., & Ceci, S. (1993). Heredity, environment, and the question "how" – A first approximation. In R. Plomin & G. E. McClearn (Eds.), *Nature, nurture and psychology* (pp. 313–24). Washington, DC: American Psychological Association.

Bruner, J. S. (1973). Organization of early skilled action. *Child Development*, *44*, 1–11.

Bruner, J. S. (1982). The organization of action and the nature of adult–infant transaction. In

M. Cransch & R. Harré (Ed.), *The analysis of action* (pp. 280–96). New York: Cambridge University Press.

Bruner, J. S. (1990). *Acts of meaning.* Cambridge, MA: Harvard University Press.

Bruner, J. S., & Koslowski, B. (1972). Visually preadapted constituents of manipulating action. *Perception, 1,* 3–14.

Campos, J. J., & Bertenthal, B. I. (1987). Locomotion and psychological development in infancy. In F. Morrison, K. Lord, & D. Keating (Eds.), *Advances in applied developmental psychology* (Vol. 2) (pp. 11–42). New York: Academic Press.

Campos, J. J., Bertenthal, B. I., & Kermoian, R. (1992). Early experience and emotional development: The emergence of wariness of heights. *Psychological Science, 3,* 61–4.

Case, R. (1985). *Intellectual development: Birth to adulthood.* New York: Academic Press.

Changeux, J-P. (1985). *Neural man: The biology of mind* (L. Garey, Trans.) New York: Oxford University Press. (Original work published 1983)

Changeux, J.-P., & Danchin, A. (1976). Selective stabilization of developing synapses as a mechanism for the specification of neuronal networks. *Nature, 264,* 705–12.

Cook-Gumperz, J. (1981). Persuasive talk – The social organization of children's talk. In J. L. Green & C. Wallat (Eds.), *Ethnography and language in educational settings* (pp. 13–24). Norwood, NJ: Ablex.

Dawkins, R. (1976). *The selfish gene.* New York: Oxford University Press.

Diamond, A. (1991). Neuropsychological insights into the meaning of object concept development. In S. Carey & R. Gelman (Eds.), *The epigenesis of mind: Essays on biology and cognition* (pp. 67–110). Hillsdale, NJ: Erlbaum.

Diaz, R. M., Neal, C. J., & Amaya-Williams, M. (1990). The social origins of self-regulation. In L. C. Moll (Ed.), *Vygotsky and education: Instructional implications and applications of sociohistorical psychology* (pp. 127–54). New York: Cambridge University Press.

Dunn & Plomin (1990). *Separate lives: Why siblings are so different.* New York: Basic Books.

Edelman, G. M. (1987). *Neural Darwinism.* New York: Basic Books.

Emde, R. N., Gaensbauer, T. J., & Harmon, R. J. (1976). Emotional expression in infancy: A biobehavioral study. New York: International Universities Press.

Erikson, E. H. (1964). *Childhood and society.* New York: Norton.

Fischer, K. W. (1980a). A theory of cognitive development: The control and construction of hierarchies of skills. *Psychological Review, 87,* 477–531.

Fischer, K. W. (1980b). Learning and problem solving as the development of organized behavior. *Journal of Structural Learning, 6,* 253–67.

Fischer, K. W. (in press). From alien dynamics to constructive dynamics: Explaining hierarchical cognitive development. In E. Amsel & K. A. Reninger (Eds.), *Mechanisms of change in development.* Hillsdale, NJ: Erlbaum.

Fischer, K. W., & Ayoub, C. (1993). Affective splitting and dissociation in normal and maltreated children: Developmental pathways for self in relationships. In D. Cicchetti & S. Toth (Eds.), *Rochester symposium on developmental psychopathology: Vol. 5. The self and its disorders* (pp. 149–222). Rochester, NY: The University of Rochester Press.

Fischer, K. W., & Bidell, T. R. (1991). Constraining nativist inferences about cognitive capacities. In S. Carey & R. Gelman (Eds.), *The epigenesis of mind: Essays on biology and cognition* (pp. 199–235). Hillsdale: NJ: Erlbaum.

Fischer, K. W., & Bullock, D. (1981). Patterns of data: Sequence, synchrony, and constraint on cognitive development. In K. W. Fischer (Ed.), *Cognitive development* (New Directions for Child Development, no. 12, pp. 69–78). San Francisco, CA: Jossey-Bass.

Fischer, K. W., Bullock, D. H., Rothenberg, E. J., & Raya, P. (1993). The dynamics of competence: How context contributes directly to skill. In R. Wozniak & K. W. Fischer (Eds.), *Development in context: Acting and thinking in specific environments* (pp. 93–117). Hillsdale, NJ: Erlbaum.

Fischer, K. W., & Elmendorf, D. (1986). Becoming a different person: Transformations in personality and social behavior. In M. Perlmutter (Ed.), *Minnesota symposium on child psychology* (Vol. 18) (pp. 137–78). Hillsdale, NJ: Erlbaum.

Fischer, K. W., & Farrar, M. J. (1987). Generalizations about generalization: How a theory of skill development explains both generality and specificity. *International Journal of Psychology*, 22, 643–77.

Fischer, K. W., Hand, H. H., & Russell, S. L. (1984). The development of abstractions in adolescence and adulthood. In M. Commons, F. A. Richards, & C. Armon (Eds.), *Beyond formal operations* (pp. 43–73). New York: Praeger.

Fischer, K. W., & Hogan, A. E. (1989). The Big picture of infant development: Levels and variations. In J. J. Lockman & N. L. Hazen (Eds.), *Action in social context: Perspectives on early development*. New York: Plenum Press.

Fischer, K. W., & Knight, C. C. (1990). Cognitive development in real children: Level and variations. In B. Presseisen (Ed.), *Styles of learning and thinking: Interaction in the classroom* (pp. 43–67). Washington, DC: National Education Association.

Fischer, K. W., & Rose, S. P. (1994). Dynamic development of coordination of components in brain and behavior: A framework for theory and research. In G. Dawson & K. W. Fischer (Eds.), *Human behavior and the developing brain* (pp. 3–66). New York: Guilford Press.

Fischer, K. W., Shaver, P., & Carnochan, P. (1990). How emotions develop and how they organize development. *Cognition and Emotion*, 4, 81–127.

Fischer, K. W., & Silvern, L. (1985). Stages and individual differences in cognitive development. *Annual Review of Psychology*, 36, 613–48.

Gardner, H. (1985). *The mind's new science: A history of the cognitive revolution*. New York: Basic Books.

Geertz, C. (1973). *The interpretation of cultures*. New York: Basic Books.

Gelman, R. (1991). Epigenetic foundations of knowledge structures: Initial and transcendent constructions. In S. Carey & R. Gelman (Eds.), *The epigenesis of mind: Essays on biology and cognition*. Hillsdale, NJ: Erlbaum.

Gelman, R., & Baillargeon, R. (1983). A review of some Piagetian concepts. In P. H. Mussen (Ed.), *Handbook of child psychology: Vol. 3. Cognitive development* (J. H. Flavell & E. M. Markman, Eds.) (pp. 167–230). New York: Wiley.

Gesell, A. (1946). The ontogenesis of infant behavior. In L. Carmichael (Ed.), *Manual of child development*. New York: Wiley.

Gottlieb, G. (1970). Conceptions of pre-natal behavior. In L. R. Aronson (Ed.), *Development and evolution of behavior*. San Francisco: W. H. Freeman.

Gottlieb, G. (1983). The psychobiological approach to developmental issues. In M. M. Haith & J. J. Campos (Eds.), *Infancy and developmental psychobiology* (Vol. 2) (pp. 1–26). New York: Wiley.

Gottlieb, G. (1992). *Individual development and evolution: The genesis of novel behavior*. New York: Oxford University Press.

Gottlieb, D. E., Taylor, S. E., & Ruderman, A. (1977). Cognitive bases of children's moral judgments. *Developmental Psychology*, 13, 547–56.

Granott, N. (1991). Puzzled minds and weird creatures: Phases in spontaneous process of knowledge construction. In I. Harel & S. Papert (Eds.), *Constructionism*. Norwood, NJ: Ablex.

Hall, G. S. (1904). *Adolescence: Its psychology and its relations to physiology, anthropology, sociology, sex, crime, religion, and education* (2 vols.). New York: Appleton.

Herman, J. (1992). *Trauma and recovery*. New York: Basic Books.

Hirsch, J. (1990). A nemesis for heritability estimation. *Behavioral and Brain Sciences*, 13, 137–8.

Ho, M-W. (1983). Effects of successive generations of ether treatment on penetrance and expression of the biothorax phenocopy in *Drosopila melanogaster*. *Journal of Experimental Zoology, 225*, 357–68.

Ho, M-W., & Saunders, P. T. (1984). *Beyond neo-Darwinism: An introduction to the new evolutionary paradigm*. London: Academic Press.

Hubel, D. H., & Wiesel, T. N. (1977). Functional architecture of macaque monkey visual cortex. *Procedures of the Royal Society, London, Series B, 193*, 1–59.

Kagan, J., Kearsley, R., & Zelazo, P. R. (1978). *Infancy: Its place in human development*. Cambridge, MA: Harvard University Press.

Kellman, P., & Spelke, E. S. (1983). Perception of partly occluded objects in infancy. *Cognitive Psychology, 15*, 483–524.

Krus, D. J. (1977). Order analysis: An inferential model of dimensional analysis and scaling. *Educational and Psychological Measurement, 37*, 587–601.

Kuleck, W., & Fischer, K. W. (1989). *Partially ordered scaling of items*. Cambridge, MA: Cognitive Development Laboratory.

Lamborn, S. D., & Fischer, K. W. (1988). Optimal and functional levels in cognitive development: The individual's developmental range. *Newsletter of the International Society for the Study of Behavioral Development*, No. 2, Serial No. 14, 1–4.

Landau, B., Spelke, E. S., & Gleitman, H. (1984). Spatial knowledge in a young blind child. *Cognition, 16*, 225–60.

Lerner, R. M. (1980). Concepts of epigenesis: Descriptive and explanatory issues. *Human Development, 23*, 63–72.

Lerner, R. M. (1991). Changing organism-context relations as the basic process of development: A developmental contextual perspective. *Developmental Psychology, 27*, 27–32.

Lerner, R. M. (1993). *The demise of the nature–nurture dichotomy. Human Development, 36*, 119–24.

Levins, R., & Lewontin, R. (1985). *The dialectical biologist*. Cambridge, MA: Harvard University Press.

Mandler, J. M. (1992). Foundations of conceptual thought in infancy. *Cognitive Development, 7*, 273–85.

Marler, P. (1991). The instinct to learn. In. S. Carey & R. Gelman (Eds.), *The epigenesis of mind: Essays on biology and cognition*. Hillsdale, NJ: Erlbaum.

Mehan, H. (1979). *Learning lessons: Social organization in the classroom*. Cambridge, MA: Harvard University Press.

Molenaar, P. C. M., Raijmakers, M. E. J., & Hartelman, P. (1994 June 2–4). *Some extensions of the paradigm for non-linear epigenetic process*. Paper presented at the Twenty-Fourth Annual Symposium of the Jean Piaget Society, Chicago.

Overton, W. F. (1973). On the assumptive base of the nature–nurture controversy: Additive versus interactive conceptions. *Human Development, 16*, 74–89.

Overton, W. F. (1994, June 2–4). *Toward a relational theory of change: Healing the biological split*. Paper presented at the Twenty-Fourth Annual Symposium of the Jean Piaget Society, Chicago.

Oyama, S. (1985). *The ontogeny of information: Developmental systems and evolution*. Cambridge: Cambridge University Press.

Perry, M., Church, R. B., & Goldin-Meadow, S. (1988). Is gesture/speech mismatch a general index of transitional knowledge? *Cognitive Development, 3*, 359–400.

Piaget, J. (1952). *The origins of intelligence in children* (M. Cook, Trans.). New York: International Universities Press. (Originally published 1936)

Piaget, J. (1954). *The construction of reality in the child* (M. Cook, Trans.). New York: Basic Books. (Originally published 1937).

Piaget, J. (1960). *The psychology of intelligence* (M. Piercy & D. E. Berlyne, Trans.). Totowa, NJ: Littlefield, Adams & Co. (Originally published 1947)

Piaget, J. (1977). The role of action in the development of thinking (H. Furth, Trans.). In W.

Overton & J. M. Gallagher (Eds.), *Knowledge & development, Vol. 1: Advances in research and theory*. New York: Plenum.

Plomin, R., Corley, R., DeFries, J. C., & Fulker, D. W. (1990). Individual differences in television viewing in early childhood: Nature as well as nurture. *Psychological Science, 1*, 371–7.

Resnick, L. (1985). Constructing knowledge in school. In L. Liben (Ed.), *Development and learning: Conflict or congruence* (pp. 19–50)? Hillsdale, NJ: Erlbaum.

Roberts, R. J., Jr. (1981). Errors and the assessment of cognitive development. In K. W. Fischer (Ed.), *Cognitive development* (New Directions for Child Development, no. 12, pp. 69–78). San Francisco, CA: Jossey-Bass.

Rogoff, B. (1990). *Apprenticeship in thinking*. New York: Oxford University Press.

Rogoff, B., Baker-Sennett, J., & Matusov, E. (1994). Considering the concept of planning. In M. M. Haith, J. B. Benson, R. J. Roberts, Jr., & B. F. Pennington (Eds.), *The development of future oriented processes*. (pp. 353–73). Chicago: University of Chicago Press.

Sameroff, A. (1975). Transactional models in early social relations. *Human Development, 18*, 65–79.

Schlesinger, M., & Langer, J. (1994 June 2–4). *Perceptual and sensori-motor causality in 10-month-old infants*. Paper presented at the Twenty-Fourth Annual Symposium of The Jean Piaget Society, Chicago.

Selman, R. L., & Schultz, L. H. (1990). *Making a friend in youth*. Chicago: University of Chicago Press.

Siegler, R. S. (1981). Developmental sequences within and between concepts. *Monographs of the Society for Research in Child Development, 46*(2, Serial No. 189).

Singer, W. (1979). Neuronal mechanisms in experience-dependent modification of visual cortex functions. In M. Cuenod, F. Bloom, & G. Kreutzberg (Eds.), *Progress in Brain Research*. Amsterdam: Elsevier Sequoia.

Spelke, E. S. (1988). Where preceiving ends and thinking begins: The apprehension of objects in infancy. In A. Yonas (Ed.), *Perceptual development in Infants* (The Minnesota Symposia on Child Psychology, Vol. 20). Hillsdale, NJ: Erlbaum.

Spelke, E. S. (1991). Physical knowledge in infancy: Reflections on Piaget's theory. In S. Carey & R. Gelman (Eds.), *The epigenesis of mind: Essays on biology and cognition*. Hillsdale, NJ: Erlbaum.

Starkey, P., & Cooper, R. G. (1980). Perception of numbers by infants. *Science, 210*, 1033–5.

Thelen, E., & Fogel, A. (1989). Toward an action-based theory of infant development. In J. Lockman & N. Hazan (Eds.), *Action in social context: Perspectives on early development* (pp. 23–63). New York: Plenum.

Thelen, E., & Smith, L. B. (1994). *A dynamic systems approach to the development of cognition and action*. Cambridge, MA: MIT Press.

Uzgiris, I. C., & Hunt, J. McV. (1987). *Assessment in infancy: Ordinal scales of psychological development*. Urbana, IL: University of Illinois Press.

van Geert, P. (1994). *Dynamic systems of development: Change between complexity and chaos*. London: Harvester Wheatsheaf.

von Bertalanfy, L. (1968). *General system theory*. New York: Braziller.

von Hofsten, C. (1984). Developmental changes in the organization of prereaching movements. *Developmental Psychology, 20*, 378–88.

von Hofsten, C. (1994). Planning and perceiving what is going to happen. In M. Haith, J. Benson, B. Pennington, & R. Roberts Jr. (Eds.), *The development of future-oriented processes* (pp. 63–86). Chicago: The University of Chicago Press.

Vygotsky, L. S. (1987). Thinking and speech. In R. W. Riever & A. S. Carton (Eds.), *The collected works of L. S. Vygotsky. Vol. 1: Problems of general psychology* (N. Minick,

Trans.). New York: Plenum. (Original work published 1934)

Watson, M. W., & Fischer, K. W. (1980). Development of social roles in elicited and spontaneous behavior during the preschool years. *Developmental Psychology, 16,* 484–94.

Weiss, P. A. (1959). Cellular dynamics. *Reviews of Modern Physics, 31,* 11–20.

Weiss, P. A. (1970). The living system: Determinism stratified. In A. Koestler & J. Smythies (Eds.), *Beyond reductionism: New perspectives in the life sciences* (pp. 3–55). New York: Macmillan.

Werner, H. (1948). *Comparative psychology of mental development.* New York: International Universities Press.

Werner, H. (1957). The concept of development from a comparative and organismic point of view. In D. B. Harris (Ed.), *The concept of development: An issue in the study of human behavior* (pp. 78–108). Minneapolis: University of Minnesota Press.

Wertsch, J. V. (1985). *Vygotsky and the social formation of mind.* Cambridge, MA: Harvard University Press.

Wertsch, J. V., McNamee, G. D., McLane, J. G., & Budwig, N. (1980). The adult–child dyad as a problem solving system. *Child Development, 51,* 1215–21.

Westerman, M. A. (1990). Coordination of maternal directives with preschoolers' behavior in compliance problem and healthy dyads. *Developmental Psychology, 26,* 621–30.

White, S. H. (1976). The active organism in theoretical behaviorism. *Human Development, 19,* 99–107.

Wood, D. J. (1980). Teaching the young child: Some relationships between social interaction, language, and thought. In D. R. Olson (Ed.), *The social foundations of language and thought* (pp. 280–96). New York: Norton.

8   A third perspective:
The symbol systems approach

Howard Gardner, Thomas Hatch, and Bruce Torff

A paradox: The triumph of hereditarian and environmental explanations

When there are competing explanations for a set of phenomena, one of two outcomes can generally be anticipated. Either evidence steadily accrues in favor of one side of the argument; in such a case, the competing position (e.g., *phlogiston* or *Lamarckianism*) is gradually abandoned or radically reformulated. Or it is determined that neither side of the debate will triumph, either because the contrasting case proves too difficult to refute, or because the apparently conflicting issues have been improperly formulated.

In the behavioral sciences, no issue has been more passionately joined than the relative contribution to human behavior of hereditary (genetic) or environmental (cultural) approaches. In one sense, the debate dates back at least to the beginnings of the modern philosophical era: Contemporary hereditarians (hereafter Hs) can date back their pedigree to Descartes and Leibnitz, while contemporary environmentalists (hereafter Es) can trace their intellectual heritage to the British empiricists, such as Locke and Hume.

In the past century, since the times of the evolutionary theorist Charles Darwin and the pioneering geneticist Gregor Mendel, the issue has been joined with ever greater sharpness, even as the pendulum has continued to swing between the rival positions. For the first several decades of the behavioral sciences, the view held sway that humans were hostages to their biological heritage. Then, with the advent of the Behaviorist and Learning Theory approaches, the contributions of the environment were reciprocally stressed. An influx of European thought to American shores contributed to a reassertion of a biologically influenced rationalist position after World War II; but the political and social events of the 1960s, as well as certain scientific findings, led many to embrace or reembrace a strong environmentalist position (Clarke & Clarke, 1977). More recently, those of a rationalist/ hereditary orientation have appeared to gain the upper hand (Degler,

1991), although new environmentalisms and contextualisms continue to be formulated (LeVine, 1991; Rogoff, 1990; Shweder & LeVine, 1984; Snow, 1991).

This breathless sweep through several hundred years of intellectual history self-consciously groups together a number of separable schools of thought. Belief in the primacy of reason (rationalism) is not identical either with an emphasis on biological factors (biological causation) or with a stress on the importance of the genetic makeup of a specific individual (hereditarianism). Belief in the importance of experience in general (empiricism) is not coequal either with an emphasis on pervasive cultural factors (cultural causation) or with a stress on the particular events of an individual's own history (environmentalism). Yet, it is typically the case that these sets of beliefs constitute a cluster; and for the purposes of this chapter, in which we seek to introduce a new perspective rather than to parse existing ones, we will continue to contrast the extreme positions of hereditarianism (H) and environmentalism (E).

What is curious about the antinomy between H and E explanations is that *both* have seemed to gain in persuasiveness over the years. On the H side, those of a hereditary/biological persuasion point to ever-increasing evidence that (1) even the most specific of traits are under genetic control; (2) a whole range of traits, from psychometric intelligence to one's temperament and even one's political attitudes, prove to be significantly heritable (Bouchard & Propping, 1993); and (3) fundamental perceptual and cognitive processes are present at birth, or unfold in infancy with ineluctable sureness and rapidity, and prove most difficult to dislodge (Bower, 1982; Carey & Gelman, 1991). As evidence mounts, many in the biological tradition have concluded that the H vs. E dispute is already resolved, with the Hs as the winners.

It is important to stress that the presence of powerful perceptual or reasoning capacities in the newborn does not establish an unambiguously and exclusively biological explanation; these prepotent responses might emerge as well because of important, if largely predictable, events that occur within the uterine environment or at the time of birth. Nonetheless, even though these positions cannot properly be conflated, none gives much comfort to the E camp.

On the E side, those of an environmental perspective have hardly abandoned their positions. Indeed, based on over a century of anthropological fieldwork, they point with confidence to the enormous differences among individuals raised in different cultures, differences that can be noted even from an early age: What better proof could exist of the potency of one's surrounding environment? Proponents cite studies within the Western cultural envelope that demonstrate clear and long-lasting effects from contrasting styles of parenting and socialization by other elders or messages conveyed by the media of communication (Bandura, 1977; Baumrind, 1971; Cole & Cole, 1989; Collier, 1994; Heath, 1983; Sternberg & Wagner, 1994; Stigler, Schweder, & Herdt, 1990; Vygotsky, 1978). They cite the power of various training methods in producing prodigious behaviors within one culture, and the fearful dampening of potential through deprivation, be it of a physiological or psychological sort (Comer, 1980; Ericsson, Krampe, & Tesch-Romer, 1993).

Finally, and most dramatically, Es seek to meet the Hs on their own ground. They show how knowledge of genetic susceptibility can itself motivate interventions that render the genetic effects weaker or nonexistent. They point to changes in an entire culture in such putatively inherited traits as physical height or psychometric intelligence. And they take careful note of emerging evidence that the development of the brain is quite flexible or plastic, particularly in early life, and that neurological regions not used for one purpose may be *rededicated* to other quite different ends (Flynn, 1992; Greenough, Black, & Wallace, 1987; Merzernich, Kass, & Wall, 1983; Neville, 1984).

And so we have a standoff, but one, we submit, of a paradoxical type. With the passage of time, *both* H *and* E positions have marshaled increasingly convincing arguments. In a sense, both have won the debate. As one commentator put it some time ago, all behavior turns out to be 100% the product of hereditarian factors and 100% the product of environmental factors as well.

Given this state of affairs, we wonder how profitable it is, in textbooks and in polemical collections, to continue to contrast the H and E positions. To quote the psychologist Robert Woodworth, writing over 50 years ago, "To ask whether heredity or environment is more important to life is like asking whether fuel or oxygen is more necessary for making a fire" (1941, p. 1). And so we raise here the possibility that another way of conceptualizing the matter and another set of mechanisms may prove more fruitful – particularly for the psychological sciences.

Desiderata for a new perspective

In developing a new approach to the issues traditionally discussed in H vs. E debates, one should have in mind the following set of features:

1. Capacity to incorporate the principal findings that are cited by proponents of the classical positions in the debate. Unless a new position is able to encompass, and to place in proper perspective, the principal concepts and data from the hereditarian and environmental positions (as broadly construed), it is unlikely to be persuasive.

2. Introduction of a framework that suggests promising new lines of work, as well as the potential for integrating the results of that work into an attractive and comprehensive position. Researchers are unlikely to be attracted to a position unless it suggests some fruitful lines of research and a new and tenable theoretical perspective.

3. Identification of new phenomena, or fresh characterization of traditional phenomena. In the end, if a framework does not add to our scientific knowledge and understanding, and promise to do so in the future, it does not deserve to survive . . . nor is it likely to do so!

4. An analysis of the ways in which the new perspective compares favorably with other competing stances, including those that have been put forth in an attempt to mediate the classical debate.

To fill any, let alone all, of these desiderata, with respect to a classical question of research, is a tall order. We do not pretend to satisfy these requirements in this chapter. Nevertheless, it is well to keep these guideposts in mind as one puts forth or evaluates a new framework.

Introducing the symbol systems perspective

The *symbol systems approach* holds that human experience – the thought and behavior of our species – is most felicitously analyzed in terms of the symbols and the systems of symbols in which human beings traffic from early in life. On an intuitive basis, the position is straightforward enough. All human beings are highly dependent on the symbol system of spoken language. Nearly all human beings (and human cultures) rely as well on other widely evolved symbol systems, such as those used in gesturing, picturing, and enumerating. In increasingly specialized contexts, individuals come to use more technical and hermetic systems, ranging from various written mathematical notations to dance notations to religious iconography. And, ultimately, most individuals also develop idiosyncratic systems of symbols, such as those that are used to organize one's personal time and space, and those that populate one's daydreams, nightmares, and other imaginative flights.

We employ the term *symbol* in the most generic sense. A symbol is any element (e.g., a mark, a pattern, a circumscribed act) that carries meaning within a culture or community. Symbols typically represent and/or express aspects of experience – aspects both internal and external to the individual. Words, numbers, rituals, and icons all function as symbols as long as their significance is acknowledged consensually within a culture or subculture. In addition, a specific element may come to function symbolically for an individual if, for example, a song represents pain or a design expresses harmony or conflict.

Symbol systems are *sets* of symbols, whose relationship to one another has been established by convention: Thus, words are combined by principles of syntax into the system of language, just as individual physical stances are choreographed so that they embody the system of classical ballet. Symbol systems are more likely to be devised by cultures over time than by individuals, but there is nothing to preclude an individual from devising a symbol system and even conveying it to others (Feldman, 1980; Gardner, 1993b).

Three lines of analysis contribute to the current consensus within the social sciences about the centrality of human symbol or semiotic use. The first line is essentially philosophical in origin. Dating back at least to the time of the American philosopher Charles Sanders Peirce (1955), who himself drew on discussions in the scholastic tradition, philosophers have probed the nature and delineation of various kinds of marks that carry meaning. Intriguing typologies have been put forth in an effort to classify systems of symbols (Cassirer, 1953–1957; Goodman, 1976; Langer, 1942), and some effort has been devoted as well to an exploration of the ways in which these symbol systems operate – their semantics, syntax, and pragmatics. While the cognitive-rationalistic approach to symbols has probably received more emphasis, there has been recognition among philosophers that symbols have important affective and emotional impact (Bachelard, 1969; Langer, 1976; Ogden & Richards, 1929; Sperber, 1975). These affective impacts are particularly likely to emerge in a highly individualized fashion, as, for example, when an individual has a strong but idiosyncratic reaction to a work of art.

A second line of analysis grows out of the cognitive revolution in the psychological sciences (Gardner, 1985). During the behaviorist period, in which any talk of reality not grounded in material entities was spurned, *ideas, images, representations*, and *schemas* were deemed terms illegitimate in scientific inquiry. The invention of the digital computer changed this state of affairs. Computers, variously, were apparatuses that operated upon strings of symbols, or were themselves conceived of as symbol systems (Newell, 1980). Given that discussion of symbol systems was unproblematic with respect to artificial mechanisms, it made little sense to withhold this locution from analyses of human activity. And so investigators in the human sciences have felt licensed to speak of human beings – as well as electronic computers – as entities that manipulate symbols (Gardner, 1982, 1993a).

Symbol systems – more broadly, representations – can be thought of in two complementary ways. In the philosophical, linguistic, and semiotic traditions, the emphasis has fallen on *external symbols*. These are marks or patterns – words, drawings, graphs, scores – that can be seen publicly and analyzed in terms of their physical appearance and their prescribed mode of operation. In the computational and psychological traditions, the emphasis has fallen correlatively on *internal symbols*. Without prejudice to the fact that human beings operate regularly with external symbols, cognitivists have explored the possibility that human mental life is best characterized as the creation, transformation, and production of various kinds of "symbols in the head."

It is worth noting that the Russian psychologist Vygotsky (1978) developed a distinctive position in which externally mediated symbols ultimately become internalized. As a consequence, individuals become able to cognize even in the absence of external tools or marks. This perspective is a promising way to synthesize the external and internal means of symbolization; additionally, when suitably expanded upon, this position may suggest ways in which cultures build upon and ultimately influence the biological underpinnings of human symbolization.

In a *language of thought* analysis (Fodor, 1975), it is assumed that individuals are either born with, or soon create, a symbolic language. Subsequent mental life consists in syntactic operations of these internal symbols. Researchers use their investigative prowess to identify these symbols and to decipher the processes by which internal symbolization operates. It remains controversial whether the internal symbols are best thought of as identical to the external symbols, as related, or as an entirely different kind of entity.

The third position recognizes, and builds upon, the robustness and contribution of the philosophical and psychological positions. Where it differs is in its scholarly agenda: an effort to identify a mission for the psychological sciences and to place the discipline centrally within the human sciences (Gardner, 1992).

Consider the way that knowledge about human nature has come to be organized in the academy during the past century. The *biological approach* is rooted in genetics and in brain study. Typically reductionist in spirit, this approach views the central points about human behavior and thought as directly reflective of our biological heritage. The more that we know about brains and genes, the more that we will know about human beings and the less we will require ancillary disciplines. Philosophy and psychology are considered at best holding operations, or way stations, en route to a fully scientific understanding of the human condition. Full-fledged hereditarians or biological determinists have little need for a psychological perspective rooted in the symbol systems approach.

Even the most determinedly biologically oriented investigators do not question the need for a second, complementary approach to the understanding of human nature. Clearly, some studies must be humanistic, contextual, and interpretive in nature. We look to historians, experts in art and literature, and cultural anthropologists to instruct us about the wide variety of experiences that human beings have had over the millennia, to help us understand the questions raised and the products executed as part of this human conversation, and to enable us to appreciate better which aspects of human experience are universal and which ones are transitory, accidental, or restricted to particular times and places (Benedict, 1946; Geertz, 1973; Rorty, 1979). Again, for those deeply rooted in the *sociocultural perspective*, there is little need for, and often deep suspicion of, a point of view that highlights processes and categories within the psyche (or the genome). That is, a psychological approach that focuses on mental representations particular to the individual is as open to the thoroughgoing cultural analyst as to the full-blown biologist.

In light of the state of affairs that emanates from this pastiche of the scholarly enterprise, the following crucial issue arises: Is it adequate to launch two parallel enterprises – the biological and the sociocultural – which tackle the human condition from disparate points of view? Or do we require a *tertium quid*, some kind of third position, one rooted in traditional psychological concerns, that allows us to span the distance between an approach rooted in the fine structure of the human genome and an approach rooted in large-scale cultural patterns?

We submit that the symbol systems approach provides a promising means for reconciling the biological and the cultural positions (cf. Piaget, 1962). Such an approach takes into account findings regarding the influence of nature and nurture on human behavior; it recognizes the importance – and interrelatedness – of both internal and external symbolization; and it provides a promising way to explore the links between scientific and humanistic approaches to the study of human nature. In short, only an approach rooted in symbol use has the capacity to "speak" to both sides, to both "cultures" (Snow, 1959) in this long-standing debate.

An aside based on frank anthropomorphizing may help to convey this point. As card-carrying Darwinian entities, both the genes and the culture exist to perpetuate themselves. Genes directly mediate the development of the body, including the brain, which must survive long enough for reproduction to occur. The genes have control only over these anatomical structures and physiological processes. They know nothing – they can know nothing – about cultures, rites, family relations, wars, works of art, and the like.

The situation is precisely the opposite from the point of view of the culture. The culture exists in the practices of the group and in the minds that apprehend these practices. To survive, the culture must ensure that its thoughts, values, behaviors, and the like are perpetuated. Such survival can be achieved only if circumstances are contrived in which the relevant practices are modeled, learned, exhibited, and passed down in reasonably faithful form to succeeding generations. The culture does not need to know anything about genes and the brain. Indeed, for most of recorded history, and in most corners of the globe, the survival of the culture occurred despite total absence of knowledge about genetics. And our own society, unprecedentedly familiar with the realm of biology, may well not have the stamina and the savvy to survive – because survival depends *not* on knowledge of gene pools or neural networks but on the capacity to ensure that competitive cultural practices endure.

Yet, despite these different interests, genes and cultures are part of the same universe. How can we best think about these strands in a unified, synergistic way? The answer, we submit, is that symbol systems provide precisely the interstitial material needed to bridge these analytic strands. The brain has developed to use symbols and is prepared to recognize them, transform them, create them. Humans will venture to virtually any length to behave symbolically. The culture, for its part, is equally poised to use and create symbols, for it is as a consequence of the careful monitoring of such symbol use that the culture's preservation is most likely to be achieved. Dramatically, by moving to the level of symbol use, mechanisms that explain how messages contained in the genetic code and messages borne by the cultural languages may be identified.

In what follows, we examine two areas – realms of cognition that have traditionally been called "domains" (Feldman, 1986; Keil, 1981) and which we in earlier work have called "intelligences" (Gardner, 1993a). One area (or domain or intelligence) is *musical*; the other is *spatial*. In each case, we begin with those properties that are most likely to belong to the human genotype – ones that we can expect to unfold in a predictable manner with little formal tutelage in all but the most severely impaired individuals. This is the territory most friendly to the biological investigator – the territory of those *biopsychological human potentials* that we term *intelligences*.

As we trace the development of this cognitive realm, we encounter the challenges that our culture supplies for young children as they begin to deal with musical and spatial representations. In both cases, children's nascent abilities lead to intuitive understandings and misconceptions that can be overcome only through experiences with symbols and symbol systems. Finally, we consider the role of formal and informal educational experiences that contribute to the mastery of the quite specific systems of meaning that have evolved in particular cultural settings – notations and compositional genres in the case of musical cognition; the conventions of the visual arts in the case of spatial cognition. While the creation and apprehension of these systems is made possible because of our biological heritage, their mastery can only be explained through immersion in the specific cultural histories and institutions of the disciplines. Throughout, we keep in mind those factors that characterize the development of intelligences in all members of a culture, as well as those factors that lead to different degrees of expertise within a population.

Our goal is not merely a summary of the biological and cultural approaches to these domains, however. We wish to show that their respective operations can best be appreciated and related to one another if one takes seriously the symbolic systems within which music and space come to be apprehended. We seek as well to provide a more felicitous framework for consideration of the issues that have classically concerned those in the H vs. E debate. These sections, then, address the first three of the four desiderata that any new conceptual framework must ultimately meet.

A symbol systems approach to music

Reputed to exist even in nonliterate and nonnumerate cultures, music exemplifies a domain in which a strong biopsychological basis or intelligence becomes intertwined with symbolic vehicles of various sorts. Music is also a domain in which hereditarians and environmentalists have often contested one another. On the one hand, conceptions of innate talent predominate in Western folk psychology and pedagogy, and a number of psychologists of music take a nativist tack (e.g., Piechowski, 1993; Seashore, 1938). On the other hand, empirically oriented researchers (e.g., Howe, 1990; Sloboda, Davidson, & Howe, 1994) downplay hereditary influences and stress environmental factors such as early experience and motivation.

We see the symbol systems approach as harboring several advantages for investigators of musical cognition. First, the symbol systems approach has the capacity to incorporate principal findings from such diverse quarters as neuropsychology (e.g., Sergent, 1993), experimental psychology (e.g., Deutsch, 1982), ethnomusicology (e.g., Blacking, 1972), and music education (e.g., Gordon, 1965). Second, this perspective raises new issues, for example, concerning the relationship between internal and external symbolic representations. Finally, encompassing universals of musical development as well as differences across individuals and cultures, the symbol systems approach provides the most comprehensive account of musical behavior.

Neurobiology of music

A number of lines of research support the claim that the human genome carries universal capacities for musical symbolization. To begin with, nearly all children reveal remarkable vocal skill early in life. The infant's babbling includes melodic and intonational as well as phonological experimentation.

At about 1½ years of age, children begin to engage in spontaneous song – individual explorations relatively untouched by environmental input. Second, infants are able to make impressive intonational discriminations; by 1 year of age, and perhaps earlier, children can match pitch at better than chance rates and are also capable of imitating intonational patterns (Kessen, Levine, & Peindich, 1978). These computational capacities reveal a basic symbol-processing skill on which later musical achievement will be constructed. Third, studies of prodigies (e.g., Feldman, 1980; Revesz, 1925) and savants with severe learning deficits (e.g., Viscott, 1970) document advanced performances not easily attributable to practice or training. While the human genome probably includes no hard-wired "ur song," it embeds an uncommitted neurobiological potential for musical symbolization that has been termed *musical intelligence* (Gardner, 1993a). Consistent with views of a modular mind, other evidence suggests that musical intelligence is largely autonomous – distinct from other intellectual competences, most notably language. Studies of music perception provide evidence that the mechanisms by which pitch is apprehended are different from the mechanisms that process language (Aiello, 1994; Deutsch, 1975, 1982). Moreover, brain-imaging research reveals that while there exists no single music center in the brain, an extensive neural network underlies the realization of musical functions, and that network is significantly different from the neural apparatus that subserves linguistic functions (Sergent, 1993). Finally, studies of individuals with brain damage reporting the existence of aphasia without amusia (Basso & Capitani, 1985) and amusia without aphasia (McFarland & Fortin, 1982) reveal a "functional autonomy of mental processes inherent in verbal and musical communication and a structural independence of their neurological substrates" (Sergent, 1993, p. 168).

Vast individual differences in musical behavior from the earliest years of life suggest that musical intelligence has a high heritability (Gardner, 1982; Piechowski, 1993; Torff & Winner, 1994). Biologically based talent *is* a factor in musical achievement, but, as we shall see, cultural and training considerations play an indispensable role as well.

Early musical development and the intuitive musical mind

Exposed regularly to recorded music and the singing of adults, most young children in the West are immersed in song. This exposure provides the catalyst for several years of musical development that occurs without much formal instruction – development that finds the child increasingly in tune with the particular musical symbol systems provided by the surrounding culture.

At about the age of $2\frac{1}{2}$, the child begins to exhibit more explicit and extended awareness of tunes sung by others (Davidson, 1994; Gardner, 1982). The child attempts to reproduce familiar nursery rhymes, initially by matching the rhythmic structure, later by following the contour of the melody, and finally by singing discrete pitches. As learned song comes to dominate, spontaneous song becomes less frequent and eventually all but disappears.

Picking up expressive as well as contextual meanings from the world around him or her, the child acquires the symbolic forms of the ambient music culture. One example is the linkage of harmonies and moods; in the course of everyday life in Western culture, the child comes to appreciate that major tonalities symbolize happiness while minor tonalities represent sadness. Another example connects music to situations: One tune denotes Sesame Street, another a birthday party or a religious ceremony.

A focus on symbol systems reveals as well that music development amounts to far more than the simple unfolding of a singular musical intelligence. Musical activities in everyday settings involve the coordination of a range of intelligences – spatial, linguistic, logical-mathematical, intrapersonal and interpersonal, as well as musical – with the particular musical activity dictating the specific blend of intelligences.

This story of spontaneous musical development draws to a close when the child reaches school age; at this point, everyday immersion in the music culture appears to lose its power to catalyze further. At least in Western culture, musical development continues beyond the age of 7 or so only in an environment that provides some form of tutelage (Gardner, 1973a; Winner, 1982). Thus, musical intelligence demonstrates the same kind of developmental plateau reached by the 5-year-old mind in other domains (Gardner, 1991).

The result is a musical mind which is at once powerful and limited. Competent to participate as consumers in a music culture, the musically untrained demonstrate impressive abilities on tests of perception (Dowling, 1994; Dowling & Harwood, 1986; Krumhansl, 1990), production (Sloboda, 1994; Swanwick & Tillman, 1986), and representation (Bamberger, 1991; Davidson & Scripp, 1988). At the same time, marked limitations also are evident. First, stereotypes may be forged; for example, research indicates that people value poems with regular meter, rhyme, and an upbeat tone – restrictions that disciplinary experts do not embrace (Richards, 1929). As children are typically oriented toward the lyric content of songs and to songs that honor a canonical form (Gardner, 1973b, 1982), similar stereotypical predilections seem to prevail in music. Second, misconceptions may emerge, such as the notion that songs must begin and end on the same note

(Davidson, Scripp, & Welsh, 1988). Finally, should formal music instruction commence, the transition to mastery may not be smooth and unproblematic (Bamberger, 1991).

Disciplinary expertise and instruction in music

Formal instruction places the biologically prepared individual in contact with the disciplinary conventions of the music culture in ways that allow most youngsters to gain considerable competence. In music classes and apprenticeships, the child confronts the concepts and moves of the disciplinary expert as he or she becomes more deeply enmeshed in symbols and symbol systems of the music of the culture.

To begin with, more differentiated symbol systems are elaborated; for example, in typical apprenticeship settings in music performance, coaching of performances yields understanding of more subtle shades of expression or of programmatic aspects of compositions (Davidson, 1989). In addition, symbols drawn from other spheres come into play; for example, linguistic symbols are seen in performance directives (e.g., "legato") and metaphor (e.g., "like a butterfly"). Symbols in the form of gestures are evident, for example, in the extended body language of the conductor.

It is important to note the existence of two musical developmental courses – a notational one, used in the West, and a nonnotational (traditional) one, employed in many cultures and in some Western subcultures (e.g., blues and rock). Notations – second-order symbol systems, which themselves denote first-order symbol systems – present the student with systems inherent in representing music on paper. Included are graphic symbols (representing pitch, articulation, and so on) and logical/mathematical symbols (particularly ones denoting rhythm and meter). Whether the music culture happens to be notational or nonnotational, as the student comes into greater awareness of these symbol systems, these systems come to dominate musical cognition. Enculturation in musical symbols fuses the individual's idiosyncratic personal development to the discipline; indeed, if one accepts the notion of the internalization of musical maneuvers and procedures (cf. Vygotsky, 1978), music instruction provides the cognitive tools through which the expert musician thinks and acts (Davidson & Torff, 1992). It may be for this reason that certain aspects of musical cognition shift the dominance of brain lateralization (Bever & Chiarello, 1976). In a perception task, right-handed subjects who are trained in music show left-hemisphere dominance, while untrained right-handed subjects favor the more typical right-hemisphere processing. These results imply that formal instruction in music may actually stimulate meaningful changes in innately specified neural organization.

Music instruction brings the symbol systems of the discipline into contact with the intuitive musical mind. In some instances, the result is a smooth path to disciplinary expertise marked by improved performance/composition, perceptual acuity, and reflective skills. In other cases, however, the transition proves to be more problematic.

In a revealing line of work, Bamberger (1991, 1994) reports a conflict between intuitive "figural" understandings embedded in particular contexts with "formal" systems such as music notation. In a paradigmatic example, trained and untrained subjects were asked to create an invented notation of the rhythmic pattern of "one, two, pick up shoes, three, four, shut the door." This pattern features two claps (grouping A) followed by three claps (grouping B), followed by a silence (a "rest") that is as long as the time between the two claps in grouping A or between the first and third claps in grouping B.

The untrained subjects typically created notations that group the claps appropriately but ignore the length of the rests:

.. ..

A person attempting to read this notation would reproduce groupings of two and three but would fail to reproduce the rests in a consistent fashion. In contrast, trained subjects follow metrical patterns, recognizing how much time is taken by both the pulses and rests:

./././...//././...//

This rendering is literally correct; however, when asked to perform the pattern itself, the trained subjects may fail to re-create the feeling of rhythmic grouping (phrasing) that is highlighted in the naive notations and performances. Crafting a notation that is metrically accurate but fails to capture the phrasing, the trained subjects produced a "correct-answer compromise" rather than a fully adequate notation. In Bamberger's terms, *formal* knowledge (the notation system) overwhelmed the intuitive *figural* knowledge of the pattern. While the expert musician is not limited to either the figural or formal interpretation, Bamberger argues, students encounter difficulty matching intuitive knowledge of "how the song goes" to its notation. The culture's formal symbol system clashed with the intuitive musical competence spawned by human evolution. Without further instruction in formal musical symbols and categories, this clash is likely to remain unresolved.

Exploring in a vein similar to Bamberger's work, Davidson, Scripp, and Welsh (1988) report that conservatory students demonstrate a conflict between perceptual knowledge and conceptual knowledge in a notation task. Attempting to notate the song "Happy Birthday" using the standard notation system of Western music, the students produced an inaccurate notation.

They opted to alter the key of the final phrase of the song, based on the misconception that songs must begin and end on the same note. Asked to sing from their notations, the students reproduced the canonical version of the song, while detecting no errors in their notations. In their notations, the students disregarded accurate perceptual knowledge of the song in favor of misconceived conceptual knowledge; in their singing, they drew on their perceptual knowledge and overlooked the inaccuracy in their notations. According to the authors, the students experienced a disjunction between perceptual knowledge of music and the formalism of music notation.

Other tensions may emerge between musical perception and production. Novice students in jazz are often given a generic pitch-set (the blues scale) as a strategy for structuring a first exercise in improvisation; while the blues scale proves effective in some settings, students often overgeneralize the strategy, imposing it in settings in which it does not work – sometimes as remote as songs in other keys. At the same time, when asked to listen to a recording in which this same overgeneralization is present, the students are able to recognize that the improvised line ill matches the accompaniment. During performance, the students subjugate perceptual knowledge as they pursue a misaligned production strategy.

The foregoing examples point up tensions that may exist between competing varieties of representation or understanding in music. The smooth developmental path implied – for different reasons – by both hereditarian and environmental theories does not obtain in music. Only an approach that addresses the encounter between biological proclivities and different culturally constituted musical systems and practices can suggest mechanisms that mediate the phenomena of musical learning and performance. The same tension can be discerned when it comes to the issue of individual talent. From all indications, youngsters differ from one another in the extent to which they are "at promise" in the domain of music. Such capacities as perfect pitch, ready mastery of tonal relations and transposition, and keen memory for meter and instrumental timbre may all be significantly under genetic control. Yet, such talent proves of no avail in the absence of thousands of hours of practice distributed over a decade or more, as the youngster gains facility in various first- and second-order musical symbol systems. Even if cultural interventions do not entirely obviate the need for powerful inborn musical analytic powers, they prove essential for mastery (Bloom & Sosniak, 1985; Sloboda, 1994).

A symbol systems approach to spatial cognition

As happens with respect to spoken language, we think of spatial abilities as developing early in life and in the absence of explicit instruction. In contrast to language (and music), however, human beings share many spatial abilities with other animals; this fact reinforces the belief that our biological/ genetic endowment, and not our cultural heritage, defines our capacities to recognize objects and perceive spatial relationships. This H-tinged belief is buttressed by two commonly held views. First, we view spatial abilities and related domains (e.g., science and mathematics) as ones in which males are more advanced than females – and, rightly or wrongly, such strong sex differences are often attributed to biological factors. Second, we view those who have reached the heights of development in certain domains – like Picasso, Einstein, and any number of chess prodigies – as having been born *gifted* in spatial intelligence. In both cases, we view spatial abilities as capacities in which effort and education are overshadowed by natural talent. Even as we may feel that natural talent is needed to excel in the spatial domain, we treat common spatial representations like maps, blueprints, and 3-D models as if they are relatively easy to use and can be employed with little or no specific tutelage.

The apparently natural course of spatial development, and the ready accessibility of certain spatial representations, belies the complexity of the spatial domain and the difficulties that both children and adults encounter in trying to master its diverse manifestations. The symbol systems approach shows how spatial ability builds both upon biological predispositions and upon adjustments to the presses of human culture. Such an approach brings together data regarding the neuropsychological bases of spatial abilities and the development of abilities to deal with such cultural constructions as 3-D models, maps, and perspective drawing. In the end, this approach reveals how achievements in areas like the visual arts, geography, or geometry may be founded upon biologically determined capacities but can only be reached through immersion in the symbolic representations, practices, and community that define the domain.

The biological bases of spatial abilities

Spatial ability is best thought of as a collection of loosely related potentials that enable humans to perceive and manipulate spatial information. Spatial abilities include the capacities to perceive the visual world accurately, to attend to specific locations in space, to manipulate objects either visually or tactually, and to organize the surrounding environment into a coherent spatial framework (Morrow & Ratcliff, 1988). In addition, although we tend to think of spatial abilities as being tied to our capacity to see, spatial abilities have been shown to develop independently of particular sensory modalities (Kritchevsky, 1988). Even blind people or those who suffer damage to the visual system can perceive spatial information (such as the location of objects) and can navigate successfully through and create representations of the world (Kennedy, 1974; Landau, 1988).

The evidence of biological bases for these abilities comes from at least three sources. First, research has shown that a variety of vertebrate and invertebrate species are predisposed to carry out specific spatial tasks (Cheng, 1986; Dyer & Gould, 1983; Gallistel, Brown, Carey, Gelman, & Keil, 1991; Margules & Gallistel, 1988). For example, infant vervet monkeys respond distinctly to alarm calls that warn of the approach of three different kinds of predators – leopards, martial eagles, and pythons (Seyfarth, Chency, & Marler, 1980). They run to or stay in a tree in response to leopard calls; they run into the bushes or look up in response to eagle calls; and they look down in response to snake calls. These actions indicate that these monkeys have innate tendencies to represent critical features of movement and shape and to respond appropriately in their spatial milieu. Such evidence suggests that human spatial ability is continuous with that of other animals, while musical or linguistic ability is more appropriately thought of as specific to the human species.

Second, research suggests that specific parts of the brain (especially posterior regions of the right hemisphere) are prepared to carry out particular spatial functions. Thus, brain injuries can disrupt a variety of spatial functions without causing serious damage to other capacities. In both monkeys and humans, damage to the right parietal lobe can lead to impairments in ability to learn a maze, while damage to the right temporal lobe limits the capacity to recognize faces. Other kinds of selective deficits in humans include difficulties learning new routes, problems producing accurate spatial representations (such as drawings or models), and the neglect of one side of the visual field. In the latter case, neglect may also apply to internalized spatial representations, such as mental images (Gardner, 1975).

There is growing evidence that, even in very young children, the right hemisphere is better prepared than the left to carry out spatial functions (Stiles-Davis, 1988; Witelson & Swallow, 1988). In addition, although there is often significant recovery of functions after damage to the right hemisphere, people who experienced such damage in childhood virtually never perform as well on spatial tasks as those who have experienced comparable injuries to the left hemisphere (Gardner, 1993a).

Third, sex differences in both humans and animals indicate that spatial abilities may depend at least to some extent on biological factors. Differences in the performances of males and females have been more widely reported on spatial tasks than on almost any other kind of task. Males have performed significantly better than females on tasks involving the rotating of mental images, the understanding of changes in the water level of a tilted container, route finding through a printed maze, and the reading of maps

(Harris, 1981; Witelson & Swallow, 1988). While such differences may be exacerbated by sociocultural influences (Harper & Sanders, 1975; Harris, 1981; Rheingold & Cook, 1975), a number of studies suggest a link between differences in performance of spatial tasks and a variety of biological factors. These factors include differences between the sexes in hormone levels, chromosomal patterns, rates of functional maturation of the two hemispheres, lateralization of functions, and rates of physical maturation (see Harris, 1981, for a review). For example, females with increased levels of adrenal androgen performed better on spatial tasks than did females with average levels, while performances of males with similarly increased hormonal levels did not differ from those of a control group. In addition, females who are missing one X chromosome have reduced levels of both testosterone and estrogen, and often perform below the normal level on tests of spatial ability (Harris, 1981; Witelson & Swallow, 1988).

Intriguing confirmation of this line of analysis comes from the fact that analogous sex differences can be found in the spatial abilities of both humans and other animals. Bever (1992) reports that although human males are better at reading maps and printed mazes, there are no differences in their ability to find their way through a neighborhood or an actual maze. He explains this finding by arguing that males focus on the configuration of space (useful for route finding on maps *and* neighborhoods) while females rely on nearby cues or landmarks (useful only in finding a route through real space). Bever documents that these same gender differences in spatial strategies can be found in male and female rats. Regardless of the specific cause of such differences, it is extremely difficult to explain the similar findings in two vastly different species without positing a role for biological factors.

The development of spatial abilities

Everywhere an infant looks or reaches, there are objects to be identified and manipulated; all around are places to go. Infants are well prepared to explore and exploit this spatial world, and during the first several years of life, the child's development includes such notable landmarks as an understanding of object permanence and the ability to navigate around complex terrains. By the time they enter school, children can manipulate spatial images mentally and appreciate how objects look from different perspectives. By adolescence, they become capable of understanding and using the rules and abstract relationships that govern disciplines such as geometry and physics (Piaget & Inhelder, 1967).

After infancy, young children will also begin to recognize and use the external (particularly second-order) symbols that human beings have created in order to make optimal use of the spatial domain. Children's initial abilities in this area may well depend on a match between genetic endowment and the culture that surrounds them. For example, quite different cognitive demands are made depending on whether a literacy system is based on ideographic or orthographic elements – and there are some who have a particular difficulty with one or another of these systems. Karmiloff-Smith (1992) argues, further, that the notational systems developed by human culture capitalize on differences in the information-processing mechanisms that have evolved in the brain; here the intriguing interaction between neurological predispositions and cultural inventions comes to the fore (Marshack, 1991).

Although young children are capable of treating spatial notations differently from others, it takes time to master the specific demands that the culture imposes on the expression of spatial knowledge. Unfortunately, as children begin to use external symbols to represent objects and spatial relationships, and as they encounter such activities as perspective drawing, they do not appear to be so well prepared. Consider the specific forms of spatial mapping (e.g., that one object can stand for or represent another). At the age of $2\frac{1}{2}$, children fail to recognize that an object hidden in a scale model can reveal to them where a similar object is located in the room that the model represents (DeLoache, 1987, 1989, 1991). Little in children's biological heritage inclines them to recognize that object can function symbolically just as words and pictures do.

Instead, there is some evidence that experience with relatively simple symbolic relationships can help young children to see and expect that objects and other kinds of spatial notations can be used for the purposes of representation. Thus, Marzolf and DeLoache (1994) show that children who are first exposed to an easy model task are significantly more likely to succeed on a more difficult task and that experience with a model task even helps the children to succeed subsequently on a task using maps. These results indicate that early interventions can have a significant positive impact in directing children's nascent abilities to comprehend and use spatial representations in culturally approved ways.

Similarly, although very young children – both sighted and blind – display some initial understanding of the spatial relationships in maps (Landau, 1988; Landau, Spelke, & Gleitman, 1984), the mastery of these means of spatial representation proves far from a simple process. For example, children can recognize that such displays as aerial photographs and maps symbolize places. However, children expect there to be a direct or iconic relationship between elements of the representations and the entities in the world to which they refer (Bluestein & Acredolo, 1979; Gregg & Leinhardt, 1994; Liben & Downs, 1991). This expectation leads to a number of common misconceptions that take time to overcome. Thus, children may conclude that a red road on a map must actually be red (Liben & Downs, 1991). Further, they may come to believe that islands float on top of the ocean and that the lines separating states and countries are actually marked on the earth (Gregg & Leinhardt, 1994; Hewes, 1982).

Such misconceptions generated about and by maps do not end in childhood. As they learn the conventional properties of graphic representations, children overgeneralize, assuming that grass in an aerial photograph must be at the bottom of the photograph while the sky must be at the top. Similarly, they often interpret overhead views as frontal views – for example, concluding that a tennis court seen from the air is actually a door (Spencer, Harrison, & Darvizeh, 1980). Even adults will assume a direct relationship between the representations in maps and the entities to which they refer. As a result, many people believe that Greenland is two to three times larger than Saudi Arabia because that is the way they appear on maps, which are commonly drawn to a Mercator projection (Liben & Downs, 1991). Only with experience do people come to understand that the Mercator projection is a means of representation that distorts area and that these two countries are actually about the same size.

Perspective in paintings and drawings provides another instance of the challenge that children face when trying to understand certain cultural constructions of the spatial domain. Single-point perspective rendering is a symbolic invention of Western culture. If human beings were born equipped to capture three dimensions on paper in an unambiguous fashion, it is hard to understand why it took until the Renaissance for artists to figure out how to represent objects and scenes from a unitary perspective (Winner, 1982). It appears that young children today are no better prepared to solve this problem than were their ancestors. Even with the invention of perspective drawing, children's initial representing capacities cannot suffice to translate three dimensions onto a two-dimensional surface. Items that are intended to be on top of one another are likely to be represented as if they are floating. Only at about the age of 9 do most children attempt to depict the third dimension, and not until adolescence are most able to render perspective somewhat properly in their art works (Willats, 1977). While more general cognitive abilities undoubtedly come into play in the achievement of perspective drawing, it is likely that some exposure to the artworks and conventional solutions developed in our culture are prerequisites as well (Gombrich, 1960; Goodman, 1976).

Spatial ability and education

While it is not possible in this survey to trace the role of education on all aspects of spatial development, the development of abilities in drawing and painting illustrates that even natural proclivities in the spatial domain depend on some formal and informal experiences in the disciplines of the visual arts. No one would mistake the writing of Shakespeare for the utterances of a toddler, but correspondences have often been discussed between the work of some of the greatest visual artists and the spontaneous drawings and pictures of young children. Picasso himself whimsically remarked, "When I was their age I could draw like Raphael, but it has taken me a whole lifetime to learn to draw like them" (Gardner, 1993b, p. 145). Such comments reinforce the view that the expression of spatial ability in paintings and drawings is a natural process that does not depend on education or training.

Such comparisons overlook the differences in the way that children and artists approach the task of representation. For the average child, drawing and painting are spontaneous acts that sometimes happen to conform to the aesthetic standards that have been developed in Western culture (Gardner, 1989). It is not until middle childhood that young children appreciate the aesthetic qualities of artwork, and it is not until about this time that they can intentionally vary their works to meet specific criteria of the artistic symbol systems of their culture (for example, they have difficulty drawing a wrecked car differently from an intact car [D. Pariser in Gardner, 1982]).

Ironically, it is at this same time that children become preoccupied with the accuracy of their representations (Gardner, 1982; Winner, 1982). This literal stage is echoed across domains as children become concerned with conforming to the standards and rules of the dominant culture; their natural talents must confront the demands of the graphic environment. In the visual arts, many adolescents and adults will never get beyond this stage (Davis & Gardner, 1993), but a few will proceed to challenge the boundaries and conventions that exist at any given time. Those who can master the cultural conventions, attend to the aesthetic qualities in graphic representations, and willfully vary their efforts to produce aesthetic effects distinguish their works both from the spontaneous productions of children and from the standardized reproductions of their peers.

In order to achieve a high level of artistry, some form of formal training and education is almost always indicated. An analysis of the career of an artist like Picasso demonstrates that, even for the most "gifted" artists, considerable learning is required (Gardner, 1993b). Both alone and in collaboration with peers and mentors, artists develop their abilities by grappling with the problems and conventions of their discipline. The same can be said of architects, geographers, surgeons, chess players, and others who traffic in spatial intelligence. Thus, while the roots of graphic and artistic mastery can be traced to our species membership, actual achievement of competence in a spatial realm presupposes a long and rigorous apprenticeship in the procedures of particular symbolic systems.

Conclusion

We have put forth the symbol systems approach as a capacious mode of analysis – one sufficiently broad as to encompass the concerns of those of a biological/hereditarian persuasion as well as the perspectives of those who embrace an environmental/cultural point of view. After setting the historical and theoretical context for such an approach, we have introduced an ensemble of examples from two disparate and yet comparable realms – musical cognition and spatial cognition. We hope to have shown, or at least to have suggested, how findings from both perspectives can be interwoven in order to yield a view that is more veridical than that attained through either perspective considered in isolation. For example, we have seen how intuitive ways of representing musical patterns, presumably reflecting biological constraints of perception and action, are gradually transformed in the face of strong dictates from cultural notational practices. By the same token, we have noted how forms of spatial mapping that come naturally to the young child must be reconfigured in the light of systematic practices entailed in specific culturally constructed mapping systems. It is not that hereditarian or environmental perspectives should completely disappear; rather, the strengths and limitations are best appreciated when the findings from the rival point of view are directly considered rather than minimized, distorted, or altogether ignored.

In an essay of this size and scope, we cannot survey this approach in its entirety, let alone subject it to adequate critical analysis. In the foregoing sections, we have indicated how disparate findings can be integrated within a common semiotic approach; and we have shown that the development of expertise can best be illuminated by taking into account the tensions between natural or intuitive forms of knowing and the requirements of disciplined work in established domains.

In conclusion, we can suggest a few ways in which our tack differs from that of other developmental psychologists' attempts at integration. While Piaget always stresses biological and cultural interaction in his writings, he does not take into adequate account the contents of specific domains, nor the requirements needed in order to master a domain or a discipline to the level expected in a specific culture or subculture. Post-Piagetian thinkers are more sensitive to the nature of different domains (Case, 1992; Karmiloff-Smith, 1992), but they fail to deal sufficiently with the ways in which our biological proclivities – our unschooled minds – may complicate the mastery of certain symbolic codes (Gardner, 1991) or clash with the particular requirements imposed by specific educational systems (Gardner, 1993a).

Whether the perspective introduced in this chapter can rise to such

challenges remains to be seen. It will be important to see whether cultural
analysts are aided in explaining the prevalence of certain practices by a
consideration of the kind of mental representations that arise naturally and
by the ways that they can (and cannot) be transformed. By the same token,
the utility of a symbol systems approach to biologically oriented investigators will be substantiated if the ways in which they analyze – and, for that
matter, identify – genetic and neurological evidence is affected by knowledge of the kinds of systems that can (and cannot) be sustained in different
cultures. Even if it proves inadequate to the full burden, the symbol systems
approach can serve a useful purpose in bringing certain scientific literatures
into contact with one another, in raising consciousness about certain
conundra that have not yet been adequately explained, and in suggesting a
continuing useful role for psychology at a time when it stands in danger of
being cannibalized by other disciplines (Gardner, 1992).

Acknowledgments

Preparation of this chapter was supported in part by the McDonnell Foundation, the Spencer Foundation, and the Shouse Education Fund. Thanks
are due to Mindy Kornhaber and the editors of this volume for their helpful
comments.

References

Aiello, R. (1994). Music and language: Parallels and contrasts. In R. Aiello & J. Sloboda
(Eds.), *Musical perceptions*. New York: Oxford University Press.
Bachelard, G. (1969). *Poetics of space*. Boston: Beacon Press.
Bamberger, J. (1991). *The mind behind the musical ear*. Cambridge: Harvard University Press.
Bamberger, J. (1994). Coming to hear in a new way. In R. Aiello & J. Sloboda (Eds.), *Musical
perceptions*. New York: Oxford University Press.
Bandura, A. (1977). *Social learning theory*. Englewood Cliffs, NJ: Prentice Hall.
Baumrind, D. (1971). *Current patterns of parental authority*. Washington, DC: American
Psychological Association.
Basso, A., & Capitani, E. (1985). *Journal of Neurology, Neurosurgery, and Psychiatry, 48*, 407–
12.
Benedict, R. (1946). *Patterns of culture*. New York: Mentor Books.
Bever, T. (1992). The logical and extrinsic sources of modularity. In M. Gunnar & M. Maratsos
(Eds.), *Modularity and constraints in language and cognition*. Hillsdale, NJ: Erlbaum.
Bever, T., & Chiarello, R. (1976). Cerebral dominance in musicians and nonmusicians. *Science,
185*, 537–9.
Blacking, J. (1972). *How musical is man?* Seattle: University of Washington Press.
Bloom, B., & Sosniak, L. (1985). *Developing talent in young people*. New York: Ballantine
Books.
Bluestein, N., & Acredolo, L. (1979). Developmental changes in map-reading skills. *Child
Development, 50*, 691–7.

Bouchard, T., & Propping, P. (Eds.). (1993). *Twins as a tool of behavioral genetics.* Chichester: John Wiley.

Bower, T. G. R. (1982). *Development in human infancy.* New York: W. H. Freeman.

Carey, S., & Gelman, R. (1991). *The epigenesis of mind.* Hillsdale, NJ: Erlbaum.

Case, R. (1992). *The mind's staircase.* Hillsdale, NJ: Erlbaum.

Cassirer, E. (1953–1957). *The philosophy of symbolic forms* (Vols. 1–3). New Haven, CT: Yale University Press.

Cheng, K. (1986). A purely geometric module in the rat's spatial representation. *Cognition, 23,* 149–78.

Clarke, A. M., & Clarke, A. D. (1977). *Early experience: Myth and evidence.* Riverside, NJ: Free Press.

Cole, M., & Cole, S. (1989). *The development of children.* New York: Freeman.

Collier, G. (1994). *The social origins of mental ability.* New York: Wiley.

Comer, J. (1980). *School power.* New York: Free Press.

Davidson, L. (1989). Observing a Yang Chin lesson. *Journal for Aesthetic Education, 23*(1), 85–99.

Davidson, L. (1994). Songsinging by young and old. In R. Aiello & J. Sloboda (Eds.), *Musical perceptions.* New York: Oxford University Press.

Davidson, L., & Scripp, L. (1988). Young children's musical representations. In J. Sloboda (Ed.), *Generative processes in music.* Oxford: Oxford University Press.

Davidson, L., Scripp, L., & Welsh, P. (1988). "Happy Birthday": Evidence for conflicts of perceptual knowledge and conceptual understanding. *Journal of Aesthetic Education, 22*(1), 65–74.

Davidson, L., & Torff, B. (1992). Situated cognition in music. *World of Music, 34,* 3.

Davis, J., & Gardner, H. (1993). The arts and early childhood education. In B. Spodek (Ed.), *Handbook of early childhood education.* New York: Macmillan.

Degler, C. (1991). *In search of human nature.* New York: Oxford University Press.

DeLoache, J. (1987). Rapid change in the symbolic functioning of very young children. *Science, 238,* 1556–7.

DeLoache, J. (1989). Young children's understanding of the correspondence between a scale model and a larger space. *Cognitive Development, 4,* 121–9.

DeLoache, J. (1991). Symbolic functioning in very young children. *Child Development, 62,* 736–52.

Deutsch, D. (1975). The organization of memory for a short-term attribute. In D. Deutsch & J. Deutsch (Eds.), *Short term memory.* New York: Academic Press.

Deutsch, D. (Ed.). (1982). *The psychology of music.* New York: Academic Press.

Dowling, W. (1994). Melodic contour in hearing and remembering melodies. In R. Aiello & J. Sloboda (Eds.), *Musical perceptions.* New York: Oxford University Press.

Dowling, W., & Harwood, D. (1986). *Music cognition.* New York: Academic Press.

Dyer, F., & Gould, J. (1983). Honey bee navigation. *American Scientist, 71,* 587–97.

Ericsson, K. A., Krampe, R. T., & Tesch-Romer, C. (1993). The role of deliberate practice in the acquisition of expert performance. *Psychological Review, 100,* 363–406.

Feldman, D. (1980). *Beyond universals in cognitive development.* Norwood, NJ: Ablex.

Feldman, D. (1986). *Nature's gambit.* New York: Basic Books.

Flynn, J. (1992). *The achievements of American Orientals.* Hillsdale, NJ: Erlbaum.

Fodor, J. (1975). *The language of thought.* New York: Crowell.

Gallistel, C., Brown, A., Carey, S., Gelman, R., & Keil, F. (1991). Lessons from animal learning for the study of cognitive development. In S. Carey & R. Gelman (Eds.), *The epigenesis of mind: Essays on biology and cognition.* Hillsdale, NJ: Erlbaum.

Gardner, H. (1973a). *The arts and human development.* New York: Wiley. Reprinted by Basic Books, 1994.

Gardner, H. (1973b). Children's sensitivity to musical styles. *Merrill-Palmer Quarterly, 19*, 67–77.

Gardner, H. (1975). *The shattered mind.* New York: Vintage.

Gardner, H. (1982). *Art, mind, and brain.* New York: Basic Books.

Gardner, H. (1985). *The mind's new science.* New York: Basic Books.

Gardner, H. (1989). *To open minds.* New York: Basic Books.

Gardner, H. (1991). *The unschooled mind.* New York: Basic Books.

Gardner, H. (1992). Scientific psychology: Should we bury it or praise it? *New Ideas in Psychology, 10*(2), 179–90.

Gardner, H. (1993a). *Frames of mind: Tenth anniversary edition.* New York: Basic Books.

Gardner, H. (1993b). *Creating minds: An anatomy of creativity as seen through the lives of Freud, Einstein, Picasso, Stravinsky, Eliot, Graham, and Gandhi.* New York: Basic Books.

Geertz, C. (1973). *The interpretation of cultures.* New York: Basic Books.

Gombrich, E. H. (1960). *Art and illusion.* Princeton: Princeton University Press.

Goodman, N. (1976). *Languages of art.* Indianapolis: Hackett.

Gordon, E. (1965). *Musical aptitude profile manual.* Boston: Houghton Mifflin.

Greenough, W. T., Black, J. E., & Wallace, C. S. (1987). Experience and brain development. *Child Development, 58*, 539–59.

Gregg, M., & Leinhardt, G. (1994). Mapping out geography: An example of epistemology and education. *Review of Educational Leadership, 64*, 311–61.

Harper, L., & Sanders, K. (1975). Preschool children's use of space. *Developmental Psychology, 11*, 119.

Harris, L. (1981). Sex related variations in spatial skill. In L. Liben, A. Patterson, & N. Newcombe (Eds.), *Spatial representation and behavior across the lifespan.* New York: Academic press.

Heath, S. B. (1983). *Ways of worldmaking.* New York: Cambridge University Press.

Hewes, D. (1982). Pre-school geography. *Journal of Geography, 81*, 94–7.

Howe, M. (1990). *The origins of exceptional abilities.* Cambridge, MA: Blackwell.

Karmiloff-Smith, A. (1992). *Beyond modularity.* Cambridge: MIT Press.

Keil, F. (1981). Constraints on knowledge and cognitive development. *Psychological Review, 88*, 197–227.

Kennedy, J. (1974). *The psychology of picture perception.* San Francisco: Jossey-Bass.

Kessen, W., Levine, J., & Peindich, K. (1978). The imitation of pitch in infants. *Infant Behavior and Development, 2*, 93–9.

Kritchevsky, M. (1988). The elementary spatial functions of the brain. In J. Stiles-Davis, U. Bellugi, & M. Kritchevsky (Eds.), *Spatial cognition: Brain bases and development.* Hillsdale, NJ: Erlbaum.

Krumhansl, C. (1990). *Cognitive foundations of musical pitch.* New York: Oxford.

Landau, B. (1988). The construction and use of spatial knowledge in blind and sighted children. In J. Stiles-Davis, U. Bellugi, & M. Kritchevsky (Eds.), *Spatial cognition: Brain bases and development.* Hillsdale, NJ: Erlbaum.

Landau, B., Spelke, E., & Gleitman, H. (1984). Spatial knowledge in a young blind child. *Cognition, 16*, 225–60.

Langer, S. (1942). *Philosophy in a new key.* Cambridge: Harvard University Press.

Langer, S. (1976). *Mind: An essay on human feeling.* Baltimore: Johns Hopkins University Press.

LeVine, R. (1991, September). *The new environmentalism: Social and cultural influences on child development.* Paper presented at the Centennial of Education at Harvard, Cambridge, MA.

Liben, L., & Downs, R. (1991). The role of graphic representations in understanding the world. In R. Downs, L. Liben, & D. Palermo (Eds.), *Visions of aesthetics, the environment, and development.* Hillsdale, NJ: Erlbaum.

Margules, J., & Gallistel, C. (1988). Heading in the rat: Determination by environmental shape. *Animal Learning and Behavior, 10*, 404–10.

Marshack, A. (1991). *The roots of civilization*. Mount Kisco, NY: Moyer Bell Ltd.

Marzolf, D., & DeLoache, J. (1994). Transfer in young children's understanding of spatial representations. *Child Development, 65*, 1–15.

McFarland, H., & Fortin, D. (1982). *Archives of Neurology, 39*, 725–7.

Merzernich, M., Kaas, J. H., & Wall, J. T. (1983). Topographic reorganization of somatosensory cortical areas 3b and 1 in adult monkeys following restricted deafferentation. *Neuroscience, 8*, 33–55.

Morrow, L., & Ratcliff, G. (1988). The neuropsychology of spatial cognition. In J. Stiles-Davis, U. Bellugi, & M. Kritchevsky (Eds.), *Spatial cognition: Brain bases and development*. Hillsdale, NJ: Erlbaum.

Neville, H. (1984). Effects of early sensory and language experience on the development of the human brain. In J. Mehler & R. Fox (Eds.), *Neonate cognition: Beyond the blooming buzzing confusion*. Hillsdale, NJ: Erlbaum.

Newell, A. (1980). Physical symbols system. *Cognitive Science, 4*, 135–83.

Ogden, C. K., & Richards, I. A. (1929). *The meaning of meaning*. London: K. Paul.

Peirce, C. S. (1955). *Selected philosophical writings*. New York: Dover.

Piaget, J. (1962). *Play, dreams, and imitation*. New York: Norton.

Piaget, J., & Inhelder, B. (1967). *The child's conception of space*. London: Routledge and Kegan Paul.

Piechowski, M. (1993). "Origins" without origins. *Creativity Research Journal, 6*(4), 465–9.

Revesz, G. (1925). *The psychology of a musical prodigy*. Freeport, NY: Books for Libraries Press.

Rheingold, H., & Cook, K. (1975). The content of boys' and girls' rooms as an index of parents' behavior. *Child Development, 46*, 459–63.

Richards, I. A. (1929). *Practical criticism*. New York: Harcourt Brace.

Rogoff, B. (1990). *Apprenticeship in thinking*. New York: Oxford University Press.

Rorty, R. (1979). *Philosophy and the mirror of nature*. Princeton, NJ: Princeton University Press.

Seashore, C. (1938). *Psychology of music*. New York: Dover.

Sergent, J. (1993). Music, the brain and Ravel. *Trends in Neurosciences, 16*, 5.

Seyfarth, R., Cheney, D., & Marler, P. (1980). Monkey responses to three different alarm calls: Evidence of predator classification and semantic communication. *Science, 210*(14), 801–3.

Shweder, R., & LeVine, R. (1984). *Culture theory*. New York: Cambridge University Press.

Sloboda, J. (1994). Musical performance: Expression and the development of excellence. In R. Aiello & J. Sloboda (Eds.), *Musical perceptions*. New York: Oxford University Press.

Sloboda, J., Davidson, J., & Howe, M. (1994). Is everyone musical? *The Psychologist, 7*, 349–54.

Snow, C. (1991, September). *The new environmentalism: Social and cultural influences on language and communicative development*. Paper presented at the Centennial of Education at Harvard, Cambridge, MA.

Snow, C. P. (1959). *The two cultures and the scientific revolution*. New York: Cambridge University Press.

Spencer, C., Harrison, N., & Darvizeh, Z. (1980). The development of iconic mapping ability in young children. *International Journal of Early Childhood, 12*, 57–64.

Sperber, D. (1975). *Rethinking symbolism*. New York: Cambridge University Press.

Sternberg, R. J., & Wagner, R. (Eds.). (1994). *Mind in context*. New York: Cambridge University Press.

Stigler, J. W., Shweder, R. A., & Herdt, G. (Eds.). (1990). *Cultural psychology: Essays on comparative human development*. New York: Cambridge University Press.

Stiles-Davis, J. (1988). Spatial dysfunctions in young children with right cerebral hemisphere injury. In J. Stiles-Davis, U. Bellugi, & M. Kritchevsky (Eds.), *Spatial cognition: Brain bases and development*. Hillsdale, NJ: Erlbaum.

Swanwick, K., & Tillman, J. (1986). The sequence of musical development: A survey of children's composition. *British Journal of Music Education, 3*, 3.

Torff, B., & Winner, E. (1994). Don't throw out the baby with the bathwater: On the role of innate factors in musical accomplishment. *The Psychologist, 7*, 361–2.

Viscott, D. (1970). A musical idiot savant. *Psychiatry, 33*, 494–515.

Vygotsky, L. (1978). *Mind in society*. Cambridge: Harvard University Press.

Willats, J. (1977). How children learn to represent three-dimensional space in drawings. In G. Butterworth (Ed.), *The child's representation of the world*. New York: Plenum Press.

Winner, E. (1982). *Invented worlds*. Cambridge: Harvard University Press.

Witelson, S., & Swallow, J. (1988). Neuropsychological study of the development of spatial cognition. In J. Stiles-Davis, U. Bellugi, & M. Kritchevsky (Eds.), *Spatial cognition: Brain bases and development*. Hillsdale, NJ: Erlbaum.

Woodworth, R. S. (1941). *Heredity and environment: A critical survey of recent material on twins and foster children*. A report prepared for the Committee on Social Adjustment, Social Science Research Council, New York, New York.

9 A cultural-psychology perspective on intelligence

Joan G. Miller

Those psychologists who do work with human beings may fail entirely to see the relevance of culture. . . . [P]sychologists accept that while everyone has culture, it is mainly relevant elsewhere where it produces certain exotic effects that anthropologists study. It is as if others have culture while we have human nature. (Schwartz, 1992, p. 329)

This chapter examines the role of culture in intelligence and the implications that a cultural perspective has for understanding hereditary and environmental influences on intelligence. This discussion builds on work being conducted in the interdisciplinary perspective of cultural psychology (Bruner, 1990; Cole, 1990; Miller, 1994a; Shweder, 1990; Shweder & Sullivan, 1993). From this perspective, psychological processes are viewed as, in part, contextually and culturally constituted, if not also as contextually and culturally variable. Culture is seen as influencing which behaviors are considered to be intelligent, the processes underlying intelligent behavior, and the direction of intellectual development. It thus is essential to any theory that is oriented to understanding the respective contributions of hereditary and environmental factors in intelligence.

Traditionally, cultural considerations have tended to be taken into account in theories of intelligence solely from an ecological perspective, if at all (e.g., Berry, 1976; Dasen, Berry, & Witkin, 1979). From this ecological perspective, cultural forms are treated exclusively in functional terms as adapted to objective constraints given in the physical setting or in the social structure. While not denying the importance of an ecological view of culture, this chapter argues that cultural factors need to be understood as influences on intelligence that vary independently of ecological constraints. The view will be forwarded that to understand the development of intelligence requires consideration not only of the hereditary and environmental factors traditionally taken into account in biological models, but also attention to cultural meanings and practices that mediate the impact of these factors.

The first section of the chapter considers some of the reasons why cultural considerations are frequently ignored in theories of intelligence and in

269

psychology more generally. Attention is given in the second section to the assumptions of cultural psychology and to respects in which work within this tradition challenges various objections to cultural accounts that have been raised by psychological investigators. The third section centers on the role of culture both in defining what is considered intelligent behavior and as a constituent of such behavior. Finally, in the last section, general conclusions are drawn regarding the relationship of hereditary and environmental influences in intelligence. Challenging behavior-genetic formulations, it is maintained that intelligence must be viewed as culturally grounded and the development of intelligence understood to be an open process.

Common objections to cultural approaches

Discussion focuses on arguments made by various cognitive researchers for downplaying the impact of cultural factors on intelligence or for claiming that cultural factors are already adequately taken into account as the environmental component of biological models. The concerns raised in these arguments are widely shared among psychologists and contribute to the relative neglect of cultural considerations in psychology.

Redundancy of collective and individual information

One argument offered by cognitive theorists who downplay the impact of cultural considerations on intelligences is that cultural information is redundant with information that the individual can obtain through self-constructive processes. It is maintained that the world presents determinate logical and empirical structures that may be discerned through deductive or inductive processing of information (Kohlberg, 1981; Piaget, 1981; Turiel, 1983). To account for the development of individual cognitive processes, then, it is argued, merely requires consideration of the objective information available to the individual and of the individual's abilities veridically to process this information.

Culture, from this perspective, is treated as primarily representational in nature and as formed through the same processes of individual induction and deduction as in the case with individual knowledge. As Wells (1981), for example, argues:

It is difficult for anyone who has raised a child to deny the pervasive influence of socialized processing that surely surfaces as causal schemata originate through secondary sources such as parents. . . . Even though socialized processing may be an important determinant of knowledge about causal forces at one level, it nevertheless begs the question. How is it that the parents knew an answer? The issue is circular. That is precisely the reason that one must consider a more basic factor – namely original processing [p. 313].

Cultural information in this view is treated as nonessential in interpreting experience or in defining objective knowledge.

Passivity of enculturation

A second related objection to viewing intelligence as culturally patterned is the argument that enculturation represents a passive process in which the individual merely reproduces, in unmodified form, understandings current in his or her culture. Cultural accounts, it is maintained, fail to appreciate the constructive nature of cognitive processes (Nucci, in press; Turiel, 1983, 1994).

This assumption that cultural learning represents a passive process explains, at least in part, the downplaying of cultural considerations in cognitive-developmental theory and in related constructivist approaches (Piaget, 1966; Turiel, 1983; Turiel, Smetana, & Killen, 1991). In these views, cultural approaches are regarded as similar to behaviorist approaches in that they are assumed to portray the individual as merely accommodating to externally presented information. It is argued alternatively that development proceeds independently of cultural learning, with individuals unable to understand socially transmitted information until they have already constructed this information for themselves.

Failure to account for contextual variation

A third objection raised in regard to cultural perspectives is that they portray cultural meaning systems as homogeneous in nature and fail to either anticipate or explain contextual variation in behavior. It is seen as stereotypical for theorists to claim that cultures can be distinguished in terms of their contrasting meaning systems. Interpreted as a claim that behavior is uniform within a given cultural tradition, the assertion that cultures differ in their views of self (e.g., Markus & Kitayama, 1991; Miller, 1994b; Shweder & Bourne, 1982) is regarded as contradicted by evidence of contextual variation in behavior. Turiel (1983), for example, makes such an argument in dismissing assertions that Americans' social judgments reflect individualistic cultural values:

We have seen that social judgments in the United States do not fall into categories reflecting holistically integrated cultural schemes. Instead, it has been found that the social cognition of individuals reflects differing orientations, whose manifestation depends, at least, on the issue and domain in question. . . . The same individuals are relativists and universalists. They are concerned with group solidarity and individual rights. They can be absolutistic and flexible. . . . The research suggests that a coherent cultural scheme is not the direct source of the social judgments and behaviors of individuals [p. 214].

The charge that a cultural analysis is unable to account for contextual variation in behavior, in turn, leads to a questioning of the explanatory force of culture as a construct. Cultures are seen as so heterogeneous and nonintegrated that they cannot be reliably distinguished from each other but rather include essentially the same elements. This stance, then, leads to a view of context as an influence on cognition that is not only independent of culture but which has greater predictive power.

The existence of culture-free psychological processes

A fourth assumption that has led certain cognitive researchers to downplay cultural considerations is the assumption that intelligence can be understood in terms of psychological processes that are unaffected by culture. Thus, it is maintained that cultures may differ in the value placed on different intellectual operations or may differ in the contexts in which these operations are applied, but the basic properties of these operations can be understood in individual psychological terms without reference to cultural phenomena. As Salovey and Mayer (1994), for example, assert in arguing for such a position:

Intelligence, as defined by Western psychology, is the property of the individual, and that individual, idiocentric or allocentric, can have his or her intelligence gauged by abilities at manipulating symbols. Certainly intelligence can exist at a group level. ... But this collective intelligence is disparate in its meaning from individual intelligence [p. 310].

This type of assumption – that "the individual person is the sole unit of analysis" (Shweder, 1984, p. 3) – has led psychological investigators to treat cultural content as unwanted error variance to be controlled (1) in the search for culture-free psychological processes, and (2) in the measurement of individual differences in psychological functioning. Thus, effort has been directed to the construction of intelligence tests which, by employing procedures that are understood in similar ways in different cultural populations, are assumed to make it possible to tap psychological processes and abilities that are uncontaminated by cultural noise.

Importantly, the present stance gives privilege to a natural science model of explanation that is oriented toward the formulation of universal generalizations. Theories that purport to achieve such predictive power have tended to be embraced, whereas theories that indicate that psychological functioning may be more culturally or contextually bound have typically met with more resistance. As Ceci (1990) observes, for example, evidence of contextual variability in responses to psychometric measures of intelligence has typically led to a questioning of the specific theory of intelligence being tested and not to a rethinking of the nature of intelligence itself.

Problems inherent in relativism

An additional problem associated with cultural perspectives is that they tend to be relativistic in nature and thus are associated with various problems viewed as inherent in relativism. Such problems include the failure of relativistic stances to provide a standard from which to evaluate diversity and the logical incoherence of relativistic arguments.

Cultural approaches, it is charged, lead to an unchecked proliferation of views of intelligence, since they provide no criteria other than consensus for determining what is adaptive or socially valued in a given cultural group. As Salovey and Mayer (1994) assert:

> By defining intelligence as adaptation, we require that its measurement be culture-bound, which is fine, so long as there is broad agreement about what is adaptive in a particular cultural context. For example, does protesting against an authoritarian government in China represent high or low intelligence from the adaption point of view? An adaptation view of intelligence is not sufficiently worked out to answer such a question [p. 310].

Relativistic definitions are seen as particularly problematic when moral issues are involved, since they provide no criteria for comparative evaluation of practices. As critics have charged, relativism appears to lead to the absence of any moral standards and by default to the condoning of moral abuse (Hatch, 1983; Turiel, 1983).

Finally, relativistic approaches are rejected as logically incoherent and as leading to a solipsistic and nihilistic stance that is judged to be incompatible not only with comparison but also with science (Kohlberg, 1971). Articulating this type of concern, Spiro (1990) claims that a relativistic stance ironically impedes rather than enhances understanding of diversity:

> The only solution to the cultural subjugation of Third World peoples by Western anthropology, so it is claimed, is to create a non-Western anthropology – one that is informed not by Western values, but by the radically different values of Third World cultures. When these cultures are conceptualized by that alternative – their own – ethnoscience, they still, of course, remain Other vis-à-vis Western culture, but they achieve a status (at the very least) of equality with it. In sum, the strange can only become the equal of the familiar not, paradoxically enough, by being made familiar but by remaining strange [pp. 53–4].

Notably, this objection assumes that only two extreme positions are possible: an *ethnocentric approach*, in which behaviors are compared in terms of the standard of one cultural group (commonly that of the American psychologist or anthropologist), or a *culturally embedded approach*, in which behaviors are analyzed in terms of indigenous categories and comparison or translation is impossible. Of these two extremes, the second relativistic approach is seen as more seriously flawed, and the first universalistic approach as preferred.

Cultural psychology: Assumptions and responses to objections

In this section, discussion focuses on the perspective of cultural psychology, presenting both the key assumptions of this viewpoint and the ways in which it addresses some of the objections to cultural approaches described in the first section. The term *cultural psychology* is applied in this chapter as an umbrella concept to characterize a loosely defined emerging tradition characterized by a variety of shared assumptions and goals (e.g., Bruner, 1990; Cole, 1990; Markus & Kitayama, 1991; Miller, 1994a, in press; Pepitone, 1976; Shweder, 1990; Shweder & Sullivan, 1993). It encompasses work associated with a wide range of perspectives including but not limited to situated cognition, action theory, psychological anthropology, sociolinguistics, and cultural-social psychology. What defines an approach as part of the perspective of cultural psychology is its theoretical commitments and not whether or not use is made of cross-cultural (as contrasted with single-culture) research designs. Thus, in this view, research that involves more than one comparison group may be part of cultural psychology. Not all theorists whose work is considered as within this tradition explicitly use the label *cultural psychology* or necessarily share all of the assumptions identified with this perspective in this chapter.

Roots in cross-cultural psychology

Assumptions of cross-cultural psychology. Cultural psychology may be briefly distinguished from the field known as *cross-cultural psychology*, which represents only one of its roots (e.g., Munroe, Munroe, & Whiting, 1981; Triandis & Lambert, 1980; Triandis & Lonner, 1980). Cross-cultural psychology shares more of the goals and assumptions of mainstream psychology than does cultural psychology. As in mainstream psychology, it is assumed that the ultimate goal of psychological research is to uncover universal generalizations. Also, as in mainstream psychology, culture is conceptualized primarily in functional terms, as adapted to the objective ecological constraints existing in a given setting.

The major difference that distinguishes cross-cultural psychology from much of mainstream psychology is its assumption that the contexts to which humans adapt are culturally variable rather than universal on a functional level. The aim of research in cross-cultural psychology, then, has been not only to test the universality of existing psychological theories but also to identify cultural parameters that affect the operation of presumed universal psychological processes.

An example of work in this tradition is early research that tested the universality of Piagetian theory (Dasen, 1972, 1977). This research proceeded initially by administering standard cognitive developmental tasks in different cultural contexts, after these tasks were translated into the local languages. The results were interpreted as supporting the universality of Piagetian theory. They also suggested, however, that in Western urbanized populations the rate of development is more rapid and higher endpoints are obtained than in other cultural populations.

Toward cultural psychology. One of the most important challenges to early research on culture and cognitive development came from investigators who were sensitive to the contrasting meaning of experimental contexts for different populations. Cole, Gay, Glick, and Sharp (1971), for example, demonstrated that when cognitive abilities are assessed utilizing procedures that employ familiar content or that are motivationally engaging, it may be possible to demonstrate the presence of cognitive competencies among individuals from traditional cultures that are not observed utilizing experimental assessment measures in their standard form. Thus, it is shown that individuals from traditional cultures who fail standard Piagetian classification tasks, which involve geometric shapes, tend to succeed on such tasks or even to do better than highly educated Americans when tested on content items, such as types of rice, which have greater ecological validity. Such results, it may be noted, find parallels in research undertaken in cognitive-developmental psychology that has been sensitive to the contrasting meaning of experimental tasks for different age groups (e.g., Donaldson, 1978; Gelman, 1978; Gelman & Baillargeon, 1983). The cross-cultural findings not only challenge claims that individuals from traditional cultures progress at a slower rate of development than do those from modernized cultures but also call into question claims that cognitive processes are highly generalized. As Cole and Scribner (1974) concluded in relation to the implications of this type of comparative experimental research, a "major implication of this view for cross-cultural work is that *we are unlikely to find cultural differences in basic component processes*" (p. 193).

Interestingly, this early challenge to stage models of culture and cognitive development, while innovative in the context of psychology, was accorded a somewhat lukewarm welcome by many anthropological investigators. The assumption that individuals in all cultures are cognitively competent at the skills demanded by their communities was considered unsurprising. As Schwartz (1981) commented in relation to this type of research:

An anthropologist will not always be satisfied will the work of their extradisciplinary associates who seem often to be proving repetitively what our experience has led us to take as axiomatic – that people everywhere are more or less competent in their own cultures and will acquire the cognitive skills demanded by the tasks which their culture requires of them. People will demonstrate this competence when tasks are presented in culturally familiar forms. We would further wish to understand the variation in cognitive tasks and the competence to meet them required by different cultures in the cultural implementation of the intellect [p. 5].

In this regard, Shweder (1977) also cautions that the conclusion that there are no differences in cognitive capacities between different populations is as unwarranted as is the conclusion that such differences exist. As he notes, the unmet challenge in supporting such a claim is to ensure that the contexts in which behavior is assessed have the same meaning to the different cultural groups being compared. Importantly, from this perspective, the possibility is left open that there may be contrasting intellectual capacities developed within different cultural communities just as there are individual differences in intellectual capacities that develop within any given cultural group.

Major assumptions of cultural psychology

Cultural psychology is emerging or, it may be argued, reemerging (Jahoda, 1993) as a perspective that builds on this concern with understanding culturally patterned diversity in psychological functioning. The perspective of cultural psychology rejects the assumption that a dichotomy exists between psychological structures and processes, which are universal, and cultural content and contexts, which are variable – a dichotomy assumed in mainstream psychology (Miller, 1988). Rather, the fundamental premise of cultural psychology is monistic. Psychological processes are seen as, in part, culturally constituted, if not also as culturally variable, with culture viewed as necessary to the development of individual capacities. As Geertz (1973) argues:

There is no such thing as a human nature independent of culture. . . . We are, in sum, incomplete or unfinished animals who complete or finish ourselves through culture – and not through culture in general but through highly particular forms of it [p. 49].

From a cultural-psychology perspective, it is assumed that psychological processes depend on cultural meanings that are symbolic in character and thus which do not bear a one-to-one causal relationship to genetic factors. Although psychological processes must conform to biological constraints, such constraints are viewed as representing only broad limits on human functioning rather than as causal determinants of such functioning. As Bruner (1990) argues:

The biological substrate, the so-called universals of human nature, is not a cause of action, but, at most, a *constraint* upon it or a *condition* for it. The engine in the car does not "cause" us to drive to the supermarket for the week's shopping, any more than our biological reproductive system "causes" us with very high odds to marry somebody from our own social class, ethnic group, and so on [p. 21].

Making a similar point about the importance of shared meaning systems in human behavior, Wahlstein and Gottlieb note that "genetic information does not specify what we think or remember, any more than the colors of the artist's palette dictate the artistic quality of the painting" (Wahlstein & Gottlieb, chapter 6, this volume).

Importantly, the thrust of a cultural-psychology perspective is to highlight the hierarchical relationship obtaining between biological constraints and psychological processes (Sahlins, 1976a,b). Thus, for example, a biologically imposed limit on short-term memory, such as that identified by Miller (1956) in his principle of "seven plus or minus two," is overcome through the cultural symbol systems that enable individuals to organize information in a way that transcends this limitation (e.g., Ericsson, Chase, & Faloon, 1980). Similarly, in a somewhat different type of example, it has been shown that newborns have an elaborate biologically based capacity to detect categorical distinctions in sound (Werker, 1989). Cultural meanings, however, have the effect of narrowing this biological propensity, with individuals losing the capacity to make certain auditory discriminations that they are not exposed to in their native language. Equally, although mating is a biological universal, it is culture that determines the specific institutions of free love, dating, arranged marriage, and so forth that embody mating in a particular context and lead to the very salient differences in experiences and in individual subjectivity through which individual lives are lived.

While not denying the existence or importance of psychological universals, the perspective of cultural psychology rejects the idea of psychic unity as a premise. As Jahoda (1993) notes, the dominance of this idea in psychology represents a relatively recent historical occurrence. It developed into the prevailing perspective at the turn of the 19th century partly in response to the popularity at that time of racial interpretations of cultural diversity – interpretations generally invoked to provide a biological grounding for the assumed superiority of European culture and European minds. With the popularity of natural-science models of explanation, the postulate of psychic unity has continued to dominate the discipline throughout the last century.

It is argued from the perspective of cultural psychology that this postulate may have outlived its usefulness and that many of the behavioral phenomena of most interest in psychology are likely to be culturally variable – in that they are premised on culturally diverse meaning systems and practices. Retention of the postulate of psychic unity, it is cautioned, contributes to the neglect of cross-cultural research – a neglect that, in turn, threatens the integrity of claims to universality:

The disinclination to conduct cross-cultural research has the long-term effect of protecting hypotheses against falsification. . . . To presume universality from intraindividual mechanisms and processes when one is really observing the script of a Western or even a single culture is to manufacture an illusory or pseudoscience of human social behavior [Pepitone & Triandis, 1987, pp. 476, 495].

Theory and research undertaken from a cultural-psychology perspective is addressing this perceived gap in current understandings of psychological processes. In investigations conducted in a wide range of content domains, including cognition, self, motivation, emotion, morality, personality, and health, investigators are uncovering the implicit cultural grounding of many existing psychological theories as well as documenting culturally based variation in psychological functioning.

Response to critiques of cultural approaches

The cultural dependence of objective knowledge.   From the perspective of cultural psychology, the claim that individual and collective knowledge are redundant is criticized for failing to recognize the interpretive element entailed in descriptive inference (Bransford & McCarrell, 1974; Goodman, 1972) and for failing to recognize that cultural meaning systems include prescriptive and constitutive functions rather than only representational functions (D'Andrade, 1984). On a prescriptive level, cultures may be seen to specify norms that may or may not be reflected in actual patterns of behavior and thus that may not be derived merely from observation of such patterns. In terms of their constitutive functions, cultural meaning systems serve to define reality, with concepts such as marriage, human rights, witches, or even emotions reflecting cultural meanings contributed to experience rather than merely or even necessarily observable natural facts (Schneider, 1968, 1976). How a parent may have knowledge to impart to his or her child cannot be explained merely in terms of original processing (Wells, 1981, p. 313) but reflects the impact of cultural beliefs and standards (Sabini & Silver, 1981).

Active and creative nature of enculturation.   The perspective of cultural psychology offers a theory of enculturation that differs markedly from the behaviorist models to which it has been linked by some critics. It rejects the oversocialized conception of the person as merely conforming to the dictates of the culture – a perspective that informed some early research in culture and personality (Shweder, 1979). Equally, it rejects FAX models of cultural transmission, with their views that cultural information is transmitted to individuals in unmodified form – an assumption that until relatively recently was made in much work in psychological anthropology.

Alternatively, although individual subjectivity is seen as always culturally influenced, it is not viewed as part of a one-to-one relationship with cultural meanings (Markus & Kitayama, 1994; Miller, 1994a). Cultural schemata are also viewed as selectively drawn upon and transformed by individuals while developing their conceptual understandings in a process that is cognitively complex and active (D'Andrade, 1984; D'Andrade & Strauss, 1992).

The present view of cultural learning treats both the endpoint and course of development as culturally patterned, if not also as culturally variable. It thus may be seen to be less deterministic and to give greater weight to human agency than do constructivist viewpoints, with their assumption that development conforms to a universal sequence given in the structure of reality.

Culturally interpreted contexts. Cultural psychology is premised on a contextual view of psychological functioning. Importantly, however, the effects of culture are assumed to be dependent on cultural meanings rather than culture-free. Such a process is illustrated in a recent cross-cultural investigation of person attribution in which contextual factors were observed to have opposite effects among Japanese as compared with Americans (Cousins, 1989). The research tapped self-descriptions on both a decontextualized task, which asked subjects to respond to the question "Who am I?", and on a contextualized task, which asked subjects to describe themselves, in turn, when at home, at school, and with close friends. Whereas Japanese were found to be more abstract on the contextualized than the decontextualized person-description task, Americans were found to be more abstract on the decontextualized task. These effects were interpreted as reflecting the contrasting cultural views of self emphasized in each culture. In assuming that behavior is naturally contextually embedded, Japanese felt less need to qualify their responses on the contextualized task. In contrast, reflecting their assumption that behavior is naturally generalized, Americans experienced difficulty in communicating their abstract views of themselves on the contextualized task.

The perspective of cultural psychology not only views cultural interpretations of behavior as compatible with context effects but also rejects the charge made by critics that the effects of culture are so contextually variable that it is impossible to identify consistent patterns of group differences between cultural communities. As Bruner (1990) notes, an attention to cultural meanings promises less rather than more subjectivity in psychology, with even such private phenomena as secrets conforming, in ways, to publicly interpretable and ordered patterns.

Cultural grounding of psychological processes. From the present cultural perspective, as noted already, individual psychological processes are viewed as always situated in a cultural context and as, in part, culturally constituted. It is assumed that psychology is inherently culturally influenced and that the major need is to make an awareness of this cultural grounding overt rather than covert, as it presently is in much of the field (Pepitone & Triandis, 1987; Markus & Kitayama, 1991). A theory is not considered to be flawed if it is discovered to be culturally bound. Rather, it is in the understanding of this cultural grounding that a fuller process of understanding psychological functioning may be achieved.

Pragmatic relativism. From a cultural-psychology perspective, extreme relativism, with its postulate of radical incommensurability, is rejected as an untenable position. However, in contrast to the position discussed earlier, it is also asserted that the alternative to extreme relativism should not be universalism, with its attendant ethnocentrism and parochialism. Rather, the argument is made for a more moderate or pragmatic relativism. It is maintained that whereas there are no culture-free or observer-independent ways of viewing experience, a move toward a more relativistic perspective leads to a greater appreciation of the other.

The present position, then, is compatible with evaluating the beliefs and practices of different cultures in terms of standards that are not based merely on indigenous criteria. However, it encourages a greater reflection and self-awareness on the part of the observer in making such judgments. As Bruner (1990) describes such a stance, it leads not to inaction but to a more informed and less ethnocentric stance from which to appraise diversity:

> The best we can hope for is that we be aware of our own perspective and those of others when we make our claims of "rightness" or "wrongness". . . . Asking the pragmatist's questions – How does this view affect my view of the world or my commitments to it – surely does not lead to "anything goes." It may lead to an unpacking of presuppositions, the better to explore one's commitments. (p. 27)

This type of relativistic approach assumes that (1) there is the possibility of gaining increasing, although never complete, understanding of alternative beliefs and practices, and (2) from such understanding comes a basis for building a less parochial psychology.

Culture and intelligence

Intelligence as abstract thought

The most traditional and widely accepted definition of *intelligence* is in terms of the g factor of assumed general intelligence. From this perspective, intelligence is accorded a determinate content as the specific constellation of abstract abilities assessed by intelligence tests. Much of the appeal of IQ measures is in their ease of measurement and apparent predictive power, with correlations demonstrated between IQ scores and a range of desired outcomes, including school achievement, work productivity, mental health, and so on (e.g., Gottfredson, 1986; Humphreys, 1994).

This psychometric view of intelligence has been challenged in recent years by investigators who offer models that give greater weight to the impact of context, knowledge, and motivational factors on IQ and call into question its predictive power (e.g., Ceci, 1990, 1991; Sternberg, 1985). It is shown that IQ performance is closely related to schooling, parenting style, occupational status, and other related variables, with these factors, in turn, often constituting better predictors of outcome than IQ. Evidence also indicates that performance on IQ tests is affected by language, background knowledge, interest, and a range of variables other than abstract cognitive abilities. Research from a cultural-psychology perspective builds on the same types of concerns raised in these latter accounts in (1) questioning the adequacy of abstract cognitive abilities as a criterion of intelligence, and in (2) highlighting the cultural dependence of abstract cognitive behavior (Irvine & Berry, 1988).

Cultural variation in valuing of abstract thought. Although both abstract and concrete modes of cognition are valuable for differing purposes, cross-cultural research reveals that (1) abstract modes of thought tend to be treated as prototypical in modern Western cultures, and (2) concrete modes as prototypical in traditional cultures. This difference is noted, by anthropologists in a distinction between religious orientations associated with each type of culture:

Traditional religions attack problems opportunistically as they arise in each particular instance ... employing one or another weapon chosen on grounds of symbolic appropriateness, from their cluttered arsenal of myth and magic.... The approach ... is discrete and irregular.... Rationalized religions ... are more abstract, more logically coherent and more generally phrased.... The question is no longer, ... to use a classical example from Evans-Pritchard, "Why has the granary fallen on my brother ... ?" but rather "Why do the good die young and the evil flourish as the green bay tree?" [Geertz, 1973, p. 172].

Similarly, Ramanujan observes that Hindu Indian culture contrasts with Western culture in its valuing of what he characterizes as *context-sensitive* (as compared with *context-free*) cognitive orientations (Ramanujan, 1990). In another example, evidence suggests that among both the Thai as well as the Kpelle of North Africa, contextualization rather than universalization is accorded greater cultural value (Goodnow, 1976).

Despite the fact that many cultural groups do not share the idealization of abstract thought found in modern Western cultures, this mode of thought has tended to be portrayed as the natural outcome of development. As Goodnow (1976) asserts:

In our culture, for instance, efforts towards generalization are usually regarded as "good" and we may easily slide into regarding generalization as something to be expected in the course of normal development, something intrinsic to the nature of thinking itself. From this point of view, it may be no accident that Bruner (1973) describes thinking as "going beyond the information given," and Bartlett (1958) describes it as "gap filling" [p. 173].

This claim that development proceeds toward increasing abstraction is made in theories forwarded to explain societal as well as individual development. On a societal level, theorists link technological development to the emergence of the values and practices associated with Western culture (Triandis, 1990), whereas on an individual level, cognitive developmental theorists portray a universal age-related transition as occurring from "context-bound" and "egocentric" orientations to orientations that are increasingly "abstract" and "objective" (Piaget & Inhelder, 1969; Werner, 1948). In turn, in certain theories that link societal development to individual development, exposure to modernizing influences, such as written language and formal schooling, is seen as necessary and sufficient for the emergence of abstract thought (Luria, 1976).

Recent cultural analyses reveal, however, that modernization and technological development is not inevitably associated with Westernization in cultural values (Yang, 1988). As illustrated in accounts of the response in Hindu India to the imposition of British colonial laws (O'Flaherty & Derrett, 1978) and of Japanese responses to American popular culture (Tobin, 1992), modern Western cultural institutions may be radically transformed and indigenized at the same time they are adopted. Practices such as formal schooling, then, do not appear to be embodied in the same form in all cultural contexts but rather to exist in culturally specific forms. Thus, for example, the hierarchically organized classrooms and emphasis on public criticism found commonly in Japanese elementary schools may be seen to reflect the interdependent cultural view of self that is stressed in Japanese culture, just as the more decentralized classrooms and emphasis on individual praise found commonly in American elementary schools reflect the individualistic cultural view of self that is stressed in modern American culture (Stigler & Perry, 1990).

Cultural dependence of abstract thought.   In terms of ontogenetic development, available cross-cultural developmental research provides some indication that a tendency to utilize more abstract modes of cognition is not an inevitable consequence of development. It is shown, for example, that the tendency to focus on abstract modes of person attribution is dependent on childrens' acquisition over development of a particular cultural view of self rather than on an age-linked gain either in the capacity for abstract thought or information about social experience. With increasing age, Americans tend to place greater emphasis on abstract psychological dispositions and to describe others in a more impersonal way – an age change not observed among Hindu Indians. In contrast, with increasing age, Hindu Indians tend to place greater emphasis on aspects of the social context and to describe others in a more self-involved way – an age change not observed among Americans (see also Bond & Cheung, 1983; Shweder & Bourne, 1982). The pattern of developmental change found among Americans appears to result from their highly individualistic cultural views of the self, whereas the pattern of developmental change found among Hindu Indians appears to result from their more relational cultural views of the self. As the person–situation debate in personality theory implies (Mischel, 1968, 1973), the abstract person schemas acquired as the endpoint of development among Americans may be seen to be more adequate than those of younger children only in terms of their congruence with an individualistic cultural world view and not necessarily in terms of their having greater predictive power.

Including work in situated cognition and in the sociohistorical tradition of Soviet psychology (e.g., Lave & Wenger, 1991; Valsiner, 1988; Wertsch, 1991), a focus of current cultural research is also on understanding respects in which cognitive performances are adapted to culturally specific contexts (Lave, 1988; Rogoff & Wertsch, 1984) as well as respects in which cultures provide tools for thought that lead their members to become experts in particular cognitive performances (e.g., Stigler, 1984; for reviews of research in this tradition, see, e.g., Cole, 1988, 1990; Laboratory of Comparative Human Cognition, 1983; Rogoff, 1990). In contrast to the assumption made in psychometric approaches that generality in cognitive performance can be expected, only limited cognitive transfer is assumed. Furthermore, this generality is viewed as dependent more on the similarity that exists in the meaning and structure of particular contexts than on the existence of generalized individual cognitive capacities. Thus, as Scribner and Cole (1981) illustrate, the literacy skills that are developed in the practice of composing letters facilitate performance on a referential communication task that draws on related skills but do not enhance performance on more remotely related cognitive tasks. It is concluded that although cultural symbols and meanings provide powerful tools for thought, they do not result in a general amplification in cognitive performance such that their effects remain in situations in which they are no longer available to the tool user (Cole & Griffin, 1980).

Intelligence as everyday adaptive competence

In recent years, there has been increased interest in understanding intelligence in terms of the range of competencies that enable successful adaptation to everyday environments (e.g., Gardner, 1983; Sternberg, 1985; Sternberg & Kolligian, 1990; Sternberg & Wagner, 1986). It is recognized that although intelligence tests predict performance in school, they are much less successful in predicting performance in contexts that present contrasting demands. Importantly, this concern with linking intelligence to everyday problem-solving behavior leads to an expansion in the range of competencies associated with intelligence, from the relatively narrow list of abstract analytical competencies assessed on standard intelligence test measures, to a broad range of social, emotional, and motivational competencies considered to be important in adaptation.

Approaches to intelligence as everyday adaptation have been presented in both universalistic and in more relativistic forms. In universalistic forms, the assumption is made that the contexts to which individuals adapt can be considered functionally similar and thus that a common set of intellectual processes can be identified as adaptive for all populations. As noted in the first section of this chapter, this type of assumption is made commonly among cognitive researchers and contributes to the downplaying of cultural considerations in their work.

In contrast, in more relativistic positions, it is recognized that cultural and subcultural groups promote the development of different competencies and thus that the behaviors that constitute intelligence may be culturally relative (Charlesworth, 1979; Ruzgis & Grigorenko, 1994; Sternberg, 1988). For example, as Sternberg (1988) illustrates, whether or not a particular ability, such as music, comprises part of intelligence depends on the practices of the particular culture under consideration. In a culture that forbade musical expression, he observes, no opportunity would be provided to exercise and thus to develop such an ability. This type of contextually sensitive approach to intelligence captures an important thrust of a cultural-psychology perspective in its concern with the role of cultural meanings and practices in defining what is adaptive and promoted in a given cultural setting.

Symbolic elements in adaptation.   One way in which a cultural-psychology perspective contributes to an understanding of the adaptive nature of intelligence is in highlighting the nonrational elements entailed in adaptation. From this perspective, individuals are seen as adapting not only to objective constraints, given in the physical and social ecology, but also to arbitrary culturally defined symbolic constraints. These cultural presuppositions underly everyday practices and are frequently not cognitively available to the natives, who may justify their practices exclusively by reference to means–ends considerations. In an illustration of such a process, LeVine (1984) reports that whereas the Nyakyusa of Kenya interpret the local practice of segregating age groups into separate villages solely in functional terms, the practice also reflects arbitrary cultural presuppositions:

When Monica Wilson (1952) was told by a Nyakyusa informant that it was necessary to segregate adult generations into separate villages because otherwise the intergenerational avoidance taboos would cause inconvenience in daily life, the explanation was a rational one – but only if one assumes that the avoidance taboos represent a high priority for all concerned. . . . But the assumption that avoidance itself is necessary is not rational in the sense of a response to environmental contingencies [p. 79].

As Sahlins (1976a) similarly illustrates in a case drawn from modern American culture, although Americans consider their food practices to be based largely on functional considerations, these practices also reflect nonrational cultural presuppositions. For example, it is the American cultural concept of personhood that affords particular animals the status of pets that protects them from being categorized as potential food.

Some implications of this attention to arbitrary cultural meanings for understanding cultural influences on intelligence may be seen in a recent comparative study of parenting conducted by Harkness and Super (1992). They report that for the Kipsigis of Western Kenya, childhood intelligence is conceptualized primarily in terms of competence in carrying out family responsibilities, with doing chores as the single largest occupier of children's time from about age 3 onward. In contrast, for Americans, childhood intelligence is conceptualized primarily in terms of verbal and social skills associated with schooling, with American children occupied almost exclusively in play activity and doing few chores throughout childhood. Super and Harkness suggest that, from an ecological perspective, this cultural variation can be understood in terms of factors such as the greater poverty and associated need for child labor existing among the Kipsigis than among the Americans. Importantly, however, this functional interpretation is not sufficient to account for the observed cultural variation. Rather, it is shown that the differences arise at least in part from nonrational presuppositions held in each culture about the nature of persons, with the Americans concerned with developing children who have a strong psychological sense of themselves and the Kipsigis concerned with developing children who show the culturally valued properties of obedience, respect, and an absence of pride.

Cultural dependence of intelligent behavior. A second way in which a cultural-psychology perspective contributes to an understanding of the adaptive nature of intelligence is in highlighting the cultural grounding of adaptive competencies that have been identified as aspects of everyday intelligence. The following examples – in the areas of motivation, emotion, and morality – are considered to illustrate this contribution.

Motivation: In the area of motivation, the capacity to exercise personal control over the environment in order to achieve desired outcomes represents a social competence that has been identified as an important component of intelligence (e.g., Baltes & Baltes, 1986; Bandura, 1977; Deci & Ryan, 1987, 1990; Rotter, 1966; Seligman, 1975). As Maciel, Heckhausen, and Baltes (1994) argue in identifying personal control as an aspect of intelligence:

> The extent to which individuals attain their personal optimum largely hinges on their mental and physical abilities. The other major component of adaptiveness is the individual's motivation to act and exert control over his or her environment. . . . The realization of one's personal potential depends on abilities and on core features of the self, such as self-ascribed capacities and perceived personal control [p. 63].

A vast body of research conducted with Americans has demonstrated that having an internal as contrasted with external locus of control, as well as experiencing one's actions in a self-determined as compared with controlled way, is associated with more positive adaptive outcomes, such as higher self-esteem, more satisfying interpersonal relationships, and greater success in school and work contexts (for reviews, see Deci & Ryan, 1987, 1991; Lepper, 1983). Cross-cultural research utilizing Rotter's Internal–External Locus of Control scale (Rotter, 1966) has been typically interpreted as supporting the universality of this view of motivation, with individuals in non-Western cultures found to place greater weight on external relative to internal control orientations than do Americans (Dyal, 1984; Hui, 1982).

Recent cultural research that has paid more attention to indigenous cultural conceptions of the person, however, is documenting the cultural dependence of personal-control motivational orientations. It has been shown, for example, that Japanese tend to display a preference for alignment with others and with groups rather than for the self-reliance emphasized among Americans (Kojima, 1984; Weisz, Rothbaum, & Blackburn, 1984). Whereas among Americans a more autonomous style of control is associated with greater maturity and independence, among Japanese a shared locus of control is considered the more mature and adaptive motivational orientation (Azuma, 1984). In another example, research indicates that Hindu Indians associate well-being with maintaining an interdependent rather than a controlling relationship with the environment (Sinha, 1990) and that, in contrast to trends observed among Americans, individual satisfaction is perceived to be as great when behavior is constrained by interpersonal social norms as when it is undertaken under less controlling conditions (Miller & Bersoff, 1994a). In still another example, it has been shown that Korean adolescents associate greater perceived parental control with greater perceived parental warmth and less perceived parental neglect (Rohner & Pettengill, 1985; see also Weisz, 1990) – a finding that contrasts with the negative relationship commonly found among American populations between perceived social control and positive adaptive outcomes (Deci & Ryan, 1990).

The importance of this type of cultural research, it should be emphasized, is not merely to highlight the culturally bound character of present theories of motivation or, as Grolnick, Ryan, and Deci (1991) acknowledge in discussing certain results observed among Americans, "the cultural relativity of our model and findings" (p. 516). Rather, the contributions of the research are theoretical. On the one hand, the findings imply that the adaptive nature of personal-control orientations for Americans and for other Western populations depends, at least in part, on the individualistic cultural view of self that is emphasized in modern Western cultures. On the other hand, the findings raise the possibility that present dichotomous formulations, such as that of internal versus external locus of control (Rotter, 1966) or of self-determined versus controlled motives (Deci & Ryan, 1990), may be inadequate to account for motivation in various non-Western cultural groups that emphasize more relational cultural views of self. Rather, it is argued that more monistic constructs such as "shared" or "distributed" locus of control may be needed to capture the forms of motivation that are adaptive in the latter types of cultures (Miller & Bersoff, 1994a; Misra & Gergen, 1992).

Emotion: In the area of emotion, the capacity to interpret emotional states and to use this information in regulating one's behavior has been identified by various theorists as a form of intelligence that is critical in adaptation. Gardner (1983), for example, portrays "personal intelligence" as involving having "access to one's own feeling life – one's range of affects or emotions; the capacity instantly to effect discriminations among these feelings and, eventually, to label them, to enmesh them in symbolic codes, to draw upon them as a means of understanding and guiding one's behavior" (p. 239). Salovey and Mayer (1994) offer a similar view in their construct of emotional intelligence, which is defined as "the ability to monitor one's own and others' feelings and emotions, to discriminate among them, and to use this information to guide one's thinking and actions" (p. 312). Consonant with claims that emotional knowledge and self-regulation is an important component of intelligence, developmental research conducted among Americans documents that an increase with age occurs in childrens' abilities to identify facial expressions of emotions and to regulate their emotional expressions (e.g., Profyt & Whissell, 1991; Saarni & Harris, 1989).

Recent cultural research on emotion, however, calls into question the universality of such views of emotional intelligence. In contrast to previous cross-cultural research on facial categorization, this latter work examines emotions in terms of indigenous rather than only English-language categories, pays greater attention to the varied connotations of emotion terms, and emphasizes the importance of social interaction and not merely of internal states in understanding emotions (Lutz & White, 1986; Kitayama & Markus, 1994).

Ethnographic research reveals that many cultural groups do not share the views of emotion assumed in most psychological theories, with various non-Western cultural groups linking what might be considered to constitute emotional elicitors only to physical illness and not to affective experiences, making no distinction between thoughts and feelings and failing to objectify emotions (Lutz, 1988; Potter, 1988; Rosaldo, 1984; Shweder, 1993). The existence of this type of cultural difference in emotional experience is supported in a recent experimental investigation conducted among Americans and among the Minangkabau of West Sumatra (Levenson, Ekman, Heider, & Friesen, 1992). Results reveal that among both cultural groups, posing the face to mimic emotional expressions results in systematic autonomic changes. However, whereas these autonomic changes are interpreted by Americans in affective terms, they are not associated with a change in subjective feelings among the Minangkabau.

Recent comparative research on emotion is also demonstrating that the tendency to focus on personal affective reactions as a guide to regulating one's behavior is culturally variable. It has been shown that Japanese tend to report lower affect levels than do Americans (Akiyama, 1992) and tend to be less accurate than are Americans in recognizing emotions (Matsumoto, 1992). It has also been demonstrated that whereas, for Americans, knowledge about the self is more elaborated and distinctive than is knowledge about others, for Hindu Indians, knowledge about others is relatively more elaborated and distinctive (Markus & Kitayama, 1991). Available cross-cultural developmental research reveals that this type of cultural difference in whether individual subjectivity or the interpersonal context constitute domains for elaboration is reflected in everyday socialization practices. Thus, whereas among Americans, caregiver responses to a child's crying involve reorienting the child's attention to his or her subjective experience, among the Kipsigis, caregiver responses to this type of behavior involve redirecting the child's attention to an external focus (Harkness & Super, 1985).

The cross-cultural work reviewed implies that emotional intelligence, as presently defined, is adaptive for Americans and for other Westerners only against the background of their independent cultural definition of selfhood. As Markus and Kitayama (1994) conclude:

Americans tied to an independent cultural framework will highlight and emphasize their individual feeling states precisely because such feelings are self-definitional. . . . This fits with the culturally shared imperative to create, or define, or objectify the self. . . . Knowing that one feels, what one feels, and that one can instrumentally control one's emotions is extremely important in Anglo-American culture [p. 109].

In contrast, the findings suggest that in cultures emphasizing more interdependent cultural views of self, emotional intelligence, as defined in present theories, may not represent a central adaptive competence. Rather, in such cultures, as Markus and Kitayama (1994) note, "it is the intersubjectivity that results from interdependence and connection that receives a relatively elaborated and privileged place in the behavioral process" (p. 102).

Morality: The currently dominant framework for understanding interpersonal morality is the morality of caring framework developed by Gilligan (1982). In contrast to approaches to morality in the cognitive developmental and distinct domain traditions (e.g., Kohlberg, 1981; Turiel, 1983), Gilligan treats morality as integrally related to problems in adaptation and thus, in this sense, as a form of everyday intelligence. The moralities of justice and of caring are assumed to exist in the same form in all cultures and to reflect a universal adaptive tension between independence and attachment. As Gilligan and Wiggins (1987) argue:

The different dynamics of early childhood inequality and attachment lay the groundwork for two moral visions – one of justice and one of care. . . . Although the nature of the attachment between child and parent varies across individual and cultural settings and although inequality can be heightened or muted by familial and societal arrangements, all people are born into a situation of inequality and no child survives in the absence of adult connection. Since everyone is vulnerable both to oppression and to abandonment, two stories about morality recur in human experience [p. 281].

In the only known attempt to examine the morality of caring cross-culturally, research that I and my colleagues have undertaken among Americans and Hindu Indians reveals marked cross-cultural variation in such a morality, with this variation unrelated to gender. We have found that Indians tend to treat responsibilities to family, friends, and other in-group members in moral terms (Miller, Bersoff, & Harwood, 1990; Miller & Luthar, 1989). In contrast, with the exception of cases involving life-threatening needs, Americans tend to treat such responsibilities as matters for personal decision making. Also, as compared with Americans, Indians consider greater self-sacrifice as morally required (Miller & Bersoff, 1995), give greater priority to interpersonal responsibilities relative to justice considerations (Miller & Bersoff, 1992), consider interpersonal responsibilities as less contingent on feelings of personal affinity and liking (Miller & Bersoff, 1994b), and place greater weight on contextual factors in moral judgment (Bersoff & Miller, 1993).

These cross-cultural differences are illustrated in the following responses of a Hindu Indian and an American adult to a situation in which an adult son failed to care for his elderly parents in his own home and instead sent them away to be cared for elsewhere. Regarding this as a moral breach, an Indian subject justified her response as follows:

Because it's a son's duty – birth duty – to take care of his parents. It's not only money that matters. It's being near your dear ones which counts more. So even though they were being looked after elsewhere, they were not with their own children [Miller & Luthar, 1989, p. 253].

In contrast, the following argument was offered by an American subject, who also disapproved of this behavior but regarded it as a matter of personal choice:

It's up to the individual to decide. It's duty to the parents versus one's own independence and, I guess, one's self interest. He's fulfilling the minimal obligations to his parents. It wasn't a life and death situation and their needs were being taken care of. Beyond that it's a personal choice [Miller & Luthar, 1989, p. 253].

Like the research reviewed earlier on motivation and emotion, the present work on interpersonal morality is important in highlighting the implicit cultural dependence of Gilligan's theory and the need for alternative frameworks to account for psychological functioning in various non-Western cultural groups, if not in contrasting ethnic groups within the United States. In particular, the cross-cultural differences noted have been interpreted as implying that the type of morality emphasized among Americans and captured, in part, by Gilligan's theory represents an individualistic form of a morality of caring (Miller, 1994b). Reflecting certain Western individualistic cultural views of self, such a moral orientation is distinguished by the centrality that it gives to individual needs, desires, and volition, its emphasis on individual responsibility for action, and its dualistic view of a core psychological self as underlying social role performances. In contrast, the type of morality of caring emphasized among Hindu Indians, it is maintained, represents a duty-based orientation that reflects a relational cultural view of self. Such a code is distinguished by its emphasis on the obligatory yet natural quality of interpersonal responsibilities, with doing one's duty regarded both as meeting social requirements and as realizing one's nature.

This work on cultural variation in the morality of caring also suggests possible cultural boundaries in the model of attachment that informs

Gilligan's approach as well as attachment theory more generally (Miller & Bersoff, 1995). The cross-cultural research raises the possibility that the tension between autonomy and connection, presumed in this view of attachment, may be most reflective of problems for the self in Western individualistic cultures, which emphasize the importance of autonomy and the voluntary nature of interpersonal commitments. In contrast, it is argued that such a tension may not be central in collectivist cultures, which stress interdependence rather than individual autonomy, and tend to treat interpersonal commitments as relatively taken-for-granted matters of duty. It is proposed that alternative theoretical conceptions may be needed to account for the salient adaptive stresses in the latter types of cultures. In particular, in collectivist cultures, adaptive tensions appear to center on regulating interpersonal interchange and on displaying the accommodation and equanimity necessary to function well, as part of a social whole, rather than on effecting a balance between potential loss of individuality, resulting from too much connection, and potential isolation, resulting from too much autonomy.

Conclusions

The cross-cultural evidence on intelligence discussed may be seen to highlight certain limitations of the interpretations of intelligence forwarded in evolutionary accounts. Evolutionary approaches focus on the functional significance of behavior and portray the environment of human development in ecological terms, as presenting certain objective adaptive constraints. Cross-cultural research suggests, however, that in many cases, functionality represents a secondary formation, with the utility of a practice only apparent against the background of cultural values or presuppositions that are themselves nonfunctional in character. Equally, it must be recognized that functionality is to some extent a culturally relative value and that in all cultures, many practices have more to do with other types of considerations, such as perceived appropriateness, spiritual merit, truth, and so forth rather than with utility.

Cultural research on intelligence also highlights the need to recognize that cultural contexts do not merely provide opportunities for intellectual development but affect the form in which cognitive potentials are realized. Research indicates that certain forms of intelligence do not arise in cultures in which these forms are not valued or in which they are unintelligible in terms of indigenous cultural understandings. It has also been seen that some of the types of intelligence presently considered as universal are identified as such through analyses that treat cognitive processes in abstraction from the contexts in which they are embodied or through a failure to undertake the type of comparative research necessary to make apparent the implicit cultural underpinnings of these forms of intelligence.

Implications for understanding heredity and environmental influences on intelligence

Theory and research from a cultural-psychology perspective challenge many of the assumptions made in behavior-genetic views of intelligence. This cross-cultural work may be seen to call into question the possibility of formulating culture-free indices of intelligence. Research from a cultural-psychology perspective rejects the deterministic stance assumed in behavior-genetic models and documents the need to recognize that the development of intelligence represents an open and culturally variable process.

Intelligence as culturally grounded. A striking feature of behavior-genetic work on intelligence is its assumption that intellectual abilities can be isolated from prior knowledge and that IQ measures provide culture-free indices of intelligence (e.g., Jensen, 1981, 1984, chapter 2, this volume; Loehlin, 1992; Plomin, 1986, 1994, chapter 3, this volume; Scarr, 1981, 1992, 1993, chapter 1, this volume). The capacity for learning is assumed to be a universal biologically based capacity that is tapped by IQ measures and on which there are stable individual differences. This is distinguished from the content of what is acquired or achieved, which is assumed to be culturally variable.

Work from a cultural-psychology perspective challenges the claim that IQ performance represents a culturally independent measure of aptitude. Empirically, this cultural work lends support to the charge of critics that much of the apparent predictive power of IQ measures results from method variance (e.g., Ceci, 1990, 1994). Thus, just as it has been shown that IQ measures show weak correlations with measures of nonacademic intelligence (e.g., Ceci, 1990; Sternberg, 1985; Sternberg & Wagner, 1986), cross-cultural research reveals that individuals in other cultures who perform poorly on IQ tests may display superior levels of performance in everyday contexts (e.g., Hutchins, 1980; Rogoff & Lave, 1984; Scribner & Cole, 1981). Furthermore, cultural evidence indicates that many of the cognitive behaviors that are assessed on IQ measures do not have adaptive value in other cultures (Berry, 1974; Goodnow, 1976; Irvine & Berry, 1988; LeVine, 1990; Ramanujan, 1990). Work in the cultural-psychology tradition leads to the conclusion that individual and subgroup variation in performance on IQ measures needs to be seen as resulting, at least in part, from differences in culturally dependent knowledge. These differences also cannot be assumed necessarily to remain stable across contexts but to shift in ways that reflect individual and group variation in experiences and in acquired expertise in different content domains (Baumrind, 1993; Jackson, 1993).

It does not follow from these conclusions that no use should ever be made of psychometric measures. However, it does follow that the meaning of scores on psychometric measures must be interpreted differently than in behavior-genetic accounts. In particular, such measures need to be understood as having relatively circumscribed predictive power, with differences in culturally based knowledge and experiences recognized to represent an important source of individual and group differences in IQ performance. It also may be expected that for many purposes, IQ measures are likely to prove neither necessary nor sufficient in tapping intelligence, with the need arising to supplement, if not to supplant, the use of IQ measures with measures that are more closely tied to everyday cognitive performances of particular theoretical or applied concern.

More generally, it must be recognized that all measures of intelligence are culturally grounded, with performance dependent, at least in part, on culturally based understandings. Thus, even in the controlled conditions of the experimental laboratory, intellectual performances reflect individuals' interpretations of the meaning of situations and their background presuppositions, rather than pure *g*. The present considerations suggest that there is likely to be little theoretical yield from behavior-genetic attempts to hold culture constant when assessing intelligence or from attempts in this tradition to partition causation into genetic and environmental components. Rather, it may be concluded that future theoretical approaches need to be premised on a view of intelligence as inherently culturally grounded and of the role of heredity and environment in intelligence as that of inseparable rather than of discrete causal factors (e.g., Wahlsten & Gottlieb, chapter 6, this volume).

Development as an open process. Behavior-genetic theories portray development in deterministic terms, with environmental influences on development assumed to be patterned by genetic propensities both through various direct effects and through the indirect effect of influencing individuals' selection of environments (Jensen, chapter 2, this volume; Plomin, 1986; Plomin & Bergeman, 1991; Scarr & McCartney, 1983). It is assumed that the impact of environments on phenotypic behavior may be predicted on a probabilistic basis. This idea is captured in the concept of *reaction range*, which is defined as the distribution of observed phenotypical reactions associated with a given genotype. From a behavior-genetic perspective, there is assumed to be little empirical need for the concept of *reaction norm*, which is defined as the range of potential phenotypic reactions associated with a given genotype. Under extreme conditions, environments can have the effect of impeding development, while under normal conditions, development is seen as relatively uniform. As Scarr (1993), for example, argues:

> Consider psychometric intelligence in humans. . . . Along the environmental axis, the reaction norm for intelligence is flat across a wide range of normal family environments but drops off drastically at the very low end of the environmental continuum. . . . Only extremely poor environments (or impaired organisms) prevent normal species-typical development by depriving the young of opportunities that environment must afford if development is to occur [p. 1339].

From this perspective, claims that development should be understood as an open process are rejected as purely hypothetical in nature, since there appears to be no empirical evidence of environmental variation in nonimpaired intellectual outcomes. The indeterminate view of development implied by the concept of reaction norm is also rejected as incompatible with the predictive goals of science.

As critics have pointed out, however, "the norm of reaction for a particular individual or group can be ascertained only by attempting to expand it" (Baumrind, 1993, p. 1303). It is this type of expansion, both in awareness of the diversity of intellectual outcomes and in enrichment of psychological theory, which work from a cultural-psychology perspective is making possible. The claim of behavior-genetic theorists that there is no significant environmental variation in normal intelligence is questioned by the cultural-psychology approach. Such a claim is seen as based on a research tradition that precludes finding such variation. In particular, by defining intelligence in terms of the circumscribed range of responses assessed on IQ instruments, work in the psychometric tradition is viewed as failing to take into account alternative cognitive competencies that comprise intelligence in different cultural communities.

Research from a cultural-psychology perspective, in contrast, may be seen to make variation in intellectual outcomes more visible through its alternative conceptual assumptions and methodological commitments. In particular, this latter work is premised on a view of intelligence that includes competencies beyond those tapped on IQ measures. It also not only taps populations of varied cultural and subcultural background but takes into account the cultural understandings and practices of these populations in theory development. Comparative research of this type is highlighting the need for multiple models to account for psychological functioning in different cultural contexts, with cultural variability shown to exist in the nature of intellectual processes and not merely in the contexts in which such processes are displayed or in the value which they are accorded (e.g., Markus & Kitayama, 1991; Miller, 1994b; Shweder, 1990).

Although incompatible with a natural science model of explanation that is focused on the identification of invariant laws of behavior, the perspective of cultural psychology is compatible with the type of historically and culturally grounded explanation that, it may be argued, has always existed in psychology and always will exist in psychology (e.g., Bruner, 1990; Kessen, 1990; Pepitone, 1976). As Boesch points out, "it is the dilemma of psychology to deal as a natural science with an object that creates history" (Boesch, cited in Cole, 1990, p. 279). In explaining the behavior of intentional agents whose understandings reflect historically and culturally variable meanings, psychological theory necessarily may be expected to offer generalizations that are relative to a particular time and context. Thus it must be recognized that certain theories of intelligence that are adequate to account for forms of intelligence among American populations differ from those that are required to account for forms of intelligence among more traditional cultural populations (e.g., Markus & Kitayama, 1994; Miller, 1994b).

In conclusion, the present examination of cultural influences on intelligence highlights the importance of recognizing that culture is part of human experience and needs to be an explicit part of psychological theories that purport to predict, explain, and understand that experience. As Jahoda (1993) muses, in arguing for the importance of recognizing the mutually constitutive role of culture and mind: "Perhaps the time will come when a psychology that treats humans as isolated, timeless organisms, and fails to take account of culture and history, will seem like a *Hamlet* with the Prince of Denmark as the only character" (p. 194).

Acknowledgments

Appreciation is expressed to Robert Sternberg and to Elena Grigorenko for their helpful comments on an earlier version of this chapter.

References

Akiyama, H. (1992, June). *Measurement of depressive symptoms in cross-cultural research.* Paper presented at the International Conference on Emotion and Culture, University of Oregon, Eugene.

Azuma, H. (1984). Secondary control as a heterogeneous category. *American Psychologist, 39,* 970–1.

Baltes, M. M., & Baltes, P. B. (Eds.). (1986). *The psychology of control and aging.* Hillsdale, NJ: Erlbaum.

Bandura, A. (1977). Self-efficacy: Toward a unifying theory of behavioral change. *Psychological Review, 84,* 191–215.

Bartlett, F. C. (1958). *Thinking.* London: Allen & Unwin.

Baumrind, D. (1993). The average expectable environment is not good enough: A response to Scarr. *Child Development, 64,* 1299–317.

Berry, J. W. (1974). Radical cultural relativism and the concept of intelligence. In J. W. Berry & P. R. Dasen (Eds.), *Culture and cognition: Readings in cross-cultural psychology* (pp. 225–9). London: Methuen.

Berry, J. W. (1976). *Human ecology and cognitive style.* New York: Sage-Halsted.

Bersoff, D. M., & Miller, J. G. (1993). Culture, context, and the development of moral accountability judgments. *Developmental Psychology, 29,* 664–76.

Bond, M. H., & Cheung, T. (1983). College students' spontaneous self-concept: The effect of culture among students in How Kong, Japan, and the United States. *Journal of Cross-Cultural Psychology, 14,* 153–71.

Bransford, J. D., & McCarrell (1974). A sketch of a cognitive approach to comprehension: Some thoughts about what it means to comprehend. In W. B. Weiner & D. S. Palermo (Eds.), *Cognition and the symbolic processes* (pp. 189–229). New York: Wiley.

Bruner, J. S. (1973). Going beyond the information given. In V. Anglin (Ed.), Beyond the information given: Studies in the psychology of knowing (pp. 218–38). New York: Norton.

Bruner, J. S. (1990). *Acts of meaning.* Cambridge, MA: Harvard University Press.

Ceci, S. (1990). *On intelligence . . . more or less: A bio-ecological treatise on intellectual development.* Engelwood Cliffs, NJ: Prentice Hall.

Ceci, S. (1991). How much does schooling influence general intelligence and its cognitive components?: A reassessment of the evidence. *Developmental Psychology, 27,* 703–22.

Ceci, S. (1994). Education, achievement, and general intelligence: What ever happened to the *psycho* in *psychometries? Psychological Inquiry, 5,* 197–201.

Charlesworth, W. (1979). An ethological approach to studying intelligence. *Human Development, 22,* 212–16.

Cole, M. (1988). Cross-cultural research in the sociohistorical tradition. *Human Development, 31,* 137–57.

Cole, M. (1990). Cultural psychology: A once and future discipline? In J. Berman (Ed.), *Cross-Cultural Perspectives, 37, Nebraska Symposium on Motivation* (pp. 279–335). Lincoln: University of Nebraska Press.

Cole, M., Gay, J., Glick, J., & Sharp, D. (1971). *The cultural context of learning and thinking.* New York: Basic Books.

Cole, M., & Griffin, P. (1980). Cultural amplifiers reconsidered. In D. R. Olson (Ed.), *The social foundations of language and thought* (pp. 343–64). New York: W. W. Norton.

Cole, M., & Scribner, S. (1974). *Culture and thought.* New York: Wiley.

Cousins, S. (1989). Culture and self-perception in Japan and the United States. *Journal of Personality and Social Psychology, 56,* 124–31.

D'Andrade, R. G. (1984). Cultural meaning systems. In R. A. Shweder & R. A. LeVine (Eds.), *Culture theory: Essays on mind, self and emotion* (pp. 88–119). New York: Cambridge University Press.

D'Andrade, R. G., & Strauss, C. (Eds.). (1992). *Human motives and cultural models.* New York: Cambridge University Press.

Dasen P. R. (1972). Cross-cultural research: A summary. *Journal of Cross-Cultural Psychology, 3,* 29–39.

Dasen, P. R. (Ed.). (1977). *Piagetian psychology: Cross-cultural contributions.* New York: Gardner.

Dasen, P. R., Berry, J. W., & Witkin, H. A. (1979). The use of developmental theories cross-culturally. In L. Eckensberger, W. Lonner, & Y. H. Poortinga (Eds.), *Cross-cultural contributions to psychology.* The Netherlands: Swets Publishing Service.

Deci, E. L., & Ryan, R. M. (1987). The support of autonomy and the control of behavior. *Journal of Personality and Social Psychology, 53,* 1024–37.

Deci, E. L., & Ryan, R. M. (1991). A motivational approach to self: Integration in personality. In R. A. Dienstbier (Ed.), *Perspectives on Motivation, 38*, Nebraska Symposium on Motivation (pp. 237–87). Lincoln: University of Nebraska Press.

Donaldson, M. (1978). *Children's minds*. New York: W. W. Norton.

Dyal, J. A. (1984). Cross-cultural research with the locus of control construct. In H. M. Lefcourt (Ed.), *Research with the locus of control construct* (Vol. 3, pp. 209–306). New York: Academic Press.

Ericsson, K. A., Chase, W. G., & Faloon, S. (1980). Acquisition of a memory skill. *Science, 208*, 1181–2.

Gardner, H. (1983). *Frames of mind*. New York: Basic Books.

Geertz, C. (1973). *The interpretation of cultures*. New York: Basic Books.

Gelman, R. (1978). Cognitive development. *Annual review of psychology, 29*, 297–332.

Gelman, R., & Baillargeon, R. (1983). A review of some Piagetian concepts. In P. Mussen (Ed.), *Manual of child psychology: Vol. 3. Cognitive development* (J. H. Flavell & E. M. Markman, Eds.) (pp. 167–230). New York: Wiley.

Gilligan, C. (1982). *In a different voice: Psychological theory and women's development*. Cambridge, MA: Harvard University Press.

Gilligan, C., & Wiggins, G. (1987). The origins of morality in early childhood relationships. In J. Kagan & S. Lamb (Eds.), *The emergence of morality in young children* (pp. 277–305). Chicago: University of Chicago Press.

Goodman, N. (1972). Seven strictures on similarity. In N. Goodman (Ed.), *Problems and projects* (pp. 437–46). New York: Bobbs-Merrill.

Goodnow, J. J. (1976). The nature of intelligent behavior: Questions raised by cross-cultural studies. In L. B. Resnick (Ed.), *The nature of intelligence* (pp. 169–88). Hillsdale, NJ: Erlbaum.

Gottfredson, L. S. (1986). Societal consequences of the *g* factor in employment. *Journal of Vocational Behavior, 29*, 379–410.

Grolnick, W. S., Ryan, R. M., & Deci, E. L. (1991). Inner resources for school achievement: Motivational mediators of children's perceptions of their parents. *Journal of Educational Psychology, 83*, 508–17.

Harkness, S., & Super, C. (1985). Child–environment interactions in the socialization of affect. In M. Lewis & C. Saarni (Eds.), *The socialization of emotions* (pp. 21–36). New York: Plenum Press.

Harkness, S., & Super, C. (1992). Parental ethnotheories in action. In I. E. Sigel (Ed.), *Parental belief systems: The psychological consequences for children* (pp. 373–91). Hillsdale, NJ: Erlbaum.

Hatch, E. (1983). *Culture and morality: The relativity of values in anthropology*. New York: Columbia University Press.

Hui, C. C. H. (1982). Locus of control: A review of cross-cultural research. *International Journal of Intercultural Relations, 6*, 301–23.

Humphreys, L. G. (1994). Intelligence from the standpoint of a (pragmatic) behaviorist. *Psychological Inquiry, 5*, 179–92.

Hutchins, E. (1980). *Culture and inference*. Cambridge: Harvard University Press.

Irvine, S. H., & Berry, J. W. (1988). The abilities of mankind: A reevaluation. In S. H. Irvine & J. W. Berry (Eds.), *Human abilities in cultural context* (pp. 3–59). New York: Cambridge University Press.

Jackson, J. F. (1993). Human behavioral genetics, Scarr's theory and the views on interventions: A critical review and commentary on their implications for African American children. *Child Development, 64*, 1318–32.

Jahoda, G. (1993). *Crossroads between culture and mind*. Cambridge, MA: Harvard University Press.

Jensen, A. R. (1981). *Straight talk about mental tests*. New York: Free Press.

Jensen, A. R. (1984). Test bias: Concepts and criticisms. In C. R. Reynolds & R. T. Brown (Eds.), *Perspectives on bias in mental testing* (pp. 507–86). New York: Plenum Press.

Kessen, W. (1990). *The rise and fall of development.* Worcester, MA: Clark University Press.

Kitayama, S., & Markus, H. (Eds.). (1994). *Emotion and culture: Empirical studies of mutual influence.* Washington, DC: American Psychological Association.

Kohlberg, L. (1971). From is to ought: How to commit the naturalistic fallacy and get away with it in the study of moral development. In T. Mischel (Ed.), *Cognitive development and epistemology* (pp. 151–232). New York: Academic Press.

Kohlberg, L. (1981). *The philosophy of moral development: Moral stages and the idea of justice: Vol. 1. Essays on moral development.* New York: Harper & Row.

Kojima, H. (1984). A significant stride toward the comparative study of control. *American Psychologist, 39,* 972–3.

Laboratory of Comparative Human Cognition (1983). Culture and cognitive development. In P. Mussen (ed.), *Manual of child psychology: Vol. 4. History, theory and method* (W. Kessen, Ed.) (pp. 295–56). New York: Wiley.

Lave, J. (1988). *Cognition in practice; Mind, mathematics and culture in everyday life.* New York: Cambridge University Press.

Lave, J., & Wenger, E. (1991). *Situated learning: Legitimate peripheral participation.* New York: Cambridge University Press.

Lepper, M. R. (1983). Social-control processes and the internalization of social values: An attributional perspective. In E. T. Higgins, D. N. Ruble, & W. W. Hartup (Eds.), *Social cognition and social development* (pp. 294–330). New York: Cambridge University Press.

Levenson, R. W., Ekman, P., Heider, K., & Friesen, W. V. (1992). Emotion and autonomic nervous system activity in the Minangkabau of West Sumatra. *Journal of Personality and Social Psychology, 62,* 972–88.

LeVine, R. A. (1984). Properties of culture: An ethnographic view. In R. A. Shweder & R. A. LeVine (Eds.), *Cultural theory: Essays on mind, self and emotion* (pp. 67–87). New York: Cambridge University Press.

LeVine, R. A. (1990). Infant environments in psychoanalysis: A cross-cultural view. In J. W. Stigler, R. A. Shweder, & G. Herdt (Eds.), *Cultural psychology: Essays on comparative human development* (pp. 454–74). New York: Cambridge University Press.

Loehlin, J. C. (1992). *Genes and environment in personality development.* (Individual Differences and Development Series, Vol. 2). Newbury Park, CA: Sage.

Luria, A. R. (1976). *Cognitive development: Its cultural and social foundations.* Cambridge, MA: Harvard University Press.

Lutz, C. (1988). *Unnatural emotions: Everyday sentiments on a Micronesian atoll and their challenge to Western theory.* Chicago: University of Chicago Press.

Lutz, C., & White, G. M. (1986). The anthropology of emotions. *Annual Review of Anthropology, 15,* 405–36.

Maciel, A. G., Heckhausen, J., & Baltes, P. B. (1994). A life-span perspective on the interface between personality and intelligence. In R. Sternberg & P. Ruzgis (Eds.), *Personality and intelligence* (pp. 61–103). New York: Cambridge University Press.

Markus, H., & Kitayama, S. (1991). Culture and the self: Implications for cognition, emotion and motivation. *Psychological Review, 98,* 224–53.

Markus, H., & Kitayama, S. (1994). The cultural construction of self and emotion: Implications for social behavior. In S. Kitayama & H. Markus (Eds.), *Emotion and culture* (pp. 89–130). Washington, DC: American Psychological Association.

Matsumoto, D. (1992). American-Japanese cultural differences in the recognition of universal facial expression. *Journal of Cross-Cultural Psychology, 23,* 72–84.

Miller, G. A. (1956). The magic number seven, plus or minus two: Some limits on our capacity for processing information. *Psychological Review, 63,* 81–97.

Miller, J. G. (1984). Culture and the development of everyday social explanation. *Journal of Personality and Social Psychology, 46*, 961–78.

Miller, J. G. (1987). Cultural influences on the development of conceptual differentiation in person description. *British Journal of Developmental Psychology, 5*, 309–19.

Miller, J. G. (1988). Bridging the content-structure dichotomy: Culture and the self. In M. Bond (Ed.), *The cross-cultural challenge to social psychology* (pp. 266–81). Newbury Park, CA: Sage.

Miller, J. G. (1994a). Cultural psychology: Bridging disciplinary boundaries in understanding the cultural grounding of self. In P. K. Bock (Ed.), *Handbook of psychological anthropology* (pp. 139–70). Westport, CT: Greenwood Publishing Group.

Miller, J. G. (1994b). Cultural diversity in the morality of caring: Individually oriented versus duty-based interpersonal moral codes. *Cross-Cultural Research, 28*, 3–39.

Miller, J. G. (in press). Theoretical issues in cultural psychology. In J. W. Berry, Y. Poortinga, & J. Pandey (Eds.), *Handbook of cross-cultural psychology: Theoretical and methodological perspectives, Vol 1.* (rev. ed.). Boston: Allyn & Bacon.

Miller, J. G., & Bersoff, D. M. (1992). Culture and moral judgement: How are conflicts between justice and interpersonal responsibilities resolved? *Journal of Personality and Social Psychology, 62*, 541–54.

Miller, J. G., & Bersoff, D. M. (1994a). Cultural influences on the moral status of reciprocity and the discounting of endogenous motivation. In D. Miller & D. Prentice (Eds.), The self and the collective: Groups within individuals [Special issue]. *Personality and Social Psychology Bulletin, 20*(5), 592–602.

Miller, J. G., & Bersoff, D. M. (1994b). *Culture and the role of personal affinity and liking in interpersonal morality.* Manuscript submitted for publication.

Miller, J. G., & Bersoff, D. M. (1995). Development in the context of everyday family relationships: Culture, interpersonal morality and adaptation. In M. Killen & D. Hart (Eds.), *Morality in everyday life: A developmental perspective* (pp. 259–82). New York: Cambridge University Press.

Miller, J. G., Bersoff, D. M., & Harwood, R. L. (1990). Perceptions of social responsibilities in India and in the United States: Moral imperatives or personal decisions? *Journal of Personality and Social Psychology, 58*, 33–47.

Miller, J. G., & Luthar, S. (1989). Issues of interpersonal responsibility and accountability: A comparison of Indians' and Americans' moral judgements. *Social Cognition, 7*, 237–61.

Mischel, W. (1968). *Personality and assessment.* New York: Wiley.

Mischel, W. (1973). Towards a cognitive social learning reconceptualization of personality. *Psychological Review, 80*, 252–83.

Misra, G., & Gergen, M. (1992, July). *Issues of gender and control: Cross-cultural perspectives.* Paper presented at the meeting of the International Association of Cross-Cultural Psychology (IACCP), Liege, Belgium.

Munroe, R. H., Munroe, R. L., & Whiting, B. B. (Eds.). (1981). *Handbook of cross-cultural human development.* New York: Garland.

Nucci, L. (in press). Morality and personal freedom. In E. Reed & E. Turiel (Eds.), *Values and cognition.* Hillsdale, NJ: Erlbaum.

O'Flaherty, W. D., & Derrett, J. D. (1978). *The concept of duty in South Asia.* India: Vikas.

Pepitone, A. (1976). Toward a normative and comparative biocultural social psychology. *Journal of Personality and Social Psychology, 34*, 641–53.

Pepitone, A., & Triandis, H. (1987). On the universality of social psychological theories. *Journal of Cross-Cultural Psychology, 18*, 471–98.

Piaget, J. (1966). Need and significance of cross-cultural studies in genetic psychology. *International Journal of Psychology, 1*, 3–13.

Piaget, J. (1981). *The psychology of intelligence.* Totowa, NJ: Littlefield, Adams & Co.

Piaget, J., & Inhelder, B. (1969). *The psychology of the child.* New York: Basic Books.

Plomin, R. (1986). *Development, genetics, and psychology.* Hillsdale, NJ: Erlbaum.

Plomin, R. (1994). *Genetics and experience: The interplay between nature and nurture.* Thousand Oaks, CA: Sage.

Plomin, R., & Bergeman, C. S. (1991). The nature of nurture: Genetic influences on "environmental" measures. *Behavioral and Brain Sciences, 14,* 373–427.

Potter, S. H. (1988). The cultural construction of emotion in rural Chinese social life. *Ethos, 16,* 181–208.

Profyt, L., & Whissell, C. (1991). Children's understanding of facial expression of emotion: I. Voluntary creation of emotion-faces. *Perceptual and Motor Skills, 73,* 199–202.

Ramanujan, A. K. (1990). Is there an Indian way of thinking? An informal essay. In M. Marriott (Ed.), *India through Hindu categories* (pp. 41–58). New Delhi: Sage Publications.

Rogoff, B. (1990). *Apprenticeship in thinking: Cognitive development in social context.* New York: Oxford University Press.

Rogoff, B., & Lave, J. (Eds.). (1984). *Everyday cognition: Its development in social context.* Cambridge, MA: Harvard University Press.

Rogoff, B., & Wertsch, J. V. (1984). *Children's learning in the "zone of proximal development."* San Francisco: Jossey-Bass.

Rohner, R. P., & Pettengill, S. M. (1985). Perceived parental acceptance–rejection and parental control among Korean adolescents. *Child Development, 56,* 524–8.

Resaldo, M. Z. (1984). Toward an anthropology of self and feeling. In R. A. Shweder & R. A. LeVine (Eds.), *Cultural theory: Essarys on mind, self and emotion* (pp. 137–55). New York: Cambridge University Press.

Rotter, J. B. (1966). Generalized expectancies for internal versus external control of reinforcement. *Psychological Monographs, 80* (1 Whole No. 609).

Ruzgis, P., & Grigorenko, E. L. (1994). Cultural meaning systems, intelligence and personality. In R. J. Sternberg & P. Ruzgis (Eds.), *Personality and intelligence* (pp. 248–70). New York: Cambridge University Press.

Saarni, C., & Harris, P. L. (Eds.). (1989). *Children's understanding of emotion.* New York: Cambridge University Press.

Sabini, J., & Silver, M. (1981). Introspection and causal accounts. *Journal of Personality and Social Psychology, 40,* 171–9.

Sahlins, M. (1976a). *Culture and practical reason.* Chicago: University of Chicago Press.

Sahlins, M. (1976b). *The use and abuse of biology.* Ann Arbor: The University of Michigan Press.

Salovey, P., & Mayer, J. (1994). Some final thoughts about personality and intelligence. In R. J. Sternberg & P. Ruzgis (Eds.), *Personality and intelligence* (pp. 303–18). New York: Cambridge University Press.

Scarr, S. (1981). *Race, social class, and individual differences in IQ.* Hillsdale, NJ: Erlbaum.

Scarr, S. (1992). Developmental theories for the 1990s: Development and individual differences. *Child Development, 63,* 1–19.

Scarr, A. (1993). Biological and cultural diversity: The legacy of Darwin for development. *Child Development, 64,* 1333–53.

Scarr, S., & McCartney, K. (1983). How people make their own environment: A theory of genotype–environment effects. *Child Development, 54,* 424–35.

Schneider, D. (1968). *American kinship: A cultural account.* Englewood Cliffs, NJ: Prentice-Hall.

Schneider, D. M. (1976). Notes toward a theory of culture. In K. Basso & H. Selby (Eds.), *Meaning in anthropology* (pp. 197–220). Albuquerque: University of New Mexico Press.

Schwartz, T. (1981). The acquisition of culture. *Ethos, 9,* 4–17.

Schwartz, T. (1992). Anthropology and psychology: An unrequited relationship. In T. Schwartz, G. White, & C. Lutz (Eds.), *New directions in psychological anthropology* (pp. 324–49). New York: Cambridge University Press.

Scribner, S., & Cole, M. (1981). *The psychology of literacy.* Cambridge, MA: Harvard University Press.

Seligman, M. E. P. (1975). *Helplessness: On depression development and death.* San Francisco: Freeman.

Shweder, R. A. (1977). Culture and thought. In B. B. Wolman (Ed.), *International encyclopedia of psychiatry, psychology, psychoanalysis and neurology* (Vol. 3, pp. 457–61). New York: Aesculapius Publishers, Inc.

Shweder, R. A. (1979). Rethinking culture and personality theory. Part II: A critical examination of two more classical postulates. *Ethos, 7,* 279–311.

Shweder, R. A. (1984). Preview: A colloquy of culture theorists. In R. A. Shweder & R. A. LeVine (Eds.), *Cultural theory: Essays on mind, self and emotion* (pp. 1–24). New York: Cambridge University Press.

Shweder, R. A. (1990). Cultural psychology – What is it? In J. W. Stigler, R. A. Shweder, & G. Herdt (Eds.), *Cultural psychology: Essays on comparative human development* (pp. 1–43). New York: Cambridge University Press.

Shweder, R. A. (1993). The cultural psychology of the emotions. In M. Lewis & J. M. Haviland (Eds.), *Handbook of emotions* (pp. 417–31). New York: Guilford Press.

Shweder, R. A., & Bourne, E. (1982). Does the concept of the person vary cross-culturally? In A. J. Marsella & G. White (Eds.), *Cultural conceptions of mental health and therapy* (pp. 97–137). Boston: Reidel.

Shweder, R. A., & Sullivan, M. A. (1993). Cultural psychology: Who needs it? *Annual Review of Psychology, 44,* 497–527.

Sinha, D. (1990). The concept of psycho-social well-being: Western and Indian perspectives. *National Institute of Mental Health and Neuroscience Journal, 8,* 1–11.

Spiro, M. E. (1990). On the strange and the familiar in recent anthropological thought. In J. W. Stigler, R. A. Shweder, & G. Herdt (Eds.), *Cultural psychology: Essays on comparative human development* (pp. 47–61). New York: Cambridge University Press.

Sternberg, R. J. (1985). *Beyond IQ: The triarchic theory of intelligence.* New York: Cambridge University Press.

Sternberg, R. J. (1988). A triarchic view of intelligence in cross-cultural perspective. In S. H. Irvine & J. W. Berry (Eds.), *Human abilities in cultural context* (pp. 60–85). New York: Cambridge University Press.

Sternberg, R. J., & Kolligina, J., Jr. (Eds.). (1990). *Competence considered.* New Haven, CT: Yale University Press.

Sternberg, R. J., & Wagner, R. K. (1986). *Practical intelligence: Nature and origins of competence in the everyday world.* New York: Cambridge University Press.

Stigler, J. W. (1984). "Mental abacus": The effect of abacus training on Chinese children's mental calculation. *Cognitive Psychology, 16,* 145–76.

Stigler, J. W., & Perry, M. (1990). Mathematics learning in Japanese, Chinese, and American classrooms. In J. W. Stigler, R. A. Shweder, & G. Herdt (Eds.), *Cultural psychology: Essays on comparative human development* (pp. 328–53). New York: Cambridge University Press.

Tobin, J. J. (Ed.). (1992). *Re-made in Japan: Everyday life and consumer taste in a changing society.* New Haven, CT: Yale University Press.

Triandis, H. C. (1990). Cross-cultural studies of individualism and collectivism. In J. Berman (Ed.), *Cross-cultural perspectives, 37, Nebraska symposium on motivation* (pp. 41–133). Lincoln: University of Nebraska Press.

Triandis, H. C., & Lambert, W. W. (Eds.). (1980). *Handbook of cross-cultural psychology: Vol. 1. Perspectives.* Boston: Allyn & Bacon.

Triandis, H. C., & Lonner, W. J. (Eds.). (1980). *Handbook of cross-cultural psychology: Vol. 3. Basic processes*. Boston: Allyn & Bacon.

Turiel, E. (1983). *The development of social knowledge: Morality and convention*. New York: Cambridge University Press.

Turiel, E. (1994). Morality, authoritarianism and personal agency in cultural contexts. In R. J. Sternberg & P. Ruzgis (Eds.), *Personality and intelligence* (pp. 271–99). New York: Cambridge University Press.

Turiel, E., Smetana, J. G., & Killen, M. (1991). Social contexts in social cognitive development. In W. M. Kurtines & J. L. Gewirtz (Eds.), *Handbook of moral behavior and development: Vol. 2. Research* (pp. 307–32). Hillsdale, NJ: Erlbaum.

Valsiner, J. (Ed.). (1988). *Child development within culturally structured environments: Vol. 2. Social co-construction and environmental guidance in development*. Norwood, NJ: Ablex.

Weisz, J. R. (1990). Development of control-related beliefs, goals, and styles in childhood and adolescence: A clinical perspective. In J. Rodin, C. Schooler, & K. W. Schaie (Eds.), *Self directedness: Cause and effects throughout the life course* (pp. 103–45). Hillsdale, NJ: Erlbaum.

Weisz, J. R., Rothbaum, F. M., & Blackburn, T. C. (1984). Standing out and standing in: The psychology of control in America and Japan. *American Psychologist, 39*, 955–69.

Wells, G. (1981). Lay analyses of causal forces on behavior. In J. H. Harvey (Ed.), *Cognition, social behavior and the environment* (pp. 309–24). Hillsdale, NJ: Erlbaum.

Werker, J. (1989). Becoming a native listener. *American Scientist, 77*, 54–69.

Werner, H. (1948). *The comparative psychology of mental development*. Chicago: Wilcox & Follett.

Wertsch, J. V. (Ed.). (1991). *Voices of the mind: A sociocultural approach to mediated action*. New York: Cambridge University Press.

Wilson, M. H. (1952). *Good company: A study of the Nyakyusa age villages*. London: Oxford University Press.

Yang, K. (1988). Will societal modernization eventually eliminate cross-cultural psychological differences? In M. H. Bond (Ed.), *The cross-cultural challenge to social psychology* (pp. 67–85). Newbury Park: Sage.

10 A bio-ecological model of intellectual development: Moving beyond h^2

Stephen J. Ceci, Tina Rosenblum, Eddy de Bruyn, and Donald Y. Lee

In her 1958 Presidential Address to APA, Anne Anastasi urged psychologists to move beyond the question of studying "how much" variance was accounted for by genetics and environment and instead to address the question of "how" genotypes were translated into phenotypes. Despite the passage of nearly 40 years, Anastasi's challenge has not been taken up. Today, we know much more than we once did about the role of genetics in intelligence, the distribution of alleles at neurologically relevant sites, and the molecular chemistry of brain functioning, but we still know as little as ever about the mechanisms through which genetics get translated into phenotypic intellectual expression.

In this chapter, we endeavor to say something about the process that translates genes into phenotypes and, in so doing, say something about the environmentally loaded nature of heritability and the genetically loaded nature of the environment. While acknowledging the important role that genetics plays in intellectual development, we shall nevertheless take issue with the traditional "percent of variance accounted for" model of behavior genetics. In its place, we will suggest that traditional heritability measures do not inform us about the manner in which genes exert their influence, nor even about the organism's genetic potential. Rather, they tell us what proportion of individual differences in *already actualized* genetic potential has been brought to fruition by the prevailing environment. Left unknown, and unknowable from the traditional model, is (1) the amount of the organism's unactualized genetic potential, and (2) the absolute level of intellectual competence around which individual variation occurs.

We conclude this chapter with a model for understanding and predicting intellectual development that complicates the standard psychometric analysis; by adding specific environmental factors that we posit to moderate the workings of genes, we claim to better understand and predict intellectual development.

Heritability and norms of reaction. While heritability (h^2) is a measure of covariation among relative ranks and is, therefore, indexed by correlation, it can tell us nothing about the overall mean levels of performance that can be attained, a point originally made by Woodsworth (1941) and repeatedly raised by others (see Bronfenbrenner & Ceci, 1994). Even when h^2 is extremely high, as it is for some physical traits like height, the environment still can exert a very powerful influence on the absolute levels reached, as attested to by the nutritionally based surge in the height of the second-generation Japanese-American sons of tall Japanese fathers, who were taller than both the American-born sons of short Japanese fathers and the Japanese-born sons of all Japanese fathers:

Even though h^2 among this group remained over .90 . . . the American-reared offspring were over five inches taller than they would have been if they had been reared in Japan (Greulich, 1957). And Tanner (1962) showed that both American and British teenagers were a half-foot taller, on average, than their predecessors a century earlier. Finally, Angoff (1988) reported that the heights of young adult males in Japan were raised by about three and a half inches since the end of World War II, an enormous gain in such a brief period of time! If something as highly heritable as height can fluctuate so dramatically in such a relatively short period of time, then surely traits like intelligence can be altered, too [Ceci, 1990b, p. 142].

The traditional additive behavior-genetic model assumes that different genotypes have different norms of reaction, and the phenotypic differences are potentially greatest when environmental resources are also greatest. In Figure 10.1, this is exemplified in the regression lines associated with Genotypes A and C, which both increase by similar amounts in their phenotypic expression as a linear function of the increasing value of the environment. However, it is not only possible but probable that the norms of reaction are nonmonotonic, with the error about the regression line being variable and irregular, a point made recently by Bors (in press): "The same maximum mean phenotypic values for all genotypes could be produced by different intersections of multiple environmental parameters" (p. 17). Thus, a demonstration that different phenotypes are produced by the same environment does not minimize the importance of the latter, because, as suggested in Figure 10.1, the maximal phenotypic expression for each of the four genotypes could be related to different levels of environmental resources. As we will argue later, the *same* environmental factors can account for different phenotypic expressions under identical levels of environmental resources, just as the important role of genetics can account for the same phenotypic expressions in *different* environments. There may be multiple gene systems, each with its own norms of reaction and developmental trajectories. If we are correct about this, then both of the above differences in phenotypic expression may be systematically related to the environment

Bio-Ecological Variousness

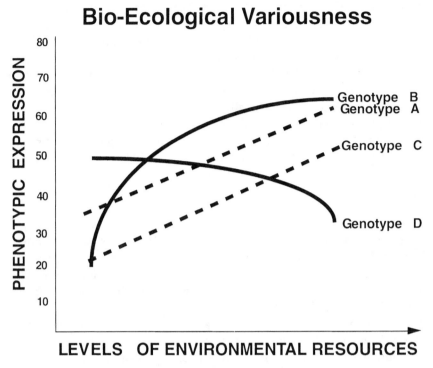

Figure 10.1. Hypothetical depiction of four different genotypes unfolding in response to increasing environmental resources.

despite the differential expressions at any given level of environment between two differing genotypes.

Heritability.   General intellectual functioning, often indexed by full-scale IQ, is one of the most highly heritable measures known to psychologists, commonly accounting for between 50% and 80% of the variance among individuals' IQ scores (Ceci, 1990a; Plomin et al., in press). Over the years, there have been debates about the meaning of heritability, and there has been a general acknowledgment that the next step in our understanding will require breakthroughs at the molecular biological level. Barring such breakthroughs, one can claim, as Kamin (1974) and others once did, that heritability is simply epiphenomenal, the result of colinearity among related variables.

Recently, Plomin and his colleagues have attempted to provide the missing link in the construct validation for the claim that heritability estimates are not simply epiphenomenal but rather reflect the biological substrate responsible for intellectual functioning. Their approach has been to show

that DNA markers, from genes thought to be relevant to neural functioning, taken from individuals with high and low IQ, have different allelic
associations. Specifically, the allelic frequency for two genetic markers appears to be different for high- and low-IQ individuals, thus supporting
the notion of a direct link between IQ and its heritability (see Plomin
et al., in press). This is exciting work and promises to open up the debate
to new levels of analysis. Presently, however, it is a promissory note that
will require far more research to document, as Plomin and his colleagues
note. Crucial reliability and measurement work await this argument,
lest the few reliable associations that have been discovered prove to be
spurious.

Assumptions of traditional psychometric models

The traditional psychometric models of intelligence (e.g., Humphreys,
1962, 1979; Jensen, 1979; Marshalek, Lohman, & Snow, 1983; Spearman,
1904) contain three assumptions. First, there is the assumption of a singular
pervasive ability, called *general intelligence*, or *g*. The evidence for this
assumption is that whenever a battery of diverse mental tests is administered to a diverse sample of subjects, and their scores are intercorrelated,
the result is a positive manifold of correlations. That is, the performance of
individuals tends to be consistent across diverse tests. The assumption is
that consistent performance on such seemingly different tests as verbal
reasoning, cultural knowledge, mathematics, and spatial ability reflect the
fact that each of these measures is saturated with a common factor –
namely, general intelligence, or *g*. This is why they correlate with each
other.

   The second assumption of traditional psychometric models is that general intelligence is biologically based. Many researchers have reached this
conclusion because measures of *g* are correlated with h^2 and other constructs of heritability, as well as with a variety of physiological measures
such as cranial blood flow, evoked potential recordings, central nerve conductance velocity and oscillation, and, as described earlier, different
allelic distributions in high- and low-IQ groups (e.g., Eysenck, 1988;
Hendrickson & Hendrickson, 1982; Jensen, 1992; Plomin et al., in press;
Schafer, 1987).

   The third assumption of traditional psychometric models is that IQ tests
are good measures of biologically based general intelligence and that general intelligence permeates most intellectual endeavors, from verbal reasoning, reading comprehension, and cultural knowledge to quantitative,
spatial, and mechanical abilities. Construct validation research has attempted to show that IQ tests are more potent predictors of a range of

outcomes such as job success, school grades, and training scores than are measures of specific cognitive abilities, motivation, or relevant background experience (e.g., Barrett & Depinet, 1991; Gottfredson, 1986; Henderson & Ceci, 1992; Hunter, Schmidt, & Rauschenberger, 1984; Hunter, Schmidt, & Rauschenberger, 1984). It is argued that there are lower correlations with these specific factors because these are less perfect markers of *g* than are IQ tests, which are alleged to be highly saturated with *g*.

Taken together, the above three assumptions and their related categories of evidence have persuaded many to view psychometric test scores as reflections of a singular, biologically based resource pool that permeates virtually all intellectual feats and that is responsible for the consistency of individual differences in many real-world outcomes.

As readers of the journal *Intelligence* know, there is an enormous research literature that is in accord with the traditional psychometric view, such as the relationship between task complexity and *g*, the relationship between h^2 and *g*, the relationship between IQ and central nerve conductance velocity, and so on. Those who support this view do not deny that the environment plays some role in performance on *g*-based tests like IQ, but they do underscore the relative ascendancy of biology, in some cases pointing out that environments themselves are genetically loaded. That is, the environment is an important contributor to individual and developmental differences by exerting its influence through genetically determined forces and through gene–environment correlations and/or interactions (Plomin & Bergeman, 1991; Plomin & Neiderhiser, 1992):

Labelling a measure "environmental" does not make it a pure measure of the environment. People make their own environments. . . . The ways in which people interact with their environments – their experiences – are influenced by genetic differences [Plomin & Neiderhiser, 1992, p. 163].

Sandra Scarr's work (1992) provides a paradigmatic example of the biologically oriented interactionists' claim that environmental effects are moderated by genetic influences. Scarr reconceptualized findings that had been commonly cited as evidence for large environmental influences on children's IQ. On this basis, she argued that the actual cause of variation in children's IQs was not the putative differences in the child's immediate or proximal environment (e.g., in parental disciplinary styles) but rather variations in mothers' IQs – a presumed genetic marker. Scarr (1989, 1992) argued that the relationship between maternal disciplinary style and offspring's IQ could be accounted for by the fact that mothers who employed positive types of discipline had higher WAIS vocabulary scores than did mothers who employed punitive disciplinary techniques. Thus, what appears to be an environmental influence (type of discipline) on their offspring's IQ is claimed to be the result of a gene–environment correlation;

the driving force behind such ostensibly environmental variation (related to parental disciplinary style) is, in actuality, genetics. Scarr (1992) asserted that as long as some minimum (nonabusive) environment is provided, variations in children's intelligence will be unrelated to variations in their environment:

Being reared within one family, rather than another, within the range of families sampled makes few differences in children's personality and intellectual development. These data suggest that environments most parents provide for their children have few differential effects on the offspring [p. 5]. . . . The important point here is that variations among environments that support normal human development are not very important as determinants of variations in childrens' outcomes [p. 6]. . . . For children whose development is on a predictable but undesirable trajectory and whose parents are providing a supportive environment, interventions have only temporary and limited effects. Should we be surprised? Feeding a well-nourished but short child more and more will not give him the stature of a basketball player. Feeding a below-average intellect more and more information will not make her brilliant. . . . The child with a below-average intellect . . . may gain some specific skills and helpful knowledge of how to behave in specific situations, but [her] enduring intellectual and personality characteristics will not be fundamentally changed [pp. 16–17].

Taking this argument to the next stage, supporters of the psychometric position attempt to estimate, through heritability formulae (for a description and interpretation of this concept, see Ceci, 1990a; Falconer, 1989; Plomin, DeFries, & McClearn, 1990), the degree to which genetics and environment influence mental development. Reports of very high heritabilities are now commonplace for cognitive performance. For example, the psychometrist J. B. Carroll (1992) has recently claimed that at least 50% of the variance in cognitive ability is genetic in origin, and Plomin and his colleagues have reported heritability estimates for specific groups of individuals that range in excess of .80 (Pedersen, Plomin, Nesselroade, & McClearn, 1992). If these claims are correct, then one might question how much the developing organism's ecology can influence mental development, because at least half, and possibly upward of 80%, of the variation in intelligence test scores would seem to be accounted for solely by genetic variation.

Taking issue with some psychometric claims

The biological basis of intelligence.   Recently, some of us have suggested that the evidence recruited to support the three assumptions of the traditional model of intellectual development is open to alternative explanations (Bronfenbrenner & Ceci, 1994). We have argued, for example, that the interpretation given to h^2 is misleading, because h^2, as it is commonly computed (e.g., see Cavalli-Sforza & Bodmer, 1971), can only refer to the proportion of actualized genetic variance; the amount of variance that is unactualized as a result of insufficiencies in the developing organism's ecology is unknown. This means that even if $h^2 = .80$, it is incorrect to infer that the environment can, at best, influence only 20% of the individual differences, because we have no way of knowing how much unactualized genetic potential exists.[1]

In the next section, which outlines the bio-ecological model of intellectual development, we show how traditional measures of h^2 are highly labile due to their failure to take into consideration a variety of assumptions involving environmental factors, which could either increase or decrease heritability estimates. Before describing the bio-ecological model of intellectual development, a caveat is in order: None of the foregoing is meant to claim that traditional assumptions about *heritability* are decidedly incorrect. Rather, the claim is that the evidentiary basis which underlies traditional assumptions allows different interpretations; that is, both psychometric test performance and intellectual development result from a concatenation of cognitive, social, and biological factors (e.g., Sternberg, 1985, 1990). We believe that biology is important but not in the manner that is implied by the concept of heritability.

The bio-ecological model of intellectual development

We turn first to a discussion of some major characteristics of the bio-ecological model. According to the bio-ecological view, the data related to intellectual development necessitate a four-pronged framework for their explication – namely, (1) the existence of not one type of intellectual resource but multiple, statistically independent resource pools, (2) the interactive and synergistic effect of gene–environment developments, (3) the role of specific types of environmental resources (e.g., interactions, called *proximal processes*, as well as more distal environmental resources such as family educational level) that influence how much of a genotype gets actualized in what type of an environment, and (4) the role of motivation in determining how much one's environmental resources aid in the actualization of their potential. We now review these four prongs and, in doing so, describe in greater detail how the bio-ecological model accounts for the bedrock data that underpin traditional models while going beyond them in predictions and explanations.

First, based on the research and arguments presented in Ceci (1990b, 1993), we assume that intelligence is a multiple resource system. Making this assumption gets around the thorny problem of domain specificity and low cross-task correlations when the same cognitive operation is involved (see Ceci, 1990a). It also accords with the analyses by Detterman and his colleagues, showing that independent cognitive processes make unique predictions compared to g-based measures (see Detterman et al., in press).

Second, the bio-ecological view is inherently developmental and interactionist. Like all interactionist perspectives, it asserts that from the very beginning of life, there is an interplay between biological potentials and environmental forces. In order to understand how individuals could begin life possessing comparable intellectual potentials but differ in the level of intelligence they subsequently manifest, the bio-ecological view posits an interaction between various biologically influenced cognitive potentials, such as the capacity to store, scan, and retrieve information, and the ecological contexts that are relevant for each of their unfoldings. At each point in development, the interplay between biology and ecology results in changes that may themselves produce other changes until a full cascading of effects is set in motion.

Although biology and ecology are interwoven into an indivisible whole, their relationship is continually changing; with each change, a new set of possibilities is set in motion until soon even small changes produce large effects. Hence, developmental change is not always or even usually linear but rather is synergistic and nonadditive. A small environmental influence on a protein-fixing gene may initially result in only tiny changes, but over time the chain of events may produce a magnification of effects on other processes. In addition, certain epochs in development can be thought of as sensitive periods during which a unique disposition exists for a specific "cognitive muscle" to crystallize in response to its interaction with the environment. During such periods, neurons within specific compartments rapidly *arborize*, spreading their tentacle-like synaptic connections to other neurons. Even though some of the arboreal connections laid down during these periods of brain spurts will not be used at that time, they can be recruited to enable future behaviors to occur, provided they are not "pruned" because of atrophy or disuse. Siegler (1989) concluded that "the timing of the sensitive period seems to be a function of both when synaptic overproduction occurs and when the organism receives relevant experience" (p. 358). It appears that while some neural processes are more fully under maturational control, others are responsive to the environment, and synapses are formed in response to learning that may vary widely among humans. Similar contextual roles have also been found in the case of various animals' cognitive skills (e.g., Lickliter & Hellewell, 1992; Smith & Spear, 1978).

Third, the bio-ecological model stipulates that proximal processes, which depend, in part, on distal resources in the child's environment, are the engines of intellectual development. *Proximal processes* are defined as reciprocal interactions between the developing child and other persons, objects, and symbols in its immediate setting. In order to qualify as a proximal process, an interaction must both be enduring and lead to progressively more complex forms of behavior (for a detailed justification of these twin assumptions, see Bronfenbrenner & Ceci, 1994). The efficiency of a proximal process is determined to a large degree by the distal environmental resources. Take the proximal process of parental monitoring as an example; this refers to keeping track of your children, knowing if they are doing their homework, who they associate with after school, where they are when they are out with friends, and so forth. Parents who engage in this form of monitoring tend to have children who score higher in school (Bronfenbrenner & Ceci, 1994). However, the process of monitoring is not enough to ensure high scores; parents must also know enough about the content of their child's lessons to help them when they study, and such parental knowledge is what we mean by *distal environmental resources.* Proximal processes are the engines that actually drive the outcome but only if the distal resources can be imported into the process to make it effective.

Appropriate proximal processes differ as a function of the developmental status of the organism. For example, in infancy a proximal process might be an activity between a caregiver and an infant that serves to maintain the infant's attention or encourage her to slightly exceed her proximal zone of potential; for an adolescent, an appropriate proximal process might be parental monitoring that occurs during a homework assignment. As we will argue, proximal processes are the engines that drive development; they are the mechanisms that translate genes into phenotypes. By adding them to our models, we can understand when heritability measures will be high and when they will be low.

Based on the reanalysis of two different data sets, Bronfenbrenner and Ceci (1994) have argued that when proximal processes are at high levels in a child's environment, then heritability estimates are high, and yet at the same time individual difference *may* be attenuated. The first part of this claim is no different from that of many in the psychometric community (e.g., Herrnstein, 1973; Humphreys, 1989) because it has always been recognized that genetic variance becomes relatively greater as a consequence of decreasing environmental variance, which is where the second part of the claim comes in: High levels of proximal processes decrease environmental variance because they serve to supply interactive experiences to children who otherwise might not have them (i.e., in some poor environments, children may still have high levels of proximal processes). This has the effect of not only increasing h^2 but also leveling group differences.

According to the bio-ecological model, it is also important to assess the dimensions of the child's ecology, or distal environment, because in two ways it provides limits on the efficiency of proximal processes. First, the distal environment contains the resources that need to be imported into proximal processes for the latter to work maximally. The total environment is therefore a combination of distal processes (e.g., books, toys, parents' education level, etc.) and proximal processes (e.g., reciprocal interactions with caregivers). For example, it is not enough for parents to engage their adolescents in reciprocal interactions that sustain their children's attention while studying algebra (an example of the proximal process called *monitoring*) if the parents themselves cannot effectively explain the relevant algebraic concepts. Thus, the distal environment can sometimes set limits on the efficacy of proximal processes. Resources from the distal environment need to be imported into the proximal environment, as in the importation of knowledge of algebra into the proximal process of parental monitoring. It simply is insufficient to train parents to monitor their children's homework in algebra if the parent does not know enough algebra to help out.

A second reason for the importance of the distal environment is that it provides the stability necessary to benefit from proximal processes. A large literature (see Bronfenbrenner & Ceci, 1994) illustrates that the less stable the distal environment, the worse the developmental outcome, regardless of social class, ethnicity, or ability levels. Frequent changes in daycare arrangements, parental partners, schools, or neighborhoods, for example, are associated with adverse outcomes, and this is presumably independent of the level of proximal processes.

Measures of heritability are extremely sensitive to secular trends, generally dropping during times of economic scarcity and climbing during times of plenty (see Bronfenbrenner & Ceci, 1994). Such trends are consistent with the view that the ecology brings to fruition differing levels of genetic potential, and h^2 will fluctuate by a factor of three in conjunction with economic fluctuations. We assume that in times of economic scarcity, the levels of proximal processes are reduced because caregivers' attention is directed externally (at staying alive) rather than internally (at reciprocal interactions with their children).

The fourth prong of the bio-ecological view is the incorporation of *motivation* as a key ingredient in its explanation of empirical findings. Briefly, an individual must not merely be endowed with some biological potential for a given cognitive resource, or merely be exposed to an environment that facilitates the expression of this cognitive resource; the individual must also be motivated to benefit from exposure to such an environment. Men in Ceci and Liker's (1986) study, who demonstrated highly complex forms of reasoning at the racetrack, did not exhibit the same degree of complex reasoning in other domains. Had they been exposed to environments that were conducive to, say, learning science or philosophy, and motivated to take advantage of such environments, they would have undoubtedly acquired the ability to think more complexly in those domains.

Figure 10.2 is a partial schematic of the bio-ecological trajectory, and it summarizes and elaborates some of the main points just made. As can be seen, the flow starts at the bottom with parental genes giving the child its early impetus and direction, ultimately influencing various forms of cognitive processing that are fairly independent of each other (depicted as four independent arrows emanating out of the child). The manifestations of these multiple gene-based resource pools are influenced by multiple forms of activities that caregivers engage in with children. One such activity – namely, proximal processes – is postulated to actualize genetic potentials. Thus, variations in the level of resources in the distal environment (dichotomized in this figure for ease of illustration) along with variations in the level of proximal processes will lead to different levels of h^2.

There are several aspects about Figure 10.2 that are worth noting. One is that proximal processes are more than expressions of parental genetics; otherwise, there would be no basis for making differential predictions about the size of h^2 because all that we would need to know is the parental genotype. Yet we shall argue that there are differential predictions about the size of h^2 *within the same levels of consanguinity* when differing levels of proximal processes exist.

Perhaps not obvious in the figure is a core assumption underlying the bio-ecological model – namely, the influences of genetics and environment are never wholly separable. From the moment of conception, the actualization of inherited cognitive dispositions for embryological development do not occur in a vacuum but are differentially responsive to the intrauterine environment (as well as to the intercellular environment, including interactions among hormones, inducers, enzymes, and proteins). This power of inherited propensities is not diminished after birth, because as the child interacts with persons, symbols, and objects in her environment, the latter becomes genetically loaded, as the active organism selects, changes, and hence constructs her own environment.

Third, this figure illustrates the stipulation that proximal processes are the engines of intellectual development, with higher levels of proximal processes associated with increasing levels of intellectual competence.

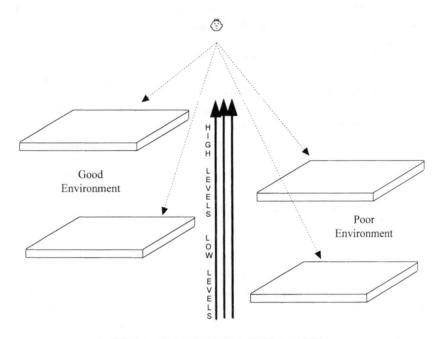

Figure 10.2. The bio-ecological model. Distances between platforms represent the differential effects of proximal processes and more global environmental resources.

Thus, it is predicted that increases in proximal processes not only increase h^2 and reduce group differences, but they will produce more competent organisms across the board.

Finally, Figure 10.2 portrays some of the major predictions regarding the inclusion of proximal processes into a model of intellectual development. Proximal processes are posited to exert a more potent influence on the various cognitive operations than does the distal environment (e.g., SES) in which they operate. Accordingly, we predict that the differences in intellectual outcomes (and their corresponding h^2 estimates) between poor and good environments will be systematically smaller than differences associated with low versus high levels of proximal processes. (This prediction is revealed in the comparison of the distances in the figure; distances between high and low proximal processes are greater than distances between good and poor distal environments.)

In addition to the prediction that the highest magnitudes of h^2 are to be found under good–high conditions (i.e., good distal environmental resources–high proximal processes),[2] the bio-ecological model makes a corollary prediction; the largest *differences* in the magnitudes of h^2 for competent outcomes are to be found between children in the high–good condition and the low–poor condition. In Figure 10.2, this is indicated by the largest physical distance between these two conditions.

There have been several attempts to provide a limited test of these predictions. For example, nearly 20 years ago, Bronfenbrenner (1975) reanalyzed the then available data on monozygotic (MZ) twins reared apart. While there were no measures of proximal processes available, the resources in the distal environment probably differed markedly as a function of the ecologies in which the twins were raised. Bronfenbrenner reported that intraclass correlations for IQ reached the high 0.80s when separated twins (MZAs) were reared in similar ecologies but that they plummeted to 0.28 when MZAs were reared in vastly different ecologies (e.g., agricultural/mining towns vs. manufacturing towns). These data are consistent with the bio-ecological model's prediction that h^2 will be both far higher and far lower than previously reported when the levels of proximal processes and distal environmental resources are systematically varied.

An implication of the foregoing argument is that the greater the genotypic *dissimilarity*, the greater the impact of proximal processes in increasing phenotypic *dissimilarity* but only in good environments. This is why the difference between the dizygotic (DZ) and MZ intraclass correlations is almost always greater in good environments than in poor ones (see Bronfenbrenner & Ceci, 1993, 1994). Thus, according to the bio-ecological view, the reason that h^2 is so high in such studies is not because the cognitive phenotypes of MZ pairs are unaffected by the environment – the interpretation that many have given to Bouchard's provocative findings of MZs reared apart – but because the cognitive phenotypes of the DZ pairs are so amply affected by the environment (e.g., Fischbein, 1982). In other words, the DZs become less alike when there are good environmental resources; parents with such resources can treat their children differentially, bringing to fruition one child's musical talent through lessons, and another child's language skills through enrollment in tutoring, special camps, and so on. As resources enable the parents to treat their genetically different twins dissimilarly, the DZ twins become phenotypically less alike, their intraclass correlations get smaller, and the difference between their correlations and those of MZs gets bigger (hence, h^2 gets bigger).[3] Thus, the bio-ecological model finds support for the interplay between ecology and biology in some of the very findings that have served as cornerstones of the behavior-genetic and psychometric traditions.

The results of reanalyses with more recent data are consistent with predictions from our bio-ecological model, although much more work clearly needs to be done before this can be accepted as definitive (Bronfenbrenner & Ceci, 1993, 1994). If we are correct about the critical role of proximal processes, then under differing levels of proximal processes, the size of the heritability estimate will change, possibly dramatically, because h^2 reflects only the proportion of actualized genetic potential, leaving unknown the amount of genetic potential that remains unactualized due to insufficient proximal processes.

Comparing the bio-ecological model with other approaches

Having described the basic features of the bio-ecological model of intellectual development, it now can be contrasted with the biological, environmental, and psychometric views of intelligence on three grounds. First, the bio-ecological view proposes multiple cognitive abilities rather than one pervasive general factor, and these multiple abilities are at best only imperfectly gauged by tests of so-called general intelligence. Detterman (1986) has made a related argument – namely, that the cognitive system can be conceptualized so that any observed interrelatedness among disparate cognitive tasks can be due to bottlenecks in the system rather than because the disparate tasks are g-saturated. To take one possibility, if the ability to retrieve information depends on the prior ability to store information, then any disruption of the latter ability will also have an impact on the former ability. This is not because these are the same abilities but merely because their efficiencies are linked. Under experimental conditions, their independent functioning can be demonstrated.

Second, the bio-ecological view differs from traditional nature–nurture interactionist views in its conception of the interaction between biological and environmental factors. The traditional interactionist view is that "genes encode phenotypes," and it is the eliciting power of the environment that releases the phenotypes. Thus, the traditional interactionist view gives primacy to the importance of biological factors. The bio-ecological view departs from this preformationist view of the genotype and gives the environment a much more significant role. Genes do not encode pheno-types; rather, they manufacture proteins and enzymes that influence the expression of neighboring genes as well as interaction among themselves. If this view is correct, it implies that the genetic interactions of proteins and enzymes are governed by physical and chemical laws independent of the strand of DNA from which they originated (see chapters in Subtelny & Green, 1982). A model of the translation of genotypes into phenotypes requires that we consider not just the proteins that genes manufacture but also the developmental role that such proteins play. This is essential, because most of the hormones, inducers, and inhibitors are connected in complex ways with the activity of multiple gene systems. Thus, the resultant morphology is only indirectly related to genes, making it impossible to explain phenotypes or morphological change exclusively or even primarily in terms of genes. As the evolutionary biologist Pere Alberch wrote in 1983:

Even if we knew the complete DNA sequence of an organism, we could not reconstruct its morphology. We need to know about the epigenetic interactions that generated the phenotype [p. 862].

Third, like other interactionist views of development, the bio-ecological view argues that the efficiency of cognitive processes depend on aspects of the context. However, according to Ceci (1990a), context is not an adjunct to cognition but a constituent of it. Unlike traditional cognitive science, which has assumed that context is merely a background for cognition, the bio-ecological view regards context as an inextricable aspect of cognitive efficiency. In this chapter, *context* is defined broadly to include not only external features of the near and far environment and their motivational properties but internal features of the organism's mental representation, such as the manner in which a stimulus or problem is represented in memory. Thus, speed in recognizing letters and numerals depends on how those stimuli are represented in memory, with more elaborate representa-tions leading to faster recognition rates (Ceci, 1990b). This explains why the same cognitive ability, no matter how basic, often operates inconsistently across diverse contexts (Ceci, 1990b). The same individuals who are slow at recognizing a stimulus in one domain may recognize it in another domain more quickly if its representation in the latter domain is more elaborate. In short, cognition-in-context research has shown that context, including the mental representation or mental context of a task, helps to determine the efficiency of cognition.

Fourth, the bio-ecological view assumes that there are noncognitive abilities that are highly important for subsequent intellectual development and that are inherited. For example, a child may inherit various types of temperament (e.g., restless, impulsive), physical traits (e.g., skin color, facial shape), and "instigative characteristics" (e.g., Bronfenbrenner's, 1989, reward-seeking type) that may influence later learning and development. While these traits are themselves influenced by gene systems, and can be shown to exert direct as well as indirect effects on subsequent IQ performance and school success, they are not cognitive in nature. So, these noncognitive characteristics and abilities can account for heritability patterns (e.g., IQs that run along consanguinity lines) without claiming that this is a consequence of the inheritance of a central nervous system with a determinate signal-to-noise ratio that limits processing capacity. Family members who share these characteristics may perform similarly as a result of their noncognitive dispositions rather than because they share the same rate-limiting nervous systems (Lehrl & Fischer, 1988, 1990). Importantly, accounts of rate-limiting cognitive functioning that are based on EEG power, spectral density measures (e.g., Weiss, 1990a,b), blood glucose levels in the brain, central nerve conductance velocity and oscillation (Jensen, chapter 2, this volume), and heritability analyses (e.g., Pedersen et al., 1992) cannot distinguish between cognitive (i.e., inherited limits on CNS functioning) and noncognitive bases of performance. This is not to deny the importance of genes in intellectual performance but merely to signal a caution: Not all that is *intellectual* is genetic in origin, and not all that is *genetic* is intellectual in nature.

Finally, the bio-ecological view departs from traditional behavior-genetic models regarding the nature and meaning of h^2. According to the bio-ecological view, h^2 reflects the proportion of actualized genetic potential, leaving unknowable the amount of unactualized genetic potential (Bronfenbrenner & Ceci, 1994). As we have argued earlier, nothing can be said about an individual's potential for success or failure without knowing about the level of proximal processes and more distal resources that exist in the child's environment. Proximal processes are the engines that drive development; they are the mechanisms that translate genes into phenotypes (Bronfenbrenner & Ceci, 1994). If there are insufficient proximal processes in one's life, then h^2 will reflect only that portion of one's potential that can be brought to fruition by the limited level of proximal processes.

In sum, current research on intelligence and intellectual development points in a direction that emphasizes the role of context in the formation and assessment of an individual's manifold cognitive potentials. Although traditional measures of general intelligence possess good predictive validity in school, job, and training situations, this does not necessarily reflect the fact that a singular resource pool underpins a significant portion of the prediction, or that the size of heritability estimates reflects the amount of variation due to purely genetic processes, or that only cognitive factors are involved, or that only the residualized variance (i.e., that portion which is not accounted for by genetics) remains to be explained by the environment.

A possible, and perhaps even probable, misunderstanding of this last claim is that because published heritability estimates for some characteristics are often as high as 0.6–0.8 (Jensen, 1992), this implies that existing levels of proximal processes in our society are already high enough to ensure that 60–80% of variance in genetic potential for developmental competence is being realized. Such a conclusion, however, does not follow if h^2 expresses only that portion of the total variance that is accounted for by linear, additive, variations in *actualized* genetic potential. The effect of environments on genotypes is not ascertainable at a single linear point; rather, these twin forces interact at multiple developmental levels and across and along many environmental dimensions. Thus, current data on phenotypic variation merely indicate the range of relevant environments to which genotypes have been exposed. The same genotype can develop dramatically different phenotypes in different environments than what were available in those studies reporting h^2 estimates, just as surely as different genotypes can exhibit different norms of reaction for variation in the same environment. Focusing on the latter while ignoring the former runs the risk of ignoring the possibility that vast reservoirs of genetically coded resource pools (cognitive and social) exist that are not actualized, or are actualized but which are expressed with little or no variation across individuals. Thus, the absence of proximal processes to bring nonactualized genes to fruition cannot be assessed by a measure of h^2, which only reflects actualized genetic potential. This is precisely why we have departed from the common practice of using h^2 interchangeably with the term *heritability*. It would be a mistake to assume that 60% or more of genetic potential is actualized in these studies or that existing levels of proximal processes are adequate to actualize all of one's genetic potential.

For the above reasons, *prediction* and *explanation* can be fundamentally disjunctive enterprises in science. The bio-ecological view of intellectual development aims to fulfill the promissory note of the interactionist perspective by being cognizant of the biological and cognitive findings that have been reported but recasting them in a new developmental-contextual light.

Notes

1. Everyone accepts the argument that the environment is genetically loaded, as certain personality types tend to gravitate toward certain environments. We would like to suggest that heritability estimates are environmentally loaded, as heritability estimates are influenced by whether one's environment contains the resources necessary to actualize an individual's genetic potential.

2. There are some circumstances where h^2 will be higher in poor environments (see Fischbein's 1982 data on arithmetic scores). But these are exceptions, and we will refrain in this chapter from delving into them for ease of exposition. The interested reader can consult Bronfenbrenner and Ceci (1994) for the reasoning.

3. For the sake of simplification, we consider only the simple case of heritability analysis where the contrast between MZ and DZ twins is given by the formula: $h^2 = 2$ (the intraclass correlations for MZ twins minus the intraclass correlations for DZ twins).

References

Alberch, P. (1983). Mapping genes to phenotypes, or the rules that generate form. *Evolution*, *37*, 861–3.

Anastasi, A. (1958). Heredity, environment, and the question "how". *Psychological Review*, *65*, 197–208.

Angoff, W. H. (1988). The nature–nurture debate, aptitudes, and group differences. *American Psychologist*, *43*, 713–20.

Barrett, G. V., & Depinet, R. L. (1991). A reconsideration of testing for competence rather than for intelligence. *American Psychologist*, *46*, 1012–24.

Bors, Douglas A. (1993). The factor-analysis approach to intelligence is alive and well – Human cognitive abilities: A survey of factor-analytic studies. *Canadian Journal of Experimental Psychology*, *47*(4): 763–66.

Bronfenbrenner, U. (1975). Nature with nurture: A reinterpretation of the evidence. In A. Montague (Ed.), *Race and IQ*. New York: Oxford University Press.

Bronfenbrenner, U. (1989). Ecological systems theory. In R. Vasta (Ed.), *Annals of child development: Six theories of child development: Revised formulations and current issues* (pp. 185–246). Greenwich, CT: JAI Press.

Bronfenbrenner, U., & Ceci, S. J. (1993). Heredity, environment, and the question "how"?: A first approximation. In R. Plomin & G. McClearn (Eds.), *Nature, nurture, and psychology* (pp. 313–24). Washington DC: American Psychological Association.

Bronfenbrenner, U., & Ceci, S. J. (1994). Nature–nurture reconceptualized in developmental perspective: A bio-ecological model. *Psychological Review*, *101*, 568–86.

Carroll, J. B. (1992). Cognitive abilities: The state of the art. *Psychological Science*, *3*, 266–70.

Cavalli-Sforza, J., & Bodmer, J. (1971). *The genetics of human populations*. San Francisco, CA: Freeman.

Ceci, S. J. (1990a). On the relationship between microlevel and macrolevel cognitive processes: Worries over current reductionism. *Intelligence*, *14*, 1–19.

Ceci, S. J. (1990b). *On intelligence . . . more or less: A bio-ecological treatise on intellectual development*. Englewood Cliffs, NJ: Prentice Hall (Century Psychology Series).

Ceci, S. J. (1993). Contextual trends in intellectual development. *Developmental Review*, *13*, 403–35.

Ceci, S. J., & Liker, J. (1986). A day at the races: A study of IQ, expertise, and cognitive complexity. *Journal of Experimental Psychology: General*, *115*, 255–66.

Detterman, D. K. (1986). Human intelligence is a complex system of separate processes. In R. J. Sternberg & D. Detterman (Eds.), *What is intelligence?* (pp. 57–61). Norwood, NJ: Ablex.

Detterman, D. K., Mayer, J. D., Caruso, D. R., Legree, P. J., Conners, F. A., & Taylor, R. (in press). The assessment of basic cognitive processes in relationship to cognitive deficits. *American Journal of Mental Retardation*.

Eysenck, H. J. (1988). The biological basis of intelligence. In S. Irvine & J. Berry (Eds.), *Human abilities in cultural context* (pp. 87–104). New York: Cambridge University Press.

Falconer, D. S. (1989). *Introduction to quantitative genetics* (3rd ed.). Harlow, U.K.: Longman.

Fischbein, S. (1982). IQ and social class. *Intelligence*, *4*, 51–63.

Gottfredson, L. S. (1986). Societal consequences of the *g* factor in employment. *Journal of Vocational Behavior*, *29*, 379–410.

Greulich, W. W. (1957). A comparison of the physical growth and development of American-born and Japanese children. *American Journal of Physical Anthropology*, *15*, 489–515.

Henderson, C. R., & Ceci, S. J. (1992). Is it better to be born rich or smart? A bio-ecological analysis of life-course outcomes (pp. 705–75). In K. R. Billingsley, H. U. Brown, & E. Derohanes (Eds.), Scientific Excellence in Supercomputing: The 1990 IBM Contest Prize Papers. Athens: The Baldwin Press, University of Georgia.

Hendrickson, A. E., & Hendrickson, D. E. (1982). The psychophysiology of intelligence. In H. J. Eysenck (Ed.), *A model for intelligence* (pp. 151–228). New York: Springer-Verlag.

Herrnstein, R. J. (1973). *IQ and the meritocracy*. Boston: Little, Brown.

Humphreys, L. (1962). The organization of human abilities. *American Psychologist*, *17*, 475–83.

Humphreys, L. (1979). The construct of general intelligence. *Intelligence*, *3*, 105–20.

Humphreys, L. (1989). Commentary. In R. Linn (Ed.), *Intelligence: Measurement, theory, and public policy* (pp. 29–73). Urbana, IL: University of Illinois Press.

Hunter, J., Schmidt, F., & Rauschenberger, J. (1984). Methodological, statistical, and ethical issues in the study of bias in psychological tests. In C. R. Reynolds & R. T. Brown (Eds.), *Perspectives on bias in mental testing* (pp. 41–97). New York: Plenum.

Jensen, A. R. (1979). *g*: Outmoded theory or unconquered frontier? *Creative Science Technology*, *2*, 16–29.

Jensen, A. R. (1980). *Bias in mental testing*. New York: Free Press.

Jensen, A. R. (1992). Commentary: Vehicles of *g*. *Psychological Science*, *3*, 275–8.

Kamin, L. J. (1974). *The science and politics of IQ*. Hillsdale, NJ: Erlbaum.

Lehrl, S., & Fischer, B. (1988). The basic parameters of human information processing: Their role in the determination of intelligence. *Personality and Individual Differences*, *9*, 883–96.

Lehrl, S., & Fischer, B. (1990). A basic information psychological parameter for the reconstruction of concepts of intelligence. *European Journal of Psychology*, *4*, 259–86.

Lickliter, R., & Hellewell, T. B. (1992). Contextual determinants of auditory learning in Bobwhite quail embryos and hatchlings. *Developmental Psychobiology*, *17*, 17–31.

Marshalek, B., Lohman, D., & Snow, R. (1983). The complexity continuum in the radex and hierarchical models of intelligence. *Intelligence*, *7*, 107–27.

Pedersen, N. L., Plomin, R., Nesselroade, J., & McClearn, G. E. (1992). A quantitative genetic analysis of cognitive abilities during the second half of the life span. *Psychological Science*, *3*, 346–53.

Plomin, R., & Bergeman, C. S. (1991). The nature of nurture: Genetic influence on "environmental" measures. *Behavioral and Brain Sciences*, *14*, 373–427.

Plomin, R., DeFries, J., & McClearn, G. (1990). *Behavior genetics: A primer* (2nd ed.). New York: Freeman.

Plomin, R., McClearn, G., Smith, Vignetti, Chorney, Kasarda, Thompson, L., Detterman, D. K., Daniels, Owne, & McGuffin, P. (1994). DNA markers associated with high versus low IQ. *Behavior Genetics*, *24*, 107–18.

Plomin, R., & Neiderhiser, J. M. (1992). Genetics and experience. *Current Directions*, *1*, 160–3.

Scarr, S. (1989). Protecting general intelligence: Constructs and consequences for interventions. In R. L. Linn (Ed.), *Intelligence* (pp. 74–115). Chicago: University of Illinois Press.

Scarr, S. (1992). Developmental theories for the 1990s: Development and individual differences. *Child Development*, *63*, 1–19.

Schafer, E. W. P. (1987). Neural adaptability: A biological determinant of *g*-factor intelligence. *Behavioral & Brain Sciences*, *10*, 240–1.

Siegler, R. S. (1989). Mechanisms of cognitive development. *Annual Reviews of Psychology*, *40*, 353–79.

Smith, G. J., & Spear, N. E. (1978). Effects of home environment on withholding behaviors and conditioning in infant and neonatal rats. *Science*, *202*, 327–9.

Spearman, C. (1904). General intelligence objectively determined and measured. *American Journal of Psychology*, *15*, 206–21.

Sternberg, R. J. (1985). *Beyond IQ: The triarchic theory of intelligence*. New York: Cambridge University Press.

Sternberg, R. J. (1990). *Metaphors of mind*. New York: Cambridge University Press.

Subtelny, S., & Green, P. B. (1982). *Developmental order: Its origin and regulation*. New York: Alan R. Liss, Inc.

Tanner, J. M. (1962). *Growth at adolescence: With a general consideration of the effects of heredity and environmental factors upon growth and maturation from birth to maturity* (2nd ed.). Springfield, IL: C. C. Thomas.

Weiss, V. (1990a). From short-term memory capacity toward the EEG resonance code. *Personality and Individual Differences*, *10*, 501–8.

Weiss, V. (1990b). The spatial metric of brain underlying the temporal metric of EEG and thought. *Gegenbauers Morphology Jahrb. Leipzig*, *136*, 79–87.

Woodsworth, R. S. (1941). *Heredity and environment*. New York: Social Science Research Council.

11 An interactionist perspective on the genesis of intelligence

Edmund W. Gordon and Melissa P. Lemons

Introduction and historical overview

Despite ubiquitous claims to scientific objectivity in the production of knowledge, few if any of our efforts at empirical or theoretical science are free of bias. All of our research efforts tend to be influenced by prior knowledge and beliefs concerning the phenomena that we study. Many of our efforts toward rigor in design are intended to control or compensate for sources of bias. Nevertheless, since it is impossible to eliminate human judgment and decision, biases continue to influence our work (see Gordon, Miller, & Rollock, 1990). Such is the case in the continuing pursuit of understanding relevant to issues concerning the relationships between intelligence, heredity, and environments. Our conceptions of each of these constructs are influenced by differential perspectives, which are often influenced by differences in the respective social position held by the investigator. One need only look at the long history of political contamination of this debate to see how much our efforts at explaining these relationships have been shaped by concern for differences in status attributable to class, race, and sex. In approaching this set of issues, we make no claims to objectivity. The authors of this chapter are conceptually biased in favor of epigenetic and interactionist conceptions of the origin and nature of human behavioral development. We are politically biased in favor of egalitarian and humanistic values as guides to inform social organization. As social scientists, we are nonetheless committed to conceptual clarity, logical consistency, and scientific rules of evidence. We seek to bring some measure of objectivity to bear on our work, but we do not claim to be unbiased.

It is no longer fashionable to debate questions concerning the relationships between intelligence, heredity, and environment from either the nature or the nurture perspective. Almost all well-informed persons now concede that such debates are dysfunctional to an adequate understanding of these relationships. Instead, one is more likely to encounter debates concerning the relative contributions of each to our understanding of the genesis of human behavior. This is clearly an advance over the old either–

323

or debates, but generic conclusions concerning the relative weight of either may be equally dysfunctional. Human behavioral adaptation and development (phenotype) are best thought of as the products of dynamic and dialectical interactions between genetic phenomena (genotype) and environmental encounters. Thus, questions concerning how much of one and how much of the other can be answered only in reference to specific interactions in which the reciprocal components are known.

Today, most of us know a great deal more about behavioral genetics than the experts in this field knew in the 1950s. Genes, in their environmental interactions, are considered the determinants of phenotype. Although most of us will acknowledge a critical role for environments, even those of us who take a critically interactionist position know very little about how environments operate to influence the functions of genes. Yet, it is these very environments that determine just where, along each gene's wide range of reaction, the phenotypic manifestation of the gene will fall. It is, perhaps, the absence of such knowledge that makes it easier to assign importance to the genetic material than to the environmental encounters, or vice versa, depending on one's perspective or bias.

Eysenck (1971) and Jensen (1969)[1] have emerged as the principal spokespersons for heredity as the primary determinant of the quality of intelligence. In the past year, the late Herrnstein was joined by Murray (1994) in the rehashing of much of their data and that of others to assert this position. Jensen took the lead in conducting elaborate statistical analyses of IQ and other test data that show African American and low socioeconomic status (SES) persons to score about one standard deviation lower than European American middle-class persons. These data were used to raise questions about the relative forces of heredity and environment in the formation of intelligence, and the degree to which measured intelligence is mutable or fixed. Jensen makes reference to Hunt's *Intelligence and Experience* (1961), which tends to support an argument counter to the fixed intelligence notion, but he identifies himself with the less optimistic view of the mutability of intelligence advanced by Bloom in the *Stability of Human Characteristics* (1964). It should be noted that Bloom argues that IQ, as with other developmental characteristics, is quite plastic early in life, before age 4 or 5, but becomes stable as the individual matures.

Jensen reports the distribution of IQ scores in several populations and concludes that except for certain systematic departures, intelligence test scores form a normal distribution as represented by the bell-shaped curve. Exceptions at the extreme lower end of the continuum, he claims, are generally linked to genetic abnormalities. Almost no children from higher socioeconomic families have abnormally low IQ scores, unless they also have neurological abnormalities. From this and other evidence cited,

Jensen suggests that the heritability of intelligence is quite high; that is, genetic factors are far more important than environmental forces in producing IQ differences among individuals, as well as between populations. His evidence included studies of selective breeding for intelligence in rats and other animals, investigations of the chromosomal abnormality called *Turner's syndrome* on specific forms of intellective functioning, studies of mental retardation, and research concerning the high degree of assortative mating in our society and its relationship to significant differences in the intelligence of the offspring of such mates.

The primary argument advanced by Jensen and those who agree with him is that the proportion of population variance that is due to genetic factors as opposed to environmental interactions that form intelligence is separable from other sources of variance, and its special contribution to the total variance can be determined.

Thus, Jensen attempted to determine the proportion of phenotypic variance that is due to genetic variance. He offered as data in support of this assertion a rather comprehensive review of studies of unrelated persons reared together and reared apart, collateral relations including siblings, dizygotic and monozygotic twins (reared together and apart), and direct-line relations including grandparent, grandchild, parent, and child. From his review of these data, he concluded that approximately 20% of the variance in IQ is attributable to environmental influences and as much as 80% is attributable to heredity. As a result of these arguments, much of the work that emerged in the 1960s and 1970s was focused on statistical analyses in support of the hypothesis that the greater portion of the variance in IQ scores can be attributed to heredity.[2]

While much of the variance in IQ scores has been historically attributed to genetic factors and statistically supported by the comparison of test scores across populations, little discussion of the characteristics of genes or their expression has accompanied that conclusion. Our bias toward epigenetic and interactionist conceptions of development requires the examination of the characteristics of the agents and actors involved in organism–environment encounters.

Epigenetic and interactionist approaches to behavioral development have three basic characteristics. (1) Behavior derives from both social and biological origins. Biology, in this relationship, includes genetic materials and other organic resources that make up the body and bodily functioning of an organism. *Social* refers in this context to both face-to-face interactions with other individuals or groups and delayed interactions between individuals using a variety of cultural artifacts and technologies. (2) The interactions between biological and social factors that produce behavior are complex and multiply determined. (3) The interactions in question result in bidirectional transformation, influencing both the social environment and the biosocial behaviors that are involved.

The nature of intellective behavior

Webster's Dictionary (1980, p. 732) defines intelligence as the "ability to learn or understand from experiences; the ability to acquire and retain knowledge, mental ability" and "the ability to respond quickly and successfully to a new situation, the use of the faculty of reason in solving problems." *Intellective behavior,*[3] then, refers to the cumulative capacity for processing and storing information and for making adaptive responses to familiar and novel circumstances. When Cole, Gay, Glick, and Sharp (1971) concluded that all groups of human beings appear to represent in their developed abilities a wide range of intellective competencies, they were, no doubt, thinking of intelligence as the capacity to adapt to one's environment and to use past adaptational experiences in response to similar and novel environmental encounters.

In a discussion of the nature of intelligence, Weinberg (1989) examines Mayr's (1982) categorization of intellective styles as "lumpers and splitters." *Lumpers* are identified with the notion of intelligence as a "general unified capacity for acquiring knowledge, reasoning, and solving problems that is demonstrated in different ways (navigating a course without a compass, memorizing the Koran, or programming a computer)." Even though Weinberg (1989) introduces some diversity in his "different way," there is a narrowness in the lumperians approach that not only refers to an overall summative ability; this ability is referred to as being manifested in universalist conceptions of intellective function (i.e., abstract reasoning and recall).

In contrast, the *splitters* seek to isolate (at least for study) different types of intellective ability. Gardner (1983), for example, has called attention to linguistic, manual, and five other specific intelligences, some of which would at least benefit from heterogeneous contexts for their assessment. What we see in the possible differences in the potential for revealing the quality of intellect is one conception that favors communicentric[4] indicators of ability, and another that favors more heterogeneous indicators (Gordon, 1988). It is this tenuousness and protean character of the construct that contributes to confusion concerning intelligence as it is manifested in people whose life experiences differ.

Unfortunately, the variety of manifestations of intellective behavior is often ignored, resulting in failure to recognize or accept some forms of practice as intelligent. Intelligence can become synonymous with the privileged practices of the dominant culture. When intelligence is so narrowly defined, it privileges some forms of symbolization, representation, and categorization and denigrates alternative expressions of situationally desirable cognitive characteristics. The particular form that cognitive abilities or bodies of knowledge come to possess is dependent on the experiences of individuals within their environments. The development of perspective and practice, as with other forms of development, results from multiple and interdependent interactions.

Whether we are focused on practical problem solving, logical reasoning, abstract symbolization, or social judgment, if the frames of reference by which criteria are drawn are too narrow, evaluative judgment with reference to any one or all of these categories of intellective function is likely to be flawed. The adaptive or intellective status assigned to behavior should be determined with reference to the current and historical contexts of behavioral expression. The failure to judge behavior by its adaptive functions at numerous levels of expression appears to be a common problem in prevailing notions of human intelligence. Most of the psychological research on intelligence and differences in intellective function has utilized a unitary notion of intelligence and a hegemonic cultural context for its assessment.

It is the position of the authors of this chapter that to improve pedagogy and assessment, it is important to define intelligence as complex and a composite of aptitudes and developed or developing abilities and achievements (Anastasi, 1980). It must be recognized that these are expressed through behaviors that are often subjectively weighed and defined socially and culturally. Now when we turn to evaluative judgments concerning the quality of adaptive abilities in different groups of human beings, the variations in manner, level, and context of expression must be factored into the equation.

Conceptualization and explanation of the mechanisms of gene–environment interaction effects are still emerging as various researchers and theorists try to represent the nature and operation of the complex processes of intellectual development. Interactionist or constructivist perspectives on the nature of intellectual development appear under many headings and vary in their emphasis. However, the approaches presented in this chapter share a common focus on the complexity of the interactions of material, ideological, and phenomenological factors in development.

Scientists have been more successful and more willing to engage in discussion about the influence of observable, physical aspects of the environment. Such factors are amenable to scientific methods of observation, documentation, and interpretation. Psychological or ideological forces that lie somewhere between the individual mind and the physical environment are less tractable. However, because psychological factors are based on consensual agreement about the nature of reality or its meaning, they are more easily studied than phenomenological factors. By these we mean existential or individually meaningful experiences and perspectives. While we believe that these protean forces are essential in directing human development, we are as yet unable to define clearly or predict patterns of outcomes based on differential phenomenological experience.

The common mechanism identified among interactionist researchers is culture. The expression of intelligence is largely a function of cultural experiences. Sources of common cultural experiences include national, ethnic, class, and gender identifications. Individuals who share one or more cultural-category associations often share historical and environmental conditions under which cultural traits evolve. Cultural circumstances and behaviors of a people are a product of the function of common experiences and expectations. It is these experiences and expectations that influence the particular manifestations of intellective behavior.

Culture is the omnibus construct that refers to the circumstances of development and performance. Culture has many definitions and connotations that are often confused and used interchangeably. We use *culture* in several senses within this chapter. (1) Culture is the organic and structural resources in which organisms grow. It is the "stuff" from which organisms build and maintain themselves. It is the source of both material and social resources used in the development of humans. As a ubiquitous and constantly changing source of raw materials, it is the root of both beneficial and harmful resources. (2) A more common sense of culture is a particular set of practices and beliefs of a particular group of people. This notion of culture is often viewed as a fixed, invariant characteristic of members of a particular group. Even when we use *culture* in the sense of features generally associated with a particular group, we do not adopt this notion of invariant characteristics. The cultures we refer to are dynamic and dialectical phenomena that are constantly changing, expanding and contracting, making it difficult to delineate an exhaustive list of specific components.

Appiah (1993) suggests that it is not so much the components of a particular culture that are important in shaping behaviors as it is the cultural identity that one develops. *Cultural identity* refers to the ideational representation of the culture in the minds of individuals. It is this symbolic representation of one's cultural referents and the meanings attached to them that are likely to shape mental behavior. By this we mean that cultural identity is a lens through which various levels of environmental encounters and personal experiences gain meaning and value. This is not to suggest that the cultural identity or the meaning placed on it is invariant, because as products of experience within constantly changing proximal and distant circumstances, the development of identity and meaning is an ongoing process.

Biosocial determinants of behavior

In an effort to examine thoroughly the dynamics of phenotypic variation in behavioral development, we must discuss the nature of genes and heredity, environments, the interactions between the two, and related heritability measures.

Genes, heredity, and heritability

According to Hirsh (1976), a genome, one's genetic makeup, is a mosaic of information. Development is the expression of *one* of the many possible alternative phenotypes. Hirsh's description illustrates that any offspring is the product of the arrangement of genetic components that are derived from its parents, including both the similar and dissimilar features that are expressed in the development of one's siblings. Siblings make readily observable a small portion of the variety of ways in which inherited components can be arranged. Siblings, who tend to share environmental resources, also reflect the variety of expression that can be observed in similar environmental circumstances.

Just as each gene has a range of possible phenotypic expression along which it may fall, each phenotype has a range, or norm, of reaction for a given genotype. Thus, the limits of heredity are plastic at every stage of expression, including genotype, phenotype, and their integrated expression in the behavior of individuals. The range of variability may change over the life span. Individuals with similar genotypes can thrive in a variety of environments. Because expression of the genotype depends on environmental influences, individuals with different genetic makeups, who share similar environmental circumstances, may express similar phenotypes (behaviors, traits, etc.). The specific outcome (phenotype) within the range of possible reactions is not predictable a priori from genetic information alone.

Individual variability in expression means that heredity is not a simple translation of genetic material into physical outcomes. Fuller and Thompson (1960) caution that "heritability is a property of populations and not of traits." By this they mean that heritability measures are estimates of the proportion of variance in trait expression within a population (Hirsh, 1976; Rowe & Waldman, 1993). The estimates do not predict the likelihood that an individual within a given gene pool will display a particular trait. Populations are gene pools, the sources of genetic information and variability. The range of possible expression depends on the number of segregating alleles (one of the genes that can appear at a given locus) of independently acting genes within a population. If only one allele of a gene is present in the gene pool, there may be little variation in trait expression.

This simple example of independently acting genes makes a strong point. Many complex behaviors, as with intelligence, depend on the presence of multiple genes, and the expression of those genes depends on multiple levels of interactions with the environment. This intricate sequence of interactions suggests that differences in phenotypes for complex behaviors cannot and should not be thought of in terms of abstract trait expressions.

Traits represent distributions of realized phenotypes among alternative expressions, but distributions represent trait expression within a population at a given time (Hirsh, 1976). The distribution of trait expression will change as genetic and environmental factors within the gene pool change (Plomin, DeFries, & Fulker, 1988). Interpretation and application of heritability measures depend on the nature of sampling and the type of measurements taken. For example, heritability estimates vary depending on which relatives are compared (parent–child, siblings, cousins, etc.), and depending on whether or not researchers base their scores on observed or self-reported environmental and behavioral measures.

Baumrind (1993) cautions that knowledge about the proportion of genetic variation between individuals in a population tells nothing about how or why individuals differ in development – nor does it suggest how to nurture that development, or how genes and environments interact. Heritability estimates cannot provided a measure of the potential impact that alterations in the environment will have on the development of a given genotype. Heritability measures are not indices of the efficacy of targeted interventions (environmental manipulations) designed to alter trait expression positively (Baumrind, 1991).

Heritability measures can be thought of as similar to census taking. A well-conducted census can estimate the number of people living in a given area and estimate the percentages of people who fall into various classification categories (gender, ethnicity, socioeconomic status, occupation, etc.) within that locale at a specific time. However, the census does not usually tell you how the people came to be there, when they arrived, or how long they plan to stay. This type of aggregate data helps to predict general trends but not individual behavior. The general trends may not apply across populations, locales, or time periods.

The necessity of gene–environment interaction in development complicates any discussion of trait inheritance. The heritability of acquired characteristics has been debated among evolutionists and geneticists since the inception of these fields. A simple understanding of interactional forces involved in all aspects of development could have reduced the century-long debate about inheritance. According to Cannon's (1960) interpretation of Lamarckian genetics,[5] Lamarck's lessons about inheritance may be instructive for researchers seriously concerned with discovering the conditions under which phenotypes are expressed and behaviors learned. Lamarck tried to demonstrate that when children grow up in environments similar to those of their parents, the children are likely to demonstrate similar behaviors and phenotypes. Simply put, if children inherit both the genes and environment of their parents, they are more likely to behave like those parents.

Socha (1991) argued that there is a difference between talking about genetic determination and discussing how genes and environments work together to construct behavior patterns. He said,

Transgenerational continuity or "fixation" of newly originated behavioral or sociocultural patterns that drive the organism into a new environment or allow it to construct its own new specific behavioral environment need not be genetically predetermined. . . . The repeated generation or "fixation" of various transgenerational changes (so-called generons) in sociocultural patterns – for example, the transmission of new song patterns in birds, or of human native languages – in the sequence of generations can be ensured by nongenetic mechanisms. The production of hereditary (transgenerational) characters cannot be reduced to a purely genetic affair, and characters repeatedly generated in the flow of generations nongenetically cannot be treated as nonhereditary [p. 407].

Phenotype variability

Variation in individual development is what psychologists and educators would like to predict from behavior-genetic studies. However, as explained above, heritability measures only provide information about general trends in populations (the domain of behavior genetics), not individual outcomes (the domain of psychology) (Lipp, 1990). Note that this is not a trivial distinction. Confounding these levels of analysis may lead to statements similar to that of Scarr's (1991), who claimed that changes in minimally supportive environments do not lead to significant differences in development. Similar confusions may have lead to Herrnstein and Murray's (1994) claims that environmental interventions have been unsuccessful. Scarr (1991) argued that because behavior-genetic research shows greater variation within families than between them, the "environments most parents provide for their children have few differential effects on the offspring" (p. 4). (See S. J. Gould's 1981 *The Mismeasure of Man* for a discussion of the limitations of both within-group and between-group statistical comparisons.)

Scarr goes on to say that children themselves construct reality from the experiences of their rearing environments. We agree that children are active constructors and interpreters of their environments, and as such, they change the world and themselves by simply acting in it. However, it is unclear to us why the majority of the responsibility for creating environ-ments is attributed to the child in this case. Scarr's earlier work showed an effect of parental impact in studies of African American infants adopted by white parents (Scarr & Weinberg, 1976). These data suggested that the responsibility for creating nurturing and culturally important environments is shared by children, caregivers, and the society at large, which helps to build rearing environments. (See Baumrind [1993] and Hoffman [1991], who provide excellent critiques of both the conceptual and statistical foun-dations of Scarr's interpretation of her 1991 data.)

Emde, Gaensbauer, and Harmon (1976) defined *development* as a process of constant change and reorganization based on biological maturation and its interaction with behavior. Cole and Cole (1993) stress the role of social forces implied by Emde et al.'s formulation. Cole and Cole refer to developmental milestones as *biosocial-behavioral shifts*. The de-scriptive label recognizes that every biological-behavioral interaction in-volves changes in an individual's relationship with the social world. Changes in behavior or biological makeup result in changes in both experi-ence of the social world and the ways in which individuals are treated by others.

Parents and other caregivers actively shape both the physical and social environments in which children are raised (Rutter, 1991). Caregivers and children themselves engage in a process of selecting among alternative environments. The selection process occurs even when individuals are not responsible for constructing the environments. Environmental resources may be within the organism or external to it, as in its physical surrounds and economic, social, cultural, or psychological conditions. The daily choices that individuals make result in changes in one or many levels of environ-mental resources. Social and cultural sources are distinguished to illustrate the differences between various groupings of individuals and the rules that govern social interactions. Social encounters include face-to-face and de-layed interactions between individuals who vary in age, gender, education, economic and political power, and numerous other dimensions. The unique configuration and characteristics of those encounters are social. The knowl-edge that one has about the interaction and the value placed on those interactions are cultural. Cultural perspectives of social interactions vary across individuals, nations, history, and so forth.

Bronfenbrenner's (1991) review of the developmental literature reveals common features in the mechanisms through which human potentials are thought to be realized. He argues that development involves "progressively more complex, mutually responsive interaction with the immediate environment" on a regular basis and over an extended period of time. In the West, several typical patterns of social interaction and changes of interaction occur over time. Early social interactions usually involve older persons, such as caregivers. As the individual develops, opportunities for interaction with same-age and younger people increase. At the same time, individuals interact increasingly more with the physical and symbolic features of particular settings that invite or permit exploration, elaboration, and restructuring of the environment. Bronfenbrenner stresses that the outcomes of interactions depend on the genetic characteristics of involved persons and the environmental conditions and events, which may be independent of genetic characteristics. This statement is in contrast to those that assert a genetic influence on environmental conditions. See Plomin and Bergman (1991) for a discussion of the genetic influences on environment.

When cultural factors greatly influence developmental outcomes, conceptions of environment must be extended to include influence from the past and from expectations and intentions for the future. Cole's (1992) cultural-historical approach to development recognizes that moment to moment (microgenetic) changes are a function of the organism's biological makeup, human history, the cultural heritage of social agents, and the organism's present and future expectations. An example based on a father's statement about his newborn daughter (Macfarlane, 1977) illustrates Cole's perspective. The statement "It can't play rugby" influences the immediate handling of a newborn girl whose parents grew up in an era when little girls did not play rugby, and whose parents planned to raise her according to those norms. The treatment of the infant at any given moment is a function of the expression of values from her parents' cultural history and their desires for her future.

Environmental interactions

According to Mendelian genetics, everything outside of the organism as well as that which is within the cell that is not part of the genetic code itself is a source of environmental variation (Hirsh, 1976). Environmental conditions are subject to systematic variations, both within and between social groups, in both the availability of time and places for regular interaction. Variations are observed in social structures, living conditions, lifestyles, shared belief systems, and competing environmental demands.[6]

The degree of variation detectable depends on the level of environment varied. Major differences in early nutritional intake may result in readily observable cognitive deficits. Differences in mother tongue may have neg-ligible influence on the same cognitive skills. Caspi (1991), following Moos (1973), stresses that environments should be analyzed in terms of their functional properties, including psychosocial and organizational climates. Caspi's review describes several variations for which there is evidence of differential individual educational outcomes. These include emphasis on school success, encouragement, and available social support. Social support also varies in terms of salient interests, abilities, attitudes, and personalities of individuals in one's environment.

Moos (1973) lists other variations in environment that influence behavior. They include organizational-structure, behavioral-settings, and ecological dimensions. *Organizational-structure dimensions* refer to fea-tures of the setting, population density, and caregiver continuity. Behavioral-setting dimensions involve spatial and social properties that regulate activities among group members. Ecological dimensions include geography, meteorology, and human-made features of the environment. Characteristics of these three categories are less likely to show genetic influence (Caspi, 1991), but they do have an impact on the range of experi-ences available to individuals.

The recognition that human development proceeds through a series of continual and bidirectionally transformative processes involving gene–environment interactions does not simplify or resolve the nature–nurture debate. A single gene is rarely responsible for the expression of complex behaviors. The presence of several genes and supportive environmental circumstances are needed for the development of a variety of sensory, motor, and cognitive skills. The coordination of various sensory, motor, and attentional skills is required for mundane though complicated behaviors, such as walking and talking.

Recent work in brain development indicates that interactions of chemi-cals on the surface of molecules that occur during cell migration establish the basis for synaptic connections and cytoarchitectural organization (Casaer, 1993). The occurrence of cell death during neural development is genetically determined. However, the pattern of synaptogenesis, pruning, and death is influenced by functional experiences of the organism, including both motor activity and inactivity.

There is substantial evidence for the influence of enriched environments and learning activities on neuronal and nonneuronal tissue development in rats (Greenough et al., 1992). Rats placed in enriched environments at the time of weaning (25–30 days) have more blood vessels per neuron and greater numbers of synapses in the visual cortex than do rats raised in impoverished or standard laboratory environments. Rats show reduced responsiveness to the effects of enriched environments with age. However, some adult rats showed increased synapses per neuron in enriched learning conditions. In other instances, adult rats show increased blood supply to the involved neurons but no increases in synapses. The increased blood flow was shown not to be simply a response to hormonal fluctuations, metabolic processes, or general activity.

Other examples from animal data include the inducement of male-like song in female canaries. Devoodg, Nixdorf, and Nottebohm (1985) reported that systemic testosterone in adult females induced male-like song and doubled the size of the forebrain, which is known to control song. The treatment resulted in a 53% increase in the number of synapses formed on involved neurons. The authors suggest that the formation of new synapses on existing neurons is important for the acquisition of new behavior. The behavioral and anatomical changes were greater when treatment was given under housing conditions that simulated spring rather than fall. The findings suggest that seasonal cues from the environment also mediate the development of new song behavior.

Mechanisms and conditions of intellective behavior

A focus on culture and expectations for the future is found in many interactionist and sociocultural approaches to development. Other researchers stress the role of experience within a cultural-historical framework. They argue that it is the experience of developing organisms and their social networks that have the greatest impact on phenotype. Experience reflects both the environment and the experiencing person's makeup (McGue, Bouchard, Lykken, & Finkel, 1991). Immediate experience, or the result of microgenetic development, influences subsequent experiences. McGue et al. call this influence of experience the *third* factor in phenotypic variance, with genes and environment being the other two factors.

Bronfenbrenner and Ceci (1993) highlight the importance of local and cumulative influences. Ceci (1993) stresses the importance of context in performance. He states, "If basic psychological and biological processes are the 'engines' that drive intellectual development, then context provides the fuel and steering wheel to determine how far and in what direction it goes" (p. 404). Accounts of individual differences and the outcome of various levels of interaction yield greater explanatory evidence when they are grounded in physical, social, historical, and mental contexts (Ceci, 1993). Acontextual-processing accounts of performance are unable to explain differences due to circumstances "existing at the time processes are initially acquired as well as later when they are deployed in the service of mentation" (p. 405). Ceci does not question the role that biological and intellectual resources play in cognitive performance; rather, he highlights the mediating role of context for those factors.

This mediating role is realized through "proximal processes" (Bronfenbrenner & Ceci, 1993) "through which genotypes are transformed into phenotypes." The notion of proximal process is based on current research and theory in behavior genetics and a bioecological perspective on development. Heritability can be shown to vary as a direct function of the magnitude of proximal processes and the quality of the environments in which the processes occur. Proximal processes are mechanisms that connect individual properties and potentials with outside factors in a two-way process that occurs over time. The processes are not self-propelling or self-directed.

Their form, power, content, and direction . . . vary systematically as a joint function of the characteristics of the developing person and the environment (both immediate and more remote) in which the processes are taking place and the nature of the developmental outcomes under consideration [Bronfenbrenner & Ceci, 1993, p. 317].

This notion of proximal processes subsumes the interaction of previously mentioned factors (genes, environment, experience, context) within a socially, culturally, and individually meaningful framework as the basis for the development of intellect.

Recent work in neurology and neuroscience suggests that cognitive deficit and sparing of function that follows brain trauma depends on an individual's personal history as much as on which part of the brain was damaged. Interpreting function from dysfunction that follows specific damage continues to be a difficult task. Recognition that the brain has a modular structure as opposed to a hierarchical structure was an important conceptual breakthrough. However, we are as yet unable to explain how a conscious individual "emerges from the cooperative, coherent activity of neurons in many brain modules" (Rose, 1994). We do know, however, that cognition alone is not the defining feature of intelligent human functioning. Some integration of emotional and cognitive responses, which depend on various electrochemical responses, is the more likely dominant force.

The interpretation assigned to a given task or environmental circumstance has an impact on both the cognitive strategies and the resources that may be applied. For example, children's performance on cognitive tasks was shown to vary as a function of their perception of the task as a game or a laboratory test (Ceci, 1993). The task required children to discover the correct algorithm for combining several sources of information (color, shape) about an object to predict the next move of that object in terms of direction (left or right, and up or down) and distance. Children failed to complete the task when it involved only colored geometric shapes. However, when the task was embedded in a meaningful cultural context (i.e., when the task was designed to look like a video game involving attempts to capture flying animals – birds, bees, or butterflies), 10-year-olds were about 90% accurate on average.

The influence of context was further supported by the fact that children could transfer their game algorithm to abstract tasks similar to the one involving geometric shapes, but only if tested within a few hours of the game experience and with the same laboratory equipment. Ceci and colleagues showed similar limitations on learning transfer across contexts for adult gamblers.

These examples illustrate the importance of meaningful, emotionally inscribed, and personally relevant circumstances to the expression of intellective behavior. Thus, even if the cognitive components of intelligence were solely programmed by genetics, the expression of the potential that resides in such components would be influenced by affective phenomena, which most of us agree are environmentally and experientially determined. The argument being advanced is that for such organized behaviors as intellective function, an interactionist perspective provides the greatest explanatory power.

Conclusions

Based on the inconclusiveness of historical debates concerning the heritability of complex organized human behaviors, and reference to selected recent contributions to related research, we have asserted the inherent fallacy in overemphasizing the genetic contribution to the quality of intellective behavior independent of environmental interactions. Using the lessons learned from decades of unproductive debate, we have argued in favor of the potential importance of environmental manipulations in fostering and changing a variety of expressions of intellective behavior. We have argued further that the conception of complex behavior as an expression of the interaction between genetic material and environmental phenomena provides greater explanatory power than does either a nature or nurture perspective.

Our interactionist perspective supports several general conclusions concerning the genesis of intellective behavior. (1) All intellective behavior is 100% heritable and 100% environmentally determined. (2) The "nature–nurture" and the "what percent of nature and nurture" debates were not only misguided but are distracting to researchers because both provide incomplete and misleading explanations. (3) The interactionist perspective

better captures modern conceptions of the dynamic and dialectical character of developmental processes. (4) Since human intelligence is both adaptive and transformative, it is best viewed as a dynamic, continually
emergent, and protean phenomenon that cannot be explained adequately
by static processes.

Notes

1. These authors do not currently hold such strong views. Please see their chapters in this
   volume for their most recent approaches.
2. Jerry Hirsh (1990) briefly examines the mathematical and logical analysis of interaction
   effects and proportion of variance assignment in "A nemesis for heritability estimation."
3. *Intellective* is used to distinguish our concerns – the variety of cognitive and emotional
   processes that are integral to daily functioning and problem solving – from both *intelligent*
   and *intellectual* behavior. We fear that the term *intelligent* is too closely associated with
   intelligence, which is too often thought of as that which is measured by IQ tests. Intellectual
   behavior seems too easily confused with the work or habits of intellectuals and professional
   scholars.
4. *Communicentric* refers to the tendency for members of a community of belief or practice to
   place special value or emphasis on their own method and perspective, often to the exclusion
   or devaluation of alternative approaches.
5. Cannon's text is more than 30 years old. His discussion of modern genetics is dated, but the
   logical arguments made about genetic inferences and cross-generational heritability issues
   are still instructive.
6. See *The Dialectical Biologist* by Richard Levins and Richard Lewontin (1985, Cambridge,
   MA: Harvard University Press) for a discussion of environment as the product of human
   activity.

References

Anastasi, A. (1980). Abilities and the measurement of achievement. In W. B. Schroder (Ed.),
    *Measuring achievement: Progress over a decade. New directions for testing and measure-
    ment* (1–10). San Francisco: Jossey-Bass.
Appiah, K. A. (1993). *In my father's house.* New York: Oxford University Press.
Baumrind, D. (1991). To nurture nature. *Behavioral and Brain Sciences, 14*(3), 386–7.
Baumrind, D. (1993). The average expectable environment is not good enough: A response to
    Scarr. *Child Development, 64*(5), 1299–317.
Bloom, B. S. (1964). *Stability and change in human characteristics.* New York: Wiley.
Bronfenbrenner, U. (1991). The nurture of nature. *Behavioral and Brain Sciences, 14*(3),
    390–1.
Bronfenbrenner, U., & Ceci, S. J. (1993). Heredity, environment, and the question "How?" –
    A first approximation. In R. Plomin & G. E. McClearn (Eds.), *Nature, nurture and
    psychology* (pp. 313–23). Washington DC: American Psychological Association.
Cannon, H. G. (1960). *Lamark and modern genetics.* Springfield, IL: Charles C. Thomas.
Casaer, P. (1993). Old and new facts about perinatal brain development. *Journal of Child
    Psychology and Psychiatry, 34*(1), 101–9.
Caspi, A. (1991). Cleaning up the environment. *Behavioral and Brain Sciences, 14*(3),
    391–2.
Ceci, S. J. (1993). Contextual trends in intellectual development. *Developmental Review, 13*,
    403–35.

Cole, M. (1992). Context, modularity, and the cultural constitution of development. In L. T. Winegar & J. Valsiner (Eds.), *Children's development within social context, Vol. 2: Research and methodology* (pp. 5–31). Hillsdale, NJ: Erlbaum.

Cole, M., & Cole, S. R. (1993). *The development of children*. New York: W. H. Freeman.

Cole, M., Gay, J., Glick, J. A., & Sharp, D. W. (1971). *The cultural context of learning and thinking*. New York: Basic Books.

Devoodg, T. J., Nixdorf, B., & Nottebohm, F. (1985). Synaptogenesis and changes in synaptic morphology related to acquisition of a new behavior. *Brain Research, 329*, 304–8.

Emde, R. N., Gaensbauer, T. J., & Harmon, R. J. (1976). Emotional expression in infancy: A behavioral study. *Psychological Issues Monograph Series, 10* (1, Serial No. 37). New York: International Universities Press.

Eysenck, H. J. (1971). *The IQ argument: Race, intelligence and education*. New York: Library Press.

Fuller, J. L., & Thompson, W. R. (1960). *Behavior genetics*. New York: Wiley.

Gardner, H. (1983). *Frames of mind: The theory of multiple intelligences*. New York: Basic Books.

Gordon, E. W. (1988, April). *Coping with communicentric bias in knowledge production through educational research*. Paper presented at the American Educational Research Association, New Orleans.

Gordon, E. W., Miller, F., & Rollock, D. (1990). Coping with communicentric bias in knowledge production in the social sciences. *Educational Researcher, 19*(3), 14–19.

Gould, S. J. (1981). *The mismeasure of man*. New York: Norton.

Greenough, W. T., Alcantara, A., Hawrylak, N., Anderson, B. A., Karr, T., & Weiler, I. J. (1992). Determinants of brain readiness for action: Experience shapes more than neuronal form. *Brain Dysfunction, 5*, 129–49.

Herrnstein, R. J., & Murray, C. (1994). *The bell curve: Intelligence and class structure in American life*. New York: Free Press.

Hirsh, J. (1976). Behavior-genetic analysis and its biosocial consequences. In N. J. Block & G. Dworkin (Eds.), *The IQ controversy* (pp. 156–78). New York: Pantheon Books.

Hirsh, J. (1990). A nemesis for heritability estimation. *Behavioral and Brain Sciences, 13*(1), 137–8.

Hoffman, L. W. (1991). The influence of the family environment on personality: Accounting for sibling differences. *Psychological Bulletin, 110*(2), 187–203.

Hunt, J. M. (1961). *Intelligence and experience*. New York: Ronald Press.

Jensen, A. R. (1969). How much can we boost IQ and scholastic achievement? *Harvard Educational Review, 39*(3), 449–83.

Lipp, H. P. (1990). Flechsig's rule and quantitative behavior genetics. *Behavioral and Brain Sciences, 13*(10), 139–40.

Macfarlane, A. (1977). *The psychology of childbirth*. Cambridge, MA: Harvard University Press.

Mayr, E. (1982). *The growth of biological thought: Diversity, evolution and inheritance*. Cambridge, MA: Belknap Press.

McGue, M., Bouchard, T. J., Jr., Lykken, D. T., & Finkel, D. (1991). On genes, environment, and experience. *Behavioral and Brain Sciences, 14*(3), 400–1.

Moos, R. H. (1973). Conceptualizations of human environment. *American Psychologist, 28*(8), 652–65.

Plomin, R., & Bergman, C. S. (1991). The nature of nurture: Genetic influence on "environmental" measures. *Behavioral and Brain Sciences, 14*(3), 373–85.

Plomin, R., DeFries, J. C., & Fulker, D. W. (1988). *Nature and nurture during infancy and early childhood*. New York: Cambridge University Press.

Rose, S. (1994). Who is at home in our heads? Neuroscience is still hard at work trying to find out. *New York Times*, Book Review, Sunday, September 11, p. 38.

Rowe, D. C., & Waldman, I. D. (1993). The question "How?" reconsidered. In R. Plomin & G.

E. McClearn (Eds.), *Nature, nurture and psychology* (pp. 355–73). Washington, DC: American Psychological Association.

Rutter, M. (1991). Origins of nurture: It is not just effects on measures and it is not just effects of nature. *Behavioral and Brain Sciences, 14*(3), 402–3.

Scarr, S. (1991, April). *Developmental theories for the 1990s: Development and individual differences.* Presidential address to the biennial meeting of the Society for Research in Child Development, Seattle, Washington.

Scarr, S., & Weinberg, R. A. (1976). IQ test performance of Black children adopted by White families. *American Psychologist, 31*(10), 726–39.

Socha, R. (1991). Problems with the "environment as phenotype" hypothesis. *Behavioral and Brain Sciences, 14*(3), 407–8.

Webster's Dictionary. (1980). Springfield, MA: Merriam.

Weinberg, R. A. (1989). Intelligence and IQ: Landmark issues and great debates. *American Psychologist, 44*(2), 98–104.

Part III

Specific issues in the nature–nurture controversy

12 Educating intelligence: Infusing the Triarchic Theory into school instruction

Robert J. Sternberg

Suppose that 100% of the variation in scores on intelligence tests was genetic in a particular population. We could then conclude that

(a) efforts to teach children to be more intelligent are a total waste of time;

(b) the environment can have no effect on intelligence;

(c) to know one's parents would be to know one's IQ;

(d) all of the above;

(e) none of the above.

The correct answer is (e). Given this fact, why do people care so much about what proportion of the variance in intelligence is heritable? Well, there are a number of reasons, but none of them relate to the topic of this chapter: educating people to optimize their intelligence.

Consider the well-worn example of phenylketonuria. Variation in whether people are susceptible to phenylketonuria is 100% genetic. If a person inherits the right gene, he or she will suffer from the disease – no doubt about it. And the person will be mentally retarded – but only if phenylalanine is not removed from the person's diet. If it is immediately removed upon birth, the person can lead a more or less normal life, aside from watching diet extremely carefully.

The example of phenylketonuria is one of many that make the same point: The question of heritability is separate from the question of mean differences. Even though height, for example, is very highly heritable, with a heritability coefficient exceeding .9, we know that in some populations, especially the Japanese, heights have increased greatly in recent years. Thus, to the extent that one's interest is in raising intellectual performance, issues of heritability are not really important. That's not to say that heritability isn't important – just that it is not important for this purpose.

The goal of this chapter is to discuss the infusion of the Triarchic Theory of Human Intelligence (Sternberg, 1985, 1988) into the school curriculum as an environmental intervention for increasing intellectual skills. Clearly, the

Triarchic Theory is not the only one that might be infused into school instruction. Gardner (1993) describes ways in which the Theory of Multiple Intelligences might be infused into the curriculum. Gardner's theory has in common with my own the goal of broadening classical conceptions of intelligence.

To a large extent, the theories are compatible: Gardner's deals with content, or, as he calls them, *symbolic domains of intelligences* (linguistic, mathematical, musical, etc.), whereas my own theory deals with domains of application of processes (analytic, creative, practical). Thus, for example, one could think in terms of linguistic intelligence as being applied analytically (as in the work of a critic), creatively (as in the work of a poet), or practically (as in the work of an advertising copywriter). Of course, one could accept one of these theories without accepting the other. For example, Gardner and I disagree as to whether what Gardner calls *musical intelligence* is truly a distinct intelligence or a cluster of abilities related to intelligence but not unitarily an intelligence.

In the past, there have been various attempts to apply theories of intelligence to the development of intelligence. Some of these programs, such as Sternberg (1986), have been based wholly on the Triarchic Theory, others in part on this theory, such as Williams, Blythe, White, Li, Sternberg, and Gardner (1996). Still other programs have been based on other theories. For example, Feuerstein's (1980) *Instrumental Enrichment* program is based on Feuerstein's own theory, and Bransford and Stein's (1993) *IDEAL Problem Solver* program is based on a combination of modern cognitive theories, as is the *ODYSSEY* program (Adams et al., 1982). There are, of course, many other programs as well (see Nickerson, Perkins, & Smith, 1985, for a review of such programs).

The large majority of these other programs differ from the one described here in that they are taught as separate courses rather than infused into already existing curricula. There are several advantages to separate courses, such as the amount of time (presumably, a whole course period several times a week) spent on thinking instruction, the inability of the material to be diluted in the press of teaching something else, the prominence given to thinking in such courses, and the tight organization that such a course can potentially have.

But there are some disadvantages as well to a separate course on thinking. For one thing, few schools have the room for yet an additional subject in their curricula. Very few schools are willing to make time for yet another subject, when administrators and teachers are already being pressured to teach so many things. For another thing, students often do not see directly the relation between what is being taught in the thinking course and what is being taught in other courses. Indeed, they may find that what is being taught in the thinking course is undermined in other courses. The last disadvantage is that many of the separate courses require teacher training beyond that which is practical for most school districts. *Instrumental Enrichment*, for example, requires very extensive and expensive training, and few school districts have the resources to invest in such training.

The procedures described here can be applied at any grade level in any subject area. In this way, they differ from the approach of Gardner, where the intelligences each more closely correspond to subject-matter areas. Although I will concentrate on our own implementation within psychology, I will also show examples of how the procedures can be applied in other disciplines. I will also argue that an advantage of the triarchic approach is the matching of instruction and assessment to students' diverse patterns of abilities.

Matching instruction and assessment with abilities

Everyone who has gone to school has, in one way or another, paid the price of the mismatch among abilities, instruction, and assessment. The only question is how much the person has paid, and in what form the payment has been extracted.

In my own case, my most memorable payment was the receipt of a grade of "C" in my college-level introductory psychology course. It was a memorize-the-book kind of course, and I have never been very good at memorizing textbooks. The grade was enough to convince me to study mathematics instead of psychology, although I later discovered that I was even worse in math and switched back to psychology. Ironically, although many courses in psychology and other sciences are memorize-the-book types of courses, as a scientist, one never has to memorize a book, and if one does not remember something, one can simply grab the book off one's shelf.

Teachers as well as students in all subject-matter areas have seen the costs of mismatching among abilities, instruction, and assessment. In science, students who are marvelously creative and ingenious in the design of experiments may receive low marks in courses requiring memorization of books. In history, students who could ably recognize the implications of past historical events for present foreign-policy conflicts find that they are unable to remember the dates or treaty names required for high grades on multiple-choice tests. In foreign-language learning, students who can easily pick up a language from context are subjected to a mimic-and-memorize method of presentation that is more foreign to them than the foreign language. In mathematics, students who could easily visualize complex relationships are forced to learn mathematical techniques algebraically, despite the availability of geometric isomorphisms. In any area of subject-matter learning, I would argue, mismatching extracts a toll.

The toll to both the individual and society is great. Potentially highly able scientists drop out of science, and people with natural foreign-language learning facility conclude that they cannot learn another language. People who might have been great historians end up doing something else, and able mathematicians are streamed into humanities or other areas that do not require math. Mismatching can waste our most precious human resource: talent.

The question addressed in this chapter is whether we can realize compatibility among abilities, instruction, and assessment; if so, how; and once we know how, what can we gain from doing so? The hypothesis proposed is that, yes, we can realize such compatibility, that it is not terribly hard to do, and that both students and teachers will gain by our doing so.

This hypothesis is far from universally accepted. Cronbach and Snow (1977) did a massive review of studies of aptitude-treatment interaction studies and concluded that, for the most part, investigators had failed to show convincingly that such interactions exist. Where they had been shown to exist, for the most part, they did not go beyond the demonstration of an interaction with general ability. Thus, the claim being made here flies in the face of much past empirical evidence. Why should we be able to achieve any success where many have failed? We believe that (1) the large majority of studies did not actually match each of aptitude, instruction (treatment), and method of assessment, and (2) having a strong and useful theory as a basis for a matching study may give us a lever that some past studies did not have.

The remainder of the chapter is divided into three main parts. First, I shall describe a theoretical basis for achieving compatibility. In doing so, I will consider both psychological and psychoeducational models that might serve as bases for achieving compatibility. Second, I will describe an ongoing study we are doing, from which we have preliminary results, that attempts to instantiate the psychological and psychoeducational models I have described. And, finally, I will briefly talk about future directions and spell out some conclusions.

Theoretical bases

Psychological theories of intelligence

An overview of various kinds of theories of intelligence.   Perhaps the first question one needs to address in attempting to deal seriously with the question of aptitude-treatment interaction is what theory of aptitude will motivate one's work. There are a number of different kinds of theories from which one can choose (see Sternberg, 1990, for a review).

The most well-known theories are certainly *psychometric* ones (called *geographic models* in Sternberg, 1990). According to these theories, planning adequate instruction and assessment depends on an understanding of the mental map that comprises abilities. For example, Spearman's (1927) theory of general ability has probably been the most popular theory of intelligence ever proposed. It states simply that among the various factors of intelligence, one stands out from all others in importance – namely, general intelligence (*g*). In planning an intervention, one would simply want to take into account different levels of this general ability. A more comprehensive attempt might be based on one of the multifactorial theories, such as those of Thurstone (1938) or of Vernon (1971). According to these theories, abilities are best understood as multiple in nature, with the exact abilities depending upon the theory under consideration. As the number of abilities specified by the theory increases, however, the question of how to match abilities, instruction, and assessment becomes more complex, and a theory as complex as Guilford's (1967) structure-of-intellect theory will almost certainly be out of the bounds of practicality. At the same time, a theory as simple as Spearman's does not give us much hope of matching in any but a rather superficial way. A problem with most of these theories is that they do not specify mental processes, at least in any detail, and to account adequately for the type of instruction one should give, it would seem to help to know what processes to teach. A second type of theory attempts to remedy this difficulty.

A second type of theory is *epistemological*. Jean Piaget actually worked in the laboratory of one of the leaders of the mental testing movement, Alfred Binet, but was convinced that Binet paid too much attention to right rather than to wrong answers. Piaget (1972) developed a theory of how children and adults think, specifying two critical processes, assimilation and accommodation, which are used to incorporate new information into existing mental structures, and to form new mental structures to incorporate new information, respectively. Piaget proposed his well-known theory of stages to specify the levels of development that result from the equilibration (balancing) of assimilation and accommodation.

Although Piaget's theory specifies mental processes, he seems to do so more at the level of competence than of performance. For example, few people actually seem to think as logically as the stage of formal operations would imply for people of roughly 12 years of age and older. A third type of theory is more oriented toward performance than toward competence.

The third type of theory is *computational*. According to this type of theory, instruction and assessment should be based on an understanding of the information processing done in actual learning and thinking. Computational theories deal with performance rather than competence. These theories range in the complexity of the information processing that they claim is central to the functioning of human intelligence. For example, Hunt (1978) proposed that verbal intelligence could be understood largely in terms of speed of lexical retrieval of information from long-term memory. He used a relatively low-level task, recognizing physical and name identities between letters (e.g., recognizing that "A" and "A" are both physically identical and identical in name, whereas "A" and "a" are identical only in name).

Some investigators have preferred to identify the basic processes at a higher level. Simon and Kotovsky (1963), for example, proposed that identification of the basic information processes of human intelligence would be better served by the consideration of higher-level tasks, such as number series. Newell and Simon (1972) as well considered high-level tasks, such as theorem proving and playing chess. In my own earlier work (e.g., Sternberg, 1977; Sternberg & Gardner, 1982), I also looked at higher-level, information-processing tasks, such as analogies, series completions, and classifications.

Although the computational theories specify processes, they have tended to be at a molar rather than a molecular level. Several theorists have suggested that a complete theory of intelligence would have to specify the biological substrate underlying human information processing. Theorists such as Arthur Jensen (1982) have attempted to provide a link between information processing and the biological substrate through a fourth approach to human abilities – namely, *biological*. Jensen, for example, has suggested that individual differences in human intelligence may be understandable in terms of individual differences in speed of neural transmission. Eysenck (1986), in contrast, has emphasized the importance of accuracy of neural transmission.

Research using the biological approach has exploded in recent years, and some of the most exciting recent findings have been through the use of this approach. I shall not consider it further, however, because it is not clear that it yet has straightforward implications for instruction or assessment of learning outcomes. We certainly don't know how to use these results either to teach or to assess what has been learned, except, perhaps, at the loose metaphorical level suggested by those who talk about left-brain and right-brain learning. It is not clear that these loose metaphorical uses of biology in fact correspond to the biology of the human nervous system.

A fifth approach is *anthropological*, according to which intelligence is largely a cultural invention. For example, Berry (1974) has suggested that we need to look at intelligence within each culture as an *-emic* construct – as something arising out of that culture. This viewpoint of intelligence as inhering within a given culture is appealing, although it does not account well for biological findings, suggesting at least some generality in mental mechanisms that cross-cut cultures. Although there may be some aspects of intelligence that are culturally specific (e.g., what is considered "smart" behavior in a given culture), there seem to be other aspects that are culturally universal (e.g., the need to define correctly the nature of a problem, taking into account that the correct definition may be culturally specific).

The sixth approach to abilities, and that which will be taken in this chapter, is a *systems* approach. This approach attempts to combine some of the best elements of the approaches described in this section. One theory under this approach, Howard Gardner's (1983) Theory of Multiple Intelligences, claims that there are at least seven relatively independent intelligences, such as linguistic, logical-mathematical, musical, and interpersonal. My own theory, the Triarchic Theory (Sternberg, 1985), specifies that three aspects of abilities are particularly important: analytical, creative, and practical. I do not claim that this theory is the only one that might be useful as a basis for matching abilities, instruction, and assessment. But I believe that it is one theory among others that may be particularly useful.

For one thing, positing three aspects of abilities is a feasible number of aspects for diversifying instruction and assessment. For another, I will argue that the implications of analytical, creative, and practical abilities for instruction and assessment across disciplines are quite clear, in contrast with the implications of other divisions of abilities. Finally, we have actually designed the instructional and assessment instruments that show the usefulness of this approach in practice as well as in theory; therefore, we shall now pursue the use of the Triarchic Theory.

The Triarchic Theory of Intelligence. Because the instructional program to be proposed is based on the Triarchic Theory, I will briefly summarize the theory. More complete information is contained in Sternberg (1984, 1985).

The theory comprises three subtheories: the componential, experiential, and contextual subtheories. The basic idea of the theory is that the information-processing components of intelligence are applied to experience so as to serve various kinds of functions in real-world contexts.

The componential subtheory specifies the kinds of processes involved in human intelligence. There are three kinds of processes, one of which is higher order, and two of which are lower order.

The first kind of process is the *metacomponent*. Metacomponents are higher-order executive processes used to plan, monitor, and evaluate one's problem-solving activities. These processes include recognizing the existence of a problem, defining the nature of the problem, deciding on the lower-order processes needed to solve the problem, deciding on the strategy into which to combine these processes, deciding on how to represent information in problem solving, deciding on how to allocate mental and physical resources in problem solving, monitoring the actual problem solving as it takes place, and evaluating the problem solving after it is completed.

The second kind of process, a lower-order process, is the *performance component*. Performance components are used to execute the instructions of the metacomponents and to provide feedback to the metacomponents. They include processes such as encoding stimuli, inferring relations between stimuli, applying rules from one stimulus to another, and responding.

The third kind of process, also a lower-order process, is the *knowledge-acquisition component*. Knowledge-acquisition components are used to learn how to solve problems in the first place. They include selective encoding, which is used to distinguish relevant from irrelevant information for a particular purpose; selective combination, which is used to decide how to piece together disparate pieces of information; and selective comparison, which is used to relate newly learned information to what one has learned in the past.

These various kinds of components are applied to experience. According to the experiential subtheory, the levels of experience most relevant for intelligence are relative novelty and automatization. In the former, one applies the components of intelligence to problems that are somewhat but not completely novel (see also Raaheim, 1974). Such problems require one to go beyond what one knows, but not so far beyond it that one really doesn't have a clue as to how to apply one's abilities and knowledge base to the problem. In the latter, one renders automatic processes that formerly were controlled, as in reading, speaking, driving, and so on. Smooth, automatic processing is needed in many domains of life, not only to facilitate information processing, as in reading, but also to allow the resources that were once devoted to these skills to be freed for other kinds of tasks.

Finally, according to the contextual subtheory, one applies the components of intelligence via experience to three life tasks: adaptation to existing environments, shaping of existing environments in order to render them better fits to what one wishes of the environments, and selection of new environments. Usually, one initially tries to adapt to an environment. As time goes on, one typically tries to shape the environment to make it more suitable for one's abilities, interests, goals, and desires. Ultimately, if one can neither adapt to nor shape the environment suitably, one may decide to select another environment.

Virtually no one will be strong in all of the abilities discussed in this chapter. The Triarchic Theory states that ultimately the person who is intelligent in his or her life is not necessarily someone who is good in everything (the traditional notion of high *g*). Rather, the intelligent person is one who figures out what his or her strengths and weaknesses are and then capitalizes upon the strengths while compensating for and remediating the weaknesses. That is, the person makes the most of what he or she does well and manages to find ways around what he or she does not do well, or manages to improve enough in these weak domains to get by. If one looks at experts in any field, they are almost never good at everything. Rather, they are experts by virtue both of doing a few things really well and of finding ways to make the things they do not do well relatively unimportant in their work and even in their lives. They can be experts in a number of ways, as specified by a three-party triarchic theory of intelligence.

The Triarchic Theory, or another psychological theory, can serve as a basis for instruction. But one also needs a psychoeducational model that links a psychological theory to instruction. Such models are discussed next.

Psychoeducational models linking abilities, instruction, and assessment

Several different psychoeducational models can be used to link abilities, instruction, and assessment. I will consider seven such models in this chapter and give an example of each.

No model. Before considering alternative models, I should note that most frequently in the schools we find the use of no model at all. The teacher teaches and tests, making no particular effort to link either her instruction or assessment to abilities. Often, the link between assessment and instruction is trivial at best, other than that the teacher attempts to assess what she teaches. But we can certainly go beyond having no model at all.

Model 1: Abilities. This model underlies much of the ability-testing movement. Abilities are measured, but test scores go into a file, and little attempt is made to use the test scores to individualize either instruction or assessment. No significant connection is made between abilities, on the one hand, and instruction and assessment, on the other.

Model 2: Instruction.   This model underlies most instructional attempts, as well as most attempts to improve instruction. A teacher decides on what to teach and how to teach it, and no significant connection is made with abilities or even assessment. Many noteworthy attempts to improve instruction, such as the movement to emphasize thinking skills or the movement to emphasize cultural literacy, draw on this model.

Model 3: Assessment.   This model underlies most thinking about assessment. For example, the multiple-choice format used in group-ability tests was not chosen in order to take into account abilities or instruction, but in order to facilitate ease of scoring and reliability of the assessments. Even the modern emphasis on performance testing and portfolios does not go beyond this model. As is so often true in education, we go from one trend (some might say "fad") to another without asking whether we are merely swinging the pendulum from one extreme to another.

Model 4: Abilities and instruction.   This model underlies traditional aptitude-treatment interaction (ATI) research. The goal here is to fit instruction to the level and often pattern of abilities of groups of students. Some teachers use this model intuitively. The problem with the model is that unless the teacher matches the assessment to the instruction, the value of the matching may never truly be known. So often in schools, for example, teachers try to teach for thinking, and then students are placed in the position of taking multiple-choice standardized achievement tests that measure little beyond rote recall.

Model 5: Abilities and assessment.   This model is at the heart of truly multifaceted ability and achievement testing, where a variety of test contents (e.g., verbal, spatial), formats (e.g., multiple-choice, short-answer, essay), and often types (e.g., group, individual) are used to assess people. The idea is to make sure that people with different ability patterns will be equally benefited by the way that testing is done. Teachers or psychologists using this approach recognize that different children benefit from different kinds of tests, and the testing reflects this fact.

Model 6: Instruction and assessment.   This model is the basis for programs that match assessment to instruction. For example, the Advanced Placement program of the College Board specifies what an introductory college course in a given discipline should contain and then provides examinations measuring this content. In the United States, many state mastery test programs specify the content to be taught in various courses and also provide tests measuring this content. Such programs almost never take into account either individual or group differences in ability and how they might influence instruction and assessment.

Model 7: Abilities, instruction, and assessment. This model is the one advocated in this chapter. It matches abilities, instruction, and assessment. It is the model that will be instantiated in the study to be described. In my view, this full model is rarely used in educational settings.

In closing the discussion of models, I should note that, ideally, Model 7 would be implemented in a way such that children of varying patterns of abilities all receive a variety of kinds of instruction and assessments. In the ideal, you do not teach or test only to a person's strengths; rather, you help him to capitalize upon strengths *and* to remediate and compensate for weaknesses. People need some instruction and assessment with which they are comfortable, but they also need to stretch themselves. As educators, we need to provide both. In our own experimental implementation, described in the following section, children of all ability patterns received all kinds of assessments, but they were randomly assigned to a single kind of instruction. This design was, we believed, ideal for experimental purposes but is not ideal in the classroom. Teachers should use a variety of instructional methods to meet varying ability needs, not just a single method of instruction.

Applying the Triarchic Theory to the model linking abilities, instruction, and assessment

In this section, I will describe a study I have conducted in collaboration with Pamela Clinkenbeard that seeks to apply the Triarchic Theory of Human Intelligence to Model 7 – the model linking abilities, instruction, and assessment. The implementation was via a summer college-level course in introductory psychology, given to bright high-school students selected and assessed in ways that measured analytic, creative, and practical aspects of performance. The course was taught during the summers of 1992 and 1993. As of the writing of this chapter, data have been analyzed for the summer 1992 data but are still being analyzed for 1993. Hence, I shall summarize only the former data. The implementations were somewhat different; thus, distinct as well as common aspects of the two implementations will be discussed.

Method

Design. The basic design was a three-way one. In the summer of 1992, students were selected for excelling in terms of either analytic, creative, or practical abilities. Students from the three identified ability groups were assigned at random to sections of an introductory psychology course that emphasized teaching for either analytical, creative, or practical thinking. Students of all three ability patterns in all instructional sections were assessed via assessments measuring analytic, creative, and practical accomplishments. Thus, the design was a $3 \times 3 \times 3$ completely crossed one, with abilities and instructional assignment between subjects and assessments within subjects. In 1993, there were two additional ability groups: a balanced group, in which students were roughly equal in analytic, creative, and practical abilities; and an above-average control group, where students were not especially high in any of the abilities.

The prediction was that students who were matched in abilities, instruction, and assessment would perform better than those not so matched.

Selection materials.   All students were selected for participation in the summer program by virtue of their scores on a research form of the *Sternberg Triarchic Abilities Test* (Sternberg, 1993), High School–College level. The test itself is divided into a *process* facet and a *content* facet. There were three levels of the process facet and four of the content facet, which, when crossed, yielded 12 subtests. The three process facets were analytic, creative, and practical; the four content facets were verbal, quantitative, figural (all multiple-choice), and essay. Each of the multiple-choice subtests contained two practice items and four or more test items, whereas each essay subtest consisted of a single essay. A brief description of each test follows.

1. *Analytic-Verbal.* "Learning from Context." Subjects receive a brief paragraph with an embedded neologism (unknown word created especially for the test). Subjects had to infer the meaning of the unknown word.

2. *Analytic-Quantitative.* "Number Series." Subjects receive a series of numbers, which they must complete.

3. *Analytic-Figural.* "Figural Matrices." Subjects receive a figural matrix with one entry missing. They must choose which of several figures belongs in the empty cell of the matrix.

4. *Analytic-Essay.* "Analytical Thinking." Subjects are presented with a problem confronted by a school and must analyze the problem systematically.

5. *Creative-Verbal.* "Nonentrenched Analogies." Subjects are presented with an analogy preceded by a counterfactual premise (e.g., "Suppose that villains were lovable," or "Suppose that sparrows played hopscotch"). Subjects need to solve the analogy as though the premise were true.

6. *Creative-Quantitative.* "Novel Number Systems." In the 1992 version, subjects were presented with number matrices into which were intermeshed

novel symbols representing a new system of enumeration. Subjects had to fill in the missing entry in the matrix with either a normal number or one in the new system of enumeration. In the 1993 version, subjects had to do mathematics problems using novel number operations (e.g., two numbers x and y are added if x is less than y, multiplied if x is greater than y, and divided if x equals y).

7. *Creative-Figural.* "Series with Mapping." Subjects had to complete a figural series. However, the rule was illustrated in a series other than the one that they had to complete, and subjects therefore had to transfer the rule from the old series to the new one.

8. *Creative-Essay.* "Creative Thinking." Subjects had to write a creative essay envisioning their ideal school.

9. *Practical-Verbal.* "Informal Reasoning." Subjects read a paragraph describing a high-school student with a life problem. They had to select which of several answer options provided the best solution to the student's problem.

10. *Practical-Quantitative.* "Everyday Math." Subjects had to solve problems using everyday math, as in following recipes, using train or bus schedules, computing costs of tickets to sporting events, and so on.

11. *Practical-Figural.* "Route Planning." Subjects were shown maps of amusement parks, portions of cities, and the like and had to plan efficient routes for getting from one place to another, given constraints.

12. *Practical-Essay.* "Practical Thinking." Students were given a practical problem faced in a school and asked how they would solve it.

The test had very liberal time limits for each subtest so that almost everyone was able to finish each subtest. Tests were sent to schools around the country (and to schools in other countries as well) and were administered by school personnel but scored by us. A person was labeled as high in an ability if his or her score in that ability exceeded his or her score in any other of the abilities by one standard deviation unit across subjects.

Instructional materials. All students received two kinds of instructional materials: the text and the lectures. The instructional materials constituted the core of an advanced-placement (college-level) course in psychology.

The text was a higher-level introductory psychology course for college students. The text, which I had written, was in preprint form. It consisted of 20 chapters that covered the standard topics of an introductory psychology course. Students in the 1992 summer session got through only 12 of the chapters in their 3-week course, whereas students in the 1993 summer session completed the text. Emphases of the text were on (1) teaching students to think like psychologists, (2) helping students to understand the dialectical evolution of ideas in psychology (and their origins in philosophy), and (3) understanding and applying scientific thinking in psychology.

All students also attended common morning lectures. The lectures were given by a Yale professor, Mahzarin Banaji, who has won an award for outstanding undergraduate teaching. The lectures were the same as she normally teaches in her introductory college lectures.

The differentiated treatment was in the afternoon, when students went to a section that emphasized either analytic, creative, or practical thinking in psychology. In the 1993 summer session, an additional type of section emphasized memory for course content, more along the lines of a standard course. These sections were taught by advanced graduate students who had experience teaching introductory psychology. In 1993, these sections were also taught by advanced-placement, high-school teachers.

The sections emphasized a discussion format. In the analytic sections, the emphasis was on analyzing theories, critiquing experiments, evaluating concepts in psychology, and the like. In the creative sections, the emphasis was on generating new theories, thinking up new experiments, imagining how theories would need to be changed if certain assumptions were changed, and the like. In the practical sections, the emphasis was on using psychological concepts, theories, and data to inform and improve everyday life. In the standard sections (1993 only), the emphasis was on remembering and understanding course material.

Assessment materials. Assessment materials were variegated according to the Triarchic Theory. There were four main assessments.

1. *Homework assignments.* The two homework assignments involved (a) reading and doing activities stemming from a study on cognitive dissonance, and (b) reading and doing activities stemming from a reading on alternative theories of depression. For each assignment, there was an analytical activity, a creative activity, and a practical activity. The analytical activity involved analyzing the experiment (assignment one) or theories (assignment two), or coming up with a new experiment (assignment one) or theory (assignment two), or discussing the relevance of the assigned experiment (assignment one) or theories (assignment two) for everyday life.

2. *Midterm.* The midterm consisted of multiple-choice test items basically emphasizing recall, plus two analytical, two creative, and two practical essays. The essays all were based on material studied in the course. Hence, the essays measured analytic, creative, or practical *use* of information.

3. *Final examination.* The final examination was parallel to the midterm except longer. It contained more multiple-choice items and three essays of each of the three kinds (analytic, creative, practical).

4. *Independent Project.* Students were all required to do an independent

project. Although they were free to choose the topic, they were required to explore analytic, creative, and practical facets of that topic.

All students in 1993 received a pretest on psychology knowledge because some of them had previously had high-school psychology courses, whereas others had not. In 1992, none had prior psychology courses. Students also received a short battery of conventional ability tests.

In the 1992 summer program, students were ungraded but received fairly extensive comments from their section leader describing their performance in the course. In the 1993 summer program, students also received these comments, and a letter grade (A–F) as well.

Subjects.   Subjects in the 1992 summer program were roughly 65 students, mostly from the United States but a few from abroad. One student was expelled from the program for behavioral reasons, leaving a final sample of 64. Subjects in the 1993 summer program were roughly 225 students, again from both around the United States and abroad. One student left voluntarily for homesickness. Three students acquired chicken pox and lost a few days for hospitalization. Students in both years were entering the junior or senior year of high school, meaning that they were roughly 16 to 17 years of age. All enrollments were voluntary, and parents generally paid for the children to attend. Generous scholarship aid was available. Students were diverse with respect to racial and ethnic groups. In 1993, in order to make the program self-supporting and still provide scholarship aid, a few full-tuition paying students were admitted who did not fit into any treatment group. They were segregated into their own section.

Students were not aware of the experimental design or even that there was any such design. They were informed that our goal was to try to use innovative methods for teaching psychology.

Main results

We have now analyzed the data for the 1992 implementation. Because we used a small sample, we did not expect strong results. Indeed, the 1992 summer program was conceived of as a pilot for the full-scale 1993 summer study. Yet the results showed interesting trends that were either statistically significant or at least marginally significant.

The results for the creative-ability group conformed to our predictions. Students identified as creative who were placed in a section emphasizing teaching for creativity and who were assessed for creative performance excelled beyond all other groups on these creativity-based assessments.

The results were similarly promising for the practical-ability group. Subjects identified as high practical who were placed in a practicality-oriented section excelled above other groups in the practicality-based assessments.

The results for the analytic-ability group were at first puzzling. Quite simply, the high analytic students did worse on all the assessments than did members of any other group. Thus, the result was a main effect rather than an interaction, and in the wrong direction! However, interviews with the students' section leaders elucidated further why we got this result.

Students in the analytic-ability group were basically those who had always done well on standardized tests and, for the most part, in school, because both tests and schooling stress analytic abilities considerably more than creative or practical ones. Most of the high-analytic students had been labeled as gifted at some point in their school careers. Moreover, many of these students were able to receive grades of "A" in their high-school courses with very little effort. The lesson they had learned was that they could "ace" a course with very little work.

When students come to Yale as regular college freshmen, many of them encounter a rude awakening. They discover, usually after a first semester of mediocre grades, that they will not be able to get by as they did in high school, with just a minimum of effort. Our high-analytic summer students went through the same rude awakening. Believing that they could do well in the course without working hard, they didn't work hard and discovered only too late that their cavalier attitude was not adequate to the demands of the course.

On the basis of these pilot results, we introduced regular letter grades in the summer of 1993 to emphasize to students the need to study hard. We also warned them that the workload and difficulty they would face were very different from those to which they were accustomed in high school.

We plan to continue the summer psychology program beyond the summer of 1993, although our funding for the project will extend for just one more year. We hope to refine our procedures, as the need arises. But we believe that the general benefits of matching extend beyond psychology and beyond a single age group. Hence, we would like to expand our work to other subject-matter areas in order to show that matching is useful not just in psychology but in other disciplines as well. Eventually, we would like also to extend the general plan to age levels other than the 16–17-year age level at which we have been working.

The results of the summer 1992 program are promising, although it remains to be seen whether the summer 1993 results replicate the 1992 results for the creative and practical groups and provide the results we have predicted for the analytic group as well. Obviously, we have a long way to go in testing our ideas, but we believe that our data provide at least tentative evidence for the feasibility and desirability of matching abilities, instruction, and assessment.

Applying the Triarchic Theory beyond psychology

In the foregoing discussion, I have shown how the Triarchic Theory of Human Intelligence can be applied to instruction in the domain of psychology. A theory that could be applied only in the domain of psychology would be of little interest, however, so I now show how the theory can be applied in any domain of instruction.

Table 12.1 shows examples of how the Triarchic Theory can be applied in six subject-matter areas: art, biology, history, literature, mathematics, and psychology. These fields, of course, are only representative of those taught in school.

When teaching and evaluating to emphasize *analytical* abilities, one is asking students to (1) compare and contrast, (2) analyze, (3) evaluate, (4) critique, (5) say why, in your judgment . . . , (6) explain why . . . , (7) explain what caused . . . , (8) evaluate what is assumed by . . . , or (9) critique. . . . Of course, there are other prompts for analytical thinking as well.

When teaching and evaluating to emphasize *creative* abilities, one is asking students to (1) create . . . , (2) invent . . . , (3) imagine what you would do if you were . . . , (4) imagine . . . , (5) design . . . , (6) show how you would . . . , (7) suppose that . . . , or (8) say what would happen if. . . . Most teachers find that their instruction and evaluation is oriented more toward analytic than creative abilities.

When teaching and evaluating to emphasize *practical* abilities, you are asking students to (1) apply . . . , (2) show how you can use . . . , (3) implement . . . , (4) utilize . . . , or (5) demonstrate how in the real world. . . . Again, these are only some of the prompts one might use to encourage students to exercise and stretch their practical-contextual abilities. Relatively little traditional instruction and evaluation are oriented toward practical abilities, which may be why children have so much difficulty applying what they learn in school to their lives outside school.

Of course, there is a fourth kind of instruction and evaluation in our schools, which is actually the kind that predominates in most classrooms. This kind asks students things like (1) who said . . . , (2) summarize . . . , (3) who did . . . , (4) when did . . . , (5) what did . . . , (6) how did . . . , (7) repeat back . . . , and (8) describe. . . . Instruction and evaluation of this kind emphasize what students know. Obviously, there is nothing wrong with this emphasis: Students need to acquire a knowledge base. But to the extent that you are interested in developing children's ability to think, you need to

Table 12.1. *Triarchic Theory applied to student instruction and assessment methods*

	Analytic	Creative	Practical
Psychology	Compare Freud's theory of dreaming to Crick's.	Design an experiment to test a theory of dreaming.	What are the implications of Freud's theory of dreaming for your life?
Biology	Evaluate the validity of the bacterial theory of ulcers.	Design an experiment to test the bacterial theory of ulcers.	How would the bacterial theory of ulcers change conventional treatment regimens?
Literature	In what ways were *Catherine Earnshaw* and *Daisy Miller* similar?	Write an alternative ending to *Wuthering Heights* uniting Catherine and Heathcliff in life.	Why are lovers sometimes cruel to each other, and what can we do about it?
History	How did events in post–World War I Germany lead to the rise of Nazism?	How might Truman have encouraged the surrender of Japan without A-bombing Hiroshima?	What lessons does Nazism hold for events in Bosnia today?
Mathematics	How is this mathematical proof flawed?	Prove: ... How might catastrophe theory be applied to psychology?	How is trigonometry applied to construction of bridges?
Art	Compare and contrast how Rembrandt and Van Gogh used light in. . . .	Draw a beam of light.	How could we reproduce the lighting in this painting in the same actual room?

keep in mind that, ultimately, what matters is not what you know but how well you can use what you know – analytically, creatively, and practically.

The interventions described here can be applied in virtually any class-room. These interventions will help students to apply their intelligence to their coursework and will also help students to develop their intelligence. I want to emphasize, in closing, that there are no miracles and no quick fixes. We can help people to develop their intelligence, but within the limits of genetic and environmental constraints. Clearly, genetic factors play an important role in intelligence. We should not let this fact, however, lead us to stand by and watch children develop rather than actively to help children make the most of what they have. No one presumably reaches their full potential. We can help children to work toward this perhaps not fully reachable goal.

Acknowledgments

Preparation of this chapter was supported under the Javits Act Program (Grant No. R206R50001) as administered by the Office of Educational Research and Improvement, U.S. Department of Education. Grantees undertaking such projects are encouraged to express freely their professional judgment. This chapter, therefore, does not necessarily represent the positions or policies of the government, and no official endorsement should be inferred.

References

Adams, M. J., et al. (1982). *Teacher's manual.* Prepared for project intelligence: The development of procedures to enhance thinking skills. Submitted to the government of Venezuela 1982.

Berry, J. W. (1974). Radical cultural relativism and the concept of intelligence. In J. W. Berry & P. R. Dasen (Eds.), *Culture and cognition: Readings in cross-cultural psychology* (pp. 225–9). London: Methuen.

Bransford, J. D., & Stein, B. S., (1993). *The ideal problem solver: A guide for improving thinking, learning, and creativity* (2nd ed). New York: W. H. Freeman and Company.

Cronbach, L. J., & Snow, R. E. (1977). *Aptitudes and instructional methods.* New York: Irvington.

Eysenck, H. J. (1986). The theory of intelligence and the psychophysiology of cognition. In R. J. Sternberg (Ed.), *Advances in the psychology of human intelligence* (Vol. 3) (pp. 1–34). Hillsdale, NJ: Erlbaum.

Feuerstein, R. (1980). *Instrumental enrichment: An intervention program for cognitive modifiability.* Baltimore, MD: University Park Press.

Gardner, H. (1983). *Frames of mind: The theory of multiple intelligences.* New York: Basic.

Gardner, H. (1993). *Multiple intelligence: The theory in practice.* New York: Basic.

Guilford, J. P. (1967). *The nature of human intelligence.* New York: McGraw-Hill.

Hunt, E. B. (1978). Mechanics of verbal ability. *Psychological Review, 85,* 109–30.

Jensen, A. R. (1982). The chronometry of intelligence. In R. J. Sternberg (Ed.), *Advances in the psychology of human intelligence* (Vol. 1) (pp. 255–310). Hillsdale, NJ: Erlbaum.

Newell, A., & Simon, H. A. (1972). *Human problem solving.* Englewood Cliffs, NJ: Prentice-Hall.

Nickerson, R. S., Perkins, D. N., & Smith, E. E. (1985). *The teaching of thinking.* Hillsdale, NJ: Erlbaum.

Piaget, J. (1972). *The psychology of human intelligence.* Totowa, NJ: Littlefield Adams.

Raaheim, R. (1974). *Problem solving and intelligence.* Oslo: Universitetsforlaget.

Simon, H. A., & Kotovsky, K. (1963). Human acquisition of concepts for sequential patterns. *Psychological Review, 70,* 534–46.

Spearman, C. (1927). *The abilities of man.* New York: Macmillan.

Sternberg, R. J. (1977). *Intelligence, information processing, and analogical reasoning: The componential analysis of human abilities.* Hillsdale, NJ: Erlbaum.

Sternberg, R. J. (1984). Toward a triarchic theory of human intelligence. *Behavioral and Brain Science, 7,* 269–87.

Sternberg, R. J. (1985). *Beyond IQ: A triarchic theory of human intelligence.* New York: Cambridge University Press.

Sternberg, R. J. (1988). *The triarchic mind: A new theory of human intelligence.* New York: Viking.

Sternberg, R. J. (1990). *Metaphors of mind: Conceptions of the nature of intelligence.* New York: Cambridge University Press.

Sternberg, R. J. (1993). *Sternberg Triarchic Abilities Test.* Unpublished test.

Sternberg, R. J., & Gardner, M. K. (1982). A componential interpretation of the general factor in human intelligence. In H. J. Eysenck (Ed.), *A model for intelligence* (pp. 231–54). Berlin: Springer.

Thurstone, L. L. (1938). *Primary mental abilities.* Chicago: University of Chicago Press.

Vernon, P. E. (1971). *The structure of human abilities.* London: Methuen.

Williams, W. M., Blythe, T., White, N., Li, J., Sternberg R. J., & Gardner, H. I. (1996). *Practical intelligence for school: A handbook for teachers of grades 5–8.* New York: Harper Collins.

13 Raising IQ level by vitamin and mineral supplementation

H. J. Eysenck and S. J. Schoenthaler

Introduction

It is well known that high IQ levels are important for individuals and societies alike. Social status, financial earnings, and educational success are all closely linked with high IQ; similarly, social failure, unemployment, and poverty are often associated with low IQ (Cattell, 1983; Itzkoff, 1991, 1994). Attempts to raise IQ levels through special environmental manipulation (i.e. special educational measures like Head Start) have been failures on the whole, producing short-lived improvements only, with many of the claims made for large improvements clearly fraudulent (Spitz, 1986). Such an outcome is not unexpected in view of the strong determination of IQ differences by genetic and generally biological factors (Brody, 1992; Plomin, 1993; Vernon, 1989). Alternative possibilities, making use of biological determinants, do not seem to have figured prominently in these efforts to raise IQ levels, possibly because of the environmentalistic prejudices of the prevailing behavioristic climate, with its antibiological orientation. Recent changes in this general orientation may herald a less blinkered approach.

One possible venue is through micronutrient supplementation (Schoenthaler, 1991). Micronutrients (vitamins and minerals) play an important part in our physical and mental well-being (Essman, 1987), and undisciplined eating habits may produce significant deficiencies in some important vitamins and minerals in spite of adequate calorie intake (Axelson & Brinberg, 1989). There are claims that such supplementation may have the effect of raising IQ (Dean & Morgenthaler, 1990; Dean, Morgenthaler, & Fowkes, 1993), and some at least seem well founded. A more detailed account is given by Schoenthaler (1991), who also points out strong behavioral effects of such supplementation; marked improvement of antisocial behavior in criminals has been found consequent upon micronutrient supplementation (Eysenck & Eysenck, 1991). Along a different line, Lynn (1990) has argued that the recent increases in IQ observed

363

over the past decades (Flynn, 1987) are largely due to improved nutrition, another argument linking IQ and food intake.

The general notion that nutrition is a vital element in physical and mental well-being is of course neither new nor anything but obvious. As the German proverb says, *Man ist was man isst* – one is what one eats. Without food we would starve. The real question is one of *differential* aptitude: Can part of the *variance* in IQ be attributed to vitamin and mineral deficiencies? Nutritionists have usually given a negative answer, provided children receive a diet satisfactory as far as the child's recommended daily allowance (RDA) is concerned. This value is calculated as representing the level at which the child is free of physical symptoms of malnourishment; it differs from country to country, being larger in the USA than in the UK for reasons that are not easy to fathom. Psychological variables have not played any part in the establishment of the RDAs, it being assumed that they would fall in line with physical variables. But, of course, an assumption is not a factual statement and may be erroneous; only detailed experimental studies can settle the point. What is certain is that the nourishment required by the brain is not proportionately equivalent to what is required by the rest of the body; glucose, for instance, plays a much larger part in brain nourishment than in bodily requirements.

Of particular interest has always been memory, and many studies have been reported concerning nutritional factors affecting this part of the general intelligence area (Cherkin, 1987). Deficiency of nutrients such as niacin, cholamin, and foliate have been identified as causes of reversible dementia. Even in healthy volunteers, niacin improved short-term memory (Loriaux, Deijen, Orleheke, & De Swart, 1985). Also studied have been precursors like the amino acids tyrosine and tryptophan and acetylcholine precursors like choline and acetyl co-enzyme A (Bartus, Dean, Bear, & Lippa, 1982). Much of this work has been carried out on animals or senile patients and does not bear directly on what may be happening in the brains of growing children. More interesting are field studies of early nutrition and later achievement (Grantham-McGregor, 1987), but as the author acknowledges, "inconsistent findings, flaws in study design and differences in treatments and subjects, preclude making firm conclusions" (p. 148). The fact that children who are extremely malnourished tend to have low IQs (Stein & Kassab, 1970; Winick, Meyer, & Harris, 1975) tells us little about the influence of nutrition on intelligence because clinical malnutrition is invariably accompanied by neglect, poor schooling, poverty, and many other conditions that would be expected to depress IQ (Dobbing, 1987).

Extreme malnutrition in the absence of other factors usually associated with it does not necessarily lead to low IQs. The Dutch famine study (Hart,

1993) looked at the effects of a famine imposed during World War II by the German occupiers on approximately one-half of the Dutch population. Later, military inductees were routinely tested for IQ, and those who had been exposed to severe prenatal malnutrition were compared with those who had not been so exposed (Stein et al., 1972). There were no effects of this exposure on final IQ. Perhaps deprivation has to be continued for a longer period, perhaps pregnant women received more food than others, or perhaps the fetus receives a higher proportion of the food eaten as a biological protection device; we cannot accept these data as disproving the importance of nutrition for IQ development. Also, of course, low calorie intake does not necessarily indicate low vitamin and mineral intake. Lynn (1990) has tried to account for the secular gains in IQ over time observed by Flynn (1987) in terms of improved nutrition and supported this hypothesis by showing that in seven studies of MZ twins, the heavier twin had a higher IQ than the lighter twin, the differences in birth weight being reasonably attributable to differences in the adequacy of prenatal nutrition. It is possible that these differences might vanish once the twins reach adulthood, of course (Lloyd-Still, 1976).

A study by Sigman, Neumann, Jansen, and Baribo (1989) on Kenyan children correlated nutritional intake with IQ data. Controlling for parental socioeconomic status and literacy, positive correlations were found for animal protein and fat intake of around .30 (Brody, 1992); there was no measure of vitamin or mineral intake. In a more experimental design, Rush, Stein, and Susser (1980) used random assignment of subjects to one of three conditions, subjects being black, pregnant women thought to be at high risk for delivering low-weight babies. Conditions were (1) a high-protein liquid supplement, (2) a high-calorie liquid supplement, and (3) a control group. Rush, Stein, Susser, and Brody (1980) tested the children at age 1 on a visual habituation task known to predict later IQ; they found that the children of mothers receiving liquid-protein supplements performed significantly better than the other two groups. Again, long-term consequences are not known. In a similar design, Harrel, Woodyard, and Gates (1955) found that in New York City, women of low socioeconomic status who were given vitamin and mineral supplementation during pregnancy had children who, at 4 years of age, averaged 8 points higher in IQ than the control group of children whose mothers had been given placebos during pregnancy. Cravioto and Delicardio (1970) have reported similar results.

These studies suggest but clearly do not prove that micronutrient supplementation in normally fed children (i.e., in children not deprived or seriously malnourished) increases IQ to a significant degree. Nutritionists and pediatricians are generally doubtful (Rutter & Madge, 1976), although certain points do not seem in doubt any longer. Thus, Richter (1979) and

Heseker, Knehler, Westenhoefer, and Pudel (1990) have demonstrated that psychological symptoms are among the earliest to appear when vitamin intake falls below the desired level. Southon (1990) has pointed out that there is no doubt that micronutrients are essential for our mental well-being. Finally, Southon, Wright, Finglas, Bailey, and Belsten (1992) have shown that in normally fed children, there are great differences in uptake and level of vitamins, with some children falling below recognized levels (which, in turn, have been established using physical well-being as the criterion). These well-established facts suggest that a more serious experimental approach to the problem of IQ increases following micronutritional supplementation might be useful.

One interesting prediction from the hypothesis linking micronutrient intake and IQ would suggest that *fasting* would tend to lower IQ levels. Little has been done in this field, unfortunately, but a few studies (e.g., Rogers & Green, 1993; Laessle, Bossent, Hank, Hahlweg, & Pirke, 1990) suggest that fasting subjects perform less well on cognitive tasks than nonfasting subjects. However, in the absence of external stressors, their performance may be improved (Herman, Polioy, Pliner, Threlkeld, & Munro, 1978). Clearly, too little is known about the effects of dieting on IQ to adduce it as evidence, particularly as the effects may be indirect – that is, depending on motivation rather than on ability.

Many studies show a link between skipping breakfast and IQ, sucrose and intelligence, food additives and intelligence, toxic metals and intelligence, and general malnutrition and behavior. These studies are too numerous to be reported here and are only marginally relevant (Conners, 1989; Schoenthaler, 1991).

Another line of enquiry suggests a positive conclusion. If improvements in nutrition increase IQ levels at least in a proportion of children (presumably those showing vitamin and mineral deficiencies), then such improvements should be capable of being traced in the scholastic achievements of the children involved, and there is a close connection between IQ and scholastic achievement (Brody, 1922; Eysenck, 1979). The facts seem to bear out this hypothesis (Schoenthaler, Doraz, & Wakefield, 1986a,b). Schoenthaler et al. (1986a) studied the results of dietary modifications in the food supplied to public school students in New York, revisions taking place during the 1979–80, 1980–81, and 1982–83 academic years, with no changes being made during 1981–82. The diet change consisted of the gradual elimination of synthetic colors, synthetic flavors, and selected preservatives. At the same time, high-sucrose foods were gradually eliminated. The California Achievement Test was administered each year, and from it the national percentile rank of the school was determined. Results were very clear-cut and are shown in Figure 13.1. The New York schools aver-

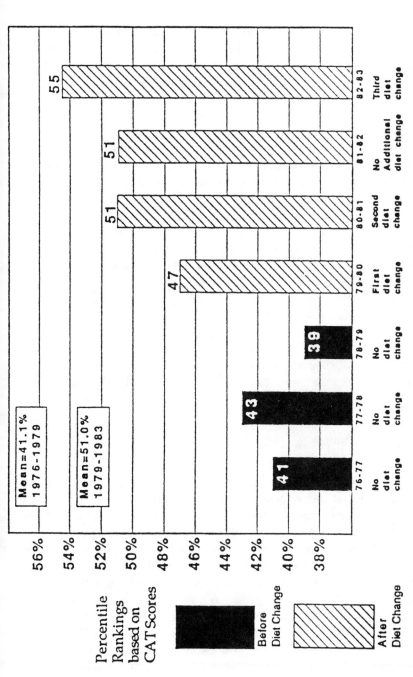

Figure 13.1. National rankings of 803 New York City Public Schools before and after diet changes. (From Schoenthaler, Doraz, & Wakefield, 1986a.)

aged around the 41st percentile in the 3 years prior to the dietary changes; they rose to the 47th percentile in the first year, and to the 51st percentile in the second year following the dietary change. There is not change in 1982, when no dietary changes were made, but in the following year, when dietary improvements were renewed, the schools achieved the 55th percentile, an overall change of 14 percentile points. (With the large numbers involved, even much smaller changes would be highly significant.)

As expected, not all children improved equally. Gains were largest in the group doing worst academically. In 1979, 12.4% of the one million student sample were performing two or more grades below the proper level, but by the end of 1983, the rate had dropped to 4.9%. (All the gains were found during 1980, 1981, and 1983; during 1982, when there were no diet changes, no improvements occurred.) The quasi-experimental, time-series design of the study does not enable us to be precise regarding specific causes of the scholastic improvement. The authors interpret the outcome as due to the fact that those foods that were partly eliminated (containing high levels of fats, sucrose, and food additives) tend to be low in the ratio of essential nutrients to calories. When the consumption of empty calories decreases, children normally eat other foods that contain a higher ratio of nutrients to calories. While this is a highly probable scenario, the experimental design obviously cannot support it directly. Nevertheless, here is another indication that a search for a connection between vitamins and mineral intake and IQ might be worthwhile.

Theory and paradigm

The broad survey given in the introduction shows that many of the experiments cited have serious weaknesses. Some of these weaknesses derive from the fact that there does not seem to be any specific theory, clearly stated, that is being tested. There is a vague feeling that there might be some cognitive benefit to be derived from micronutrient supplementation, but the nature of the benefit is not specified, and neither are the micronutrients in question. There are some studies looking at specific vitamins or minerals, and these may be exempted from this general criticism. Thus, Jordan, Bayley, and Henry (in press) have shown that administration of an iron supplement produced significant improvement in psychometric test scores; this improvement was correlated with iron status in blood samples and was noted in nonverbal reasoning tests but not in verbal ones. This is an important theoretical point that will be discussed presently.

Thiamin supplementation has also been found to raise IQ in children (Harrell, 1946). The children took either 2 mg thiamin or a placebo for a year, and the thiamin children were significantly taller, had better eyesight, quicker reaction time, and higher memory scores. Both reaction time and memory are of course related to IQ. Vitamin C administration, on the other hand, may lead to poorer reaction times (Benton, 1981), although Miller et al. (1978) found no effects after chronic administration over a 5-month period. Zinc, too, does not seem to have much connection with IQ, although Rimland and Larson (1983) found that intelligent individuals tended to have higher levels of zinc in their hair. As will be seen, the evidence is neither large nor impressive; Benton (1992b) has given a more detailed review, including studies using geriatrics and defective populations, but the picture is not much improved by their inclusion.

The weaknesses of early studies will be clear when we discuss the requirements of such studies and the theory underlying them. The first point, of course, is related to the question: Who is likely to benefit? We start with the obvious points that (1) vitamins and minerals are *essential* for proper (optimal) mental functioning, and (2) there are great individual differences in the intake and metabolism of such micronutrients. Granted that this is so, improvements following supplementation are only expected in those children whose levels of one or more relevant nutrients are below some optimal level – with the question "Which nutrients?" unanswered so far, for the most part, although we do have some ideas (Benton, 1992b). It would seem to follow that positive effects would only be expected for a (possibly small) subsection of the population – namely, those children *below* the optimal level. It is unknown how large this subsection might be, and of course, it is likely to vary from country to country and from high to low socioeconomic groups. Deficiencies are more likely to occur among poor, deprived, and single-parent families, although faulty eating habits may produce such deficiencies even among the wealthy. One important consequence of this argument is that *average* gains following supplementation will not be large, even though quite substantial gains may be found among those showing deficiencies. If we define true gains as the difference between supplementation and placebo children, and if 25% of a group of children is deficient, then a gain of 14.4 IQ points for the deficient children is hidden in a small average gain of 3.6 IQ points for the total group, which is the observed value (Table 13.1) in all 10 studies so far undertaken. This means that we have to be very careful about considerations of statistical power; large samples are needed to pick up such small mean gains. (Of course, children cannot be subdivided into deficient versus adequate; there will be a normal distribution from one extreme to the other.)

Can one give a guesstimate of the numbers involved, from the quite inadequate information available? Let us call a *significant* increase in IQ 1 of 6 points above the 9-point increase in the control group (due to repetition), typical for the Wechsler Intelligence Scale for Children–Revised (WISC–R). It seems that such an improvement would be observed in some 20% of middle-class children, 30% of working-class children, 40% or more of deprived inner-city children, and an even greater proportion of children in famine-ridden Third World countries. These are guesses at best, but they may serve to define the order of magnitude expected.

It is often argued by nutritionists that there is no evidence of subclinical micronutrient deficiency in Western countries, but the facts do not support such a view. Benton (1992a) has given an excellent survey of the available data; using conventional biochemical definitions, there is a consistent picture that the vitamin intakes of a minority of individuals are deficient or marginal. In the authoritative survey of *The Diets of British School-children* (Department of Health, 1989), there is evidence of certain low intakes (e.g., Fe, Ca, and riboflavin in teenage girls, and a generally low intake of pyridoxine). "Where the average intake for particular nutrients reached accepted levels, there were wide variations in individual consumption with evidence of a minority consuming 70% or even 50% of RDA" (Benton, 1992a, p. 298). The evidence indicates clearly that there exists a minority in Western countries that has a marginal intake of minerals and vitamins.

Should supplementation comprise only one specific element (zinc, iron, Vitamin C), or should we use a cocktail principle, combining many possible contributions to the total supply needed by the brain? There are arguments for both venues, and no doubt reseachers will follow both roads. With a single element, one feels that results can be attributed with some certainty to a single cause, as in the Jordan et al. (in press) experiment. But this certainty may be mistaken; intake of one element may be linked with another, and it may be the other that has the effect erroneously attributed to the first. Probably it is better at the present stage to demonstrate that there *is* a definite effect, and then combine all or most of the likely ingredients in a single multivitamin and mineral pill. Closer analysis of which ingredients produce the effect can come later. The search will undoubtedly be a long one, particularly in view of the fact that there are probably interactions among the ingredients that may potentiate or destroy individual effects, but these are questions to be answered after overall effectiveness has been established.

The next question is related to quantity. Just how much of each vitamin and mineral should be included in the cocktail? In the absence of any certain knowledge, one might be inclined to use maximum doses, but these may produce side-effects, and too large a dose of some minerals may have depressant effects of one kind or another. The optimum procedure might be that adopted by Schoenthaler et al. (1991) – namely, to use three different strengths equal (roughly) to 50%, 100%, and 200% RDA. This may at least give us some idea of what strength to aim at in future research, although the optimum strength for one element may differ from that for another, and quite generally, the optimum strength for one child will differ from that for another, depending on the degree of deficiency. Perhaps the requirements of the more seriously deficient children should decide this issue, there being little chance of oversupply having a negative effect on the nondeficient children within the limits described.

In view of the hypothesis that effects will only be observed in children with deficient micronutritional levels, it is obviously important to establish the *degree of deficiency* in individual children, in order to see (1) who is liable to improve, (2) whether degree of improvement is proportional to degree of deficiency, and (3) whether the micronutritional level at the end of supplementation has been restored to normal. There are two ways of accomplishing some of these things. We can keep a record of the kind and amount of food eaten by the child; this is a difficult and probably inaccurate task, particularly if we have to rely on the child's or the parent's testimony. Only in institutions can we really carry out this sort of task properly. The preferred but much more expensive and intrusive method is that of taking and analyzing blood samples, both before and after supplementation. This, too, is not perfect; some micronutrients are present in such minute quantities that no accurate measurement is possible. States of deficiency may of course be suspected in terms of socioeconomic status and other such variables, but clearly direct assessment is preferable and should generally form part of the study.

We have now established that large samples should be studied, supplements should be compared with placebos, and blood samples should be analyzed pre- and postsupplementation. Duration of supplemention should be long enough to give effects a chance to appear; 3 months appears to be a reasonable minimum, although periods of up to a year have been used. Compliance should be assured; some children pretend to take the pills but do not in fact do so. Teachers are not always conscientious in handing out pills or checking compliance. Very erroneous conclusions may be arrived at if procedures involve carelessness of this kind.

We have thus far not discussed the major point of any such enquiry. What is the variable we wish to affect, and how is this to be measured? By simply referring to intelligence, or IQ, we have avoided a more detailed discussion of this vital point. As Cattell (1963, 1980) has shown, intelligence has two aspects: *fluid ability* (g_f) and *crystallized ability* (g_c) (Eysenck, 1979). Fluid ability is the dispositional variable that allows us to solve problems, to learn, to think abstractly; crystallized ability is the *result* of such learning. Of course, the two are quite highly correlated; if a child has high g_f, he or she is able to learn more quickly and efficiently and will thus acquire more knowledge (g_c). Verbal tests are usually good measures of g_c; vocabulary is an obvious example. Nonverbal tests, such as the Raven's Progressive Matrices, or the nonverbal tests in the Wechsler scales, are good examples of g_f. It would seem to follow from the nature of these two types of intelligence that *if there is an effect of micronutrient supplementation on IQ, it should be on g_f and not on g_c*. Supplementation can affect the physiological basis of fluid ability but hardly that which has been learned in the past; in a vocabulary test, for instance, either we know the meaning of a given word, or we do not. Supplementation will not tell us. But in solving one of the Matrices problems, no prior knowledge is required. Supplementation may very well increase the speed of cortical functioning and hence improve the score on the test. This has been the theory accepted by all modern research workers in this field, and we have already mentioned one example of its successful application (Jordon et al., in press).

Would it be sufficient simply to use only nonverbal tests in our experiments? The theory makes *two* predictions, and *both* should be tested; the theory would be disproved as much by finding that supplements improved g_c as by finding that supplementation failed to improve g_f. Hence, tests like Wechsler are particularly suitable for use in this connection, measuring both g_f and g_c, with several types of test each. Another advantage is the *individual administration* of the test; group testing is less reliable and less valid. Hence, we consider the Wechsler test the gold standard of work in this field and suggest its use whenever possible.

If we predict no improvement in g_c, how can we account for the improvement in educational achievement found by Schoenthaler et al. (1986b)? Educational achievement, if anything, should be a measure of g_c, and improvement here would seem to disconfirm our theory. Not so. The theory states that improvement in g_f is shown by better problem solving, better reasoning, and better learning. Giving the IQ test twice does not furnish any opportunity for learning, as there is no knowledge of results; hence, the only improvement, shown by controls as well as experimental subjects, is through memory, test sophistication, and similar factors. But school learning is different; by improving g_f, we should improve learning ability, and that should show up as better scholastic achievement. To put the matter slightly differently, repeating the same vocabulary test after the end of the supplementation period, we would not expect any supplementation effects *because there had not been any learning*. But if the school taught word knowledge for a whole year, the improvement of the supplemented children, compared with that of control children, would be a measure of g_f, at least in part, and would be expected to be greater. In the Schoenthaler, Amos, Eysenck, Peritz, and Yudkin (1991) experiment, to be discussed later, the CTSB (Comprehensive Test of Basic Skills) was given routinely at the schools which supplied the children who took part in the experiment; hence, the 3-month period during which supplementation was applied included formal teaching and learning, and on several subjects, significant improvement was found in comparing the experimental and the control groups.

These are the major parameters of the theory considered in this chapter. There are a few additional comments that may be worth making. Would our prediction apply to adults as well as to children? Probably not; it is the growing, immature brain that is most likely to be affected. Hence, supplementation is most likely to be effective in the fetus, through the agency of the pregnant mother. Neonates and babies are next most likely to benefit, followed by toddlers, and then older children. It might not be fanciful to expect a linear regression in effect size, declining with increasing age of the children involved. Some results will be cited that support such a view. As Lucas et al. (1990) have found, "a short period of early dietary manipulation in pre-term infants had major consequences for later development, which suggests that the first weeks may be critical for nutrition" (p. 1477).

How long will results last? We have little guidance from our experiments, but it might be argued that supplementation produces certain growth and changes in the maturing brain that may remain fairly permanently. The matter clearly demands a firm experimental answer. So, of course, do many other questions, some already mentioned, others to be raised in connection with the discussion of empirical results in the next section. As we shall see, the evidence that a fair proportion of children in the USA and in the UK, none of them suffering from clinical malnutrition, is deficient in various vitamins and minerals, and can improve in fluid ability by micronutrient supplementation, is now so strong that there is little doubt about its essential truth (Benton, 1992a). This fact may encourage the ordinary business of science to proceed and to provide answers to the questions posed so far.

Nutrition and nonverbal intelligence

Substantial interest in nonverbal IQ and nutritional supplementation followed the January 1988 BBC broadcast of two controlled trials (Benton & Roberts, 1988; Schoenthaler et al., 1991). Eight subsequent controlled studies have compared nonverbal IQs of children/young adults randomly assigned to a vitamin-mineral supplement or placebo group (Benton & Butts, 1990; Benton & Cook, 1991; Crombie et al., 1990; Nelson, Naismith, Burley, Gatenby, & Geddes, 1990; Nidich, Moorehead, Nidich, Sands, &

Table 13.1. *Net IQ gains among supplemented groups in 10 independent studies*

Author and source	Note	Ages	Control group N	Active group N	Control group change	Active group change	p	Net IQ gain
Benton & Roberts, 1988	a	12–13	30	30	1.8	9.0	.01	+7.2
			26		4.0			+5.0
Nelson et al., 1990	b	7–12	105	105	12.4	15.7	none	+3.3
Crombie et al., 1990		11–13	44	42	1.5	3.9	.22	+2.4
Benton & Butts, 1990		13	87	80	2.0	5.0	.02	+3.0
Benton & Cook, 1991		6	22	22	1.2	10.8	.001	+9.6
Schoenthaler et al., 1991a		13–16	11	15	−1.0	5.0	.05	+6.0
Schoenthaler et al., 1991b	c	12–16	100	105	8.9	10.1	.01	+1.0
				105		12.5		+3.5
				100		10.3		+1.4
Nidich et al., 1993		8–9	16	18	4.9	9.8	.04	+4.9
Schoenthaler & Eysenck, 1994a	d	12–16	25	24	7.0	8.1	.27	+1.1
				30		8.5		+1.5
				21		11.1		+4.1
Schoenthaler et al., 1994	e	18–21	90	96	6.3	7.1	.12	+0.8
				90		8.9		+2.6
MEAN	f		556	883	6.16	9.68	.001	+3.5

[a] 30 controls received placebos, and 26 received no pills.

[b] The main index of nonverbal IQ (i.e. the WISC–R Coding Scale) is reported, but it is *not* a reliable measure of IQ.

[c] Three strengths of supplement were compared with placebo. As strength increased, net gain increased with the exception of the strongest formula, but the difference between the two stronger formulas was not significant and probably chance.

[d] Three strengths of supplement were compared with placebo. As strength increased, net gain increased.

[e] Two strengths of supplement were compared with placebo. The stronger formula produced the greater gains.

[f] The probability of 10 independent trials each producing greater gains in the supplement group on the primary measure of nonverbal IQ is .5 to the 10th power ($p < .001$).

Sharma, 1993; Schoenthaler et al., 1991; Schoenthaler, Amos, Eysenck, Korda, & Hudes, 1994). Table 13.1 summarizes the nonverbal IQ findings from all 10 studies. Various criticisms of these studies have been reviewed by Eysenck (1991, 1992).

The most striking pattern in Table 13.1 is that all ten studies showed supplemental subjects performing better than subjects who received placebo regardless of location, age, gender, race, formula, or research-team composition. The probability of 10 randomly selected, independent experimental groups always performing better than 10 randomly selected, independent control groups is one-half to the 10th power, or 1 in 1,024 (*p* < .001). It follows that Table 13.1 provides very robust evidence that vitamin-mineral supplements produce increases in nonverbal IQ.

Nine of the 10 studies used at least one *individually* administered test of nonverbal IQ and reported *p* values. The 10th study (Nelson et al., 1990) relied on group-administered tests of nonverbal *reasoning*, one-fifth of a valid IQ test, and no reported *p* values. These flaws make intepretation and generalizations about nonverbal IQ difficult, if not impossible, for that one study.

Notwithstanding the shortcomings of Nelson's study, 9 out of 10 reported a very narrow range of mean net gains in nonverbal IQ (i.e., from 1 to 6 points) and reported a narrow cluster of two-tailed *p* values) (i.e., .27, .22, .12, .05, .04, .02, and .01). The exception was the study involving younger children where net gains equaled 9.6 points and *p* < .001 (Benton & Butts, 1990). The *p* values in Table 13.1 are two-tailed, although the fact that the results were predicted would allow the use of one-tailed tests, thus halving the *p* values. We have preferred to err on the side of caution. In evaluating the published data, however, we would like to draw attention to the needs of addressing the problem of statistical *power*. To test a given hypothesis properly, the test must have sufficient power to give a 95% probability of achieving significance if the working hypothesis is true; otherwise, we are likely to make Type 2 errors. Calculation of a test's power is dependent on (1) sample size, (2) variance in the test instrument, and (3) the hypothetical effect size. The power of the test becomes important whenever the *p* value exceeds .05, since the absence of statistical significance may be a function of (1) inadequate sample size, (2) excessive variance in the instrument, (3) an excessive estimate of effect size, or (4) a false hypothesis.

The three studies that failed to reach significance at the .05 level using a two-tailed test were each designed in 1990 based on a hypothesized effect size of 9 points, which was believed to be present in the first published study (Benton & Roberts, 1988) before a corrected analysis showed the effect size to be 6 points. All three of these studies had an adequate sample size to have a 95% chance of producing significant results if the real effect size was 9 points, but not 6 points, nor 3 points. In fact, not one of the three studies had even a 50% chance of finding significance if the real effect size was about 3 points! In the absence of statistical significance, the proper conclusions for each of these three studies should have been that there is a 95% certainty that the real magnitude of effect on these populations is not 9 points or larger, but the possibility of smaller effect sizes remained unknown (rather than not supported, since each supplemental group did better).

It follows that the five proper conclusions from Table 13.1 are: (1) all 10 studies *individually* provided limited support for the hypothesis that vitamin-mineral supplementation may increase nonverbal IQ, because each study found that their supplemented subjects had a greater net gain in nonverbal IQ than control subjects: (2) six of nine studies that reported inferential statistics *individually* reached the traditional level of acceptable risk of Type 1 error (i.e., 5%); (3) three of nine studies did not but came fairly close with p values of .27, .22, and .12, which represent the probabilities that their net differences were coincidental; (4) there is a 95% certainty that the real magnitude of effect on these populations is not 9 points or larger based on these studies; and (5) *collectively*, all 10 studies show supplemented subjects performing better than subjects who received placebos, which is statistically significant ($p < .001$), providing no other studies exist where this did not occur.

A review of the hypothesis at the macro level, looking for more meaningful patterns across these 10 studies, produced further insight into the relationship. That analysis and summary follow.

The 1986 Benton and Roberts trial

Ninety schoolchildren, aged 12 to 13 years, were randomly assigned to one of three groups: (1) a vitamin-mineral supplement group, (2) a placebo group, or (3) a nontablet group. The children took the Cognitive Abilities Test and the Calvert Non-verbal Test at the beginning and end of the trial (9 months later), as well as a 3-day dietary record that was analyzed for nutrients. In nonverbal IQ, the supplemented children gained 7.2 points more than children on placebo and 5.0 points more than children on no tablets, with the differences being significant. However, in verbal IQ, no significant differences were found.

The trial attracted immediate criticism. For example, the statistics should have incorporated a three-group design instead of just a comparison between supplement and placebo groups. This was corrected and made no difference; the results remained significant. Others noted that the diet analysis was not adequate to determine individual nutrition, which would be essential to isolate why the supplements had an effect. However, knowledge about individual nutrition is not necessary to determine whether supplements will raise IQ; thus, this criticism does not negate the primary finding. Rather, the lack of these data limits the ability to speculate which nutrients may be involved. In fact, there were no fatal flaws in design or implementation. The legitimate concerns were (1) the lack of assessment of the children for their blindness to treatment, (2) the absence of multiple measures of IQ, (3) adequate assessment of nutrition, (4) randomization based on presupplementation scores to ensure initial similarity, and (5) the need for independent replication, points with which Benton concurred.

The 1986 Schoenthaler trial

Although broadcast on the same BBC program as Benton's work, it is unfortunate that the editor of *Lancet* declined an invitation to publish both together, since that could have provided (1) independent replication, (2) two additional measures of IQ, (3) excellent nutrition data due to the use of 7-day diet analyses, blood assays, and clinical examinations, (4) assessment of blindness, and (5) initial similarity on IQ, the same five issues upon which the Benton study was sttacked.

Twenty-six schoolchildren, aged 13 to 16 years, were randomly assigned to a vitamin-mineral supplement group or a placebo group for 13 weeks. The children had taken the verbal and nonverbal halves of the WISC–R, over a year before the beginning of the trial and again at the end of the trial. Each completed a battery of tests for nutrition. Brain Electrical Activity Mapping (BEAM) was conducted on 10 subjects with no history of neurological disorders but a history of chronic conduct disorders of unknown etiology; 6 of the 10 BEAM pretests were abnormal for no known reason. In nonverbal IQ, the supplemented children gained 5.0 points, while children on placebo fell 1.0 points, with the differences being significant. In contrast, no significant differences in verbal IQ were found, which supports the Benton and Roberts study on both types of IQ.

The study went beyond Benton and Roberts in three important areas. First, the gains in IQ were not evenly spread throughout the subjects. Five on supplements gained 9, 15, 18, 21, and 25 points in nonverbal IQ, and one on placebo gained 13 points; these amounts are clinically significant because they equal or exceed twice the 4.5-point standard error of measurement on the WISC-R when it is given far enough apart to produce no test–retest gains. Second, it was possible to predict that only these subjects would produce the clinically significant gains in nonverbal IQ by looking for low preintervention blood concentrations that had been corrected by the posttest blood assays. Third, of the six children with abnormal BEAM tests, all four on supplements improved significantly with three or four becoming normal, while neither of the two subjects on placebo with abnormal BEAM tests exhibited any improvements at all. This study suggests that the supplements were correcting undiagnosed low blood-nutrient concentrations, correcting selected brain function problems, and improving nonverbal IQ in about one-third of the experimental subjects (i.e., 5 of 15).

The 1988 Nelson and Naismith trial

This study claimed to be a replication of the Benton and Roberts study. It used 227 schoolchildren, aged 7 to 12 years, who were randomly assigned to a vitamin-mineral supplement or placebo group for 28 days. The children aged 7 to 10 years took the HEIM AHIX nonverbal reasoning test, while the older children, aged 11 and 12 years, took the HEIM AH4 test of verbal and nonverbal reasoning. All the children completed the WISC–R Digit Span and Coding scales, with the HEIM being given in groups and the WISC–R administered individually. A valid 7-consecutive-day diet analysis was also recorded for each child. On retesting, there were no significant differences in verbal or nonverbal IQ on the four measures.

This study was not a valid replication because of the shorter period of supplementation, the use of different IQ tests, and the use of a different formula. However, there exist more serious methodological shortcomings that are fatal in terms of testing the nutrition and IQ hypothesis.

First, there are important distinctions between intelligence quotient tests (better known as IQ tests) and tests of nonverbal or verbal reasoning. The individually administered WISC–R measures the former, while the group-administered HEIM measures the latter. Although correlated, they are not synonymous and should not be treated as such. Thus, the WISC–R Digit Span scale became the sole measure of verbal IQ and the Coding scale the sole measure of nonverbal IQ, with the HEIM measuring reasoning.

Second, the manner in which the WISC–R was used is not a valid measure of IQ because just one of the five nonverbal scales was used to measure nonverbal IQ and just one of the five verbal scales was used to measure verbal IQ. The WISC–R instructor's manual is quite clear that in order to produce a valid measure of nonverbal and verbal IQ, one should test the subject on five verbal subscales and 5 nonverbal subscales but that four on each type of IQ is the minimum to retain validity. For these nutritionists to ignore the instructor's manual and make up their own definitions of verbal and nonverbal IQ is a fatal flaw from which there can be no recovery.

Third, in addition to the conceptualization errors on the nature of the WISC–R and HEIM, problems exist with the statistical nature of the HEIM and using these two WISC–R subscales alone. More specifically, if one compares the predicted magnitude of gain in nonverbal IQ to the variance on these tests, the power of these tests is not adequate. For the HEIM, one would need over 1,000 subjects per group and 280 on the Coding scale per group to have a 95% chance of finding a relationship if it existed. Accordingly, it is hardly surprising that no significant differences were found. The flaws, however, do not stop here.

Fourth, the experimental and control groups were not properly matched. They were matched by sex, age, height, and then nonverbal test score on the HEIM AH4 nonverbal scale. The primary dependent variable, the individually administered WISC–R, was not used. This is an elementary methodological mistake; assignment to group should be either random or made on the basis of the primary and secondary dependent variables (the WISC–R and the AH4) rather than potential covariates such as sex, age, and height.

The effects of this error are all too apparent to anyone trained on the WISC–R. According to table 4 of the study, the children on supplements started out at about 2.0 raw scores above the children on placebo. This translates into about a 1.0 higher scaled score, which represents about 5.0 points in the total scaled score and translates into a 6-point IQ difference! The same error was made with the AH4; although the raw scores on the supplement group were only about 1.0 point higher, this translated into about *4* IQ points. Although neither was significantly different due to the aforementioned variance problems associated with those two measures, the magnitude of these differences during the pretest exceeded the hypothesized differences at posttest and should have alerted the researchers to a serious problem.

Fifth, age differences must be considered in IQ assessment in children aged 7 to 12 years. The WISC–R manual instructs investigators to scale the raw scores, controlling for age in 4-month units. Thus, 18 different age-related scales each spanning one 4-month unit should have been used, but the study employed only two age-related scales (in 2- and 4-*year* units).

These errors may explain two anomalies in the data: (1) the subjects' reported average initial raw score of about 51 points translates into an average initial IQ score of about 120 points (which is most unlikely in a "normal school population"); and (2) the subjects' IQ gains in the Coding test are reported as between 4.5 and 7.1 points (far higher than the average 2.0 raw-point gain predicted in the instructor's manual). The weight of all these anomalies leads to the conclusion that the Nelson and Naismith IQ data are *not* reliable.

However, if one wished to argue that they were reliable and one focused on all the children who had not been already taking supplements before the study began, the 0.9 greater gain in the Coding scale among children on supplements translates into about a 0.5 scaled-score difference, and a 3.5 greater net gain in nonverbal IQ among the supplement group in the

Nelson data. This is about the same magnitude as the average net gain in the other nine studies.

The 1990 Crombie and Todman trial

Eighty-six schoolchildren, aged 11 to 13 years, were randomly assigned to a vitamin-mineral supplement group or a placebo group for 7 months. The children took the Cognitive Abilities Test, the Calvert Non-Verbal Test, the Cattell Culture-Fair Test, AH4 (Parts I and II), and the Raven's Progressive Matrices as measures of verbal and nonverbal IQ and reasoning at the beginning and end of the trial. A 7-day dietary record was analyzed for nutrients. On nonverbal IQ, the supplemented children gained 2.4 points more than the children on placebo on the primary measure, the Calvert Non-Verbal Test of IQ. No significant differences were found in any of the other five tests.

In marked contrast to the Nelson trial, this study corrected for most of the shortcomings in the Benton and Nelson studies. Crombie improved on Benton's design by including (1) an assessment of the children for their blindness to treatment, (2) multiple measures of IQ and reasoning, and (3) adequate assessment of nutrition. Crombie improved on Nelson's design by (1) using almost as long a period of supplementation as Benton, (2) including the same IQ tests as Benton plus others, (3) using the same formula as Benton, (4) not making up intelligence tests like Nelson, and (5) conducting a power analysis to determine the adequacy of the instruments unlike Nelson.

The Crombie trial failed to reject the null hypothesis that children on supplements would not produce significantly greater gains in nonverbal IQ than children on placebo, because the 2.4-unit greater net gain among children on supplements produced a p value of .22. This means that there is a 22% probability that *if* one uses the data to accept the working hypothesis, the likelihood of these results being due to chance is 22%. On the other hand, there is also a 16% chance of being wrong if one rejects the working hypothesis based on a power analysis, which predicts a 6-point net gain. In short, this study standing by itself should be viewed as inconclusive if everything else were in order. However, a closer review of the paper suggested that this was not the case for three main reasons.

First, the power analysis was based on the faulty assumption that net gains of 6 or 9 points should occur rather than 3 points. As a result, failure to find statistical significance may have been due to inadequate sample size alone.

Second, the random assignment to group is not good. Proper

randomization requires that assignment be done on the dependent variable (i.e., nonverbal IQ). However, the Crombie team randomized on the basis of *verbal* IQ only. This error may appear trivial, but in fact, there are often wide differences in verbal and nonverbal IQ in the same individual. Their data show that before the experiment started, there was already a substantial difference between the placebo and experimental groups; the latter had a mean IQ 1.9 points higher, a 2.0-point larger standard deviation, and a 20% larger standard error, because five of six children with nonverbal IQs over 120 points were in the supplement group. The average IQ of these six children places them in the top 1% and creates serious problems. Given that a corollary of the primary hypothesis is that children with the highest IQs are likely to be the best nourished (and therefore not likely to respond to supplementation), allowing 5 of 42 subjects to be this bright in the experimental group unfairly disadvantages testing of the hypothesis. Furthermore, the samples cannot be normally distributed with 6 of 86 subjects in the highest percentile. This is critical since using an ANOVA on change in IQ when the data are not normally distributed falsely elevates the value of p.

Third, the statistical procedures contained two errors. Since this was a replication study that predicted direction, the authors conceded that a one-tailed test should be preferred to the two-tailed test they used. Furthermore, the data were not normalized before conducting an ANOVA, which invalidates the reported value of p.

Just as the Medical Research Council has endorsed the design used in this study, the Research Oversight Committee of the California Legislature endorsed the following ANCOVA statistical procedure on nutrition and IQ data. The pre- and posttest IQ distributions should be normalized first using $\log 10$ transformations; the dependent variable should be posttest IQ; the covariate should be pretest IQ; and then any other variables of concern may serve as potential covariates when doing a two-tailed test at the .05 level.

If these statistical procedures are used, the results approach significance ($p = .06$) in spite of the fact that randomization placed the supplement group at a substantial disadvantage. By eliminating the four subjects who had very high initial IQs from the supplement group, leaving the one subject (IQ 135) who was closest to the very bright subject on placebo (IQ 133), the net difference rises to 3.5 points.

Fourth, a less critical anomaly was resolved after examining the raw data. Their abstract states, "This direction of effect was not consistently seen with three other tests of non-verbal reasoning." More precisely, the Calvert Non-Verbal test, Raven's Progressive Matrices, and AH4 part II (nonverbal) all produced greater net gains in the supplement group (about +2.4,

+1.0, and +6.9 IQ points, respectively) but not the Cattell Culture-Fair test (−0.6). The raw data showed that the Cattell was posttested not once as implied by the paper but *four* times, with the children who received placebos more likely to have taken all four posttests. Thus, the controls were more likely to experience greater test–retest gain. The hypothesis that multiple retests affected subsequent learning was examined by comparing the four retests by cohort, but the absence of the predicted direction is more likely due to poor administration of this test; both groups fell about 14 IQ points at the first retest, gained 28 IQ points at the second retest, surged ahead 4 more at the third, and fell back 12 IQ points at the last test, leaving both groups with about a 6-point IQ net gain. Such massive swings call the skill of the person supervising this test into question. Last, but not least, the Raven's matrices also failed to pass test–retest reliability requirements ($p = .55$).

In short, if one limited a review of this literature to the Crombie data, one could either reject the null hypothesis with a 10% chance of creating a Type 1 error (an erroneous acceptance of the working hypothesis) or fail to reject the null with a 60% chance of creating a Type 2 error in spite of a real net gain of 3 points.

The 1990 Benton and Butts trial

One hundred and sixty-seven Belgian schoolchildren, aged 13 years, were randomly assigned to a vitamin-mineral supplement group or a placebo group for 5 months. Flemish-speaking children took the Calvert Non-Verbal Test and the Differentiele Geschiktheidsbatteri, while French-speaking children took Les Examens Otis-Ottawa d'Habileté Mentale at the beginning and end of the 5-month trial. Fifteen-day dietary records were analyzed for nutrients that allowed children to be placed in a better or poorer diet category.

As before, no effect was found on verbal IQ. As expected, no difference was found among children who were consuming a better diet. No significant effect was found on nonverbal IQ for females who were consuming a poorer diet either. In contrast, among male children who consumed a poorer diet, nonverbal IQ rose 4 points for supplemented male children and fell 2 points for males given placebos, which was statistically significant ($p < .02$).

Although well-designed and implemented, the statistical analysis was somewhat suspect due to the separation of males and females without a priori theoretical grounding. The more conservative approach would have been to combine males and females utilizing one ANCOVA among the entire sample or just the 81 poorer nourished children. This would have produced a greater net gain of about 3 or 4 points respectively among supplemented children. Such an analysis would just miss being statistically significant, making the study much like the Crombie study – that is, suggestive of a 2- to 4-point net gain but inconclusive by itself.

The 1991 Benton and Cook trial

Forty-seven British schoolchildren, aged 6 years, were randomly assigned to a vitamin-mineral supplement group or a placebo group for six to eight weeks. The children took four sub-scales of the British Ability Scale at the beginning and end of the trial. In non-verbal IQ, the supplemental children gained 7.6 points while children who received placebos declined 1.7 points. This was the first study where net differences were significant at the .001 level.

This was only the second trial to receive no criticism other than the first Schoenthaler trial. Unfortunately, both of these superior studies have been largely ignored.

The 1992 Schoenthaler, Amos, Eysenck, Peritz, and Yudkin trial

Four hundred and ten American schoolchildren, aged 12 to 16 years, were randomly assigned to one of three different-strength vitamin-mineral supplement groups (equal to, roughly, 50%, 100%, or 200% RDA) or a placebo group for 3 months. The children took the full WISC–R, Raven's progressive matrices, and the Matrix Analogies Test as measures of verbal IQ, nonverbal IQ, and reasoning at the beginning and end of the trial. Blood samples were assayed for nutrients. In nonverbal IQ, the supplemented children gained 3.5 points more than children on placebo on the primary measure, the nonverbal half of the WISC–R. No significant differences were found in any of the other five tests.

The study was criticized on several grounds: (1) the blood data were not published with the original study (see the next section); (2) the two nonverbal tests of reasoning were not significant, making the significant finding appear coincidental; and (3) children receiving the strongest formula did not gain as much as children receiving the middle-strength formula.

At the time of publication, the blood had not been fully analyzed since it had been archived until the authors were sure of a significant effect; it made sense to first determine if a significant effect existed before undertaking a very expensive examination of blood to determine why it occurred. The lack of significance on Raven's Progressive Matrices and the Matrix Analogies Test is consistent with other researchers using similar nonverbal tests of reasoning; the lack of significance may be due to their relatively large variance or the nature of the tests not adequately measuring nonverbal intellignce. Once the blood was assayed and analyzed, the anomaly surrounding the strongest formula became apparent. In the original study, which had no controls for children being at risk of malnutrition, the 200% group's net gain of 1.4 points over placebo was not significant. When the reanalysis was limited to children who were at risk of nutritional deficiencies (as measured by low concentrations of vitamins in the blood), the difference between placebo and 200% rose to 5.1 points and became significant. The difference between placebo and the 100% group rose to 8.1 points, but the difference between 100% and 200% was not significant. In contrast, the difference between placebo and 50% was still not significant. This suggests that very low dose supplements may not work but that a supplement set at 100% of the USRDA is adequate (USRDA = United States Recommended Daily Allowance).

The 1992 Schoenthaler and Eysenck trial (1996a)

This British study was supposed to be a replication of the previous American trial using the same three strengths of supplements and placebos but with fewer blood tests. Unfortunately, the number of subjects who completed the trial was only 100, one-quarter the number called for in the design due to financial and other problems. The 21 children who received the 200% formula gained 4.1 points more than 25 children who received placebos. The 30 children who received the 100% formula gained 1.5 points more than children who received placebos, and the 24 children who received the 50% formula gained 1.1 points more than children who received placebos. Although the magnitude of the gain increased as the daily average intake increased, the results were not significant ($p = .27$), and the power of the test was low (32% chance of finding significance if, in fact, it did exist) due to low sample size.

The 1993 Nidich trial

Thirty-four children, aged 8 to 9 years, were randomly assigned to a placebo or supplement group for 3 months and took the Calvert nonverbal test of IQ at the beginning and end of the study. The 18 children who received supplements produced a 4.9-point greater net gain in IQ than the 16 who received placebos ($p = .04$).

The 1994 Schoenthaler, Amos, Eysenck, Korda, and Hudes trial

Two hundred and sixty-six male corrective institution inmates, aged 18 to 25 years, were randomly assigned to one of two different-strength vitamin-mineral supplement groups or a placebo group for 15 weeks. The males took the nonverbal half of the Wechsler Adult Intelligence Scale–Revised (WAIS–R), as a measure of nonverbal IQ at the beginning and end of the trial. Blood samples were assayed for nutrients. In nonverbal IQ, the 90 males on the 300% USRDA formula gained 2.6 points more than 90 males on placebo, and the 96 males on the 100% USRDA formula gained .8 points more than the males on placebo. Although the magnitude of net gain increased as the daily average intake increased, the results did not quite reach significance ($p = .12$).

These 10 studies show a number of interesting patterns other than the fact that the mean gain in nonverbal IQ for subjects in each study who received placebos never equaled their experimental group counterparts.

First, the *group* that was administered tests of nonverbal *reasoning* consistently failed to produce significant differences six times. More specifically: (1) Crombie and Schoenthaler independently failed to find significant differences using Raven's Progressive Matrices; (2) Crombie and Nelson independently failed to find significant differences using the AH4 part II; and (3) the two Schoenthaler research teams failed to find significant differences using the Matrix Analogies Test (MAT). This pattern may be partially, or totally, due to the relatively large amounts of variance/error associated with group-administered tests; the average variance of these group-administered tests is about 10 times larger than individually administered tests and would necessitate a much larger sample to attain statistical power. On the other hand, it is also possible that the consistent lack of significance of these nonverbal tests of reasoning is due to their not being really analogous to nonverbal tests of IQ; the latter control for age and measure a broader spectrum of mental skills, which might explain why *p* values of these six tests never reached significance.

Second, none of the studies that measured verbal IQ using individually administered IQ tests found any significant effect on verbal IQ, and none of the group-administered tests did either, which suggests that the short-term effects are limited to nonverbal IQ with a hypothesized long-term effect on both types of intelligence.

Third, children aged 6 years showed the largest net gains: 9 points. Young adult males, mostly aged 18 to 21 years, produced the smallest mean net gain: about 2 points. The mean gains of children aged 12 to 16 years were in between, with the younger gaining more. Thus, an inverse correlation exists between net gain and age of participants. Do improvements in IQ last? The only relevant study is the Schoenthaler et al. (1991b) trial. One year after its completion, the children were retested and were found to have retained their enhanced status. Whether this will be maintained indefinitely is not known.

Blood analysis

The data examined so far support our view that micronutrient supplementation increases nonverbal IQ in selected children, but while it seems likely that the children who benefit are those who start out with relatively low levels of vitamins and minerals, direct proof of this requires direct measurement of vitamin and mineral status at the beginning and end of the experiment. Only the Schoenthaler et al. (1991a) study provides such evidence, if only on a very small sample. However, the recent Schoenthaler et al. (1991b) study included blood taking and led to a detailed blood analysis (Schoenthaler & Eysenck, 1996). In this study, as mentioned in the preceding section, a sample of 410 California schoolchildren received placebos or 50%, 100%, or 200% of the USRDA for 13 weeks. Concentrating on the results obtained from the WISC–R, the placebo group improved by 9 points of IQ, based on retesting; the 50% supplementation group exceeded that by 1.0 point, the 100% group by 3.5 points, and the 200% group by 1.4 points. Supplementation overall had a significant effect, but only the 100% group produced individually significant improvement.

The hypothesis requires that children who receive the supplementations but do not improve in IQ above the 9 points measured at retesting (nonresponders) should have significantly higher blood vitamins and mineral levels than those who show a significant improvement (responders). The nonresponders, with retest improvements of 9 points or less, numbered 37 in all, out of 310 children (i.e., 12%). Responders were defined as follows: The standard error of the estimate (SEE) for the nonverbal WISC–R scale is 4.5 points. Responders were defined in terms of twice the SEE – that is, 9 points over and above the test–retest gain of 9 points (i.e., 18+ points). There were 66 such responders (i.e., 21% of the total group).

The basis for defining a child to be at risk substantially was a listing of the means and the ranges of blood assay values for the various vitamins and minerals used in the supplementation formula for the nonresponders. Using this listing, a subject would receive a zero for each nutrient *within* this range, and a one for each nutrient below the bottom end of this range. The sum of the 10 vitamin scores became each subject's vitamin malnutrition index, and the sum of the 8 mineral scores become the mineral malnutrition index. As predicted, the 60 responders were at greater risk of vitamin malnutrition than the 37 nonresponders ($p = .012$). However, no significant difference was observed for minerals ($p = .362$). With the exception of magnesium and iron, low mineral concentrations were not more common among responders; it is impossible to say whether this pattern would occur again in a replication study. It appears that perhaps the total effect of micronutrients may be due to vitamins, with minerals playing little part. Iron may be an exception (Jordan et al., in press).

Among children with low blood-vitamin concentrations, the net gain in nonverbal IQ above placebo was 8.1 points ($p < .001$) for the 100% group, 5.1 points for the 200% group ($p < .01$), and 3.0 points for the 50% group (not significant). It is possible that for groups of deprived children, even higher on the vitamin malnutrition index than our responder sample, supplementation in the 200% USRDA range might give better results than 100%; for our groups, there was no significant difference between 100% and 200%.

The theory under test assumes that the normalization of low vitamin concentration corrects brain function, so that if we compare responders and nonresponders, both with low pretest blood concentrations, we would expect those who showed high posttest concentration to be responders and those with low posttest concentration to be nonresponders. It was found that responders had significantly fewer low vitamin concentrations at the posttest than nonresponders ($p < .002$). In fact, the elimination of low vitamin concentration by the posttest was a slightly better predictor of IQ gain than group assignment ($p < .006$). Reasons for failure to benefit from supplementation are not clear. It may signify noncompliance (failure to take the pills regularly) and/or bio-individuality.

These data show that the second part of our hypothesis is supported – it is children with low vitamin concentrations who benefit from supplementation, and they do so *pari passu* (in equal proportion) with improvement in their vitamin concentration. Our data suggest that roughly 20% of normal children, not deprived but regarded as fed in a normal fashion, are in fact two standard errors of the estimate below the bottom of a normal sample (nonresponders) nutritionally (i.e., indexed in terms of vitamin and mineral concentration). This proportion is likely to rise drastically in inner-city areas and among deprived children generally. It seems likely that for such groups, increases after 100% or 200% supplementation would exceed, possibly markedly, the gains of 8.1 IQ points over the placebo group achieved by the 100% group. Certainly, further experiments should be carried out to test this hypothesis.

Summary and conclusions

This chapter enables us to record some conclusions with a certain degree of confidence.

1. Inadequate levels of vitamins and minerals in the blood stream reduce a child's IQ below the optimum level for that child.
2. Supplementation of the child's standard diet by vitamin and mineral pills can raise the child's nonverbal IQ significantly.

3. Supplementation affects only *fluid* intelligence (nonverbal tests) but not *crystallized* intelligence (verbal tests).
4. The younger the child, the greater the effects. There is little effect beyond the teens.
5. Supplementation has no effect on children with an adequate level of vitamins and minerals.
6. Vitamin deficiencies seem to be more important than mineral deficiencies, with the possible exception of magnesium and iron.
7. Approximately 20% of a normal diet group of American children are responders; that is, they react to supplementation with IQ increases of 9+ points over and above test–retest increases in a placebo group.
8. This percentage would be likely to be greater in inner-city and other deprived groups.
9. The *mean* over all the children tested is likely to be relatively small (3.5 IQ points is the average for all 10 studies so far reported), because for the majority of children, there is no or little effect. This does not deny the possibility of large effects for children low in vitamins and minerals.
10. Effects of micronutrient supplementation have so far been shown to continue for 1 year and may last even longer.

Why have these conclusions met with considerable, if unjustified, criticism by nutritionists? It is only natural that experts in a given field should be hostile to outsiders, as psychologists are perceived to be, particularly if these outsiders question the traditional methods of arriving at estimates of the recommended daily allowance (RDA) and argue that the agreed RDA is grossly underestimated. RDAs are established solely on the basis of absence of physical symptoms of malnutrition; clearly, using psychological functioning as the criterion gives different values and constitutes an important innovation. Nutritionists, as their published studies reviewed earlier show, have little expertise or understanding in this field.

The practical effects are most likely to be in the public field. For mothers of well-fed children, recourse to supplementation would only be recommended if the children refuse to eat fruit, vegetables, and other material containing vitamins and minerals and concentrate on junk food, sugar, and so forth, as many apparently do. But parents of poorly fed children, presumably most at risk, are unlikely to be aware of these studies and unable to pay the cost. For these children, provision of supplementation at school meals might be a solution. But, essentially, the practical applications of these findings require democratic discussion, widespread publication, and community action; it is not our purpose to frame the necessary rules. To allay one frequently voiced fear, overdosing is not a danger if doses like RDA 100% are not exceeded; even RDA 200% presents no danger. At the moment, we cannot identify specific vitamins as being specially involved, but no doubt future research will tackle this problem.

At the beginning of this chapter, we mentioned the strong genetic component that produces differences in IQ. Vitamin and mineral intake clearly is essentially an environmental determinant, so that if we equated such intake, roughly at least, for all children, the importance of genetic factors would increase. However, as Plomin (1994) has emphasized, the environment that a person experiences is not independently given from outside; people select their environments, partly in accord with their genetic predisposition. Thus, a child's preferences for junk food, sugary sweets, and cola might themselves be inherited. Here is another interesting area for research; at the moment, little is known about these matters.

Clearly, this whole area of micronutrient supplementation requires much further research, and in particular, extension to less adequately fed groups. Extensions are also needed in the area of supplementation of prospective mothers, and babies, where IQ increases are most likely to be found. Duration of effects is another important area. So is determination of the most successful formula to be used, both in general and for particular individuals, depending on blood analyses. Last but not least, it is important to discover why supplementation in some deficient children does not produce increased vitamin and mineral levels or IQ increments. There is much still to be learned, but the fundamental fact of IQ increases after vitamin and mineral supplementation seems firmly established.

References

Axelson, M. L., & Brinberg, D. (1989). *A social and psychological perspective on food-related behaviour*. London: Springer-Verlag.

Bartus, R. T., Dean, R. L., Bear, B., & Lippa, A. S. (1982). The cholinergic hypothesis of geriatric memory dysfunction. *Science, 217*, 408–17.

Benton, D. (1981). The influence of large doses of Vitamin C on psychological functioning. *Psychopharmacology, 75*, 98–9.

Benton, D. (1992a). Vitamin-mineral supplements and intelligence. *Proceedings of the Nutrition Society, 51*, 295–302.

Benton, D. (1992b). Vitamin and mineral intake and human behaviour. In A. P. Smith & D. Jones (Eds.), *Handbook of human performance* (Vol. 2) (pp. 25–47). Cambridge: Cambridge University Press.

Benton, D., & Butts, J. (1990). Vitamin-mineral supplementation and intelligence. *Lancet, 335*, 1158–60.

Benton, D., & Cook, R. (1991). Vitamin and mineral supplements improve the intelligence scores and concentration of six-year-old children. *Personality and Individual Differences, 12*, 1151–8.

Benton, D., & Roberts, G. (1988). Effect of vitamin and mineral supplementation on intelligence of a sample of school-children. *Lancet, 1*, 140–3.

Brody, N. (1992). *Intelligence*. New York: Academic Press.

Cattell, R. (Ed.). (1983). *Intelligence and national achievement*. Washington, DC: The Institute for the Study of Man.

Cattell, R. B. (1963). Theory of fluid and crystallized intelligence: A critical experiment. *Journal of Educational Psychology, 54*, 1–22.

Cattell, R. B. (1980). The heritability of fluid, g_f, and crystallized g_c, intelligence, estimated by a least squares use of the MAVA method. *British Journal of Educational Psychology, 50,* 253–65.

Cherkin, A. (1987). Interaction of nutritional factors with memory processing. In W. B. Essman (Ed.), *Nutrients and brain function* (pp. 72–94). London: Karger.

Conners, K. (1989). *Feeding the brain: How foods affect children.* New York: Plenum Press.

Cravioto, J., & Delicardio, E. (1970). Mental performance in school-age children: Findings after recovery from severe malnutrition. *American Diseases of Childhood, 120,* 404–10.

Crombie, I., Todman, J., McNeil, C., Florey, C., Menzies, I., & Kennedy, R. (1990). Effect of vitamin and mineral supplementation on verbal and non-verbal reasoning of school-children. *Lancet, 335,* 744–7.

Dean, W., & Morgenthaler, J. (1990). *Smart drugs and nutrients.* Santa Cruz: B. & J. Publications.

Dean, W., Morgenthaler, J., & Fowkes, S. (1993). *Smart drugs II: The next generation.* Mento Park, CA: Health Freedom Publications.

Dobbing, J. (Ed.). (1987). *Early nutrition and later achievement.* New York: Academic Press.

Essman, W. B. (Ed.). (1987). *Nutrients and brain function.* London: Karger.

Eysenck, H. J. (1979). *The structure and measurement of intelligence.* New York: Springer Verlag.

Eysenck, H. J. (1991). IQ and vitamin supplements. *Nature, 751,* 263.

Eysenck, H. J. (1992). Fact or fiction? *The Psychologist, 4,* 409–11.

Eysenck, H. J., & Eysenck, S. B. G. (Eds.). (1991). Improvement of IQ and behaviour as a function of dietary supplementation. *Personality and Individual Differences, 12,* 329–65.

Flynn, J. R. (1987). Massive IQ genius in 14 natives: What IQ tests really measure. *Psychology Bulletin, 101,* 171–91.

Grantham-McGregor, S. (1987). Field studies in early nutrition and later achievement. In J. Dobbing (Ed.), *Early nutrition and later achievement* (pp. 128–53). New York: Academic Press.

Harrell, R. F. (1946). Mental responses to added thiamine. *Journal of Nutrition, 31,* 283–98.

Harrell, R. F., Woodyard, E., & Gates, A. I. (1955). *The effects of mother's diet on the intelligence of offspring.* New York: Bureau of Publications, Teachers' College, Columbia University.

Hart, N. (1993). Famine, maternal nutrition and infant mortality: A re-examination of the Dutch hunger winter. *Population Studies, 47,* 27–46.

Herman, C. P., Polioy, J., Pliner, P., Threlkeld, J., & Munro, D. (1978). Distractibility in dieters and non-dieters: An alternative view to "externality." *Journal of Personality and Social Psychology, 36,* 536–48.

Heseker, H., Knehler, W., Westenhoeker, J., & Pudel, V. (1990). Psychische Veraenderungen als Fruezeichen einer suboptimalen Vitaminversorgung. *Ernaehrungs-Umschau, 37,* 87–94.

Itzkoff, S. W. (1991). *Human intelligence and national power.* New York: Peter Lang.

Itzkoff, S. W. (1994). *The decline of intelligence in America: A strategy for national renewal.* Westport: Praeger.

Jordan, T., Bayley, P., & Henry, J. (in press). Effects of iron supplementation on attention and cognitive performance of school-children. *Personality and Individual Differences.*

Laessle, R., Bossent, S., Hank, C., Hahlweg, K., & Pirke, K. (1990). Cognitive performance in patients with bulimia nervosa: Relationship to intermittent starvation. *Biological Psychiatry, 27,* 549–51.

Lloyd-Still, J. P. (Ed.). (1976). *Malnutrition and intellectual development.* Lancaster: Medical and Technical Publications.

Loriaux, S. M., Deijen, J. B., Orleheke, J. F., & De Swart, J. H. (1985). The effects of nicotonic acid and xanthinol nicotinate on human memory in different categories of age. *Psychopharmacology, 87,* 390–5.

Lucas, A., Morley, R., Cole, T., Gore, S., Lucas, P., Crowle, P., Pease, R., Boon, A., & Powell, R. (1990). Early diet in pre-term babies and developmental status at 18 months. *Lancet, 3352*, 1477–81.

Lynn, R. (1990). The role of nutrition in secular increases in intelligence. *Personality and Individual Differences, 11*, 273–85.

Miller, J., Gauce, W., & Kang, K. (1978). A co-twin control study of the effect of vitamin C. *Twin Research: Clinical Studies, 2*, 151–6.

Nelson, M., Naismith, D., Burley, V., Gatenby, S., & Geddes, N. (1990). Nutrient intakes, vitamin-mineral supplementation, and intelligence in British school-children. *British Journal of Nutrition, 64*, 13–22.

Nidich, S., Moorehead, P., Nidich, R., Sands, D., & Sharma, H. (1993). The effect of the Maharishi student Rasayana food supplement on non-verbal intelligence. *Personality and Individual Differences, 15*, 599–602.

Plomin, R. (1994). *Genetics and Experience*. London: Sage.

Richter, M. C. (1979). Psychische Auswirkung sub-Klinischer Vitaminmangel Zustaende. *Ernaehrungs-Umschau, 263*, 381–84.

Rimland, B., & Larson, G. (1983). Hair mineral analysis and behavior: An analysis of 51 studies. *Journal of Learning Disabilities, 16*, 279–85.

Rogers, P., & Green, M. (1993). Dieting, dietary restraint and cognitive performance. *British Journal of Clinical Psychology, 32*.

Rush, D., Stein, Z., & Susser, M. (1980). *Diet in Pregnancy*. New York: Alan R. Liss, Inc.

Rush, D., Stein, Z., Susser, M., & Brody, N. (1980). Outcome at one year of age: Effects of 2 somatic and psychological measures. In D. Rusch, Z. Stein, & M. Susser (Eds.), *Diet in pregnancy*. New York: Alan R. Liss, Inc.

Rutter, M., & Madge, N. (1976). *Cycles or disadvantage*. London: Heineman.

Schoenthaler, S. (1991). *Improve your child's IQ and behaviour*. London: BBC Books.

Schoenthaler, S., Amos, S., Doraz, M., Kelly, M., & Wakefield, J. (1991a). Controlled trial of vitamin-mineral supplementation on intelligence and brain function. *Personality and Individual Differences, 12*, 343–50.

Schoenthaler, S., Amos, S., Eysenck, H., Korda, D., & Hudes, M. (1994). Controlled trial of vitamin-mineral supplementation on non-verbal intelligence of incarcerated males aged 18 to 25 years. Manuscript submitted for publication.

Schoenthaler, S., Amos, S., Eysenck, H., Peritz, E., & Yudkin, J. (1991b). Controlled trial of vitamin-mineral supplementation: Effects on intelligence and performance. *Personality and Individual Differences, 122*, 351–62.

Schoenthaler, S., Doraz, W., & Wakefield, J. (1986a). The impact of low food additive and sucrose diet on academic performance in 803 New York City public schools. *The International Journal of Biosocial Research, 7*, 189–95.

Schoenthaler, S., Doraz, W., & Wakefield, J. (1986b). The testing of various hypotheses as explanations for the gains in national standardized academic test scores in the 1978–1983 New York City Nutrition Policy Modification Project. *International Journal of Biosocial Research, 8*, 196–203.

Schoenthaler, S., & Eysenck, H. (1996a). The impact of different strengths of supplements on non-verbal intelligence among British school-children: A pilot study. Manuscript submitted for publication.

Schoenthaler, S., & Eysenck, H. (1996b). The effect of vitamin-mineral supplementation on non-verbal IQ of school-children, controlling for low nutrient concentrations in blood. Manuscript submitted for publication.

Sigman, M., Neumann, C., Jansen, A. A., & Baribo, N. (1989). Cognitive abilities of Kenyan children in relation to nutrition, family characteristics, and education. *Child Development, 60*, 1463–74.

Southon, S. (1990). Micronutrients and IQ. *Journal of Micronutrient Analysis, 7*, 179–91.

Southon, S., Wright, A., Finglas, P., Bailey, A., & Belsten, J. (1992). Micronutrient intake and psychological performance of school-children: Consideration of the value of calculated nutrient intake for the measurement of micro-nutrient status in children. *Proceedings of the Nutrition Society, 51*, 315–24.

Spitz, H. (1986). *The raising of intelligence.* Hillsdale, NJ: Erlbaum.

Stein, Z. A., & Kassab, H. (1970). Nutrition. In J. Worlis (Ed.), *Mental retardation* (Vol. 2) (pp. 269–82) New York: Grune & Stratton.

Stein, Z., Susser, M., Saenger, G., & Marolla, F. (1972). Intelligence test results of individuals exposed during gestation in the World War II famine in the Netherlands. *T. Soc. Geneesk., 50*, 766–74.

Vernon, P. A. (Ed.). (1993). *Biological approaches to the study of human intelligence.* Norwood, MJ: Ablex.

Winick, M., Meyer, K., & Harris, R. (1975). Malnutrition and environmental enrichment by adoption. *Science, 190*, 1173–5.

14　The resolution of the nature–nurture controversy by Russian psychology: Culturally biased or culturally specific?

Elena L. Grigorenko and Tatiana V. Kornilova

The historical context in which the individual objects, large or small, . . . appear in their true relative meaning is itself a whole, in terms of which every individual thing is to be fully understood in its significance, and which in turn is to be fully understood in terms of these individual things [Gadamer, *Truth and Method*, 1984, p. 156].

Any theory, even in the natural sciences, is inevitably influenced by culture. A Russian atomic physicist conducting an experiment with elementary particles (what could be more culturally neutral?) at an American university joked that Russian and American atomic physics are like cousins – they share some genes in common, making them similar, but they were raised in different homes, leading to dissimilarities.

It has often been noted that although science is thought of as the dispassionate pursuit of facts, in reality it is much more than that. As Le Vay, a gay scientist who discovered a link between human brain structure and homosexuality, stated in an interview, "Everyone has some place they are coming from; every scientist is a human being" (Marshall, 1992).

We all view the world from our own perspectives. According to Stephen Jay Gould (Barinaga, 1992), it is a pervasive fact of human existence that we as social beings find it extraordinarily difficult to step outside our own convictions and see them through the eyes of a detached observer. The same sentiment has been expressed with particular passion and persistence by psychologists. Indeed, it is becoming clear that it is difficult to discuss psychological constructs and create psychological theories without reference to the cultural environment of both behavior and the theory created to explain it (e.g., Ruzgis & Grigorenko, 1994). Nowhere is this more evident than in discussions of the nature and nurture of intelligence.

The issue of hereditary and environmental contributions to intellectual development has a special political connotation due to its relation to issues of education and life success. Every culture[1] raises the question of the relative contributions of nature and nurture in intellectual development and provides some type of answer to this question. At present, a range of theoretical positions exists, and considerable experimental data on the relation between genes and environment in intellectual development has been accumulated by psychologists in different countries, with little agreement on this issue.

Our purpose in writing this chapter is to point readers' attention to the existence of cultural and political biases in each particular resolution of the nature–nurture dilemma. Neither Russian, American (see Scarr, chapter 2, and Gordon & Lemons, chapter 11, this volume), or any other psychology is free of these biases. To make an even stronger claim, the way the issue is resolved in a given theory is influenced by the cultural context in which the theory appears (Kessen, 1990; Packer & Addison, 1989; Wertsch & Youniss, 1987). However, should a theory be considered flawed if it is found to be culturally bound? Lately, more and more psychologists have suggested that only by comprehending this cultural grounding can a fuller understanding of psychological functioning be achieved (Markus & Kitayama, 1991; Miller, chapter 9, this volume; Pepitone & Triandis, 1987).

To illustrate this statement, we will present an overview of the original work of Soviet/Russian[2] psychologists studying the issue of biological and social determinants of intellectual development. A brief historical account of the events that determined the character of Soviet/Russian psychology for three quarters of this century, and a discussion of key concepts developed, will provide the reader with an understanding of the cultural context and the categorical apparatus of the theories discussed in the main body of the chapter. In addition, we will canvass the major meta-theoretical assumptions underlying mainstream studies of thinking[3] in Soviet/Russian psychology. The second section is a summary of research targeted to understanding the role of biological, specifically genetic, determinants of intellectual development. And, finally, the last section deals with studies of the role of social-historical/cultural and microsocial influences on the development of thinking.

Thinking as generalized and mediated cognition of reality: Major concepts and general ideological-philosophical bases for research on thinking in Russian/Soviet psychology

The historical circumstances of the development of the Soviet psychology of thinking and intelligence

Soviet psychology, as well as many other sciences in the former USSR, developed under a heavy load of ideological attitudes and requirements.

Scientific programs quite often were evaluated and redirected through discussions in the mass media and at different party meetings, or by state decrees and resolutions.

This tradition was established soon after the Civil War of 1918–1922. In 1922–1923, a large group of Russian intellectuals, among whom were famous psychologists and philosophers Nikolai Berdiaev, Sergei Bulgakov, Nikolai Lossky, Ivan Lapshin, and Semen Frank, were briefly imprisoned and subsequently exiled to the West. They were charged with opposing the dominant ideology based on Marxism-Leninism by their "idealistic" theories. Former colleagues of the exiled scholars did not raise any question concerning their fate; it was as if they had never existed.[4] As Kozulin wrote, "The exile of the psychologists was in some sense just one more act of the 'class struggle,' but it provided a clear sign of the erosion of scientific ethics. While their colleagues were exiled, the rest of the psychological community remained silent . . ." (1984, p. 13).

This silence had a compelling cause: By the early 1920s, a psychologist could either follow the Marxist orientation or be exiled or subdued (Etkind, 1990). Remember, though, that the spirit of time was such that Marxist ideas were everywhere; they were easy to learn and accept. To paraphrase Karl Popper's words about Marx, one cannot do justice to young Soviet psychologists without recognizing their sincerity. Driven by the Marxist idea of creating a new type of a person – the liberated proletarian, with new morals, culture, and rules of conduct – Soviet psychologists stormed old bourgeois ideological and philosophical barricades on their way to building this new Soviet man.

No scientific struggle existed between Marxist and non-Marxist psychological theories: No philosopher-psychologists remained in the country who could argue with enthusiasts of Marxist approaches. According to Kozulin (1984), "The new Soviet psychology from its very beginning abandoned dialogue and therefore lost the invigoration that comes with pluralism. Later on it started to 'discover' idealists within its own number and treated them accordingly" (p. 14).

In the early years, struggles were waged between rival groups of Soviet psychologists. Each of these groups claimed that its methodology was the most scientific and the most purely Marxist. Because of the ideological importance of psychology as a science for raising a new type of a man, the party was closely involved in these debates. In 1930, Boris Anan'ev gave a talk at the Congress on Human Behavior that essentially ended these discussions. In closing his presentation, Anan'ev made a statement that became the basis of the meta-theoretical and theoretical work of Soviet psychologists for the next half a century: "The real founders of Soviet psychology as a dialectical-materialist discipline are neither schools nor trends ... but the founders of Marxism-Leninism" (cited from Kozulin, 1984, p. 21). This statement corresponded to the Party line, and from that time until recent years, it was accepted that Soviet psychology was to derive the categories of consciousness and behavior directly from the works of Marx, Engels, and Lenin. Marxism, as interpreted by Soviet psychologists, was a deterministic social doctrine that alone was "correct" as a philosophy of the proletariat. Psychologists who deviated from the Party line were forced to admit their "mistakes" in the early 1930s.

In 1936, educational psychologists' work came under fire. A decree by the Central Committee of All-Russia Communist Party (Bol'shevikov), known as the *Decree of Pedology*,[5] criticized the social impact of their work, condemning those psychologists who had been engaged in pedological studies and testing. Given that almost all work in educational psychology in the 1920s was called *pedology*, one can imagine the consequences of this decree. All forms of intelligence testing were forbidden.

Numerous trumped-up charges, including cosmopolitism (a tendency to analyze and refer to Western theories), idealism, and being an enemy of the people, complicated or even interrupted the lives of some psychologists and geneticists who had high official positions.[6] Along with pedology, social psychology, and psychotherapy, behavior-genetic research was prohibited in the USSR for more than 40 years. Marxism was imposed on all research as the only possible philosophical underpinning for any science, including the study of thinking.

A mosaic of approaches

Although the majority of researchers accepted the Marxist approach, emphasizing the social determination of the development of the psyche in general, and thinking in particular, their positions varied significantly. It would be wrong to state that this strong philosophical inclination, characteristic of Soviet psychology, could be explained only by ideological pressure from the state. In philosophical debates on the bases of Soviet psychology, one can see a spectrum of different explanatory principles and ideas related to mechanisms of social determination.

There are four good reasons for the variation of ideas and opinions in Soviet/Russian psychology, which existed regardless of state and party attempts to unify all approaches and merge them into one easily controllable theory. First, Soviet psychology incorporated various ideas developed in Russian philosophy, characterized by its humanistic ideals (for example, Losev's [1991] ideas of the role of sign, symbol, and myth in development; the philosophical and religious ideas of Berdiaev [1990] and Florensky [1991]) and scientific findings (the scientific ideas of Pavlov [1953] and

Sechenov [1952]), despite the exile of pre-Revolutionary Russian psychologists and philosophers. Vygotsky's theory, for example, appeared at the close of the Silver century of Russian culture, when there were no strict borders between sciences, arts, philosophy, and theology.[7] Vygotsky, a true man of his time, was a philosopher, literary critic, meta-theoretician, scientist, and psychologist. All of these aspects of Vygotsky's professionalism were expressed in his cultural-historical theory. The most important characteristic of his theory was its integration of knowledge about man with different approaches and methods of understanding human development.[8]

Second, traditionally, the Russian intelligentsia was very open to foreign ideas, and Russian scientists themselves never divided science into "ours" and "theirs." Even though a concept of bourgeois science appeared in the late 1920s, many Soviet psychologists who were forced to contrast their work to this false bourgeois science, did so with scientific sensitivity, rigor, and tact.[9] Such constant attention to psychological developments and events abroad also guaranteed variability in opinions among Soviet psychologists.

Third, differences among theoretical approaches were also due to the implicit division of psychological research into periods: (a) 1917–1936, the period prior to the 1936 State Decree on Pedology; (b) 1936–1950 (Dubinin, 1988), the period ending with Pavlov's session, where many Soviet psychologists were charged with cosmopolitism and a tendency to depart from Marxism; (c) the early 1950s to late 1980s, a relatively quiet and stable period in which Marxist-oriented mainstream psychology was still dominant but connections were established with Western psychology,[10] and some approaches deviating from Marxism began to develop; and, finally, (d) the current period, in which no ideological limitations exist, and Soviet/ Russian theories and ideas can be compared with foreign approaches without fear of life threat. These historical stages of Soviet/Russian psychology determined the nature of the theories. Thus, in the late 1950s, when the atmosphere in the country changed dramatically with Stalin's death, many psychologists digressed from the required earlier crass meta-theoretical principles (Zinchenko & Morgunov, 1994). For example, Rubinstein not only deviated from the principle of the unity of consiousness and activity and the principle of determinism (Rubinstein, 1957) but also inverted his own formulas "the external through the internal," arguing the internal to be dominant (Rubinstein, 1958). Thus, scientists tended to evaluate critically their own theories and to modify them, deviating from the official dogmas, adding to the variability in positions.

Fourth, experimental results that were obtained in studies implementing traditional methods in the psychology of thinking – such as concurrent verbal report, establishing physiological correlates of thinking processes, and concept formation – accidentally led to the discovery of unconscious determinants of thinking (e.g., insight). Findings in this area disturbed the clear theoretical pictures based on the social determination of thinking, because researchers' interpretations did not always correspond to the widely accepted psychological theories. As previously discussed, these official theories were often forced to flirt with Marxism (e.g., Leont'ev, 1975).[11] It became obvious that other theoretical schemes were necessary to incorporate these newly obtained experimental results in the body of psychological theory, once again bringing variability.

In sum, certain historical and political circumstances surrounding the development of psychology in Russia – in particular, the victory of Marxism-oriented psychology and the Decree on Pedology – led to many consequences, some of which are relevant to the subject of this chapter. First, IQ tests were prohibited; that is, intelligence, in its Western-like definition, was not present in psychological research. Second, it led to the appearance of the psychology of thinking, which defined thinking (according to Marxist philosophy – that is, as the highest form of mental reflection) as its main object of study. And, finally, the majority of Soviet/Russian psychologists recognized an unquestionable dominance of the social in the development of thinking. As a result, the concepts of intelligence and thinking have been studied in different theoretical and empirical contexts that were quite incomparable to the Western tradition. However, regardless of the heavy ideological pressure, Soviet/Russian psychology remained a rich scientific field, integrating many different approaches. A precise overview of the study of thinking in Soviet/Russian psychology would take a great many pages. Moreover, such attempts have already been made (e.g., Mattaeus, 1988). Our overview is selective and reflects the priorities formed in our own thinking via the influence of various social factors.[12]

The concepts of intelligence and thinking in
Soviet/Russian psychology

The term *intelligence* was introduced to Soviet psychology in the early 1920s and attracted considerable attention from professionals. A decade later, by will of the State and the Communist party, this concept was no longer mentioned or studied in mainstream psychology. Due to the official prohibition on intelligence testing, announced in 1936, the term *intelligence* was excised from the official psychology. As a result, until the late 1970s, the concept of intelligence was used mostly in two contexts: (1) research and theories on the development of the psyche in philogenesis (mostly in relation to animal intelligence), and (2) condemnations of Western tests of intelligence.

However, some related concepts remained in the terminology of Soviet/ Russian psychologists. For example, Vygotsky (1984) used the terms *intellectual development* and *cognitive development* interchangeably in his theory of higher mental functions.[13] It was, in part, due to such terminology, that Vygotsky was criticized for the "intellectualism" of his theory (Brushlinsky, 1968). Another psychological tradition uses the concept of *intellectual strategy* (Gurova, 1976; Kornilova & Tikhomirov, 1990). The broad meaning of this term includes a reference to the regulating role of an individual in his cognitive efforts and actions. Mostly, the researchers using this concept allude to: (1) the selective direction of problem solving and decision making; and (2) the goal-oriented nature of thinking activity and the correction of subjective plans in the process of problem solving. The concept *intellectual solution* is used to refer to a type of decision making that is mediated by thinking, in contrast to other forms of decision making (Kornilova, Grigorenko, & Kuznetsova, 1991; Kornilova & Tikhomirov, 1990).

Intelligence testing was also implicitly present in clinical psychology. Clinical psychologists in their diagnostic practice always used experimental methods, which, in reality, were analogous to subtests of intellectual assessments (classification, analogies, vocabulary, etc.). The interpretation of these results, however, was done in the context of understanding generalization and categorization processes, and motivational and goal-oriented regulation of activity (Rubinstein, 1970; Zeigarnik, 1962).

The prohibition on intelligence testing did not mean that Soviet psychologists did not use diagnostic tasks of children's intellectual development. Such tasks were designed as criterion tasks or educational tasks, not as level-oriented tests (Gurevich & Gorbacheva, 1992; Podgoretskaia, 1980; Talyzina, 1981; Zaporozhets, 1986).

The term *intelligence* reappeared in Soviet/Russian psychology in the 1960–1970s, when the first study using intellectual tests was performed at Sanct-Petersburg State University (Palei, 1974). Since then, Soviet/Russian psychologists have adapted and created many psychological tests of intelligence, and this concept is widely used in experimental psychological research. Intelligence, however, has not yet become popular as a theoretical construct. Its previous cultural image has not yet been restored, and the concept has not been integrated in the current conceptual apparatus (Zinchenko & Morgunov, 1994).

The term *thinking* was defined on the basis of Engels and Marx's work as the highest level of human cognition. In response to the demands of the party and state officials regarding the "true Soviet psychology," a number of Soviet psychologists (e.g., Rubinstein and Leont'ev) ventured to derive psychological categories directly from the works of Marx and Lenin. Although sensations were considered the sole source for thinking, according to Leont'ev (1978), thinking transcended the limitation of direct sensory reflection and enabled the human being to receive knowledge about objects, qualities, and relations of the real world that cannot be sensed directly. *Thinking* in its broad, philosophical, definition was viewed as the subject matter of logic, psychology, and neurophysiology. In the context of experimental research, psychologists developed a narrower definition of thinking as a process of problem solving.

The following meta-theoretical ideas, formulated in the context of studying thinking, were regarded as the basis of any theoretical or experimental study of thinking in Soviet psychology. Soviet psychologists stressed the primary role of *external* (social) sources and activity in mediating individual thinking schemes (Zinchenko & Smirnov, 1983). The principle of *historicism* (the principle of noninterrupting change in the societal conditions of development) was treated as one of the most important philosophical ideas in Soviet psychology. Based on this principle, psychologists criticized the "mistaken nature" (Smirnov, 1975, p. 253) of naturalistic and sociological theories.

In addition, when elaborating ideas regarding the social determination of thinking and its development, researchers stated that individual *mastery* of social-historical experience cannot be explained only by the accumulation of individual experience. Such processes as interiorization, translation of activities, internalization, and others were suggested to explain how thinking develops. The idea that mental formations evolve as individuals develop and have a material basis was represented by the concept of *functional systems of brain processes*.

Determinants of thinking were perceived to vary for: (1) philogenesis, (2) the cultural-historical development of the human psyche, (3) ontogenesis, and (4) the functional unfolding of an individual's thinking. Soviet psychologists studying philogenesis treated the mediated nature of thought by sign (or, more precisely, by the meaning of words) as the most crucial factor in the "humanization" of behavior. The process of mastering signs was included in the context of the analysis of human activity and consciousness (Leont'ev, 1975; Vygotsky, 1982b; Zaporozhets, 1986). Psychologists studying ontogenesis sought the *impetus* of development in the contradictions of a child's life, in a child's activity. Examples were contradictions between new tasks and old ways of thinking, and between what a child is allowed to do and what a child wants to do. Genetic factors were treated as prerequisites, limiting the possibilities for managing and manipulating the development of children's thinking through ontogenesis. Even though the limiting nature of genetically imposed individual characteristics was acknowledged by Soviet psychologists, the genetic basis of behavior was treated as something unfortunate, something necessarily negative and restrictive of positive cultural and societal influences. The link among different theories such as those of Leont'ev (1975), Rubinstein (1958), Kostiuk (1969), Zaporozhets (1986), and Luria (1969) was the statement that child *activity* – that is, the way in which a child interacts with the environment – circumscribes the conversion of external and inner determinants into psychological regulators. The physical maturation of the organism and its neural system is absolutely necessary for mental development, but this maturation is dependent upon the child's relation to the environment. In turn, a child's relation to the environment is *mediated by the world of adults*; that is, social determination is characteristic of all forms of human activity in ontogenesis.

Even though modern readers of many of the classical pieces on the mechanisms of *socialization* written by Soviet psychologists note the lack of experimental support for some conclusions, few would question the originality of their ideas. In terms of the conceptual apparatus of Soviet psychology, its research on the sources of individual variability in thinking cannot be classified into neat dichotomies of *social* versus *biological* or *genetic* versus *environmental*. The vast majority of research on thinking was done in the context of looking for social determinants of thinking, while ignoring the biological ones.

Before discussing research on the role of biological determinants of intellectual development, we have to recognize how unequally the roles of heredity and the environment were studied in the Soviet/Russian psychology of thinking. In this tradition the *environmental* always dominated the *hereditary*. In the next section, we will show that Soviet behavior genetics, from the very moment of its appearance, was such that research was designed, conducted, and analyzed in such a way that possible social influences to the development of a given trait were always taken into account.

Searching for the nature of intelligence

The late 1920s and early 1930s were a time of extremely rapid development in the study of genetics and behavior genetics in Russia. The whole pleiad of gifted biologists and geneticists (e.g., Vavilov, Chetverikov, Filipchenko, Timofeev-Resovsky, Agol, Slepkov, and Lobashev) was very active and productive at that time (see for review Dubinin, 1988; Frolov, 1988). In addition, biometricians developed a number of mathematical models for analyses of the variability of different traits in the human population (Filipchenko, 1929; Ignat'ev, 1934). Levit played an important role in establishing the new field of behavior genetics in Russia. He was one of the first of many Soviet scientists sent by the young Soviet government to various countries around the world to establish international scientific relationships. After returning from his successful trip to the USA, Levit headed the Medical Biological Institute in Moscow, where, along with purely genetic and medical genetic research, he started the first Soviet studies of behavioral genetics (Levit, 1934). All of the behavior-genetic studies conducted in the 1930s in the Soviet Union were performed under Levit's supervision.

Mirenova (1934) implemented the control twin method (this method assumes that one twin from a pair is given a special treatment, while another is not) in order to study the effect of psychomotor training on the general level of intellectual development. The research was carried on in the twin kindergarten affiliated with the Medical Biological Institute. The results showed that the intellectual level of the trained twins rose in comparison with the control twins. Another study (Mirenova & Kolbanovsky, 1934) sought to investigate the efficiency of different methods for the development of combinative functions in preschool twins. One of the members of each pair was trained by the methods of elementary figures (E) and the other by the method of models (M). The E method required the subject to reproduce with blocks certain given figures in which all the component elements and combinations were visible. For the M method, the same figures were presented but in the form of paper-covered models, in which the construction of the parts was concealed from the subjects. In the former case, the subject copied the model passively; in the second case, the subject constructed it actively. After 2 months of such training, a control task was presented in which a more complex figure was given to both twins. Analysis of these data showed that the free constructions executed by the M subjects were more complicated in intention and more complex in their execution than the constructions of the E subjects. The authors concluded that (1) the methods of active and independent construction develop the combinative functions much more effectively than do the methods of passive copying and imitation, and (2) for pedagogical practice, the method of active stimulation is the most desirable method of teaching.

Another group of studies was devoted to determining genotype–environment contribution to variability in IQ, measured by the Stanford-Binet (Luria, 1936; Luria & Mirenova, 1936; Luria & Yudovich, 1956). The researchers worked with two groups of children: preschoolers and school-aged children. They found a relatively high difference in MZ–DZ correlations for preschool-aged twins in comparison to the small or negative difference in school-aged twins. This led Luria to conclude that IQ variability in school-aged children had little or no relation to genetic influences. The authors also stated that mental processes in the course of development change in structure and become complicated and dependent on other factors. They emphasized that the relative importance of heredity and environment is not constant. The researchers also arrived at a more general conclusion arguing that tests themselves do not measure "natural general giftedness" but rather psychological traits that are complex in their structure and social in their genesis.

While working at the Medical Biological Institute and experimenting at the twin kindergarten, Luria collected interesting anecdotal facts and observations on twins' development that later were summarized in a book on speech and mental development (Luria & Yudovich, 1956). The authors also studied and emphasized the importance of the twin environment for overall mental and language development in twins.

Because the underlying ideology of behavior-genetic studies of normal psychological functions contradicted the official policy of the omnipotence of the Soviet state and the Soviet educational system in creating a new type of a man, such studies were completely prohibited in the late 1930s. However, the twin methodology continued to be used actively in medical genetics studies. Research started under Levit's supervision was continued by Gindilis and Finogenova (1976), Gofman-Kadoshnikov (1973), Efroimson (1968), Lil'in (1975), Trubnikov (Trubnikov et al., 1993), and others (see for review Sokolova, Gofman-Kadoshnikov, & Lil'in, 1980) in the framework of medical genetics. Because of its use in medical studies, the twin methodology survived through the 1930s and 1950s and was rather quickly reintroduced to the psychology of normal behavior in the 1960s. In addition, as mentioned previously, clinical psychologists used tests that essentially were intelligence tests (tests of analogical reasoning, vocabulary, verbal fluency, etc.). Some of these tests were used in twin-concordance studies of psychiatric disorders (e.g., Alfimova & Trubnikov, in press).

Behavior-genetic studies of cognitive functions were reintroduced into Soviet psychology in the late 1960s. The motivation for studying heredity and variability in cognitive function came from questions about the neural system properties (Nebylitsyn, 1966; Teplov, 1985). If, as argued by the theory of differential psychophysiology (Nebylitsyn, 1966; Teplov, 1985), properties of the nervous system determine to a certain degree individual differences in various psychological characteristics, it seemed logical to assume that they were not only stable and innate but also related to genotype.

The emphasis of experimental studies performed in the framework of differential psychophysiology was on studying gene–environment co-

contributions to EEG and evoked potentials measured in the context of solving various cognitive tasks. EEG, evoked potentials, and other analogous parameters remained the main objects of research. Some knowledge of hereditary and environmental contribution to cognitive functioning was also obtained but was considered to be of secondary value. The shift to studying purely psychological cognitive characteristics took place slowly and was accompanied by theoretical discussions of whether the study of genotype–environment co-contributions to the variability of human psychological traits was at all possible.

The first twin studies of normal psychological functions performed after 35 years of prohibition were carried out at the Laboratory of Psychophysiology and the Development of Individuality, led by Inna Ravich-Shcherbo, a former student of Leont'ev and one of Teplov's colleagues. Ravich-Shcherbo has been credited with reestablishing behavior genetics in Russian psychology. In the early 1970s, she created a laboratory that concentrated on behavior-genetic studies and supervised many behavior-genetic dissertations (e.g., Dumitrashku, 1992; Grigorenko, 1990; Iskol'dsky, 1988; Kiriakidi, 1994; Semenov, 1982; etc.).

In the 1970s, however, experimental research always had to be accompanied by theoretical discussions of the methodological limitations of behavior genetics in psychology, because Marxist emphasis on dominance of the social over the natural was still extremely popular at that time. However, the tone of discussion had changed dramatically since the 1930s, and many scientists raised their voices, urging a reconsideration of the role of genotype in mental development (e.g., Efroimson, 1976; Frolov, 1983; Krushinsky, 1977; Los' & Faddeev, 1981; Malinovsky, 1977; Shishkin, 1979). At that time, even though the primary goal of research in the field of behavior genetics was to estimate the genotype's contribution to various psychological functions, aspects of the development of monozygotic and dizygotic twins (such as their interpair relations, role distribution in twin pairs, and conditions of living) were generally taken into account (e.g., Iskol'dsky, 1988; Semenov, 1982).

Since the middle 1980s, the main emphasis in behavior-genetic studies of psychological traits has changed; intelligence and special abilities have developed into the main object of research (Egorova, 1988). Since that time, research in this area has become well recognized and independent. Because this line of research originated in the field of differential psychophysiology, the main emphasis remains on psychological traits and the psychophysiological characteristics that mediate the relations between the genome and psychological traits. Theoretical debate also revolves around issues concerning the specificity of psychophysiological and psychological human traits. If a researcher includes these characteristics in a genetic study, is he or she inevitably treating these traits in a simplified way, disregarding their complex nature? The complexity of these characteristics may, for example, manifest itself in a situation in which a characteristic is identical in different settings at the phenotypic level but involves completely different mechanisms of regulation in these different settings – that is, differs in terms of relative genotype–environment co-contributions to its variability.

The hypothesis suggested above has been verified by behavior-genetic analyses of evoked potentials (Mariutina, 1978, 1988; Mariutina & Ivoshina, 1982). Highest heritability estimates are found for evoked potentials registered for reactions to simple sensory stimuli, such as a light flash and the image of a chess board, while lowest heritability estimates are found with evoked potentials registered for reactions to a word or a drawing that can be easily related to a word. In addition, Mariutina and her colleagues showed that attentional mobilization increases the proportion of genetic control of evoked potentials. Moreover, relative co-contributions of genetic and environmental components of variability may vary, depending on subjects' perceptual strategies and age. Specifically, a higher genetic contribution was found for 10–12-year-olds than for both older and younger age groups.

An analogous situation at a psychological rather than a physiological level of analysis was discovered by Panteleeva (1977). Studying a decision-making situation with two alternatives, she evaluated the co-contributions of genetic and environmental components to variability at the initial stages of training (e.g., at the phase of "entering the task") and at the end of the training, when decision-making skills in the situation were automated. Panteleeva found that the heritability estimates were higher for characteristics measured when the skills were automated than for the characteristics measured at the initial stage of training. Only measurements of the time it took to begin a motor action after receiving a task and of the delay between the evoked potential appearance and the actual beginning of the motion were controlled genetically to the same degree at both stages of skill formation.

These two phases of skill formation differ by the degree of conscious control of action regulation: The training stage requires more conscious control than later when the skills are automated. In other words, with this task, for all but two temporal characteristics, the co-contribution of genes and environment is mediated and changed by psychological factors.

The idea that restructuring the inner mechanisms of a mental function may lead to a shift in heritability estimates was initially formulated by Luria and Morozova in their study of natural (nonmediated) and higher (mediated) memory, performed in 1936 but not published until 1962. The main idea of this study was that "the nature of every mental function (in other words, its relation to the genotype) changes in the process of human mental development as well as its structure" (Luria, 1962, p. 16). Luria suggested that the natural mental functions are controlled by genotype to a greater degree than the higher ones. He also hypothesized that genetic control will decrease over the life span because higher, social mental functions dominate at later stages of development and adulthood. It is remarkable that Luria formulated a hypothesis about the unequal influences of heredity and environment at different stages of individual development as early as 1936. Similar research was not conducted in the West until the 1970s. Luria's justification of this hypothesis was different from the current explanation of these findings and was given in the context of Vygotsky's cultural-historical theory.

Averina (1980, 1983), who studied three groups of twins of different ages (second-, sixth-, and tenth-graders), collected more data supporting Luria's hypothesis regarding the unequal influence of genes and environments at different stages of development. She found that variability in visual memory is genetically determined in all three age groups. In addition, heritability estimates for recognition of previously observed stimuli were higher across the three ages than h^2 (a measure of heritability) for active reproduction – that is, for the reconstruction of previously viewed patterns. However, the MZ correlation for the generalized mnemonic ability was higher than the DZ correlation only in the youngest age group. Variance partitioning determined an additive and a shared environment component at the youngest stage of development, and only environmental components for both older groups. However, given small sample sizes, these findings should be interpreted with caution. These results, nevertheless, provide further evidence for the argument that it is inappropriate to talk about the heritability of a psychological trait in general. Instead, heritability estimates should be obtained for various components of the studied traits, in a variety of situations in which the trait is included in psychologically different functional structures, and at different developmental stages.

This idea, initially formulated by Ravich-Shcherbo in the late 1970s (Grigorenko & Ravich-Shcherbo, in press), determined the ideology and structure of all behavior-genetic studies in Russia from the late 1970s to the early 1990s. Researchers studied the phenotypic variance of cognitive traits in groups of twins from a wide variety of backgrounds and found differences in h^2 estimates in different groups of adult and adolescent twins for IQ and stylistic cognitive characteristics (Egorova 1988; Grigorenko, 1990; Grigorenko, LaBuda, & Carter, 1992; Iskol'dsky, 1988). In a study of 21–25-month-old infants, Kiriakidi (1994) investigated the difference in genotype–environment contributions to IQ variability in groups of twins attending child care or staying home with their grandmothers. She found that the presence of grandmother as a significant caregiver was associated with higher IQs in the studied twins. In addition, in partitioning the environmental component of the phenotypic variance, Kiriakidi demonstrated the presence of the gender effect: For boys the major component is shared family environment; for girls, unique environment.

Bogoyavlenskaia and Susokolova (1985) obtained heritability estimates for various intellectual components of such elaborate activities as playing cylindrical chess (the authors constructed a cylindrical chessboard by joining the top and the bottom of a regular chess board). Their heritability estimates were very low, suggesting a strong environmental component in the variability of the studied traits. Grigorenko (1990) showed that such different indicators of the hypothesis-making process as dynamic components (speed of creating and testing hypothesis) and contextual components (the content of hypothesis and their elaboration and flexibility) have different heritability estimates. It was found that h^2 estimates for dynamic components are higher than estimates for contextual indicators.

In a study of preschool twins, researchers (Talyzina, Krivtsova, & Mukhamatulina, 1991) in addition to measuring IQ investigated the influence of special methods of teaching on fluctuations in heritability estimates for dynamic intellectual characteristics (reaction time, mental orientation speed, etc.). The authors showed that teaching may influence heritability of intellectual characteristics, such as reaction time, which traditionally have been regarded as highly heritable.

Zyrianova (1992) compared genotype–environment contributions to the variability of phenotypically stable psychological traits at different ages. Researchers also used laboratory conditions in order to vary the structure of psychological functions while their phenotypic expression remained the same. They did this by changing the degree of mastery of an activity in which the psychological characteristic manifested itself (e.g., they introduced a new psychological function to a subject who was then trained until the function became automated). In this case, heritability estimates before the beginning of the learning process and after its completion were found to vary with different stages of mastery (Zyrianova, 1992; Malych, Egorova, & Piankova, 1993).

Dumitrashku (1992) used the family studies methodology to compare children with two or more siblings from low SES families to children from families with singletons, matched for SES and parental age. She found that children from large families tended to be behind peers from one-child families on all measures of cognitive development.[14] Moreover, older children from large families tended to do better than their younger siblings.

Only one main effect of gender was found: Boys scored higher on Perform-ance IQ measures than girls. However, these effects were mediated by parental characteristics that were present for almost all measured traits. In general, the cognitive characteristics of parents were more highly correlated with the cognitive characteristics of first-born sons than later-born sons or daughters. Correlations with fathers' cognitive scores were more highly mediated by children's birth order and gender than correlations with moth-ers' scores. Measures of child cognitive development were more likely to correlate with the father's IQ than with any other measures. The observed pattern of correlations suggested that a low paternal IQ was more detri-mental for children's cognitive development than a low maternal IQ, espe-cially for first-born children. Correlations with mothers' scores had a more general nature and were not significantly different for younger and older boys and girls. A tendency did exist, however, for maternal characteristics to be more strongly related to the characteristics of third-born children. Analyses of parental attitudes showed that in large families, at least one of the parents tended to be more authoritarian than the average parent in small families. Fathers' authoritarian style of parenting, however, corre-lated positively with children's cognitive scores, while mothers' authoritar-ian style correlated negatively. In addition, both parents in large families tended to avoid infantilization of their children's behavior. Dumitrashku concluded that in this sample, which was representative of the lower-SES population in Russia, birth order and family size were negatively related to children's cognitive development: Children from large families tended to have a lower level of cognitive abilities, and younger siblings tended to be behind their older brothers and sisters. It was also found that measures of shared and nonshared environment did not appear to mediate the detri-mental effects of family configuration.

To summarize, because of the historical and cultural peculiarities of the situation, Soviet/Russian behavior-genetic studies have been conducted in a manner that is quite different from traditional Western approaches. In particular, the main emphasis of behavior-genetic studies of cognitive de-velopment in Russia was on studying how genotype–environment contribu-tions to cognitive functions differ depending upon the internal structure of the studied functions (e.g., novel versus automated actions) and the social microconditions of their development. Early on, Russian psychologists dis-covered that heritability estimates of the same phenotype, when studied as a component of different psychological functions, vary significantly. It was also shown that heritability estimates of cognitive functions may vary de-pending upon microsocial conditions of the functions' development (e.g., h^2 for IQ measured in twin pairs with a distinct leader versus pairs with reciprocal relations), family environment (e.g., parenting styles and caregiver situation), and developmental stage (the degree to which higher mental functions have been formed). Research using the control-twin method found a high degree of flexibility in heritability estimates for cognitive functions in conditions of directed training. Thus, in Russia, even behavior-genetic studies were designed, conducted, and analyzed in such a way that the importance of the environmental component in cognitive development was always stressed.

Searching for social determinants of thinking

As stated above, the assumption that the development of thinking is socially determined underlies virtually all Russian research on thinking. The concept of social was divided into two main components (Burmenskaia, in press): (1) social-historical experience, meaning culture as a source of reproduction of human abilities, and (2) aspects of specific microsocial environments (e.g., school, familial, situational). Correspondingly, studies of the mechanisms of social determination of thinking can be classified into three groups: (1) theories dealing with the means and mechanisms of social-historical/cultural determination of thinking, (2) theories of environmental conditioning of the development of thinking, and (3) approaches to application of these various theoretical ideas in educational practice.

The social-historical/cultural determination of thinking

As examples of theoretical studies that explore the social-historical/cultural mechanisms underlying the development of thinking, we will consider: (1) the theory of practical thinking (Teplov, 1990); (2) studies of the inner and outer determination of scientific thinking and scientific creativity (Yaroshevsky, 1971, 1981); (3) ideas on "raising" of thinking (Zinchenko & Mamardashvili, 1977; Mamardashvili, 1992a,b); (4) research on the active nature of thinking (Smirnov, 1985, 1994); and, finally, (5) an approach that emphasizes the cultural-religious determination of thinking (Bratus', in press; *Psikhologiya i novye idealy nauchnosti*, 1993).

Teplov's paper, written in 1943 during World War II, was the first theoretical study to investigate issues regarding the social-historical determination of thinking. The paper was devoted to understanding practical thinking in military leaders in critical situations. In order to reconstruct the peculiarities of thinking in situations of extreme difficulty and high levels of responsibility, Teplov relied on military historical materials, military commanders' autobiographies, and literary pieces. Stressing the most remarkable traits of military commanders and the historical and cultural determination of their thinking, Teplov considered a number of personalities and historical and geographical events, describing such different military commanders as Alexander of Macedonia, Julius Caesar, Hannibal, Napoleon, Suvorov, and Kutuzov. In the first part, Teplov analyzed the situations that military commanders face and described their activities and the tasks these activities should accomplish. Then the author analyzed the psychological characteristics of military commanders, as demonstrated in their professional activities. In the third part, Teplov considered the relations between these characteristics, attempting to draw a complete psychological portrait of a military commander. In addition, in the paper, Teplov treated the social-historical situations in which every military commander acted as initial determinants and stimuli for commanders' intellectual activity.

It is also important to note that Teplov compared the intellectual, emotional, and self-regulatory characteristics of military commanders. It is trivial to say that a distinctive military commander can be characterized by a remarkable intelligence and a strong will. The question, however, is: Which is more advantageous and crucial for a talented commander? Discussing this issue, Teplov wrote: "I have never seen any discussion in which this question was ever resolved in intelligence's benefit. Usually the question itself is formulated in order to state the dominance of strong will in a military commander's activity" (1985; vol. 1, p. 288). Teplov himself, however, disagreed with the emphasis on the dominance of will, stating that practical intelligence is a unity of intellectual and volitional components. Knowing that his interpretation was contrary to most widespread ones, Teplov supported his statement with a citation from Klauzevits: "Resoluteness is obliged by its existence to a special type of mind" (Teplov, 1985; vol. 1, p. 251).

This paper established a whole school of the study of thinking devoted to the issues of how personality, reacting to a particular social-historical situation, regulates processes of thinking. However, the experimention based on this approach has not reached in its originality, deepness, and novelty the level of generalization established by Teplov.

One of the most interesting concepts, developed by Yaroshevsky (1981) in the context of his work on the mechanisms of social determination in scientific thinking, is that of the *categorical regulation* of scientific thinking. According to this theory, with a change of paradigms in any system of knowledge, the greatest shift is observed in the categorical apparatus used rather than in the logical basis of knowledge. Using as a basis for his studies the development of psychological science worldwide, Yaroshevsky showed how a shift in scientific interpretations leads to a change at the level of formal logical operations – that is, in the objective logical structure of scientists' thoughts. In this sense, the individual creativity of scientists is inevitably heavily influenced by shared, historically developing, categorical networks. Consequently, individual perceptions of reality are objectively determined, independent of the originality of the ideas developed by a creative personality (Yaroshevsky, 1971, 1981).

Merab Mamardashvili, one of the most interesting Soviet/Georgian philosophers, theorized about mechanisms for the social-cultural determination of thinking. He wrote on the objective method in psychology (Zinchenko & Mamardashvili, 1977) and in the content and forms of thinking (Mamardashvili, 1968) and on the importance of the raising of thinking (Mamardashvili, 1992a,b). Among the extremely rich psychological ideas he left behind, we mention only two. One is related to the role of science and culture in the development of thinking, and the second deals with issues of the *precision of thinking* in intellectually developed people. Science, as well as culture, is normative (Mamardashvili, 1992b). Both embed the possibility of mastering knowledge; that is, they are connected not to a person but to *a possible person*, one who can potentially develop from a given person. Mamardashvili related a person's mastery of scientific thinking to this person's overcoming his innate propensities at the moment of his "second birth" (Mamardashvili, 1992b). Scientific activity involves oscillating movements in two different directions: toward new possible structures (mastering the norms of cultures and science) and, in the opposite direction, toward destroying these structures for the sake of creating new ones. Cognition in this sense is experimenting with forms, not the forms themselves. When a person defines himself in an action of thought, while he is dependent on cultural and scientific norms of thinking, he is simultaneously free to choose or reject those norms.

Not every individual "producing thoughts" demonstrates relative freedom of individual thought from societally habitual "laws of ideas or moral structures." The first instance of modern intellectual work, according to Mamardashvili, was that of the Russian classical writer Fedor Dostoyevsky. Dostoyevsky allowed himself to think about a number of issues in a way that was not typical of the public consciousness at his time (for example, his views on the oppressed and abused in society).

Precision of thinking refers to an ability to think through an issue or a problem, from all possible angles, realizing the complete responsibility of a thinker for the level and products of his thinking. From this point of view, an individual when thinking responsibly cannot use excuses such as "I did not think," "I did not mean," "I could have not guessed" (Mamardashvili, 1992a, p. 130). When the Russian intelligentsia capitulated in the face of evil at the beginning of this century, its true betrayal was not its social sin, but its "slovenliness of thinking" (Mamardashvili, 1992a, p. 132). Russian intelligentsia failed to "create a sphere of autonomous thought, autonomous spiritual life, behind the back of which there is tradition, . . . in which you immerse yourself and for which you feel personal and professional responsibility" (Mamardashvili, 1992a, p. 132).

In this sense, *precision of thinking* can be interpreted as the moral obligation to finish thoughts, to think everything through. Consequently, social factors that determine thinking in the context of this framework are themselves shaped by the cultural traditions' ways of reproducing thought. In addition, the standards of thinking defined by culture and science imply that a person perceives them as something more highly ordered and superior to his own ways of thinking. Therefore, a thinking person develops from a person with spontaneous attitudes toward himself and others into a *possible person* – that is, a person mastering his second nature via the qualities of his thinking.

These ideas on fusing the artificial and the natural in individual thinking and on the roles of cultural systems in the social determination of thinking have not been transformed into experimental tasks. Remaining a philosopher, Mamardashvili, due to his unique personality, strength, and depth of thought, himself became the ideal possible person. In discussing his work, we wanted to show an alternative understanding of the social nature of thinking in Soviet philosophy and psychology that is receiving much attention in the current circuit in the search for criteria for rationality of human knowledge and cognition (Mamardashvili, 1984; *Psikhologiya i novye idealy nauchnosti*, 1993).

Other theorists studied the social-historical/cultural determination of thinking by focusing on the active nature of thinking (Smirnov, 1985, 1994). According to Leont'ev and his students and colleagues (e.g., Leont'ev, 1972; Smirnov, 1985), the dualism of environment and heredity, which inevitably occurs in the study of human development, can be avoided by introducing the concept of activity, which takes into account both the psychophysiological particulars of each child (determined primarily genetically) and unlimited environmental variation. Both of these factors influence rather than determine development, because the positive or negative constitutional qualities are mediated by a child's social environment from the very moment of birth. The reason activity holds such a central position is that, within this theory, activity is viewed not simply as a combination of various physical actions mediated by mental processes. Rather, object-oriented activity forms the connection between the individual and the world, and this connection is bidirectional. That is, the individual acts on and changes the environment through activity, but, as a result, the individual is also changed as he or she absorbs a wider range of experiences from the environment. According to the theory of activity, it is necessary to study neural-physiological aspects of the ontogenetic development of thinking; however, the main area of psychological research remains the content and conditions of a child's optimal interactions with the world.

Dr. Bratus', Leont'ev's student and a colleague of Mamardashvili, is developing a humanistic paradigm for the study of thinking, incorporating the ideas of Russian philosophers on the nature of man and his soul and on the concept of spirituality (*Psikhologiya i novye idealy nauchnosti*, 1993). According to this author (Bratus', in press), the Russian orthodox tradition is the most natural cultural tradition for Russian people, determining their structure of personality as well as peculiarities of thinking. Bratus' argues that the Russian consciousness is one specific manifestation of Orthodox thinking. Understanding this will save the nation from the spiritual egotism and distortion that equates Russian nationalism with religion itself. The author states that this salvation can be achieved by a unification of church, state, and science, where church is one of the most important social institutions to influence the formation of consciousness.

The environmental determination of thinking

Social environment as a determinant of the development of thinking. Following chronological order, this review of Soviet psychological theories, which view the social environment as a determinant of the development of thinking, will begin with an analysis of Kornilov's work, because Kornilov was the first psychologist to develop a theory of the interaction of heredity and environment as a basis for "Marxist differential psychology" (Kornilov, 1980). He also introduced the *biosociological concept* of reaction[15] that supposedly differed in various societies and could be developed in its "correct" form in members of a socialistic society. However, Kornilov's *reactology* was so mechanistic that even historians of psychology do not treat it as an original interpretational scheme.

A more concrete and well-developed attempt not only to build an objective psychology, but also to study mechanisms for the social determination of thinking, was Blonsky's theory, developed in the 1920s–1930s (Blonsky, 1935a,b).

According to Blonsky, the mind has an ability to gain successful individual experience – that is, the ability to learn: The mind is seen as nonemotional and noninstinctive; practical interest is its main engine. Thinking is a process, and the way one thinks shows one's mental level. Thinking as *something developing* can be represented by three stages: imaginative (primitive), visual, and systematic.

According to Blonsky, during secondary-school education, only memory reaches its developmental summit. Thinking remains immature and can be fully developed only at higher levels of education. Blonsky interpreted changes in thinking caused by learning and school experience as an increase in the volume of thinking (e.g., mastering grammar, a child learns things unrelated to his own experiences, things he does not have to hear or see for himself), and a shift from visual to abstract (systematic) thinking occurs (especially influenced by studying math). Formal logic creates the necessary background for mastering dialectics. Consequently, school systems directly influence the psychological characteristics of pupils' minds. Such a simplified interpretation of the mechanisms for the social determination of thinking evolved in the context of challenging Vygotsky's approach, in which Blonsky tried to criticize the distinction between natural and higher mental functions.[16]

Vygotsky, called a "Mozart of psychology" (Tulmen, 1981), created a cultural-historical theory that, on a surface level, does not appear to have been influenced by the forces of Soviet or worldwide psychological thought, nor the methods of his time. As a modern Russian psychologist wrote: "similar to 'Don Quixote,' cultural-historical psychology was possible but not inevitable" (Puzyrei, 1986, p. 10). In Vygotsky's work, thinking itself is regarded as possible but not inevitable. Thinking is not simply one of the functions of the human mind but is a process related to mastering sign and its function.

Vygotsky worked as a professional psychologist for only 10 years.[17] However, his writing does not seem either unprofessional or archaic and remains a focus of active debate. His ideas have influenced many modern theories to an extent frequently unrecognized by their authors because of the high degree to which Soviet/Russian psychology emerged from Vygotsky's theory.[18]

Vygotsky's scientific path was quite unusual. In 1925, he began to try to develop a concrete psychological study of consciousness. This idea seemed extremely brave and challenging in the context of the struggle between the old subjective-empirical psychology, the main object of which was consciousness, and the objective orientation of new psychology (behaviorism and reactology). As has been mentioned, with the ongoing debates on the appropriate objects of study in Marxist psychology, scientists of the 1920s and 1930s had only two options if they were to survive – to choose to study the materialistic basis of mind and thinking or to recognize the dominance of social (cultural) mechanisms in determining cognitive development. Given those options, Vygotsky chose culture (Etkind, 1993).

Vygotsky was the first Russian scientist to develop a complete Marxist-oriented psychological theory. In studying thinking and development, Vygotsky created an approach that became the basis for 50 years of experimental research.

Vygotsky's theory has attracted attention because it provides an alternative to classic psychological approaches in regard to the role of social factors in the development of thinking and the role of tools and signs (especially human language) in the formation of human psychological processes. Vygotsky's writings on thinking emphasized the use of developmental analysis and made the claim that higher (that is, uniquely human) mental functions[19] have their origin in social life and are heavily shaped by the historically evolved tools and sign systems (especially human language) that mediate them. No attempt is made to summarize Vygotsky's theory, which is presented in six volumes of his writings; instead we briefly touch upon those aspects of his writing on thinking mentioned previously (Vygotsky, 1982b).

Preoccupied with the idea of developmental analysis of higher mental functions, Vygotsky developed the historical-genetic[20] method (Vygotsky, 1982b, 1983, 1984), which allowed him to obtain results that would be unobtainable using the regular cross-sectional method. Along with this method, Vygotsky developed a new psychological hypothesis regarding the mediated nature of higher (cultural) mental functions in contrast to natural functions that differ in their structure, nature, and control mechanisms. The concept of mediation implied mastering psychological tools as a means to mastering psyche. The tools are initially connected to a situation involving communicative social interaction with a partner but are then turned by the child into a means of self-regulation. In regard to thinking, an example of such a means is a word used as a sign. The process of mastering stimuli means is, in its essence, the development of self-regulation, a transition from external signs to inner signs. Such widely different things as "making a knot in order to remember something" and a word meaning are both artificial things intentionally created by man and are both cultural elements and have dialogical natures in the sense that they can originate only in human communicative interactions.

Children do not invent, but they also do not simply learn to manipulate signs. Learning and intellectual discoveries are embedded in the history of the development of each child's operations on signs. The development of verbal thought takes place at the intersection of two roads: Speech becomes intellectualized, and intelligence becomes verbalized (after age 2). After mastering words, children discover new ways of manipulating and dealing with objects. The inner formation of sign relations occurs over a long period of time; that is, each child's thought develops through understanding the relationships between a sign and its meaning. The particular way in which a child uses a word is determined by the level of generalization and conceptualization: Initially, concepts are syncretic concepts, then complexes, then pseudo-concepts, then functional concepts, and only finally, much later, true concepts. Scientific concepts, reflecting essential qualities of objects and learned systematically at school, outstrip everyday concepts, which develop as empirical generalizations of individual experience. This finding – that the understanding of scientific concepts developed more quickly than everyday ones – led Vygotsky to formulate the concept of the zone of proximal development (Vygotsky, 1982b), a zone in which the actual level of intellectual development can be changed. Learning in this sense is ahead of development because it stimulates psychological functions (e.g., memory) that are still at the stage of maturing – that is, in the zone of proximal development. Accordingly, in the zone of proximal development, children can change their levels of achievement with a teacher's help – that is, can do with help what they cannot do alone. This potential level of mental development reflects the change of systemic connections in consciousness, which is a result of the stage development of consciousness.

Wertsch and Youniss (1987) saw the historical-cultural theory as primarily applicable to the formal school setting. According to these authors, to the extent that Vygotsky was trying to outline a psychology of pedagogy, his tendency to focus on the skills required in formal schooling contexts was legitimate. However, Vygotsky, his colleagues, and his students viewed their overall enterprise as one of constructing a theoretical approach that had broader applications than merely education. For example, Smirnov (1975), the author of a classic book on the history of theoretical discussions in Soviet psychology, wrote that the concept of the meaning of a word as the unit of analysis in thinking, developed by Vygotsky, is the key to understanding the nature of human consciousness as a whole. In this context, verbal thought is "a social-historical form of behavior" (Smirnov, 1975, p. 175). That is why the issue of the relationship between scientific and everyday concepts cannot be simplified to the issue of child-developed individual generalization versus socially given concepts, because both types of concepts are learned in a social situation involving communication with adults. The difference between scientific and everyday concepts lies in the degree of one's awareness and systematization of them. The system of mental functions, developing in a child's consciousness along with the mastery of scientific concepts, changes everyday concepts as well.

Vygotsky and his associates lived at a time when the major scientific myth shared by Soviet scientists was that they could override human nature and create a system of upbringing that would result in a new type of a man exemplifying only the best of humanity. This ethno- and historicocentrism resulted in the use of sociohistorically specific concepts in an attempt to examine the development of human consciousness in general. Vygotsky and his colleagues and students viewed themselves as creating a general theory of development, with the leading role belonging to culture and society.

Two circumstances led to the wide acceptance and use of Vygotsky's methodological and theoretical approach. Vygotsky's group, in Kharkov, survived the repression in the 1930s and upon their return to Moscow had no serious theoretical or experimental competition in the field. In addition, their theoretical approach corresponded to the overall Marxist dogma of the dominance of social influences in development. Hence, ideas formulated by Vygotsky and his methodological approaches[21] were successfully developed in a variety of different theories. For example, his ideas on the mediated nature of psyche were developed by Leont'ev in his theory of activity (Leont'ev, 1975); his semiotic ideas were used by the famous Russian producer Sergei Eisenstein (Leont'ev, 1982); and his thoughts on the systemic structure of consciousness, functional organs, the localization of higher mental functions, and the disintegration of consciousness were cultivated by Zeigarnik (1962), Luria (1969, 1979), and Rubinstein (1970).

It is important to note that even though higher mental functions in Vygotsky's theory were contrasted with natural mental functions, his approach focuses on the opposition of *social and individual* rather than *social and biological*. The *higher–natural* dichotomy was criticized by Rubinstein and his students (Brushlinsky, 1968; Gurevich & Gorbacheva, 1992; Rubinstein, 1958) when they pointed out that in this interpretation, social could be equated with external. This, according to the critics, was a wrong interpretation; according to Marx, inner conditions could not be limited to pure biological conditions, because "human nature itself is a product of social history" (Brushlinsky, 1968, p. 97).

A reinterpretation of the concepts of *higher* and *natural* functions is suggested by Puzyrei, a representative of the third generation of Vygotsky's students (Puzyrei, 1986). Puzyrei treats the act of mastering a sign as creating and building that sign, based on its particular meaning and uses to the individual. He stresses the manipulative nature of culture and of culturally determined mediating tools (e.g., symbols and signs). When mastering a stimulus, "an action occurs that is organized in a special form, the performance of which alone allows development to happen" (Puzyrei, 1986, p. 85). These actions allow the transformation of one's innate natural functions into higher ones and are the means by which humans regulate behavior, and by which the human psyche is reorganized. Such an understanding of the higher–lower or cultural–natural contradiction suggests a change in our understanding of the mechanisms of higher mental functions: The objects of study are not the natural processes of mental functioning but rather systems of actions targeted toward the transformation of mental functioning.

Puzyrei's interpretation of the concept of *individual* in Vygotsky's theory assumes the dialogical nature of human consciousness: It is possible for a man to relate to himself, as to another person, and in this process to change his self-regulation of mental functioning.

Social situations involving communication with others as a determinant of the development of thinking. Vygotsky's idea that thought is born not from a word or another thought but from the "motivating sphere" of our consciousness has been the source of many different approaches to understanding the motivation for thinking and the relations between the mental processes of a subject as an individual and as a bearer of socially determined forms of thinking.

In Russian, the notion of consciousness might be translated as "co- (or 'shared') knowledge," determined by the coexistence of an individual and human culture. This notion was widely accepted among Marxist psychologists. On the way to mastering one's culture (or social-historical experience), human thinking develops. Within the inner activity of thinking, an individual forms inner schemata which, even though they do not correspond exactly to external activity, link individual thoughts to forms of thinking developed by humanity. The transition between these forms, according to Vygotsky, is a shift from shared thinking to individual thinking but is mediated by social structures of thoughts through the semantic function of a word as a sign.

It is important to note that the most frequent criticism of Vygotsky's theory was complexity of the mediated nature of an individual's thought by systems of signs – that is, the difficulty in finding a correspondence between mediation by activity and mediation by sign in a single process of the development of individual consciousness. Brushlinsky (1968) wrote that the Achilles' heel of cultural-historical theory was the idea that signs transform interpersonal interaction into individual spiritual life. According to this author, Vygotsky's interpretation of signs led to intellectualism (the assumption that consciousness is prior in its development to activity) and, consequently, to idealism[22] (Brushlinsky, 1968).

In the context of a concrete psychological study, however, the question of the mediated nature of an individual's thought can be reformulated into a study of the internal dialogue of thinking, and into a study of internal modification of thinking through mastering higher-order psychological tools (for example, informational technologies). Initially, Kuchinsky, a Belorussian psychologist, was one of the first to explore the dialogical nature of thinking (Kuchinsky, 1983). He attempted a holistic analysis of the structural involvement of speech in pure thinking activity. It is important to note that the author did not formulate the task as one of describing the structural-functional peculiarities of thinking in communication situations – that is, the structure of thinking as a goal-directed activity. Rather, his analysis focused on the formal characteristics of dialogues between partners attempting to solve problems and in finding analogous formal structures in individual problem solving.

Kuchinsky based his analysis on ideas coming from linguistics, especially on the concept of sign in Bakhtin's theory (Bakhtin, 1979). Bakhtin analyzed the dialogical position of different meanings using literary characters (e.g., Dostoyevsky's characters). Kuchinsky (1988), using experimental data from dialogues between partners who were attempting to solve problems, separated speech components into different clusters – a cluster related to finding a solution, and a cluster related to planning steps of thinking. However, Kuchinsky focused only on studying the exteriorized (e.g., translated from the inner to the external), as expressed in speech actions, ignoring purely mental actions as components of the goal-directed activity of thinking. In this sense, Kuchinsky's theory appears to be limited.

Soviet psychologists also developed a variety of other approaches, including studies of the internal content of an action, reflection, goal creation, and the correspondence between the orientation and performance of an action (e.g., Gal'perin, 1959, 1981; Davydov, 1986). Another approach to reinterpreting the idea of internal mediation of the thinking process was suggested by Znakov (1991), who described specific forms of understanding that influence thinking. For example, he studied Afghan veterans' understanding and recollection of critical situations they had experienced. Znakov experimentally demonstrated dual determination of thinking: Thinking is determined by the structure of a situation and by an individual's orientation to implicit norms of communication.

Many authors studied how problem solving is influenced by communicating information at various times and in various ways. Sources of additional knowledge in these studies were other people and different technological systems (e.g., Brushlinsky, 1982; Kornilova, 1986; Kornilova & Tikhomirov, 1990; Urvantsev, 1974).

Mechanisms of social determination conceptualized in terms of external and inner conditions of thinking. The methods used in Soviet/Russian psychology were determined by the shared understanding of thinking as the process or activity of solving particular problems. Different theoretical schools used a variety of methods – for example, having the subject perform creative tasks that required no special knowledge, studying concept development (methods of Vygotsky-Sakharov [Vygotsky, 1982b], Bruner [1977], and Tikhomirov [1969]), organizing help for a subject in the form of

ordered clues, and associative experiments. These shared methods made
the findings obtained in different schools of thought comparable. However,
the broad Marxist definition of thinking as a generalized and mediated type
of cognition of reality allowed enough freedom for variability in theories of
thinking.

One theory of thinking, developed by Rubinstein and recognized as the
official approach to the study of thinking, overshadowed Vygotsky's and
Leont'ev's theories in the period between the 1930s and 1960s. Rubinstein's
theory was based on the assumption that the mechanism of *analysis through
synthesis* was the main mechanism of regulating human thinking
(Rubinstein, 1958). The mechanism of analysis through synthesis determines the productivity of thinking. The object of thinking is incorporated
into new connections and relations with other objects, allowing the individual to learn new sides of the object and to reflect on new characteristics
of the object in new concepts. According to Rubinstein, the principal means
of thinking are analysis, synthesis, abstraction, and generalization
(Rubinstein, 1958). Tikhomirov (Luria's student) criticized this approach,
pointing to the lack of a psychological flavor and the too general nature of
these concepts. According to Tikhomirov, these concepts could be included
easily in a variety of different frameworks and theoretical approaches, some
of which would result in contradictory arguments and predictions. He also
pointed to the fact that Rubinstein's students' favorite experiments, using
the method of reasoning out loud, failed to address the causal mechanisms
of thinking (Tikhomirov, 1975).

Rubinstein's line of research was continued by his students,
Abul'khanova (1968), Antsyferova (1988), and Brushlinsky (1979). These
scientists experimentally developed one of Rubinstein's general ideas:
namely, the idea of *transformation of the external thought through the inner*
as the mechanisms that determines of thinking as (Rubinstein, 1958).[23] The
importance of inner conditions of thinking has been recognized in many
theories. The question is: What aspects of thinking are considered to be
these inner conditions of thinking? In Rubinstein's tradition, inner conditions were perceived as different stages in the process of thinking. For
example, in order to be ready to accept a clue and to progress in finding a
solution, subjects must reach a certain stage in their thinking; for example,
they must have completed their analysis of the task and be ready to synthesize the information it provides. That is, for Rubinstein, the inner conditions
of thinking are determined by characteristics of the process of thinking
itself. In addition, the advocates of this theory assume that these inner
conditions can be influenced by the external conditions of the problem-
solving situation. In other words, subjects can be led from one stage to
another stage in their thinking under the guidance of experimenters. Inner
conditions have been treated differently in more recent experimental work

by Kornilova (Kornilova, Grigorenko, & Kuznetsova, 1991; Kornilova & Tikhomirov, 1990). In her research, such formal-dynamic constructs as cognitive styles, cognitive risk, situational and personal anxiety, and success motivation were regarded as inner conditions of thinking.

Obviously, the way in which inner conditions of thinking are understood in different theoretical frameworks determines how issues regarding the determination of thinking can be addressed. If inner conditions of thinking are defined as types and levels of generalization, manipulable by external influences, then it seems logical to search for social causes of interindividual variability on these characteristics. Then researchers can describe more or less successful clues and other pedagogical or experimental tricks and observe their manipulative role in subjects' searches for solutions. In contrast, in defining formal-dynamic stylistic characteristics as inner conditions of thinking, researchers make a reasonable case for formulating questions about social as well as nonsocial causes of individual differences in stylistic traits. They state, for example, that a subject's cognitive style is a fundamental inner condition of his thinking that mediates the effect of a clue. Thus, in order to understand the clue's influence on the process of thinking, a researcher should understand the structure of the subject's cognitive style, which may be of social as well as nonsocial nature.

In any discussion of the inner determinants of thinking, research into the motivational aspects of thinking cannot be ignored. These aspects were addressed in analyses of the subjective regulation of thinking developed in the school of Tikhomirov. He accepted Leont'ev's theory of activity and concentrated his attention on studying the structure of thinking activity, focusing on the process of goal development. Tikhomirov views intermediate goals that arise during problem solving as new products of thinking activity; as long as the task has not been accepted by a subject, he or she will not attempt to solve it. The act of accepting a task consists of connecting the task to a motivational structure that is active in the specific situation. Tikhomirov separated external and inner (purely cognitive) motivations, but he also noted that this separation on its own does not completely resolve issues regarding the emotional-motivational regulation of thinking.

The originality of Tikhomirov's approach was in pointing out the regulating aspects of goals and emotions in the activity of thinking. "The goal-making process is one of the most essential components of thinking activity, and concentrates in itself connections between thinking and other mental processes – memory, imagination, emotional-volitional and motivational aspects of personality" (*Psikhologicheskie mekhanismy tseleobrazovaniya*, 1977, p. 24). This theoretical idea was realized in experiments investigating associations between aspects of final and intermediate goals of thinking (e.g., their trivial or original nature, their concrete or general character, their hierarchy) and characteristics of thinking activity and its productivity, as well as exploring the influence of shifts in emotional-motivational evaluations on changes in characteristics of thinking activity.

A separate line of studies was devoted to research on the regulatory influence of motives on thinking activity. Researchers, working in the framework of Tikhomirov's approach (Berezanskaia, 1977; Kornilova, 1994; Kornilova & Chudina, 1990; Vasil'ev, Popluzhnyi, & Tikhomirov, 1980), developed the concept of a structural function of a motive. The authors argue that thinking as an activity has its own motive, which structures thinking – that is, defines a set of actions, constituting the activity, and determines a set of goals and operations/strategies leading to the realization of these goals. These studies, combining experimental and clinical methods, investigate different types of motives and their structuring influences on subjects' thinking. The researchers show that different motives and motivational shifts dramatically influence the structure of thinking, as well as leading to changes in subjects' plans and goals, and in types of intellectual strategies.

The most important idea in Tikhomirov's approach is that goal-making processes and motivation are directly related to the management of cognitive activity (Tikhomirov, 1977). Tikhomirov argues that thinkers are active, that they themselves produce the motivational and regulating mechanisms of their thinking activity, and that motivations in thinking are not completely determined by external influences. This active nature of the thinker was a point of a long argument between Tikhomirov's school and Gal'perin and Talyzina's theory of stage formation (Gal'perin, 1959, 1981; Talyzina, 1981, 1984). Gal'perin's approach (see next section) treats thinking as a form of orientation, which, according to Tikhomirov, restricts thinking to very limited forms from the very beginning. The active nature of a subject (at the level of goal making) is suppressed under conditions in which he or she is required to follow an explicitly formulated scheme of mental activity.

Understanding thinking as a high-level, self-regulatory process, Tikhomirov objects to Gal'perin's contention that thinking can be formed through a system of direct manipulating influences. He proposes a less direct management of thinking that involves changing a thinker's motivation by providing motivational and emotional support to problem solving (Tikhomirov, 1984).

Application of theories of the social determination of thinking in educational practice

The belief that the development of thinking is fully and completely socially determined[24] had been the basis of Soviet education throughout its 70-year history. Thus, it was assumed that all of the children attending regular Soviet schools could be taught to do virtually anything. According to the official belief, education plays an extremely important role in development. We will briefly present three psychological-educational theories as examples of how the notion of the social determination of thinking was applied to pedagogy: (1) Galperin's theory of planned formation, (2) Shchedrovitsky's approach to education, and (3) Davydov's theory of the content and structure of learning activity.

The development of thinking in children is closely related to their education. Influential situations range from spontaneous practical interactions, such as game playing, to goal-directed, controlled interactions, such as classroom lessons. Each of these types of interaction contributes to a child's development in its own way. In his theory of the zone of proximal development, Vygotsky showed that every step in development depends upon the previous steps. However, these steps are not determined by children themselves. Their education (in its broadest sense) is the result of child–adult cooperation that broadens their zone of proximal development.

Even though the mastery of thinking and, therefore, its development, takes place only through children's own activity, this activity itself should be directed and organized (Gal'perin, 1985) in order to achieve maximum impact on development. The forms and organization of activity are not arbitrary but are determined by certain conditions.

In an attempt to find an alternative to a traditional type of schooling, Gal'perin developed a method[25] that allows any action in any given child to be formed with minimum energy and maximum efficiency (Liders & Frolov, 1991). This method is based on the principle that "an action is not formed in parts, but rather as a complete and correct structure containing all necessary components in their proper relation to other concepts" (Gal'perin, 1978, p. 102). Gal'perin's theory of stage/planned formation of actions and concepts suggested that the critical and purely psychological aspect of an individual's activity is its orientational aspect (Gal'perin, 1976). That is why one of the main components of this method is the process of building an orientational scheme – that is, a scheme, containing all of the relevant information needed to create a plan of action and all of the information needed to master an action faultlessly on the very first attempt. Thus, the orientation scheme is a special type of instruction, characterized by the completeness and clarity with which it describes an action. Moreover, this instruction is created in a way that permits the transfer of a mastered action to completely different situations.

Gal'perin showed experimentally that both the process and results of education depend upon the nature and efficiency of an individual's orientation and upon how well this orientation reflects the aspects of objects and environments needed for successful action [for detailed review, see Burmenskaia (in press)]. The concept of orientational activity provides a key to understanding the mechanisms of learning and to the quality of new knowledge. In addition, orientational activity makes more concrete the complicated connections between learning and natural, spontaneous ontogenetic processes. From this point of view, the traditional form of education, widely found in schools today and embedded in many experiments, can be classified as based on the first type of orientation. This type of orientation does not provide a differentiated understanding of the full system of conditions necessary for the successful outcome of an action. In the context of such an orientation, students are forced to search for missing orientators themselves. Thus, learning is performed via trial and error, with students finally (usually only partially) discovering the products of their activity through blind groping. The efficiency of this type of education is mostly determined by a child's previously formed systems of orientation. Hence, this type of learning is subordinated to the level of mental development of the child. Piaget was, no doubt, right in his evaluation of the weak developmental potential of traditional education. However, this is not the only type of education. According to Gal'perin, education can be built on a completely different foundation.

In a second type of education, based on the second type of orientation, student activity is organized in a special way: Students are given all of the directions necessary for successful completion of a given task from the very beginning. This leads to the disappearance of a lengthy and psychologically unjustifiable period of trial and error. Children, according to Gal'perin, obtain valuable knowledge and master actions consciously, based on common sense and their sense of reality. However, researchers have found that this type of education does not have any direct influence on children's general level of cognitive activity, because it does not guarantee the wide applicability of the mastered knowledge to other situations, and it does not require students to develop the ability to establish a system of orientations for new tasks in the same knowledge domain.

These limitations can be overcome if children master not actions and concepts (or their systems) but methods of analyzing objects and developing orientations. This constitutes a third type of education, with a corresponding orientation – that is, when a child can build an orienting basis for each concrete situation based on general principles. When children master a method of creating orientational bases for their actions, they then have a real means for independent and rational research. In this process, children learn to select the main qualities, units, and connections among studied materials, revealing their underlying structure. As a result, general orientational schemes for a wide range of activities are formed, constituting, according to Gal'perin, the main condition for development. During the 1960–1980s, a set of studies was conducted that showed a close connection between the third type of orientation and the development of thinking. In particular, the third type of orientation was applied in teaching scientific concepts to preschoolers (Obukhova, 1981) and in the development of complex graphical skills (Gal'perin, 1985).

Thus, unlike many other types of education, Gal'perin's method of planned formation of actions and concepts, which models various fragments of children's general mental and cognitive development in experimental conditions, is not limited to problematic representations of the learning material. This method assumes that a full system of conditions is completely established, guaranteeing that a child can successfully perform new actions, that this action can be correctly transferred to a new context, and that step by step the child will master and learn the desired qualities of these actions. Mastery of the objective means for analyzing objects constitutes the basis for organizing a child's rational search for a problem solution. Thus, in contrast to widely spread traditional types of education that vary only at the surface level, this third type of education, based on the third type of orientation, addresses the internal mechanisms of children's cognitive development. According to Gal'perin, the heart of developmental education is teaching a child to be able to establish the third type of orientation independently in any new situation.

Another major practical application of the principle of social determination of thinking was developed in the framework of the activity of Moscow Meta-theoretical Circle (MMC).[26] These scientists believed that it was necessary to study the educational process holistically, as a system, and to restructure it accordingly.

This approach rests on a number of epistemological premises. First is a distinction between naturalistic and activity approaches to human behavior. The naturalistic view assumes that objects in the surrounding world are independent entities with which humans interact. The principal theoretical categories of the naturalistic approach are subjects and objects, where subjects are always human beings. The activity approach starts with human activity as an all-embracing principle within which objects and relations are revealed as embodiments of the activity itself. Objects in the external world now appear as secondary constructions dependent for their existence on the activity that is applied to them. As for the subject, the concept of activity is more important than the humans involved in it. This is Shchedrovitsky's major thesis: Activity should not be regarded as an attribute of the individual but rather as an all-embracing system that captures individuals and forces them to behave in a certain way [for detailed review, see Dmitriev (in press)].

Activity thus appears as a complex system, whose structure can be viewed from different perspectives and grasped by different means of analysis. What follows from this is the third of Shchedrovitsky's theses – namely, that an essential distinction exists between an object of study and its presentation in a particular scientific domain. A single complex object such as behavior might be analyzed in a number of different scientific domains, depending on the epistemological and methodological positions chosen. The scientist's task therefore includes not only the study of the object within the framework of a chosen scientific subject but also the choice of theoretical procedures that mark this subject as a distinct component of scientific knowledge.

In 1957, Shchedrovitsky and Alekseev published an article about the study of thinking, in which they argued that thinking should not be viewed as only fixed knowledge but also as a process and an activity in which knowledge about objects is formed and used. Such a formulation of the main research task implies both (1) a separation and description of the structure of processes of thinking and the creation of an "alphabet" of elementary thinking operations, and (2) a study of laws, based on which complex combinations of ways and types of thinking can be created from existing thinking operations. This approach had more than theoretical significance: The scientists involved always stressed its practical orientation. They stated that this research would "allow them to improve education in such a way that a teacher would not only translate certain types of knowledge for his students, but also would be able to consciously shape certain actions and ways of thinking in his pupils" (Shchedrovitsky & Alekseev, 1957, p. 46). Based on this approach, a new activity-oriented and content-oriented type of education was developed – one impossible to realize in the context of regular school subjects (physics, math, chemistry, history, etc.). These subjects, as taught in schools, cannot answer the question "What should students be taught?" School subjects only provide material in a form of prepackaged knowledge from which the content of education (e.g., those ways and tricks of thinking used to obtain this knowledge) has yet to be extracted through special logical analysis (Shchedrovitsky, 1964).

That is why this group of scientists perceived education as "a complex structural whole, containing a number of heterogeneous parts: along with what children should study there are also pupils' learning activities and teachers' activities, organizing this process" (Shchedrovitsky, 1963, p. 162). Therefore, along with logical and meta-theoretical studies that allowed researchers to examine the content of education (that is, the types of thinking and the activities to be mastered), two other types of studies were designed and undertaken. The first type included psychological studies of how children master the content of education (e.g., the activity that students must perform in order to master a new type of activity). In this context, special attention was given to the problem of the ontogenetic development of thinking in learning settings (Shchedrovitsky & Alekseev, 1957). The second group of studies involved purely pedagogical research that was designed to explore the types of teacher activity that required children to master new content.

Such a complex and simultaneously clear-cut formulation of the program of studies was possible in the MMC because the object of research was not an abstract process of learning or an isolated, decontextualized educational situation but rather a holistic system of education in its social environment. Education was perceived as a special social institution that had its own history and functional role in society. Its historically formed function was to provide an uninterrupted process of societal reproduction by means of implementing in new generations the abilities and types of thinking needed to perform socially significant types of activity (Shchedrovitsky, 1966). This approach not only provided a certain framework for educational research (including educational-psychological research) but also allowed the formulation of a much broader circle of research tasks related to the reorganization of the educational system based on the activity principle.

Finally, Davydov's pedagogical theory was founded on his criticism of the traditional Soviet educational psychology, which, he argued, was based on an empiricist doctrine of concept formation that prevented students from acquiring a broad-scale theoretical foundation in science and the humanities. Using Shchedrovitsky's distinction between the object of study and the scientific subject, Davydov has argued that the only coherent way to acquire new knowledge is through theoretically constructed scientific concepts. Educational psychology must therefore revise the methodological apparatus that has focused on the gradual development of quasiscientific concepts, based in the everyday experience of the child.

In the 1960s, Davydov and El'konin conducted a series of studies directed toward creating new forms of education, inspired by Vygotsky's work (Vygotsky, 1982b, 1983), which provided the theoretical basis for their efforts. Their main premise was that the role of pedagogy and of pedagogical psychology is not to search for effective new ways of schooling (as was required by traditional pedagogy) but rather to concentrate on the content of mastered knowledge. Vygotsky viewed ways of teaching as derived from the content of mastered knowledge, and he stressed that the content of knowledge has a leading influence on a child's mental development. That is why it is necessary in school settings to introduce children to a system of scientific (theoretical) concepts. These concepts cannot be mastered independently by children, simply through personal experience. Vygotsky argued that a teacher should not be simply an organizer of children's personal experience, or a translator of his/her own experience, but rather a representative of science and cultural knowledge, the mastery of which requires special settings. The traditional methods of teaching theoretical concepts in Soviet schools did not make theoretical knowledge the content of education but instead resembled the process of mastering primitive everyday (empirical) concepts (Davydov, 1972).

Davydov designed a special curriculum for primary school in which scientific subjects were presented in a way that provided students with a basis for mastering scientific concepts that meet the epistemological standards of modern science and the humanities. The special curriculum in grammar and mathematics designed by Davydov and his collaborators has shown that a 7-year-old child can in fact handle highly abstract concepts.

In addition, it has been shown that theoretical knowledge cannot be mastered if memorization and recall are the only means by which a child studies material. The mastery of theoretical concepts requires special types of activity – learning activity – that must be intentionally encouraged and developed in a child. That is why, along with developing new curricula and new ways of teaching, psychologists working within this theoretical framework conducted a series of theoretical and experimental studies aimed at exploring the process by which learning activity develops in young school-age children. The results of these studies were assembled later into the "theory of the content and structure of learning activity" (Davydov, 1972; Davydov & Vardanian, 1981; El'konin, 1974).

To sum up, in this section we have attempted to present a complex and elaborate picture of Soviet/Russian theories on mechanisms for the social determination of the development of thinking and their practical applications. The approaches reflect not only an intricate system of environmental determination of thinking, taking place at different levels of influence (historical, cultural, social, and situational), but also detailed and elaborate theoretical, experimental, and observational data that convincingly illustrate the ideas described in this chapter.

Culturally biased or culturally specific?

This chapter aimed to provide a comprehensive account of how Soviet/Russian psychology has addressed the issues of nature and nurture in the study of intelligence and thinking. In addition, we have tried to place the theories in the cultural context in which they appeared and were developed.

Within the framework of cultural psychology, culture is seen as influencing those behaviors considered to be intelligent, the processes underlying intelligent behavior, and the direction of intellectual development (Miller, chapter 9, this volume). In this essay, we attempted to demonstrate that culture is also relevant to the way the nature–nurture controversy is resolved in a given theory.

Having described the historical background and major assumptions underlying studies of cognitive development in Soviet/Russian psychology, we directed readers' attention to the fact that even though significant ideological pressure existed, Soviet/Russian psychology was never made uniform and flat. Of course, the assumption that the primary source of variability in normal cognitive development is social, rather than biological, unquestionably dominated research on cognitive development.

The major question that remains to be answered is one raised in the introduction – Should a theory be considered flawed if it is found to be culturally biased? The belief in cultural/societal dominance penetrated deeply, not only into Soviet/Russian psychology but also into pedagogy and education. Theories implemented in the school system became part of the social practice that, in its turn, may have changed the mechanisms of cognitive development, leading to the predominance of social factors in cognitive development.

Certainly, the idea of the dominance of social determination in cognitive development did correspond in its basic form to the ideological pressures of Marxism-Leninism. The crucial question is whether we can say that this idea was truly false or whether we must consider the possibility that it was true for that particular cultural situation.

We suggest that at that point in time, with that particular cultural constellation, cognitive development may, indeed, have been determined primarily by social forces. The mechanisms of cognitive development are astonishingly flexible: A redistribution of the social and the biological may have occurred in response to the picture of man adopted in this society during this historical period, with its implicit beliefs regarding the ability of society to influence individual development. In this sense, any theoretical paradigm that tries to resolve the nature–nurture controversy is not culturally biased but is culturally specific.

Notes

1. *Culture* is defined in this chapter as a system of meanings that produces a particular sense of reality and includes both values and variations in conceptions of personhood (Ruzgis & Grigorenko, 1994).
2. We are using the term *Soviet/Russian psychology* to reflect that this chapter is based on studies performed in the context of Soviet psychology prior to December 1991, when the Soviet Union fell apart, as well as the studies performed later in the territory of Russia.

3. We intentionally prefer the usage of the concept *thinking* over the more traditional (for American psychology) term *thought*. Thinking reflects the processlike nature of this psychological function, and this aspect of the concept is extremely important for understanding its meaning and usage in Soviet/Russian psychology. Leont'ev (1978, p. 708) defines thinking as "the *process* of reflection on objective reality; the highest level of human cognition." According to this author, each human being becomes a subject of thinking only by mastering language, concepts, and logic, which are products of the development of social practice. This definition of thinking serves as the basis for theories of thinking in Soviet/Russian psychology. *Thought* in Soviet/Russian psychology is treated as the product of thinking.

4. Most of the exiled scientists continued to work abroad and developed original philosophical and psychological theories (e.g., Berdiaev, Frank).

5. *Pedology* was the term used in the 1920s for research and testing in child and educational psychology.

6. For example, Sergei L. Rubinstein, one of the most famous figures in Soviet psychology, was awarded the State Premium of the USSR in 1941 but was then fired in 1951 from his position as chair of the division of psychology in the philosophy department of Moscow State University. He was charged with cosmopolitism and was forced to defend his research in front of various party and state committees (Zhdan, 1993).

   The fate of Dr. Levit is an even more unfortunate example. As director of the Medical Biological Institute in Moscow, he was severely criticized in 1936 in the first round of Vavilov-Lysenko discussions (this meeting was an example of ideologically driven struggle against cosmopolitism, where genetics was announced to be a bourgeois science and prohibited). Levit was subjected to repression in 1937, imprisoned, and perished (Dubinin, 1988).

7. There are many examples of scientists' interest in humanities and philosophers' interest in science. Even the great antipsychologist Pavlov wrote to Chelpanov when the latter opened the first Russian psychological institute: "I, who have always excluded in my laboratory work on the brain any mention of subjective conditions, congratulate your psychological institute and you as its creator and leader, from my heart and wish you full success" (Pavlov, 1993, p. 92).

8. As time progressed, Vygotsky, and later his students, thoroughly developed some aspects of the cultural-historical theory, while others were left sketchy (e.g., the concepts of symbol and myth were hardly touched by either Vygotsky or his colleagues). However, the initial image of the theory, reflecting Russian spirituality and a humanistic interest in mankind, often referred to as the *Russian tradition*, was present even in the darkest days of its development. A similar evaluation could be made of many other Soviet/Russian psychological theories (e.g., Bernstein [1966], Rubinstein [1957, 1958]). The Russian philosophical and scientific traditions, in which the majority of Soviet psychological theories were rooted, served as a vaccine against the ideological epidemics of the regime, providing the basis for varying approaches.

9. An example of the nature and style of such critical evaluation is given in an article on the crisis in psychology written in 1927 by the young Vygotsky (1982a). Analyzing in detail schools of foreign psychological thought, the author pointed to philosophical and methodological dead ends and suggested new principles of accumulating psychological knowledge. Vygotsky's idea was to build psychology using the example of hard-core experimental sciences and to use "psychopractice" as a way for the "creation of a man by himself." According to modern historians of psychology, Vygotsky himself did not accomplish these goals (Vygotsky died when he was 38); it was done instead by his students (in particular, Leont'ev [*Psikhologiya i novye idealy nauchnosti*, 1993]). In addition, due to his interest in testing, his name was put on a blacklist in 1936 after the State decree on pedology. His work was not published until the 1960s. However, Vygotsky's influence was

tremendous. He created a humanistically oriented psychological school, which survived the 1930s and early 1940s and served as the basis for Leont'ev's theory of the psychology of activity, which, in its turn, was recognized in the 1960s as an official Soviet psychological doctrine. Moreover, Vygotsky's ideas provided a fulcrum for the work of Soviet philosophers Zinov'ev (1954) and Mamardashvili (1968, 1984).

10. A turning point in the history of Soviet psychology's attitudes toward foreign psychology was the publication of Tikhomirov's book (1969), in which the researcher suggested an approach based on Bruner's theory and data (Bruner, 1977) but providing a completely different interpretation of the results.

11. For example, Leont'ev's theory was criticized for its nonpsychological nature because, according to Marx, the outer determines the inner; in Leont'ev's theory, activity plays a major role, and consciousness "is held in leash activity" (*Psikhologiya i novye idealy nauchnosti*, 1993, p. 7) – that is, it is always secondary. Thus, Leont'ev "directly" applied Marx's philosophical concepts to psychological reality, for which he was criticized.

12. Our selections were influenced, no doubt, by the fact that both of the authors are graduates of Moscow State University and belong to what is known as "Moscow Psychological School." If a similar analysis were performed by psychologists affiliated with Saint Petersburg State University, it would have included Wekker's (1970) work, for example, which is not presented here due to its lack of connection to the major figures of the Moscow Psychological School (Vygotsky, Leont'ev, Luria, Gal'perin, and others).

13. Unfortunately, in many cases, the dates associated with a particular author refer to the date of publication, sometimes many years after the text was written.

14. We use the term *cognitive development* as a general concept covering the development of intelligence, thinking, and other cognitive functions.

15. In contrast to the concept of reflex, Kornilov's reaction contained "a wealth of ideological content, which was not characteristic of a reflex" (Smirnov, 1975, p. 146).

16. Blonsky's objections to Vygotsky's theory were published only after Vygotsky's death, and a public debate did not take place.

17. Before the Revolution of 1917, Vygotsky was a student in the Law School and in the history and philology departments of Moscow State University. Then he worked in Homel (Belorussia) as a school teacher. In 1924, after his presentation at the Second Russian Psychoneurological Congress, Vygotsky was invited by Kornilov to return to Moscow, where Vygotsky accepted a position as an assistant research fellow at the Psychological Institute.

18. Here we draw an analogy to a famous statement that all Russian literature emerged from Gogol's "Greatcoat."

19. There is a distinct tradition in the Soviet/Russian psychological literature, established in the 1920s (Krogius, 1980), of using the term *mental/intellectual functions* in its broadest sense – that is, applying it to all cognitive processes – perception, thinking, attention, memory, and so on.

20. In this context, *genetic* means *developmental*; Vygotsky never referred to *genetic* in the *biological* sense of the word.

21. In addition to introducing Piaget's clinical method, the method of dual stimulation, the method of completing unfinished sentences, and the method of analyzing children's explanations, Vygotsky introduced a number of other methodological approaches. For example, in his scientific writing, he used an analysis of literary dialogues, and he suggested a method of placing a child in a group of children speaking foreign languages in order to study changes in the egocentric speech of this child.

22. This argument was addressed by Smirnov (1975). According to Smirnov, it is wrong to mix the cultural-historical context of the development of consciousness, which Vygotsky considered in talking about the development of higher mental functions, and the context of

individual development in a society that (via child–adult interactions) guides the self-directed activity of a child on his or her way to mastering a sign and its relations.

23. These experiments were mostly performed with so-called small, creative tasks. An example of such a task was the question "Will a candle light up on a spaceship (under conditions of weightlessness)?" At different stages in a subject's thinking, an experimenter gives the subject different specially formulated clues. The way in which a subject uses or ignores different clues provides information on the mediated nature of external influences via inner conditions.

24. One exception to this belief existed in work done with handicapped children. The importance of biological determination was fully recognized. Handicapped children were placed in special educational institutions where they were trained more or less successfully to overcome their deficits. However, such exceptions existed outside of the psychology of thinking and were not incorporated into the theories discussed in this chapter. Moreover, a specific area of science, independent of psychology and pedagogy, called *defectologiya*, was defined as an integrated scientific discipline that embraces the study and education of all handicapped children and adults. It is in the framework of this science that issues regarding the development of abnormal thinking are studied.

25. It is important to note that in this school, *planned formation* is understood not only as a method of teaching but also as a "method of forming new concrete mental processes and phenomena, that is, as a method of psychological research" (Gal'perin, 1978, p. 93).

26. This circle was formed on the basis of the Moscow logical circle, which appeared in 1952 in the philosophy department at Moscow State University. The founder and the real leader of this circle was Alexander Zinoviev, but in the fall of 1954, when Zinoviev was exiled from the country, the leadership switched to Georgy Shchedrovitsky.

References

Abul'khanova, K. A. (1968). *Mysl' v deistvii (Psikhologiya myshleniya)* [Thought in action (Psychology of thinking)]. Moscow: Politizdat.

Alfimova, M., & Trubnikov, V. (in press). Psychological variables and genetic predispositions to schizophrenia. *Russian and Eastern European Psychology*.

Antsyferova, L. I. (Ed.). (1988). *Kategorii materialisticheskoi dialektiki v psikhologii* [Categories of materialistic dialectics in psychology]. Moscow: Nauka.

Averina, I. S. (1980). Geneticheskaya obuslovlennost' nekotorykh osobennostei pamyati [Genetic determination of selected memory characteristics]. In D. B. El'konin & I. V. Ravich-Shcherbo (Eds.), *Psikhologiya formirovaniya lichnosti i problemy obucheniya* (pp. 59–65). Moscow: Pedagogika.

Averina, I. S. (1983). *Vozrastnaya dinamika sootnosheniya genotipa i sredy v individual'nunykh osobennostyakh pamyati doshkol'nikov* [Age-dependent dynamics of heritability in preschooler's memory]. Unpublished doctoral dissertation, NIIOPP APN SSSR.

Bakhtin, M. M. (1979). *Estetika slovesnogo tvorchestva* [Esthetics of verbal creativity]. Moscow: Arts.

Barinaga, M. (1992). Confusion on the cutting edge. *Science, 257*, 616–19.

Berdiaev, N. A. (1990). *Samopoznanie (opyt philosophskoi biographii)* [Self-cognition (philosophical autobiography)]. Moscow: International Affairs.

Berezanskaia, N. V. (1977). K analizy kritichnosti i vnushaemosti v intellektual'noi deyatel'nosti [Toward an analysis of critical ability versus suggestibility in intellectual activity]. In A. N. Leont'ev & E. D. Khomskaya (Eds.), *Psikhologicheskie issledovaniya* (pp. 50–7). Moscow: Moscow State University.

Bernstein, N. A. (1966). *Ocherki po phisiologii dvizhenii i phisiologii activnosti* [Essays on the physiology of movements and the physiology of activity]. Moscow: Nauka.

Blonsky, P. P. (1935a). *Pamyat' i myshlenie* [Memory and thinking]. Moscow: Sotsekgiz.

Blonsky, P. P. (1935b). *Razvitie myshleniya shkol'nika* [The development of thinking in preschoolers]. Moscow: Uchpedgiz.

Bogoyavlenskaia, D. B., & Susokolova, I. A. (1985). Opyt psikhogeneticheskogo issledovaniya intellectual'noi activnosti [A behavior-genetic study of intellectual activity]. *Voprosy Psikhologii, 3*, 154–8.

Bratus' (in press). The sunset of the empire through the eyes of a psychologist. In E. L. Grigorenko, & R. J. Sternberg, P. Ruzgis, (Eds.), *Russian psychology: Past, present, & future*. Commack, NY: Nova Science Publishers, Inc.

Bruner, J. (1977). *Psikhologiya poznaniya. Za predelami neposredstvennoi informatsii* [The psychology of cognition. Beyond the limits of immediate information]. Moscow: Progress.

Brushlinsky, A. V. (1968). *Kul'turno-istoricheskaya teoriya myshleniya* [Cultural-historical theory of thinking]. Moscow: Higher School.

Brushlinsky, A. V. (1979). *Myshlenie i prognozirovanie* [Thinking and prognosis]. Moscow: Thought.

Brushlinsky, A. V. (Ed.). (1982). *Myshlenie: protsess, deyatel'nost', obshchenie* [Thinking: process, activity, communication]. Moscow: Nauka.

Burmenskaia, G. V. (in press). The psychology of development. In E. L. Grigorenko, P. Ruzgis & R. J. Sternberg (Eds). *Russian psychology: Past, present, & future*. Commack, NY: Nova Science Publishers, Inc.

Davydov, V. V. (1972). *Vidy obobshcheniya v obuchenii (logiko-psikhologicheskie problemy postroeniya uchebnykh predmetov* [Types of generalizations in learning (logic-psychological issues of school subjects' curriculum]. Moscow: Pedagogika.

Davydov, V. V. (1986). *Problemy razvivauyshchego obucheniya: opyt teoreticheskogo i eksperiental'nogo issledovaniya* [Issues of developing learning: Results of theoretical and experimental psychological research]. Moscow: Pedagogy.

Davydov, V. V., & Vardanian, A. U. (1981). *Uchebnaya deyatel'nost' i modelirovanie* [Learning activity and modeling]. Erevan: Luis.

Dmitriev, D. (in press). Pedagogical psychology and the development of education in Russia. In E. L. Grigonenko, P. Reejgis, and R. J. Sternberg (Eds.). *Russian psychology: Past, present, and future*. Commack, NY: Nova Science Publishers, Inc.

Dubinin, N. P. (1988). *Genetika – stranitsy istorii* [Genetics – pages of history]. Kishenev: Shtiintsa.

Dumitrashku, T. A. (1992). *Factory formirovaniya individual'nosti rebenka v mnogodetnoi sem'e* [Factors influencing children's development in large families]. Unpublished doctoral dissertation, NIIOPP APN SSSR.

Efroimson, V. P. (1968). *Vvedenie v medetsynskuyu genetiku* [Introduction to medical genetics]. Moscow: Meditsina.

Efroimson, V. P. (1976). K bio-khemicheskoi genetike intellekta [Toward biochemical genetics of intelligence]. *Priroda, 9*, 13–20.

Egorova, M. S. (1988). Genotip i sreda v variativnosti kognitivnykh phunktsii [Genotype and environment in variation on cognitive functions]. In I. V. Ravich-Shcherbo (Ed.), *Rol' sredy i nasledstvennosti v formirovanii individual'nosti cheloveka* (pp. 181–236). Moscow: Pedagogika.

El'konin, D. B. (1974). *Psikhologiya obucheniya mladshego shkol'nika* [The psychology of primary school students]. Moscow: Znanie.

Etkind, A. M. (1990). Obschestvennaya atmosphera I individual'nyi put' uchenogo: opyt prikladnoi psikhologii 20-kh godov [The societal atmosphere and a scientist's individual path: An overview of applied psychology in the 1920s]. *Voprosy psikhologii, 5*, 13–22.

Etkind, A. M. (1993). Eshche o Vygotskom: zabytye teksty i nendidennye konteksty [Again on Vygotsky: Forgotten texts and undiscovered contexts]. *Voprosy psikhologii, 3*, 37–54.

Filipchenko, Yu. A. (1929). *Izmenchivost' i metody ee izucheniya* [Variation and methods of studying variation]. Moscow: Nauka.

Florensky, P. (1991). *Vodorazdely mysli* [Paths of thought]. Novosibirsk: Shol'nye biblioteki.

Frolov, I. T. (1983). *Perspectivy cheloveka* [Man's prospects]. Moscow: Politizdat.

Frolov, I. T. (1988). *Filosophia i istoriya genetiki. Poiski i diskussii* [Philosophy and history of genetics. Research and discussion]. Moscow: Nauka.

Gadamer, H.-G. (1984). *Truth and method.* New York, NY: The Crossroad Publishing Company.

Gal'perin, P. Ya. (1959). Razvities issledovanii po formirovaniya umstvennykh deistvii [Research on the formation of mental actions]. In *Psychological science in the USSR* (pp. 441–61). Moscow: USSR Academy of Sciences.

Gal'perin, P. Ya. (1976). *Vvedenie v psikhologiyu* [Introduction to psychology]. Moscow: MGU.

Gal'perin, P. Ya. (1978). Poetapnoe formirovanie kak metod psikhologicheskogo issledovaniya [The method of stage formation as a method of psychological research]. In P. Y. Gal'perin, A. V. Zaporozhetst, & S. N. Karpova (Eds.), *Current issues in developmental psychology* (pp. 93–110). Moscow: MGU.

Gal'perin, P. Ya. (1981). K Issledovaniyu intellektual'nogo razvitiya rebenka [Toward research on children's intellectual development]. In I. I. Il'yasov & V. Y. Lyaudis (Eds.), *Textbook on developmental psychology. Soviet psychologists' papers in 1946–1980* (pp. 198–203). Moscow: MGU.

Gal'perin, P. Ya. (1985). *Metody obucheniya i ymstvennoe razvitie rebenka* [Methods of learning and child intellectual development]. Moscow: MGU.

Gindilis, V. M., & Finogenova, S. A. (1976). *Nasleduemost' kharakteristic pal'tsevoi i ladonnoi dermatogliphiki u cheloveka* [Heritability of finger and palm structures in humans]. *Genetika, 8*, 139–59.

Gofman-Kadoshnikov, P. B. (1973). Vozmozhnost' veroyatnostnoi otsenki diagnosa zigotnosti bliznetsov, ustanavlivaemoi metodom podobiya [A possibility of probabilistic zygosity determination by a method of likeness]. *Genetika, 1*, 156–61.

Grigorenko, E. L. (1990). *Eksperimental'noe issledovanie protsessa vydvizheniya i proverki gipotez* [Experimental study of hypothesis making in the structure of cognitive activity]. Unpublished doctoral dissertation, NIIOPP APN SSSR.

Grigorenko, E. L., LaBuda, M. L., & Carter, A. S. (1992). Similarity in general cognitive ability, creativity, and cognitive styles in a sample of adolescent Russian twins. *Acta Genet Med Gemellol, 41*, 65–72.

Grigorenko, E. L., Ravich-Shcherbo, I. V. (in press). Russian psychogenetics: Sketches for the portrait. In E. L. Grigorenko, P. M. Ruzgis, & R. J. Sternberg (Eds.). *Russian psychology: Past, present & future.* Commack, NY: Nova Science Publishers, Inc.

Gurevich, K. M., & Gorbacheva, E. I. (1992). *Umstvennoe razvitie shkol'nikov: kriterii I normy* [School students' cognitive development: Criteria and norms]. Moscow: Znanie.

Gurova, L. L. (1976). *Psikhologicheskii analiz resheniya zadach* [A psychological analysis of problem solving]. Voronezh: VGU.

Ignat'ev, M. V. (1934). Opredelinie genotipicheskoi i paratipicheskoi obuslovlennosti pri pomoshchi bliznetsovogo metoda [The measurement of genotypic and paratypic influences on continuous characteristics by means of the twin methods]. In S. G. Levit (Ed.), *Trudy mediko-biologicheskogo instituta* (pp. 18–31). Moscow: Biomedgiz.

Iskol'dsky, N. V. (1988). *Vliyanie sotsial'no-psikhologicheskikh factorov na individual'nye osobennosti bliznetsov i ikh vnutriparnoe skhodstvo* [Social-psychological factors influencing twins' individual characteristics and their similarity on psychological traits]. Unpublished doctoral dissertation, NIIOPP APN SSSR.

Kiriakidi, E. F. (1994). *Genotip–sredovye sootnosheniya v individual'nosti rebenka preddoshkol'nogo vozrasta* [Genotype–environment contributions to individuality of young children]. Unpublished doctoral dissertation, NIIOPP APN SSSR.

Kessen, W. (1990). *The rise and fall of development.* Worcester, MA: Clark University Press.

Kornilov, K. N. (1980). Biogenetichesky printsip i ego znachenie v pedagogike [Biogenetic principle and its significance for pedagogy]. In I. I. Il'yasov & V. Ya. Lyaudis (Eds.), *Textbook on developmental and pedagogical psychology. Soviet psychologists' papers in 1918–1945* (pp. 12–17). Moscow: MGU.

Kornilova, T. V. (1986). Myshlenie, oposredovannoe dannymi EVM [Thinking mediated by a computer]. *Voprosy psikhologii, 6*, 123–30.

Kornilova, T. V. (1994). Risk i myshlenie [Risk and thinking]. *Psikhologicheskii zhurnal, 4*, 20–32.

Kornilova, T. V., & Chudina, T. V. (1990). Personality and situational factors influencing decision-making in dialogues with a computer. *Mind, Culture, Activity. An International Journal, 4*, 25–32.

Kornilova, T. V., Grigorenko, E. L., & Kuznetsova, O. G. (1991). Pozanavatl'naya activnost' i individual'no-stilevye osobennosti intellektual'noi deyatel'nosti [Cognitive activity and individual-stylistic characteristics of problem solving]. *Vestnik MGU, 1*, 16–26.

Kornilova, T. V., & Tikhomirov, O. K. (1990). *Prinyatie intellectual'nykh reshenii v dialoge s komp'uterom* [Intellectual decision making in dialogues with a computer]. Moscow: MGU.

Kostiuk, G. S. (1969). Printsip razvitiya v psikhologii [The principle of development in psychology]. In E. V. Shorokhova (Ed.), *Metodologicheskie i teoreticheskie problemy psikhologii* (pp. 118–52). Moscow: Nauka.

Kozulin, A. (1984). *Psychology in Utopia. Toward a social history of Soviet psychology.* Cambridge, MA: The MIT Press.

Krogius, A. A. (1981). Eksperimental'noe issledovanie intellectual'nykh functsii studentov [Experimental studies of students' intellectual functions]. In I. I. Il'yasov & V. Ya. Lyaudis (Eds.), *Textbook on developmental and pedagogical psychology. Soviet psychologists' papers in 1918–1945* (pp. 38–43). Moscow: MGU.

Krushinsky, L. V. (1977). *Biologicheskie osnovy rassudochnoi deyatel'nosti. Evolutsionnyi I foziologo-geneticheskii aspecty povedeniya* [Biological bases of mind activity. Evolutionary and physiological-genetic aspects of behavior]. Moscow: Nauka.

Kuchinsky, G. M. (1983). *Dialog i myshlenie* [Dialogues and thinking]. Minsk: Belorussian University.

Kuchinsky, G. M. (1988). *Psikhologiya vnutrennego dialoga* [The psychology of inner dialogues]. Minsk: Universitetskoe.

Leont'ev, A. N. (1972). *Problemy razvitiya psikhiki* [Issues in the development of the psyche]. Moscow: MGU.

Leont'ev, A. N. (1975). *Deyatel'nost'. Soznanie. Lichnost'.* [Activity. Consciousness. Personality.]. Moscow: Politizdat.

Leont'ev, A. N. (1978). Thinking. In *Great Soviet Encyclopedia* (a translation of the third edition), pp. 708–9. New York: Macmillan Educational Corporation.

Leont'ev, A. N. (1982). O tvorcheskom puti L.S. Vygotskogo [On Vygotsky's creativity]. In *Selected works by Vygotsky* (Vol. 1, pp. 9–41). Moscow: Pedagogika.

Levit, S. G. (1934). Nekotorye itogi I perspektivy bliznetsovykh issledovanii [Some results and prospects of twin studies]. In S. G. Levit (Ed.), *Trudy mediko-biologicheskogo instituta* (pp. 5–17). Moscow: Biomedgiz.

Liders, A. G., & Frolov, Y. I. (1991). *Formirovanie psikhicheskikh processov kak metod issledovania v psikhologii* [The formation of mental processes as a method of research in psychology]. Moscow: MGU.

Lil'in, E. T. (1975). *Bliznetsy, nasledstvennost', sreda* [Twins, heredity, environment]. Moscow: Znanie.

Los', V. A., & Faddeev, E. T. (1981). Chelovek, obshchestvo i priroda v vek NTP [Men, society and nature in the scientific revolution]. *Voprosy philosophii, 12*, 92–7.

Losev, A. F. (1991). *Philosophiya. Miphologiya. Kul'tura.* [Philosophy. Mythology. Culture.]

Moscow: Politizdat.

Luria, A. R. (1936). The development of mental functions in twins. *Character and Personality*, *1*.

Luria, A. R. (1962). Ob izmenchivosti psikhicheskikh funktsii v protsesse razvitiya rebenka [On variability in mental functions in child development]. *Voprosy psikhologii*, *3*, 15–22.

Luria, A. R. (1969). *Vyshie korkovye funcstii cheloveka I ikh nerusheniya pri lokal'nykh porazheniyakh mozga* [Higher cortical functions in man and their disturbance by local brain damage]. Moscow: MGU.

Luria, A. R. (1979). *Yazyk i soznanie* [Language and consciousness]. Moscow: MGU.

Luria, A. R., & Mirenova, A. N. (1936). Issledovaniya eksperimental'nogo razvitiya vospriyatiya metodom differentsial'nogo obucheniya odnoiyaitsevykh bliznetsov [A study of experimental development of identical twins in the context of differential teaching]. *Nevrologia I Genetika*, *2*, 18–23.

Luria, A. R., & Yudovich, F. Y. (1956). *Rech' i umstvennor razviriw rebenka* [Speech and children's mental development]. Mocow: Academy of Pedagogical Science Publishing House.

Malinovsky, A. A. (1977). *Biologiya cheloveka* [Human biology]. Moscow: Nauka.

Malych, S. B., Egorova, M. S., & Piankova, S. D. (1993). The etiology of individual differences in cognitive strategy. *Journal of Russian and East European Psychology*, *31*(3), 47–55.

Mamardashvili, M. K. (1968). *Soderzhanie i formy myshleniya* [Content and forms of thinking]. Moscow: Vyshaya Shkola.

Mamardashvili, M. K. (1984). *Klassichesky i neklassichesky idealy ratsional'nosti* [Classical and nonclassical ideals of rationality]. Tbilisi: Metsniereba.

Mamardashvili, M. K. (1992a). D'yavol igraet s nami kogda my myslim ne tochno . . . [The Evil plays with us when our thinking is not precise . . .]. In M. K. Mamardashvili (Ed.), *Kak ya ponimau philosophiyu* (pp. 126–42). Moscow: Progress.

Mamardashvili, M. K. (1992b). Nauka i kyl'tura [Science and culture]. In M. K. Mamardashvili (Ed.), *Kak ya ponimau philosophiyu* (pp. 291–310). Moscow: Progress.

Markus, H., & Kitayama, S. (1991). Culture and the self: Implications for cognition, emotion and motivation. *Psychological Review*, *98*, 224–53.

Marshall, E. (1992). When does intellectual passion become conflict of interest? *Science*, *257*, 620–4.

Mariutina, T. M. (1978). O geneticheskoi obeslovlennosti vyzvannykh potentsialov cheloveka [Heritability of human evoked potentials]. In B. F. Lomov & I. V. Ravich-Scherbo (Eds.), *Problemy geneticheskoi psikhophiziologii cheloveka* (pp. 72–93). Moscow: Nauka.

Mariutina, T. M. (1988). Priroda mezhindividual'noi izmenchivosti vyzvannykh potentsialov [The nature of variability in evoked potentials]. In I. V. Ravich-Shcherbo (Ed.), *Rol' sredy i nasledstvennosti v formirovanii individual'nosti cheloveka* (pp. 107–38). Moscow: Pedagogika.

Mariutina, T. M., & Ivoshina, T. G. (1982). Vliyanie genotipa na vyzvannye potentsialy levogo i pravogo polushary pri vospriyatii zritel'nykh stimulov [Heritability of left- and right-brain evoked potentials during visual perception]. In E. D. Khomskaya & A. R. Luria (Eds.), *Functsii lobnykh dolei mozga* (pp. 246–65). Moscow: Nauka.

Mattaeus, W. (1988). *Die sowietische Denkpsychologie* [Soviet psychology of thinking]. Goettingen: Hogrefe.

Mirenova, A. N. (1934). Psikhomotornoe obuchenie doshkol'nika i obshchee razvitie [Psychomotor training and the general development of preschool children]. In S. G. Levit (Ed.), *Trudy mediko-biologicheskogo instituta* (pp. 86–104). Moscow: Biomedgiz.

Mirenova, A. N., Kolbanovsky, V. N. (1934). Sravnitel'naya otsenka metodov razvitiya kombinatornykh funktsii u doshkol'nika [A comparative evaluation of methods for the development of combinative functions in preschool children]. In S. G. Levit (Ed.), *Trudy mediko-biologicheskogo instituta* (pp. 105–18). Moscow: Biomedgiz.

Nebylitsyn, V. D. (1966). *Osnovnye svoistva nervnoi sistemy cheloveka* [Main properties of the human nervous system]. Moscow: NIIOPP APN.

Obukhova, L. F. (1981). *Kontseptsiya Zhana Piazhe: za i proiv* [Pros and cons of Piaget's theory]. Moscow: MGU.

Packer, M. J., Addison, R. B. (Eds.). (1989). *Entering the circle*. Albany, NY: State University of New York Press.

Palei, I. M. (1974). K differentsial'no-psikhologicheskomu issledovaniyu studentov v svyazi s zadachami izucheniya potentsialov razvitiya vzroslogo cheloveka [Toward differential-psychological studies of students in the context of adult development]. *Sovremennye psikhologo-pedagogicheskie problemy vyshei shkoly, 2*, 133–43.

Panteleeva, T. A. (1977). *Analiz individual'nykh razlichy v sensomotornykh reaktsiyakh cheloveka* [An analysis of individual variability in human sensorimotor reactions]. Unpublished doctoral dissertation, NIIOPP APN SSSR.

Pavlov, I. P. (1953). *Issledovanie vyshei nervnoi deyatel'nosti* [Research on higher nervous activity]. Kiev: Gosmedizdat USSR.

Pavlov, I. P. (1993). Pis'mo Chelpanovy [A letter to Chelpanov] *Voprosy Psikhologii, 2*.

Pepitone, A., & Triandis, H. (1987). On the universality of social psychological theories. *Journal of Cross-Cultural Psychology, 18*, 471–98.

Podgoretskaia, N. A. (1980). *Izuchenie priemov logicheskogo myshleniya u vzroslych* [Studying strategies of logical thinking in adults]. Moscow: MGU.

Psikhologiya i novye idealy nauchnosti [Psychology and the new ideals of the scientific] (materials of the round table) (1993). *Voprosy philosophii, 5*, 3–42.

Puzyrei, A. A. (1986). *Kul'turno-istoricheskaya teoriya Vygotskogo i sovremennaya psikhologiya* [Vygotsky's cultural-historical theory and modern psychology]. Moscow: MGU.

Rubinstein, S. L. (1957). *Bytie i soznanie* [Being and consciousness]. Moscow: Nauka.

Rubinstein, S. L. (1958). *O myshlenii i putyakh ego razvitiya* [On thinking and methods of studying thinking]. Moscow: AN SSSR.

Rubinstein, S. Y. (1970). *Eksperimental'nye metodiki patopsikhologii I opyt ikh primeneniya v klinike* [Experimental methods in psychopathology and their applications in clinics]. Moscow: Meditsyna.

Ruzgis, P., & Grigorenko, E. L. (1994). Cultural meaning systems, intelligence, and personality. In R. J. Sternberg & P. Ruzgis (Eds.), *Personality and intelligence* (pp. 248–71). New York: Cambridge University Press.

Sechenov, I. M. (1952). *Refleksy golovnogo mozga* [Brain reflexes]. Moscow: AMN SSSR.

Semenov, V. V. (1982). *Priroda mezhindividual'noi izmenchivosti kechestvennykh priznakov individualnosti* [Nature of interindividual variability on qualitative characteristics of individuality]. Unpublished doctoral dissertation, NIIOPP APN SSSR.

Shchedrovitsky, P. G. (1963). Mesto logiki v psikhologo-pedagogicheskikh issledovaniyakh [The place of logic in psychological-pedagogical research]. *Detskaya I pedagogicheskaya psikhologiya*. Moscow: Izdatel'stvo APN RSFSR.

Shchedrovitsky, P. G. (1964). O printsipakh analiza ob'ectivnoi struktury myslitel'noi deyatel'nosti na osnove ponyatii soderzhatel'no-geneticheskoi logiki [On the principles of the analysis of objective structure of thinking activity on the basis of concepts of content-genetic logic]. *Voporsy psikhologii, 2*, 125–31.

Shchedrovitsky, P. G. (1966). Ob iskhodnykh printsipakh abaliza problemy obucheniya I razvitiya v ramkakh teorii deyatel'nosti [On basic principles of the analysis of learning and development in the context of the theory of activity]. In *Obuchenie i razvitie* (pp. 89–119). Moscow: Prosveshchenie.

Shchedrovitsky, P. G., & Alekseev, N. G. (1957). O vozmozhnykh putyakh issledovaniya myshleniya kak deyatel'nosti [On possibilities of studying thinking as activity]. *Doklady APN RSFSR, 3*, 41–6.

Shishkin, A. F. (1979). *Chelovecheskaya priroda I moral'. Istoricheskor lriticheskoe esse* [Human nature and morality. A historical-critical essay]. Moscow: Nauka.

Smirnov, A. A. (1975). *Razvitie i sovremennoe sostoyanie psikhologicheskoi nauki v SSSR* [History and the modern situation of psychological science in the USSR]. Moscow: Pedagogika.

Smirnov, S. D. (1985). *Psikhologiya obraza: problema aktivnosti psikhicheskogo otrazheniya* [Psychology of image: The active nature of mental reflection]. Moscow: MGU.

Smirnov, S. D. (1994). Intelligence and personality in the psychological theory of activity. In R. J. Sternberg & P. Ruzgis (Eds.), *Personality and intelligence* (pp. 221–47). New York: Cambridge University Press.

Sokolova, E. I., Gofman-Kadoshnikov, P. B., & Lil'in, E. T. (Eds.). (1980). *Ocherki blinznetsovykh issledovanii (v klinicheskoi meditsine)* [Essays on twin methodology (clinical medicine applications)]. Moscow: Meditsina.

Talyzina, N. F. (1981). Printsipy sovetskoi psikhologii i problema psikhodiagnostiki poznavatel'noi deyatel'nosti [Principles of Soviet psychology and issues concerning the assessment of cognitive activity]. In I. I. Il'yasov & V. Y. Lyaudis (Eds.), *Textbook on developmental and pedagogical psychology. Soviet psychologists' papers in 1918–1945* (pp. 64–86). Moscow: MGU.

Talyzina, N. F. (1984). *Upravlenie protsessom usvoeniya znanii* [Managing the process of mastering knowledge]. Moscow: MGU.

Talyzina, N. F., Krivtsova, S. V., & Mukhamatulina, E. A. (1991). *Priroda individual'nykh razlichii: opyt issledovaniya bliznetsovym metodom* [Nature of individual difference: A study of twins]. Moscow: MGU.

Teplov, B. M. (1985). *Izbrannye trudy (t. 1,2)* [Selected works (Vols. 1–2)]. Moscow: Pedagogika.

Teplov, B. M. (1990). *Um polkovodtsa* [A military commander's mind]. Moscow: Pedagogika.

Tikhomirov, O. K. (1969). *Stuktura myslitel'noi deyatel'nosti* [The structure of thinking activity]. Moscow: MGU.

Tikhomirov, O. K. (1975). Aktual'nye problemy razvitiya psikhologicheskoi teorii myshleniya [Current problems in the psychological theory of thinking]. In O. K. Tikhomirov (Ed.), *Psukhologicheskie issledovaniya tvorcheskoi deyatel'nosti* (pp. 5–22). Moscow: Nauka.

Tikhomirov, O. K. (Ed.). (1977). *Psikhologicheskie mechanismy tseleobrazovaniya* [Psychological mechanisms of goal formation]. Moscow: Nauka.

Tikhomirov, O. K. (1984). *Psikhologiya myshlenia* [The psychology of thinking]. Moscow: MGU.

Trubnikov, V. I., Uvarova, L., Alfimova, M., Orlova, V., Ozerova, N., & Abrosimov, N. (1993). Neuropsychological and psychological predictors of genetic risk for schizophrenia. *Behavior Genetics, 23,* 455–9.

Tulmen, C. (1981). Mostart v psikhologii [Mozart in psychology]. *Voprosy philosophii, 10,* 129.

Urvantsev, L. P. (1974). *Formirovanie suzhdenii v usloviyakh neopredelennosti visual'noi stimulyatsii* [The formation of reasoning across conditions that vary the certainty of visual information]. Unpublished doctoral dissertation, MGPI, Moscow.

Vasil'ev, I. A., Popluzhnyi, V. L., & Tikhomirov, O. K. (1980). *Emotsii i myshlenie* [Emotions and thinking]. Moscow MGU.

Vygotsky, L. S. (1982a). *Istorichesky smysl psikhologicheskogo krizisa* [The historical meaning of psychological crises]. Moscow: Pedagogy.

Vygotsky, L. S. (1982b). *Myshlenie i rech'* [Thinking and speech]. Moscow: Pedagogika.

Vygotsky, L. S. (1983). *Istoriya razvitiya vyshikh psukhicheskikh phunktsii* [The history of development of higher mental functions]. Moscow: Pedagogika.

Vygotsky, L. S. (1984). *Razvitie myshlenia podrostka i obrazovanie ponyatii* [The development of thinking in adolescents and concept formation]. Moscow: Pedagogika.

Wekker, L. M. (1970). *Psikhicheskie protsessy. Myshlenie i intellekt* [Mental processes. Thinking and intelligence]. Sanct-Peterburg: S-PGU.

Wertsch, J. V., & Youniss, J. (1987). Contextualizing the investigator: The case of developmental psychology. *Human Development, 30*, 18–31.

Yaroshevsky, M. G. (Ed.). (1971). *Problemy nauchnogo tvorchestva v sovremennoi psikhologii* [Modern psychological research on scientific creativity]. Moscow: Nauka.

Yaroshevsky, M. G. (1981). Katerogial'nyi apparat psikhologii [Categorical apparatus of psychology]. In *Sechenov i mirovaya psikhologicheskaya mysl'* (pp. 139–52). Moscow: Nauka.

Zaporozhets, A. V. (1986). *Izbrannye psikhologicheskie trudy (v dvykh tomakh)* [Selected psychological writings (Vols. 1–2)]. Moscow: Pedagogika.

Zeigarnik, B. V. (1962). *Patologiya myshleniya* [Thinking pathologies]. Moscow: MGU.

Zhdan, A. N. (1993). Prepodavanie psikhologii v Moscovskom universitete [Teaching psychology in Moscow State University]. *Voprosy psikhologii, 4*, 80–93.

Zinchenko, V. P., & Mamardashvili, M. K. (1977). Ob ob'ektivnom metode v psikhologii [On objective method in psychology]. *Voprosy philosophii, 7*, 109–25.

Zinchenko, V. P., & Morgunov, E. B. (1994). *Chelovek razvivaushchiisya: osherki rossiiskoi psikhologii* [A developing person: Essays on Russian psychology]. Moscow: Trivola.

Zinchenko, V. P., & Smirnov, S. D. (1983). *Metodologicheskie voprosy psikhologii* [Metatheoretical issues in psychology]. Moscow: MGU.

Zinov'ev, A. A. (1954). *Voskhozhdenie ot abstractnogo k konkretnomu* [A rise from the abstract to the concrete]. Unpublished doctoral dissertation, MGU, Moscow.

Znakov, V. V. (1991). Ponimanie kak problema psikhologii myshleniya [Understanding as a psychological problem]. *Voprosy psikhologii, 1*, 18–26.

Zyrianova, N. M. (1992). *Genotip–sredovye sootnosheniya v izmenchivosti pokazatelei kognitivnoi sphery u detei 6–7 let* [Genotype–environment contributions to variability in cognitive characteristics in 6–7-year-old children]. Unpublished doctoral dissertation, NIIOPP APN, Moscow.

15 The emerging horizontal dimension of practical intelligence: Polycontextuality and boundary crossing in complex work activities

Yrjö Engeström, Ritva Engeström,
and Merja Kärkkäinen

Introduction

In a landmark article on culture and intelligence, the collective of the Laboratory of Comparative Human Cognition (1982) pointed out that a context-specific account of the cultural basis of intelligence is crucially dependent on the formulation and employment of an adequate unit of analysis.

Concerned as we are with specifying how the outside influences the inside and vice versa, we cannot proceed leaving these two systems as independent entities. Somehow, we must deal with the problem of inside and outside together, as mutually influencing systems [p. 695].

The authors of that paper suggested the concept of activity, and the closely related notion of cultural practice, as the most promising candidates for such a unit. Object-oriented, artifact-mediated activity is the central concept of the cultural-historical approach to higher mental functions, initiated by Vygotsky (1978), Leont'ev (1978), and Luria (1978).

Activity is a molar, not an additive unit of the life of the physical material subject. In a narrower sense, that is, at the psychological level, it is a unit of life, mediated by psychic reflection, the real function of which is that it orients the subject in the objective world. In other words, activity is not a reaction and not a totality of reactions but a system that has structure, its own internal transitions and transformations, its own development [Leont'ev, 1978, p. 50; also Wertsch, 1981; Smirnov, 1994].

Activities are relatively durable collective formations, mediated by artifacts (including signs and symbol systems), social rules, and division of labor (for contemporary developments in activity theory, see Engeström, 1987, 1990; Engeström, Miettinen, & Punamäki, in press). An activity system is characterized by constant parallel processes of internalization and externalization, acquisition and creation, of cultural artifacts and meanings by the participating subjects. An activity system, such as an organized work practice, may be seen as a carrier of evolving interactive expertise, distributed among the participating subjects and their artifacts (Cole & Engeström, 1993; Engeström, 1992; for the related concept of community of practice, see Lave & Wenger, 1991).

The collective of the Laboratory of Comparative Human Cognition (1982, p. 710) concluded in their article that intelligence "will be different across cultures (and across contexts within cultures) insofar as there are differences in the kinds of problems that different cultural milieus pose." However, they added an important caveat, noting that this radical relativist conclusion "fails to consider the fact that cultures interact." We see tremendous theoretical and methodological implications in this point. Not only does it call attention to intercultural inequalities and asymmetric power relations in definitions of intelligence. Most importantly for the present investigation, interaction between cultures and contexts is in itself becoming a constitutive feature of intelligence. We will examine this claim, focusing on expertise and practical intelligence in work activities.

Expertise is understood in this chapter as working intelligence (Scribner, 1984) or practical intelligence (Sternberg & Wagner, 1986) – that is, as intelligent action and thought performed in everyday contexts of work activity. From an activity-theoretical point of view, practical or working intelligence can be seen as at the forefront of this induced evolution of intelligence. The impact of changing market forces and new technologies is most rapidly and directly experienced in work activities. Schooling, family life, and other social contexts tend to follow with a marked lag.

Collins (1990) distinguishes between two opposite approaches to expertise: "an 'algorithmic model,' in which knowledge is clearly statable and transferable in something like the form of a recipe, and an 'enculturational model,' where the process has more to do with unconscious social contagion" (p. 4). The former approach may be characterized as *top-down*, the latter as *bottom-up*. Most of mainstream experimental cognitive science and cognitive psychology have consistently represented the former approach (e.g., Chi, Glaser, & Farr, 1988; Ericsson & Smith, 1991; Hoffman, 1992). Researchers inspired by philosophers such as Heidegger, Polanyi, and the late Wittgenstein have argued for the latter approach (e.g., Dreyfus & Dreyfus, 1986; Göranzon & Josefson, 1988; Nyíri & Smith, 1988).

Both of these approaches share a vertical view of expertise. Characteristic of both is a discourse of *stages* or *levels* of knowledge and skill. Such a vertical image assumes a uniform, singular model of what counts as an *expert* in a given field. It is not accidental that such a well-bounded domain as chess, with its rigorous universal ranking system, has persistently served as the favorite example of expertise for both approaches.

In this chapter, we argue for a broader, multidimensional view of working intelligence. While the vertical dimension remains important, a horizontal or lateral dimension is rapidly becoming increasingly relevant for the understanding and acquisition of expertise. In various fields, today's practitioners operate in and move between multiple, parallel activity contexts. These multiple contexts demand and afford different and complementary yet conflicting cognitive tools, rules, and patterns of social interaction. Criteria of expert knowledge and intelligent functioning are different in the various contexts. Practitioners face the challenge of negotiating and combining ingredients from different contexts to achieve hybrid solutions. The vertical master–novice relationship, and with it, in some cases, the professional monopoly on expertise, become problematic as demands for dialogical problem solving increase.

Two central features of this newly emerging landscape of expertise may be designated as *polycontextuality* and *boundary crossing*. First we discuss these two concepts theoretically. Then we present exploratory analyses of three cases of collaborative problem solving in team environments. The three cases are based on recordings and observations conducted in 1993 and 1994 in a municipal welfare and health center, a primary school, and a factory that manufactures cabins for large ships; all three cases were located in Finland. The cases allow us to examine both the difficulties and the potentials of various types of polycontextuality and boundary crossing. Finally, we discuss the implications of our analyses for understanding emerging new aspects of working intelligence and expert cognition.

Polycontextuality and boundary crossing as theoretical concepts

Drawing on a series of *shadowing* studies on work teams in different settings, Stephen Reder (1993) notes that in all cases, the work group and its members were engaged in multiple ongoing tasks. This *polycontextuality* or *coordinated multitasking* is recognized as a challenge.

Little is known about how ongoing work group action is organized so that multiple ongoing tasks can be smoothly interrupted, suspended, later resumed, and eventually completed amidst a patchwork of other ongoing, yet intermittent tasks. Nor do we have a clear picture of the intricacies of a group "working on several things at once," even though it appears to happen frequently and may often be the norm for group (if not individual) work [p. 123].

Charles Goodwin (1990) takes the issue one step further in a case study of expert work on an oceanographic research vessel. He describes the work of a physical oceanographer as being situated within a set of "distributed interlocking participation frameworks" in that she is simultaneously interacting with both a geochemist and a winch operator, located in different parts of the ship, while also attending to the representations provided by her tools.

Each participation framework invokes its own micro-world, providing particular forms of access, structures that shape perception and talk, ways of acting, etc. Such arrangements constitute frameworks for the organization of phenomena such as talk, access, perception and action that are not lodged within the individual. However, despite their diversity, and their inherent distributed multi-party organization, this entire set of participation frameworks is relevant to the production of a single utterance/speech act . . . by a single actor [p. 46].

Both Reder and Goodwin talk about polycontextuality at the level of tasks and work actions. Leont'ev's (1978, 1981) theoretical insights help us to realize, however, that polycontextuality operates also at the level of larger collaborative activity systems. An activity system is a complex and relatively enduring *community of practice* that often takes the shape of an institution. Activity systems are enacted in the form of individual goal-directed actions. But an activity system is not reducible to the sum total of those actions. An action is discrete; it has a beginning and an end. Activity systems have cyclic rhythms and long historical half-lives (see Engeström, 1987, 1990).

Polycontextuality at the level of activity systems means that experts are engaged not only in multiple simultaneous tasks and task-specific participation frameworks within one and the same activity. They are also increasingly involved in multiple communities of practice.

A recent study by Tyre and von Hippel (1993) illuminates the activity-system level of polycontextuality. The authors analyze problem solving around new manufacturing equipment following field tests and early factory use in two industrial plants. At both sites, engineers had the primary responsibility for diagnosing and resolving problems that occurred with the new machines.

A striking feature of the adaptation process was the use of different physical settings for responding to a single problem. In most of the cases studied, engineers needed to investigate the same issue in two different locations (the plant and the lab). They often shifted repeatedly between locations before they felt they could understand and resolve the problem [p. 7].

The plant and the laboratory were very different and were separate activity systems, or communities of practice. The tools, languages, rules, and social relations of the two contexts had little in common. This very difference proved to be a decisive resource. The solution of the machine problems required that the two contexts be iteratively connected. The engineers became boundary crossers.

This is obviously but one type of boundary crossing between activity systems. Brown and Duguid (1994) and Suchman (1994) give further examples of crossing boundaries between the design and use of technologies.

Management consultants such as Lipnack and Stamps (1993) and Peters (1992) maintain that boundary crossing will be the basic mode of operation in flat, team- and network-based organizations. While such sweeping claims should be taken with a grain of salt (see Hirschhorn & Gilmore, 1993), scholarly studies of new organizational forms spreading in various industries indicate that a need for horizontal boundary crossing may indeed be emerging as a vital challenge (Alter & Hage, 1993; Harrison, 1994; Nohria & Eccles, 1992; Powell, 1990). This is particularly evident in the spheres of new product development and technological innovation (Biemans, 1992; Håkansson, 1987, 1989).

Boundary crossing is a broad and little studied category of cognitive processes. Classic studies of innovation and creative thinking emphasize the potential embedded in transporting ideas, concepts, and instruments from seemingly unrelated domains into the domain of focal inquiry (e.g., Bartlett, 1958; Ogburn & Thomas, 1922; see also Margolis, 1993). Yet such processes seem to be rare and quite demanding. As Suchman (1994) points out, "crossing boundaries involves encountering difference, entering onto territory in which we are unfamiliar and, to some significant extent therefore, unqualified" (p. 25). To overcome such a deficiency, boundary crossing calls for the formation of new mediating concepts. In this sense, boundary crossing may be analyzed as a process of collective concept formation.

Various forms of cognitive inertia and compartmentalization are powerful obstacles to boundary crossing. One of the mechanisms behind such inertia is *groupthink*, a mode of thinking that people engage in "when they are deeply involved in a cohesive in-group, when the members' strivings for unanimity override their motivation to realistically appraise alternative courses of action" (Janis, 1983, p. 9). Groupthink typically leads to an overestimation of the in-group, closed-mindedness, and stereotypes of out-groups. Another possible mechanism that prevents boundary crossing is, paradoxically, almost the opposite of groupthink – namely, fragmentation of viewpoints and lack of "shared mental models" among a community of practitioners (Cannon-Bowers, Salas, & Converse, 1993). Such fragmentation may make it impossible for experts from different contexts to "speak the same language" and exchange ideas about a problem. As Shchedrovitskii and Kotel'nikov (1988) point out:

Today, in operating the technical systems we have created, and in the process of our ever-expanding appropriation of the world around us, we continually encounter assignments and tasks whose solution is beyond the capacities of any one person and require the participation of a large team that includes representatives of different professions, different scientific disciplines, and different subjects. However, the coordinated organization of all these people into one working system has, as a rule, proved impossible: a person's thinking, organized by profession and subject, poses obstacles that are difficult to overcome, and a high level of professionalism inter-feres with, more than helps to achieve, joint team effort [p. 58].

A number of studies on expert decision making have found a pervasive tendency toward overconfidence and compartmentalization in the judg-ments of experts in various domains. Massive amounts of experience in no way guarantee an improved ability to deal with uncertainty and probabilistic reasoning tasks (Brehmer, 1980). Experts often "appear to be mainly interested in how consistent the evidence is with the hypothesis they are testing and fail to consider its consistency with an alternative hypoth-esis" (Ayton, 1992, p. 95). Sternberg and Frensch (1992) point out that "it is exceedingly difficult to break up and reorganize an automatized local processing system to which one in all likelihood no longer even has con-scious access" (p. 197). In a similar vein, Argyris (1992) talks about "skilled incompetence" as the dilemma of professionals.

In the face of such obstacles, boundary crossing seems to require signifi-cant cognitive retooling. Leigh Star's (1989) notion of *boundary object* is a useful attempt at identifying mediating artifacts that may help to overcome groupthink and fragmentation.

Boundary objects are objects that are both plastic enough to adapt to local needs and constraints of the several parties employing them, yet robust enough to main-tain a common identity across sites. . . . Like the blackboard, a boundary object "sits in the middle" of a group of actors with divergent viewpoints [p. 46].

Star identified repositories, ideal types, terrains with coincident boundaries, and forms and labels as types of boundary objects. Basically, these are different types of shared external representations of a problem or domain. In a similar vein, Schrage (1990, pp. 154–5) emphasizes the importance of playing with multiple forms and technologies of representation as a require-ment in network organizations. From a mediational viewpoint, it is also useful to consider shared mental models as internalized cognitive artifacts – that is, as a specific type of boundary object (see Norman, 1993; Rogers & Rutherford, 1992).

While representations and artifacts are important in facilitating bound-ary crossing, the realization of their potential depends on the ways they are used. In their study of the introduction of a new computer system in a hospital, Aydin and Rice (1992) found that involvement in the implementa-tion process of the new system increased employees' interaction with other departments, but involvement in computer use alone did not have such an effect. Communication-based forms of involvement in implementation were overwhelmingly most important in predicting increases in interdepart-mental interaction.

Another direction of cognitive retooling that has potential for breaking overconfidence and routinization is argumentation and dialogue (Bakhtin,

1981, 1986; Billig, 1987; Markovà & Foppa, 1990; Wertsch, 1991). Several recent studies from the Finnish school of Developmental Work Research and in Scandinavian approaches to participatory design have explored the potential of expanded discursive and argumentative repertories and *voices* for boundary crossing in expert work (Engeström, 1992; Engeström, in press; Engeström, Brown, Christopher, & Gregory, 1991; Saarelma, 1993; Sjöberg, 1994; Virkkunen, 1991). Shchedrovitskii and his colleagues in Russia have developed *organizational activity games* that attempt to combine the representational, implementational, and dialogical antecedents of boundary crossing into a unified complex intervention (Shchedrovitskii & Kotel'nikov, 1988).

The status and methodology of the case studies

In the following case analyses, we explore variations of polycontextuality and boundary crossing in order to take a step further toward understanding the horizontal aspect of expertise. The cases are taken from a large set of prolonged field observations – videotaped interactions and interviews collected in Finnish and American schools, medical centers, factories, and banks in 1993 and 1994 – within an ongoing cross-cultural project entitled "Learning and Expertise in Teams and Networks." The first two cases are from a health and social welfare center and a primary school, both settings of fairly traditional professional expertise. The third case is from an industrial plant that specializes in producing cabins for ships. It involves expertise not associated with a professional status. All three cases deal with socially distributed or collaborative cognition (Cole & Engeström, 1993).

The three cases presented are not meant to provide proof for the hypothesized increasing significance of boundary crossing, nor is our analysis aimed at establishing a full explanation of these phenomena. Boundary crossing was not our topic as we began our research. During our fieldwork, however, we have witnessed a number of incidents that seem to require conceptualization of the horizontal dimension of expertise. Instead of pushing for a fixed theoretical framework to account for these incidents, we have opted for a more data-driven and stepwise approach, resembling the *grounded theory* methodology of Glaser and Strauss (1967).

At this point, our research on this issue has an exploratory character. The examples are discussed with three aims in mind. First, we want to describe and illuminate the types of concrete incidents we find relevant at this exploratory stage. Second, we examine the uses of an activity-oriented approach to and some promising theoretical constructs in the analysis of the three cases. Finally, we discuss boundary-crossing processes as indications of an emerging horizontal dimension of practical intelligence.

Recently, Phelan and her colleagues (Phelan, Davidson, & Cao, 1991; Phelan, Yu, & Davidson, 1994) have presented a framework for analyzing adolescents' problems and pressures in terms of relations between the social worlds of family, peers, and school. The adolescent's ability to cross borders between the three social worlds is seen as an important indicator of psychological well-being. The authors use interview data to classify individual subjects into four categories: (1) congruent worlds/ smooth transitions; (2) different worlds/border crossings managed; (3) different worlds/border crossings difficult; and (4) different worlds/border crossings resisted. In a somewhat similar vein, Ancona (1991) has classified work teams into different orientations with regard to their relations to the outside world.

While such typologies may be instructive and diagnostically valuable, we find the embedded assumption of relatively stable and quickly generalizable individual or group predispositions quite problematic. In line with key principles of activity theory (e.g., Engeström, 1991a,b), instead of classifying individuals or groups into general and stable categories, we aim at capturing important *interactive processes* and *mediating artifacts* involved in boundary crossing in *specific cultural-historical activity systems*. Our assumption is that these processes and tools are constructed and employed by groups and individuals in ways that vary and develop.

Our data collection was based on the notion of problem trajectory (for a related concept of trajectory, see Strauss, 1985). After entering a team, we identified a complex problem, issue, or task that the team was about to tackle. We followed and recorded the handling and resolution of this problem or issue as closely and completely as possible. We attended team meetings, interviewed team members, and shadowed on-line work interactions with the help of video cameras and audiorecorders (for shadowing methodologies, see Reder, 1993; Sachs, 1993). One problem trajectory was typically followed for a period of 2 to 3 months. The amounts and types of data from which our examples are drawn are shown in Table 15.1. The numbers in Table 15.1 reflect differences between the workplaces and work activities. In the health and social welfare center, formally arranged meetings were the predominant form of collaboration. In the cabin production plant, informal exchanges on the shopfloor were the predominant form of collaboration. The school was somewhere between those two in this respect.

Case 1: Trying to cross the boundary between professional work and clients' everyday life

Our first case is situated in a municipal welfare and health center in a mid-size Finnish city. The center provides basic welfare and health services to

Table 15.1. *Overview of data from which the cases are drawn*

	Case 1: Welfare and health center	Case 2: Primary school teachers	Case 3: Cabin production plant
Videotaped team meetings	11	12	9
Interviews	4	6	38
Hours of recorded shadowing	4	8	30

the population and employs a broad variety of professional groups, ranging from physicians and nurses to social workers and daycare personnel. In this center, the staff had made an effort to reorganize their work into multiprofessional teams, each responsible for the population of a specific geographic area. While teams as such are not new in health care and social welfare services, in these teams, physicians, social workers, and other groups collaborated on all issues related to their work, which indeed is a new phenomenon in Finland. However, in this case, the number of employees per each area was so large that the teams were formed on a representative basis so that not every staff member could participate directly in the team meetings.

In the fall of 1993, one of the teams took on the task of creating a more organized and effective collaboration with the inhabitants of its area. During a period of 3 months, we attended and videotaped all the meetings in which this issue was discussed. In this case, the problem trajectory led to a dead end, and the team gave up on the task after a series of meetings that did not produce concrete results.

In a meeting of the leadership group of the center, a representative of the team explained their project.

Team representative: Basically, our team would be like a consultant group. So the activity would definitely be the inhabitants' own. Not so that we would take charge of all kinds of projects like hobbies for the unemployed or fitness groups or first aid training or such; we would just support it with the professional competencies we have. . . . So we don't know yet where this will lead and in what forms we'll be in it, so this is more like feeling it out. We don't have any ready-made rules, telling them that such and such nice things should happen here. We will rather take kind of a passive role, so that we are involved but by no means in charge.

The idea was to cross the boundary between professional work and the voluntary activity of lay clients. Professionals wanted their clients to take a more active role in helping themselves, yet they did not want to take the lead in shaping that role. Instead, they offered their expert support.

The initiative stemmed largely from increasingly tight budget constraints in health and welfare. The staff felt that inhabitants' volunteer work might be a solution. In a joint meeting of the team and representatives of the inhabitants' association, a team member expressed the complexity of the task.

Team member 1: Our trade unions have also not regarded positively these forms of volunteer work. So there are many contradictory pressures in the picture. But probably this new situation forces people to reassess the place of volunteer work in relation to the tasks of the bureaucracy and city officials. So perhaps our challenge is to seek the very boundary, what is the right intermediate terrain and what is the need for volunteer work, and on the other hand, the need for support.

Other team members pointed out that the term *volunteer work* implies sacrifices and should not be used.

Team member 2: It's a bad word. It must be dropped now.
Team member 1: So we must find another word. What would it be, that would be enough . . . ? What would you want us to use?
Team member 3: I haven't developed that idea.

The problematic nature of the task is evident here. The professionals wanted to activate inhabitants as volunteers, but the very notion of volunteer work seemed demotivating. At the same time, the professionals themselves wanted to remain relatively passive supporting experts.

Together with the leaders of the inhabitants' association, the team organized a town hall meeting for the inhabitants. Very few people attended the meeting. In the meeting, a social worker formulated the situation as follows.

Social worker: Our problem is that all these occasions are attended mainly by people who already know about these matters. Sometimes, like today, it seems that half of the participants are officials who largely are talking to each other. In my opinion, this is the problem, and I maintain that those who are worst off will not so much as take a step toward these meetings. They stay away, and then we who are well off are solving their problems here.

After 11 meetings, some of them conducted together with representatives of the inhabitants' organization, the team abandoned the task. The boundary was not crossed.

In this case, the notion of inhabitants and their needs remained strikingly abstract throughout the process. The professionals insisted on not taking the lead. They expected the inhabitants to come and take the initiative in town hall meetings. The concrete flesh-and-blood inhabitants that the professionals saw every day as their clients were not approached. Thus, specific needs, objects and contents of collaboration and volunteer work were never taken up.

The team relied on meetings and talk as the sole tools for boundary crossing. The meetings were attended by representatives, not by those whose problems were the initial motivation for the project. Due to their self-contained nature, holding meetings seemed to become almost an end in itself.

The discussion of the term *volunteer work* and its possible alternative may be seen as an attempt at collective formation of a theoretical concept. The discussion never became grounded in practical experimentation and instantiation – or *enactment*, as Weick (1979) would call it. Thus, the search did not lead to a viable new concept.

Case 2: Crossing the boundary between two teacher teams

Our second case is from an elementary school in Helsinki. In the fall of 1993, two groups of teachers formed teams (teams A and B), both consisting of five teachers. The aim of both teams was to plan and execute the local curriculum in a collaborative fashion. We followed the two teams for a period of approximately 2 months. During that period, team B planned and implemented a new curriculum unit, aimed at getting to know the local community. The problem trajectory proceeded from an initial search for basic principles through a period of concrete planning to the implementation of the unit in classrooms.

Two teachers in team A had collaborated closely over several years. At the beginning of the school year, they presented their model of curriculum to the team. Team B spent their two first meetings exchanging ideas of interesting themes for collaborative curriculum units, not settling for a specific model.

The next meeting was conducted jointly with the two teams. Team A wanted to present their model to team B. Team B was not willing to emulate team A but was open to an exchange of ideas. In the discussion, the division of students into groups became the trigger issue.

Team A's model was based on elective courses given to selected students during each of the six periods of the school year. These elective courses were taught in small groups, with 10 students in a group. Together with their parents, students had to select one of two alternative courses offered to them. Teacher B5 questioned the rationale of the model. For her, it offered forced alternatives, not genuine choice.

Teacher B3: So then you divide your students, you divide them into groups of appropriate size . . . ?
Teacher A1: We give them two alternatives. Yes, two alternatives, which can be, for instance, video and soccer. They are together because both have the objective of strengthening social skills. And now they get a slip which says that we have selected for your child this and this course, and for some of them, there is video or soccer, from which . . .
Teacher A5: Fill the selection slip.

Teacher B5: So they select at home one or the other?

Teacher A4: Then they select one or the other.

Teacher B5: But what is the ideology here, since the basic idea would be to increase the child's right to choose according to his or her own interests and to progress in the direction of his or her own choice? And now, however, it's like "Take this or this, but this is what you'll take."

Teacher A2: Our point of departure is . . .

Teacher A1: These elective courses are not for that purpose.

Teacher A2: Yes, they are not electives in that sense. They must be in line with the objectives. They have to serve the objectives, and the selection must be based on the teacher's familiarity with the student and on educational work done together with the parents.

Team B held its own meeting immediately after the joint meeting of the two teams. In their meeting, members of team B settled for a model of their own, based on groups of 30 students, with each group having a different theme within a shared curriculum unit. The discussion within team B in effect continued the argument with team A's ideas, although team A was no longer physically present. Notice the use of reported speech as an argumentative tool by teachers B5 and B3 in the following excerpt (see also Engeström et al., 1991).

Teacher B5: We should probably have a discussion on the principle, to agree whether it's a problem for us that we have 30 students each. Because we have had that idea so far, but now we heard that "Aha! We will only have 10."

Teacher B3: So now we should suddenly begin to change it.

Teacher B5: So we are like . . .

Teacher B3: Yes, but you know, I think it's to our advantage, although we have 30. My idea of the activity is not that we tell them that "You go there, and you go here." The students will have a strong motivation for what they'll do, so that when they come, for example, to make their own paper or when they go somewhere, they won't start to fool around although there are 30 kids. Instead, I feel that if we determined ahead of time what course they can take, then their motivation would be low and it would take a lot of time to get them interested in the first place.

The discussion led to a formulation of team B's own model as distinct from that of team A. This model meant that students could choose between five different themes within the shared broader topic of a curriculum unit. Members of team B called their model *theme teaching*.

Teacher B1: There is also this difference, theirs are clearly called *elective courses*. Yet I think in theme teaching choice comes within the unit. . . . So let's stick to the name of *theme teaching*, and it will contain choice.

Teacher B5: Yes, there really was a lot that somehow clarified things. . . .

Teacher B3: We have the freedom to make our own model.

Teacher B5: We have to reflect, what are the differences and similarities between these. We really don't have to talk about elective courses.

Teacher B3: No, we don't.

Teacher B5: We haven't made elective courses so far. And we don't have to strive to be similar.

Teacher B3: Yes, I think it's more like we will seek functional diversity, or a new way of working. Of course, we have objectives regarding the contents, too, but it's more a question of students learning to work in different ways . . .
Teacher B1: And we, too.
Teacher B3: And then to collaborate, also.
Teacher B5: Students get involved in different situations, in different groups, with different teachers and different students. So the goal is pretty social, because the contents change. Plus then, of course, the advantage of interaction between us, that we get to exchange and receive ideas as much as possible.

Here, the joint meeting of the two teams was a form of boundary crossing. It did not lead to a shared concept or action plan between the teams. To the contrary, it sharpened the differences between the views held by the two teams. Teacher B5 used the argumentative question ("But what is the ideology here?") as a key discursive tool to sharpen up the differences. This argumentative sharpening of differences was decisively important for the evolution of team B's model. Team B formulated and subsequently successfully implemented in practice the theoretical notion of theme teaching.

Argumentation is not fruitful if there is no common point of reference. In this case, the question of dividing the students into groups functioned as a temporary boundary object or "springboard" (Engeström, 1987, pp. 328–31) that enabled the two teams to compare and contrast their views. It led to the fundamental questions of ideology and choice, of individualism and social interaction, and then back to the practical pros and cons of working in groups of 30. Such stepwise alternation between theoretical principles and practical implementation seems typical to creative concept formation (Davydov, 1990).

Case 3: Crossing the boundary between two production departments in an industrial plant

Our third case is from a plant that specializes in the manufacture of modular ship cabins, bathrooms, and hotel rooms. The plant has some 170 employees, 130 of whom are blue collar. The plant's unique prefabricated modular construction process means quicker completion and lower capital costs for the customer. While the basic technology is well tested, each ship is a new project that requires specific solutions. The plant has a very flat organization that relies heavily on informal communication and rapid problem solving on the shopfloor.

For a period of 3 months in the fall of 1993 and winter of 1994, we observed and recorded the production of bathrooms for a large luxury cruise ship. This problem trajectory was a subtrajectory of the entire project of designing and producing nearly 500 cabins and bathrooms for the ship.

The bathrooms were produced first; the cabins followed. The quality standards in this case were extremely high, and the production took place under heavy time pressure. These circumstances meant that there were a tremendous number of problem-solving sequences for us to observe and record. The trajectory of bathroom production was one of constant troubleshooting and readjustment.

In this production process, a crucial transition takes place when prefabricated parts are moved from the parts production hall to the assembly hall. Problems appear when the prefabricated parts – in this case, wall panels for the bathrooms – do not meet specifications. In these cases, the assembly workers either jerryrig the parts so that they can assemble the final product in spite of faults in the parts, or they take up the issue with the parts production hall and ask them to improve the quality of the parts, sometimes even returning the faulty parts. This latter measure is seldom feasible under time pressure, so the assembly workers are often in a double bind. Assembling the product from faulty parts runs the risk of turning out a substandard product; returning the parts means delaying the production.

In a traditional functional organization, when a production department has complaints about another department, the issue is taken to the central management, which then investigates and resolves it from above. In the plant we studied, this route was seldom taken. Instead, the plant relied on boundary crossing. Representatives of the assembly hall, typically a foreman or the hall supervisor, contacted representatives of the parts production hall directly and discussed the problems without intervention from the central plant or project management.

We observed and recorded a number of such boundary-crossing interactions focused on problems with the tilesetting of the bathroom wall panels. In the project we observed, a new method of horizontal tilesetting was used. The typical sequence of these interactions was as follows. First, the foreman and possibly the hall supervisor of assembly were prompted to notice a deficiency in the tilesetting of the wall panels of one or more bathrooms that were being assembled. This was usually manifested in the form of excessively wide seams (the allowable maximum width of a seam was 3 millimeters). The foreman or the hall supervisor would then contact the foreman or hall supervisor of parts production and ask him to visit the assembly hall to observe and discuss the problem. During such a visit, the problem would be assessed, and remedies would be discussed.

An assembly worker described the problems as follows.

Assembly worker: I wonder if it will work. It seems that when they fit on one side, the other side won't fit. I don't know, they just change it, and after that, they [the panels] come down here, and then we wonder again why they don't fit. Nobody says anything, nobody asks anything. Then we go again and get the boss [foreman or hall supervisor], "Come and see for yourself." And again he calls them, and they change it, and then it comes here again. And we try it, and they run back and forth like that.

According to tilesetters and their superiors working in parts production, the problems were largely due to uneven materials and the newness of the production technique. The different contexts of the two production halls are manifest in the following excerpt. The hall supervisor of parts production comes to the assembly hall to examine a problem with the supervisor and foreman of the assembly hall.

Talk	*Nonverbal events*
Foreman: Any comments?	Parts production supervisor arrives to observe an assembled bathroom.
Parts production supervisor: This is a done deal.	
Foreman: What is?	
Parts production supervisor: This here.	Points to an assembled wall.
Foreman: What, this?	
Parts production supervisor: This repair, this change. So previously there was that, you know that.	Points at a chink in the wall.
Foreman: Previously there was a finger-wide chink in here. And he [refers to the foreman of parts production] said that he'll take more this way, and he overdid it.	Waves his hand and looks at the parts production supervisor in the eyes.
Parts production supervisor: Yes. And now everything that comes belongs to this first batch, or old system. Everything must be redone now, this second batch. And we'll change it. Did you get it?	Looks at the foreman.
Foreman: No, I sure didn't get it.	Smiles.
Parts production supervisor: Well, you see, we'll make it with the old system this time.	Points a finger at a seam in the panel.
Foreman: You mean the panel seam?	
Parts production supervisor: Yes, all those that are coming.	

This sequence demonstrates the importance of using the physical artifact, the bathroom itself, as a mediating boundary object. Without being able physically to point at and observe visually and touch the walls, negotiated problem solving of this kind would be very difficult. On the other hand, the talk about "first batch" and "second batch" and "old system" and "new system" is not mediated by any mutually accessible boundary object. This seems to be a major source of difficulty in the discourse.

In the end, this particular interaction boiled down to the question of what shall be done with the four already assembled bathrooms with deficient tilesetting.

Talk	Nonverbal events
Assembly supervisor: But we cannot use these.	Looks at parts production supervisor.
Parts production supervisor: Can't you move this seam a little bit?	Shows the seam with his hand.
Assembly supervisor: No, because we can't change that at all.	Shows with his hand.

This leads to multiple steps of examining the problem visually and probing various possible remedies. The parts production supervisor eventually comes up with something that seems to move the discussion a step closer to a solution.

Talk	Nonverbal events
Parts production supervisor: But what if you took away that tank.	All three men move behind the wall.
Foreman: This is what you get when you take away the tank.	
Parts production supervisor: So there are four walls like that. Take a bit inward there.	Shows with a hand.
Assembly supervisor: So we'll tear a little bit there, or? But we can't take away this wall in any case.	
Parts production supervisor: No, but that toward there, this wall just from inside, that edge away.	Shows with a hand.
Assembly supervisor: We might get a little bit there. But then again, the bathtub . . .	Scratches his head, bows down.
Foreman: We could try that, with these four walls.	
Parts production supervisor: Yes.	

The solution was that the assembly foreman promised to fix the four faulty units. The whole interaction is interesting in that the original idea of complaining and asking the parts production to stop sending faulty parts took a more constructive turn through dialogue. The parts production supervisor engaged in joint problem solving with the two representatives of the assembly, figuring out ways to fix the four already assembled deficient bathrooms – something obviously outside his normal duties. There is a joint realization that problems such as these will not disappear at once; they will have to be argued and negotiated.

Talk	Nonverbal events
Assembly supervisor: Well, when the guys begin to work, it will be fixed, too.	Grabs parts production supervisor on the shoulder; they begin to walk away.
Parts production supervisor: Yes.	Laughs.
Foreman: I'll fix these four. We will discuss the next one again.	Yells smilingly after them.

This acceptance of the continuous nature of the problem is a double-edged phenomenon. On the one hand, it may be interpreted as mature cognitive commitment to long-term dialogical problem solving. On the other hand, it may reflect an inability to engage in collective concept formation that could yield generalizable (i.e., theoretical) models for solving the problem rather than mere local fixes.

This case seems to be a good example of what Zuboff (1988, p. 61) calls the *action-centered* skill, based on sentient information derived from physical cues and developed in physical performance. The interaction was heavily mediated by body movement, physical pointing, and gestures. On the other hand, this characterization misses the significance and power of verbal dialogue in the interaction Zuboff (1988, p. 61) maintains that action-centered skill "typically remains unexplicated" and that "it is the individual body that takes in the situation," providing "a sense of interiority." We suspect that such a view underestimates the role of dialogical collaboration and absolutizes the role of the individual body in many forms of physical work. In case 3, verbal dialogue was intricately interwoven with the collaborative use of body movements and physical artifacts of the environment. As observers, we doubt that the knowledge that emerged was characterized by "a sense of interiority." Our strong impression is that the knowledge had much more an emergent and interactive quality, as if it were hanging in the air between the actors and their artifacts, waiting to be further tested, modified, and discussed.

Conclusion

As we noted earlier, the three case analyses presented are exploratory and tentative. Our first aim was to describe and illuminate the range of phenomena we find relevant in the elaboration of the concepts of polycontextuality and boundary crossing in expert work. Key characteristics of the three cases may be summarized with the help of Table 15.2, which gives an indication of the diversity of boundary-crossing processes. The very notion of boundary is multifaceted. In case 1, the boundary to be crossed was between professionals and lay inhabitants. In case 2, it was between two teams of the same profession. In case 3, it was between two production halls of the same plant.

Table 15.2. *Key features of boundary crossing in the three cases*

	Case 1: Welfare and health center	Case 2: Primary school teachers	Case 3: Cabin production plant
1. Who did the boundary crossing?	Multiprofessional team of staff representatives	Teacher team	Supervisors and foremen of two production halls
2. What was the boundary?	Between the professionals and lay inhabitants	Between two teacher teams	Between two production departments
3. What was the problem?	Collaboration with and volunteer work by inhabitants	Constructing a local model of curriculum	Deficiencies in prefabricated parts
4. What tools or boundary objects were used?	Meetings and talk	Argumentation; dividing students into groups as boundary object	Physical artifact as boundary object; pointing; bodily movement
5. How did the process evolve?	The inhabitants and their needs remained abstract: the initiative evaporated	Differences were sharpened: team B formulated its model in contrast to that of team A	Complaint was turned into mutual problem solving; local solution was reached (but no general model)
6. How were theory and practice related?	Theoretical search for an alternative to "volunteer work": no practical implementation	Theoretical notion "theme teaching" debated and implemented in practice	Practical problem resolved locally through physical action and artifacts; no theorizing

Case 1 demonstrates the difficulty of crossing boundaries by means of meetings only, without identifying concrete problems and engaging flesh-and-blood partners on the other side of the boundary to be crossed. Case 2 demonstrates that boundary crossing does not have to achieve mutually accepted interpretations across boundaries to be fruitful. Realization of differences and contrasts by means of argumentation may trigger significant, collective concept formation on one or both sides of the boundary. Finally, case 3 demonstrates that boundary crossing can be a mutual process of problem solving in which the initially assumed roles of the parties may be changed or reversed.

The analyses of the three cases indicate that an activity-theoretical framework, focusing on the objects and mediating artifacts of actual recorded processes of collaborative work and problem solving, may indeed be useful in the articulation of the horizontal dimension of expertise. In particular, boundary objects and argumentation seem promising in the conceptual elaboration of the mediating artifacts typical to boundary-crossing processes. On the other hand, we have only begun to work out a conceptual toolbox for effective analysis of boundary crossing in collaborative work activities.

Boundary crossing entails stepping into unfamiliar domains. It is essentially a creative endeavor that requires new conceptual resources. In this sense, boundary crossing involves collective concept formation. The three cases demonstrate variations of how concept formation and learning that attempt to combine theory and practice may be involved in processes of boundary crossing.

In case 1, a theoretical search for an alternative concept to volunteer work was initiated. But the search was confined to talk in meetings and never led to practical implementation. Thus, concept formation reached a dead end. In case 2, crossing the boundary between two teacher teams led to debate and disagreement, which seemed to facilitate concept formation, or theorizing "from the ground up," in team B. The theoretical notion of theme teaching was formulated and put into practice. In case 3, the boundary between two production halls was crossed by means of dialogue heavily relying on the use of physical artifacts and body movements. While the dialogue itself proceeded through flexible role reversals and the end result was agreement, no significant collective theorizing and concept formation seemed to take place beyond a very local and specific fix.

Some years ago, Goodnow (1986) pleaded for a research agenda that would study intelligence as working together in particular social settings. While there are some promising examples of such research (e.g., Granott & Gardner, 1994), we feel that analyses of historically evolving and newly emerging objective demands for new forms of working intelligence remain seriously underdeveloped. The implication is that psychological studies of intelligence should join forces with historical, sociological, and ethnographic analyses of collaborative work activities. Careful analyses of work and organizations in transition can teach us a lot about the future of intelligence. Horizontal boundary crossing between cultures and practices is an important component of that future.

The concepts of boundary crossing and horizontal dimension of expertise call for more focused elaboration on the social and interactive nature of practical intelligence. To understand new forms of intelligence, we need to engage in concrete analysis of interlinked activity systems and their technological and organizational mediations. While objective, activity systems cannot be understood as stable, externally given task-demand structures. Activity systems are themselves constantly reconstructed by their participants. Thus, the evolution of intelligence in cooperative activities is characterized by local innovations, some leading to new stable patterns (Engeström, in press; Fogel, 1993).

Acknowledgments

The research reported in this chapter has been conducted within the framework of the project "Learning and Expertise in Teams and Networks," funded by the Academy of Finland. We express our thanks to the informants of our various field sites for their generous collaboration. We also thank Rainer Bromme, Michael Cole, Elena Grigorenko, and Robert Sternberg for useful critical comments on earlier drafts of this chapter.

References

Alter, C., & Hage, J. (1993). *Organizations working together.* Newbury Park: Sage.
Ancona, D. G. (1991). *The changing role of teams in organizations: Strategies for survival.* The International Center for Research on the Management of Technology. Sloan School of Management, Massachusetts Institute of Technology. Working Paper # 37–91.
Argyris, C (1992). *On organizational learning.* Cambridge: Blackwell.
Aydin, C. E., & Rice, R. E. (1992). Bringing social worlds together: Computers as catalysts for new interactions in health care organizations. *Journal of Health and Social Behavior, 33,* 168–85.
Ayton, P. (1992). On the competence and incompetence of experts. In G. Wright & F. Bolger (Eds.), *Expertise and decision support.* New York: Plenum.
Bakhtin, M. M. (1981). *The dialogic imagination: Four essays by M. M. Bakhtin.* Austin: University of Texas Press.
Bakhtin, M. M. (1986). *Speech genres and other late essays.* Austin: University of Texas Press.
Bartlett, F. (1958). *Thinking: An experimental and social study.* London: Allen & Unwin.
Biemans, W. G. (1992). *Managing innovation within networks.* London: Routledge.
Billig, M. (1987). *Arguing and thinking: A rhetorical approach to social psychology.* Cambridge: Cambridge University Press.
Brehmer, B. (1980). In one word: Not from experience. *Acta Psychologica, 45,* 223–41.
Brown, J. S., & Duguid, P. (1994). Borderline issues: Social and material aspects of design. *Human-Computer Interaction, 9,* 3–36.
Cannon-Bowers, J. A., Salas, E., & Converse, S. (1993). Shared mental models in expert team decision making. In N. J. Castellan, Jr. (Ed.), *Individual and group decision making: Current issues.* Hillsdale, NJ: Erlbaum.
Chi, M. T. H., Glaser, R., & Farr, M. J. (Eds.). (1988). *The nature of expertise.* Hillsdale, NJ: Erlbaum.
Cole, M., & Engeström, Y. (1993). A cultural-historical approach to distributed cognition. In G. Salomon (Ed.), *Distributed cognitions: Psychological and educational considerations.* Cambridge: Cambridge University Press.
Collins, H. M. (1990). *Artificial experts: Social knowledge and intelligent machines.* Cambridge, MA: The MIT Press.
Davydov, V. V. (1990). *Types of generalization in instruction.* Reston, VA: National Council of Teachers of Mathematics.

Dreyfus, H., & Dreyfus, S. (1986). *Mind over machine: The power of human intuition and expertise in the era of the computer.* Oxford: Basil Blackwell.

Engeström, Y. (1987). *Learning by expanding: An activity-theoretical approach to developmental research.* Helsinki: Orienta-Konsultit Oy.

Engeström, Y. (1990). *Learning, working and imagining: Twelve studies in activity theory.* Helsinki: Orienta-Konsultit.

Engeström, Y. (1991a). Activity theory and individual and social transformation. *Multidisciplinary Newsletter for Activity Theory, 7*(9), 6–17.

Engeström, Y. (1991b). Developmental work research: Reconstructing expertise through expansive learning. In M. I. Nurminen & G. R. S. Weir (Eds.), *Human jobs and computer interfaces.* Amsterdam: Elsevier Science Publishers.

Engeström, Y. (1992). *Interactive expertise: Studies in distributed working intelligence.* University of Helsinki, Department of Education. Research Bulletin #83.

Engeström, Y. (in press). Innovative organizational learning in medical and legal settings. In L. Martin, K. Nelson, & E. Tobach (Eds.), *Theory and practice of doing: Sociocultural psychology.* Cambridge: Cambridge University Press.

Engeström, Y., Brown, K., Christopher, C., & Gregory, J. (1991). Coordination, cooperation and communication in courts: Expansive transitions in legal work. *The Quarterly Newsletter of the Laboratory of Comparative Human Cognition, 13,* 88–97.

Engeström, Y., Miettinen, R., & Punamäki, R.-L. (Eds.). (in press). *Perspectives on activity theory.* Cambridge: Cambridge University Press.

Ericsson, K. A., & Smith, J. (Eds.). (1991). *Toward a general theory of expertise: Prospects and limits.* Cambridge: Cambridge University Press.

Fogel, A. (1993). *Developing through relationships: Origins of communication, self, and culture.* Chicago: The University of Chicago Press.

Glaser, B. G., & Strauss, A. L. (1967). *The discovery of grounded theory: Strategies for qualitative research.* Chicago: Aldine.

Goodnow, J. J. (1986). A social view of intelligence. In R. J. Sternberg & D. K. Detterman (Eds.), *What is intelligence? Contemporary viewpoints on its nature and definition.* Norwood, NJ: Ablex.

Goodwin, C. (1990). *Perception, technology and interaction on a scientific research vessel.* Paper presented at the 89th Annual Meeting of the American Anthropological Association, New Orleans, LA.

Göranzon, B., & Josefson, I. (Eds.). (1988). *Knowledge, skill and artificial intelligence.* London: Springer-Verlag.

Granott, N., & Gardner, H. (1994). When minds meet: Interactions, coincidence, and development in domains of ability. In R. J. Sternberg & R. K. Wagner (Eds.), *Mind in context: Interactionist perspectives on human intelligence.* Cambridge: Cambridge University Press.

Håkansson, H. (Ed.). (1987). *Industrial technological development: A network approach.* London: Dover.

(1989). *Corporate technological behaviour: Cooperation and networks.* London: Routledge.

Harrison, B. (1994). *Lean and mean: The changing landscape of corporate power in the age of flexibility.* New York: Basic Books.

Hirschhorn, L., & Gilmore, T. (1993). The new boundaries of the "boundaryless" company. In R. Howard (Ed.), *The learning imperative: Managing people for continuous innovation.* Boston: Harvard Business School Press.

Hoffman, R. R. (Ed.). (1992). *The psychology of expertise: Cognitive research and empirical AI.* New York: Springer.

Janis, I. L. (1983). *Groupthink: Psychological studies of policy decisions and fiascoes.* Boston: Houghton Mifflin.

Laboratory of Comparative Human Cognition (1982). Culture and intelligence. In R. J.

Sternberg (Ed.), *Handbook of human intelligence*. Cambridge: Cambrige University Press.

Lave, J., & Wenger, E. (1991). *Situated learning: Legitimate peripheral participation*. Cambridge: Cambridge University Press.

Leont'ev, A. N. (1978). *Activity, consciousness, and personality*. Englewood Cliffs: Prentice-Hall.

Leont'ev, A. N. (1981). *Problems of the development of the mind*. Moscow: Progress.

Lipnack, J., & Stamps, J. (1993). *The teamnet factor: Bringing the power of boundary crossing into the heart of your business*. Essex Junction: Oliver Wight.

Luria, A. R. (1978). *The making of mind: A personal account of Soviet psychology* (Edited by M. Cole & S. Cole). Cambridge, MA: Harvard University Press.

Margolis, H. (1993). *Paradigms and barriers: How habits of mind govern scientific beliefs*. Chicago: University of Chicago Press.

Markovà, I., & Foppa, K. (Eds.). (1990). *The dynamics of dialogue*. New York: Harvester Wheatsheaf.

Nohria, N., & Eccles, R. G. (Eds.). (1992). *Networks and organizations: Structure, form, and action*. Boston: Harvard Business School Press.

Norman, D. A. (1993). *Things that make us smart: Defending human attributes in the age of the machine*. Reading: Addison-Wesley.

Nyíri, J. C., & Smith, B. (Eds.). (1988). *Practical knowledge: Outlines of a theory of traditions and skills*. London: Croom Helm.

Ogburn, W. F., & Thomas, D. (1922). Are inventions inevitable? *Political Science Quarterly, 37*, 83–98.

Peters, T. (1992). *Liberation management: Necessary disorganization for the nanosecond nineties*. New York: Knopf.

Phelan, P., Davidson, A. L., & Cao, H. T. (1991). Students' multiple worlds: Negotiating the boundaries of family, peer, and school cultures. *Anthropology and Education Quarterly, 22*, 224–50.

Phelan, P., Yu, H. C., & Davidson, A. L. (1994). Navigating the psychosocial pressures of adolescence: The voices and experiences of high school youth. *American Educational Research Journal, 31*, 415–47.

Powell, W. W. (1990). Neither market nor hierarchy: Network forms of organization. In B. M. Staw & L. L. Cummings (Eds.), *Research in organizational behavior* (Vol. 12). Greenwich: JAI Press.

Reder, S. (1993). Watching flowers grow: Polycontextuality and heterochronicity at work. *The Quarterly Newsletter of the Laboratory of Comparative Human Cognition, 15*, 116–25.

Rogers, Y., & Rutherford, A. (1992). Future directions in mental models research. In Y. Rogers, A. Rutherford, & P. A. Bibby (Eds.), *Models in the mind: Theory, perspective and application*. London: Academic Press.

Saarelma, O. (1993). Descriptions of subjective networks as a mediator of developmental diaglogue. *The Quarterly Newsletter of the Laboratory of Comparative Human Cognition, 15*, 102–12.

Sachs, P. (1993). Shadows in the soup: Conceptions of work and the nature of evidence. *The Quarterly Newsletter of the Laboratory of Comparative Human Cognition, 15*, 125–33.

Schrage, M. (1990). *Shared minds: The new technologies of collaboration*. New York: Random House.

Scribner, S. (1984). Studying working intelligence. In B. Rogoff & J. Lave (Eds.), *Everyday cognition: Its development in social context*. Cambridge, MA: Harvard University Press.

Shchedrovitskii, G. P., & Kotel'nikov, S. I. (1988). An organization game as a new form of organizing and a method for developing collective thinking activity. *Soviet Psychology, 26*(4), 57–90.

Sjöberg, C. (1994). *Voices in design: Argumentation in participatory development* (Thesis No. 436). Linköping: Linköping University, Linköping Studies in Science and Technology.

Smirnov, S. D. (1994). Intelligence and personality in the psychological theory of activity. In R. J. Sternberg & P. Ruzgis (Eds.), *Personality and intelligence.* Cambridge: Cambridge University Press.

Star. S. L. (1989). The structure of ill-structured solutions: Boundary objects and heterogeneous distributed problem solving. In L. Gasser & M. N. Huhns (Eds.), *Distributed artificial intelligence* (Vol. 2). London: Pitman.

Sternberg, R. J., & Frensch, P. A. (1992). On being an expert: A cost-benefit analysis. In R. R. Hoffman (Ed.), *The psychology of expertise: Cognitive research and empirical AI.* New York: Springer.

Sternberg, R. J., & Wagner, R. K. (Eds.). (1986). *Practical intelligence: Nature and origins of competence in the everyday world.* New York: Cambridge University Press.

Strauss, A. L., (1985). *Social organization of medical work.* Chicago: The University of Chicago Press.

Suchman, L. (1994). Working relations of technology production and use. *Computer Supported Cooperative Work, 2,* 21–39.

Tyre, M. J., & von Hippel, E. (1993). *Locating adaptive learning: The situated nature of adaptive learning in organizations* (Working Paper #90–93). Cambridge, MA: MIT, Sloan School of Management, The International Center for Research on the Management of Technology.

Vikkunen, J. (1991). Toward transforming structures of communication in work: The case of Finnnish labor protection inspectors. *The Quarterly Newsletter of the Laboratory of Comparative Human Cognition, 13,* 97–107.

Weick, K. E. (1979). *The social psychology of organizing.* Reading, MA: Addison-Wesley.

Wertsch, J. V. (1991). *Voices of the mind: A sociocultural approach to mediated action.* Cambridge, MA: Harvard University Press.

Zuboff, S. (1988). *In the age of the smart machine: The future of work and power.* New York: Basic Books.

16 Cognitive development from infancy to middle childhood

Stacey S. Cherny, David W. Fulker, and John K. Hewitt

Introduction

General cognitive ability, or general intelligence, may be the area in which we know the most from a behavior-genetic perspective. Bouchard and McGue (1981) noted over 140 studies of this domain, which yielded the largely consistent result that genetic differences account for approximately 50% of the observed variability in general cognitive ability. However, it was not until recently that we began to learn more about cognitive ability than this simple, albeit important, finding.

The conceptualization of general cognitive ability as variance shared by a number of tests of various specific abilities dates back to Spearman (1904). The development of general cognitive ability, conceptualized in this manner, from infancy through middle childhood, is the focus of this chapter. Longitudinal twin and sibling data can be used to address two aspects of the developmental process. The first aspect concerns the sources of observed variation in individual differences in general cognitive ability at each age of assessment. By comparison of correlations from identical (monozygotic) and fraternal (dizygotic) twins, as well as adoptive and biological siblings, behavior-genetic methodology allows us to partition the observed variation into variation due to differences between individuals in genetic makeup and environmental differences between individuals. These environmental differences can be further subdivided into those shared by members of the family and those nonshared environmental influences that are unique to individuals. We can then examine whether heritable and environmental contributions change across the developmental period in question.

The second and more interesting aspect of the developmental process involves the processes of continuity and change. Researchers have hypothesized major changes in the structure of mental functioning across development (e.g., McCall, 1979). If such changes were occurring, we might expect some discontinuity in cognitive ability scores over this period, with genetic variation being one salient cause. The extent that genetic and environmental sources of individual differences contribute to continuity and change in general cognitive ability can be estimated and explicitly tested.

The Louisville Twin Study (Wilson, 1972a, 1972b, 1978, 1983) was the first large-scale longitudinal investigation of the development of general cognitive ability. The study yielded the finding that genetic influences underlying cognition increase in importance throughout the period of infancy through adolescence. However, a thorough developmental analysis of these data was never performed. A fuller account of the developmental genetics of intelligence has only recently been made possible by advances in the methodology to assess continuity and change in development and to partition these developmental processes into those attributable to genetic differences between individuals, environmental differences between families (commonly referred to as the *shared environment*), and environmental influences unique to individuals (or the *non*shared environment) (Eaves, Long, & Heath, 1986; Hewitt, Eaves, Neale, & Meyer, 1988; Phillips & Fulker, 1989).

The purpose of this chapter is to illustrate this new approach by describing some aspects of the development of general cognitive ability from infancy through middle childhood. This will be done through analysis of data from three major developmental studies conducted at the Institute for Behavioral Genetics at the University of Colorado, Boulder: (1) the Colorado Adoption Project (CAP) (DeFries, Plomin, & Fulker, 1994; Plomin & DeFries, 1985; Plomin, DeFries, & Fulker, 1988); (2) the MacArthur Longitudinal Twin Study (MALTS) (Emde et al., 1992; Plomin et al., 1990; Plomin et al., 1993); and (3) the Twin Infant Project (TIP) (Benson, Cherny, Haith, & Fulker, 1993; DiLalla et al., 1990). In this chapter, we focus on the development of individual differences in general cognitive ability from ages 1 to 10. This updates previous reports of CAP and MALTS general cognitive ability data (Cardon, Fulker, DeFries, & Plomin, 1992b; Cherny & Cardon, 1994; Cherny et al., 1994; Fulker, Cherny, & Cardon, 1993), presenting, for the first time, an analysis including age 10. The two main questions we address are: (1) How important are genetic and environmental influences at each age of assessment? and (2) What is the relationship between these influences over time? The first question is no more than a simple nature–nurture question, asked at each point in time. The second question is the more important one and involves the notions of continuity and change. The extent that phenotypic differences are correlated over time implies continuity in development; the extent that they are not correlated implies change. Thus, the relationship between genetic and environmental influences across time indicates the degree to which these processes of continuity and change are driven by genetic factors and by the environment. To conclude the chapter, a discussion of future directions in behavior-genetic research on cognitive development is presented, including a recent finding of a quantitative trait locus influencing reading disability (Cardon et al., 1994).

The Colorado studies of cognitive development

At the Institute for Behavioral Genetics (IBG), University of Colorado, Boulder, there are three longitudinal studies of child development using genetically informative twin, adoption, and family-study designs. The Colorado Adoption Project (CAP), which began in 1975, is a longitudinal prospective adoption study of genetic and environmental determinants of behavioral development (DeFries et al., 1994; Plomin & DeFries, 1985; Plomin et al., 1988). It is a *full* adoption design, in that both the adoptive and biological parents were assessed, in addition to the adopted away children. Furthermore, nonadoptive control families were matched to the adoptive families for age, education, and occupational status of the fathers. However, this chapter focuses only on the sibling data from the CAP, where genetically unrelated siblings living together were obtained in many of the adoptive families, and genetically related siblings were tested in the control sample.

The MacArthur Longitudinal Twin Study (MALTS), which began in 1986, is a longitudinal study of individual differences in the development of personality and cognition, in which monozygotic (MZ) and dizygotic (DZ) twins have been thus far evaluated between 14 months and 3 years of age (Emde et al., 1992; Plomin et al., 1990; Plomin et al., 1993). The Twin Infant Project (TIP), which commenced in 1984, was a longitudinal study of continuity and change in infant intelligence (Benson et al., 1993; DiLalla et al., 1990) and also served to recruit twins to be studied later in the MALTS or the CAP, which also assess twins at ages 1, 2, 3, 4, and 7, as part of an ongoing comprehensive project of behavioral development in early childhood, now collectively referred to as the Colorado Twin Study (CTS). In all these studies, children were followed from birth prospectively, given individually administered tests of intelligence, and tested at regular intervals. Together, they constitute a unique study of the genetic and environmental determinants of cognitive development.

In this chapter, we report on general cognitive ability data obtained on 221 MZ and 192 same-sex DZ twin pairs tested at 1 year of age, of whom 50 MZ and 42 DZ twin pairs have thus far been tested at age 7.[1] In addition, 242 adopted probands – that is, individuals selected for study – and 245

Table 16.1. *Number of siblings and twins by age and relationship*

Relationship	Age (years)						
	1	2	3	4	7	9	10
MZ twins							
Twin 1	222	200	187	161	50	—	—
Twin 2	222	203	182	160	50	—	—
Maximum pairs	221	200	180	158	50	—	—
DZ twins							
Twin 1	193	178	168	145	42	—	—
Twin 2	193	179	167	145	42	—	—
Maximum pairs	192	176	166	144	42	—	—
Nonadoptive siblings							
Probands	245	228	213	213	215	213	217
Natural siblings	101	95	92	95	95	74	74
Maximum pairs	101	93	90	93	93	72	70
Adoptive siblings							
Probands	242	217	207	197	196	190	196
Unrelated siblings	90	95	87	90	86	66	56
Maximum pairs	87	88	84	87	81	59	51

Note: Dashes indicate not yet tested.

matched control probands were assessed at age 1, along with their 90 and 101 unrelated and biological siblings, respectively. These numbers decrease to 196 adopted and 217 nonadoptive probands, paired with 56 biologically unrelated and 74 natural siblings at age 10. Table 16.1 shows the number of MZ and DZ twins and nonadoptive and adoptive siblings at each age, along with the maximum number of pairings.

The tests used were the Bayley Mental Development Index (Bayley, 1969) at ages 1 and 2, the Stanford-Binet IQ test (Terman & Merrill, 1973) at ages 3 and 4, the Wechsler Intelligence Scale for Children–Revised (Wechsler, 1974) at age 7, and the first principal component from a telephone-administered cognitive test battery at ages 9 and 10. This telephone battery was designed to assess verbal, spatial, perceptual speed, and memory abilities. The tests follow those described by Kent and Plomin (1987) and have similar reliabilities to in-person versions of the same tests, which were those used in the Hawaii Family Study of Cognition (DeFries, Vandenberg, McClearn, Kuse, & Wilson, 1974). The present battery also has similar factor structure to an in-person version (Cardon, Corley, DeFries, Plomin, & Fulker, 1992a).

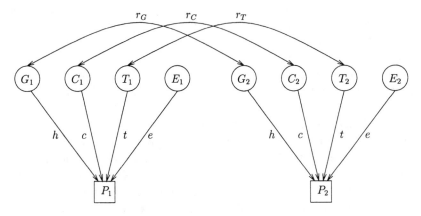

Figure 16.1. Univariate sibling model.

A behavior-genetic model for individual differences

Our approach to understanding individual differences and their continuity and change during development is based on the mathematical modeling of the variation and covariation of our subjects' test scores. This is the *structural modeling approach*, which is widely used throughout the social sciences. However, behavior-genetic models are more powerful than most. The reason for this is that structural models in behavior genetics are built on the foundations of Mendelian genetics, which sets certain parameters in the models to known theoretical values. Outside of behavior genetics, there are few other instances in the social sciences of parameters with values that are known on the basis of well-established theory. A sample of current state-of-the-art behavior-genetic modeling can be found in the first text in this area (Neale & Cardon, 1992).

The behavior-genetic model presently employed recognizes four sources of individual differences in general intelligence. In this model, shown for a pair of siblings measured on a particular phenotype, P, in Figure 16.1, G represents additive genetic differences among individuals, C represents common environmental influences shared by children reared together in the same home, T represents environmental influences shared only by twins over and above shared sibling influences, and E represents nonshared environmental influences unique to the individual. These four variables are latent variables; that is, they are not directly observable. However, use of genetically informative data allows us to estimate the magnitude of the contribution of these sources of individual differences. The genetic correlation (r_G) between genotypes for pairs of individuals, shown in Figure 16.1, varies in the present sample from 0 in the case of adoptive siblings, to 1/2 for nonadoptive siblings and DZ twins, to unity in MZ twins. For this reason, the combined twin/adoption design, in which a key theoretical parameter varies throughout the full range from 0 to 1, is optimally powerful. The correlation (r_C) between the shared environment of sibling 1 with that of sibling 2 is, by definition, unity. The correlation (r_T) between twin environments is, by definition, 1 for twin pairs and 0 for nonadoptive and adoptive siblings.

The impacts of these four sources of variation – genetic, shared environmental, additional shared twin environmental, and nonshared environmental – are shown as h, c, t, and e, respectively, in the model depicted in Figure 16.1, and the variance explained by each is the square of these quantities, h^2, c^2, t^2, and e^2. The quantity h^2 is referred to as the *narrow-sense heritability*. In the absence of Mendelian dominance or epistasis (gene × gene interaction), this parameter describes the total variation due to genetic differences between individuals. Should sources of nonadditive genetic variation be important, the study of twins and siblings will permit their evaluation. In addition, information from the parents is capable of resolving the effects of assortative mating and genotype–environment correlation, if these should prove important. The studies we are carrying out at the Institute for Behavioral Genetics are unique in this respect because they combine a variety of informative behavior-genetic designs in order to evaluate and validate better the basic models we employ in data analysis.

A behavior-genetic model for development

Given the assessment of these basic sources of variation at each age point, we then need a developmental model, in addition to the basic genetic and environmental model, to evaluate the relationships among the genetic and environmental variables over time. For the present analysis, the model that was fitted to these data was one first proposed by Eaves et al. (1986) and represents a combination of a single general factor present at all ages and a quasi-simplex model of age-to-age transmission of effects. The general factor implies a static process where influences are global across all ages. The quasi-simplex implies a more dynamic process in which new variation arises at each age and persists to the next age. The two processes, the static common factor and dynamic simplex, are illustrated in Figure 16.2. In the behavior-genetic literature, a parameterization called the *Cholesky decomposition* is also frequently used as a longitudinal developmental model (e.g., Cherny et al., 1994; Fulker et al., 1993). While the Cholesky decomposition can essentially answer most of the developmental questions that the present model can address, the simplex-factor model we present is a more precise

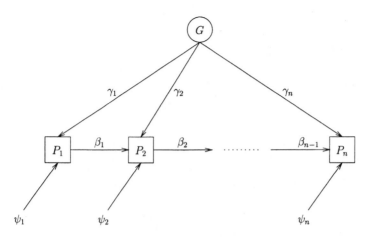

Figure 16.2. Simplex model of cognitive development. The β_i and γ_i are path coefficients, and the ψ_i are variances to be estimated.

formulation of underlying developmental processes and therefore is better suited to testing those hypotheses.

A simplex model implies a matrix of correlations between measures taken across time, where measures taken closer together are more highly correlated than measures further apart in time. The matrix would then have higher correlations nearer the diagonal and decreasing correlations the further one goes from the diagonal. This is typically observed with longitudinal data and is the case for our studies of cognitive ability. When developmental models of this kind have been applied to longitudinal cognitive data at the phenotypic level (Humphreys & Davey, 1988), a simplex was found to provide a better account of the data than a common factor from ages 1 through 9.

The full longitudinal behavior-genetic model involves a fourfold expansion of the model presented in Figure 16.2, to allow for genetic and shared, twin, and nonshared environmental levels of variation and covariation, rather than the single source depicted in the figure for simplicity. The expected covariance matrices for MZ and DZ twin pairs and nonadoptive and adoptive siblings, implied by the full model, can be derived using three parameter matrices adapted from the LISREL model (Jöreskog & Sörbom, 1989), **B**, **Γ**, and **Ψ**, at each of the genetic and shared, twin, and unique environmental levels, or 12 parameter matrices in total. The **B** matrices contain the age-to-age transmission parameters, the **Γ** matrices contain the common factor loadings, and the **Ψ** matrices contain the time-specific variances or the new variation at each age. The three parameter matrices, given for a 4 time-point model, are as follows:

$$\mathbf{B} = \begin{pmatrix} 0 & 0 & 0 & 0 \\ \beta_1 & 0 & 0 & 0 \\ 0 & \beta_2 & 0 & 0 \\ 0 & 0 & \beta_3 & 0 \end{pmatrix}, \tag{1}$$

$$\boldsymbol{\Gamma} = \begin{pmatrix} \gamma_1 \\ \gamma_2 \\ \gamma_3 \\ \gamma_4 \end{pmatrix}, \tag{2}$$

and

$$\boldsymbol{\Psi} = \begin{pmatrix} \psi_1 & 0 & 0 & 0 \\ 0 & \psi_2 & 0 & 0 \\ 0 & 0 & \psi_3 & 0 \\ 0 & 0 & 0 & \psi_4 \end{pmatrix}. \tag{3}$$

The free parameters are indicated by the appropriate Greek letter, and all other elements of the above matrices are fixed. The parameters in $\boldsymbol{\Psi}$ were constrained to be positive. The elements of these parameter matrices can be seen placed on the path diagram in Figure 16.2. The expectations for the genetic component of covariance, $\mathbf{G}$, would be given by:

$$\mathbf{G} = \left(\mathbf{I} - \mathbf{B}_G\right)^{-1}\left(\boldsymbol{\Gamma}_G\boldsymbol{\Gamma}_G' + \boldsymbol{\Psi}_G\right)\left(\mathbf{I} - \mathbf{B}_G'\right)^{-1}. \tag{4}$$

The expectations for the shared, twin, and unique environmental components of covariance, $\mathbf{C}$, $\mathbf{T}$, and $\mathbf{E}$, respectively, are obtained in an analogous manner. In summary, the model has three types of parameters to explain continuity and change: (1) the loadings of the common set of genes or environmental influences (G) influencing the measures (P_i) at all ages, symbolized by γ_i; (2) new genetic or environmental influences or innovations appearing at each age (ψ_i); and (3) age-to-age transmission of genetic and environmental influences (β_i). The transmission of genetic (or environmental) influences from age to age is what developmentalists generally refer to as *continuity*. The new variation at each age reflects developmental change. Changes in the structure of cognitive functioning (e.g., McCall, 1979) might be expected to manifest themselves as the appearance of such new genetic variation during development.

Fitting models to data from the Colorado studies of cognitive development

To test the adequacy of our models and to estimate the importance, or magnitude, of each type of influence, we compare the observed patterns of test scores, both among family members (e.g., pair of twins) and across ages, with those expected from the model. A first step is to compute the matrix of variances and covariances among family members and across ages that we expect on the basis of the model. The MZ, DZ, nonadoptive, and adoptive expected convariance matrices take a special form whereby they are partitioned into four equal quadrants. The top left and bottom right quadrants contain the within-pair variances and covariances and therefore contain the phenotypic variances and covariances. The other two quadrants contain the cross-sibling variances and covariances. These are expected to differ across the four groups of varying genetic and environmental similarity. The expected covariance matrices are estimated as:

$$\Sigma = \begin{pmatrix} \mathbf{G+C+T+E} & r \otimes \mathbf{G+C+t \otimes T} \\ r \otimes \mathbf{G+C+t \otimes T} & \mathbf{G+C+T+E} \end{pmatrix}, \tag{5}$$

where r is the genetic correlation and equal to 1 for MZ pairs, 1/2 for DZ pairs and nonadoptive siblings, and 0 for adoptive siblings, and t is equal to 1 for twin pairs and 0 for sibling pairs. The parameters of the model can then be estimated by minimizing the discrepancy between the expected and the observed covariance matrix, using optimization routines on the computer.

Due to the incomplete nature of our developmental studies, we must fit the model directly to the raw data rather than to observed covariance matrices. A Maximum-Likelihood (ML) pedigree approach (Lange, Westlake, & Spence, 1976) was employed to analyze the data in an optimal manner, allowing one to accommodate any pattern of missing data. We maximized the following ML pedigree log-likelihood (LL) function, using the MINUIT optimization subroutines (CERN, 1977):

$$LL = \sum_{i=1}^{N} \left[-\frac{1}{2} \ln |\Sigma_i| - \frac{1}{2} (\mathbf{x}_i - \boldsymbol{\mu})' \Sigma_i^{-1} (\mathbf{x}_i - \boldsymbol{\mu}) \right], \tag{6}$$

where

- $\mathbf{x}_i$ = vector of scores for sibling pair i
- Σ_i = appropriate MZ, DZ, adoptive, or nonadoptive expected covariance matrix
- N = total number of sibling pairs
- $\boldsymbol{\mu}_i$ = appropriate vector of means

and where

$$2(LL_1 - LL_2) = \chi^2 \tag{7}$$

for testing the difference between two alternative models. (Alternatively, these models could have been fitted using the freely available and highly flexible *Mx* modeling package [Neale, 1994].) The vector of means can either be modeled or, as in the present case, where we postulate no theory of mean structure, simply fixed to the observed means. This has the added advantage of being less computer intensive, which is a major concern for relatively large models of the present size. Use of this fit function, as opposed to the more common ML function used by such programs as LISREL (Jöreskog & Sörbom, 1989) and EQS (Bentler, 1989), allows all the data to be analyzed. It is assumed that the missing data are missing completely at random, which is a reasonable assumption in the present case. Use of the more common fit function for complete data would necessitate elimination of those sibling pairs from the analysis who were not measured at all time points, which would mean losing information unnecessarily. Furthermore, had this been done, we would need to assume that those cases eliminated from the analysis were eliminated at random. In the case where there is no missing data, this pedigree function yields the same results as the ML function for covariance matrices.

The data were first standardized within each age, across all individuals as a single group. This standardization procedure effectively eliminates age differences in variances, which most likely are merely a result of using different tests at different ages, while preserving MZ, DZ, adoptive, nonadoptive, sib 1, and sib 2 variance differences. Resulting parameter estimates were standardized to imply a phenotypic variance of unity, using an extension of the procedures employed by LISREL (Jöreskog & Sörbom, 1989).

The results of behavior-genetic modeling in the Colorado studies of cognitive development

Estimates of h^2, c^2, t^2, and e^2, obtained from fitting the full longitudinal genetic model to the data, are presented in Table 16.2. Heritability and shared environmental influences appear relatively constant across the developmental period under consideration, although h^2 appears to be a bit higher at age 7 years, with c^2 correspondingly lower. The influence of the shared environment and additional twin environment combined was generally substantially less than was found in the Louisville Twin Study (Wilson, 1983), with our estimates of heritability somewhat greater, on average. We can speculate that the inclusion of low birth-weight twin pairs in the Louisville Twin Study sample can be a cause of increased between-family variation.

To test which developmental processes may be operating for general cognitive ability from ages 1 through 10 and to arrive at the most parsimonious model that could explain these data, a series of model comparisons were performed, beginning with tests of the nonshared environ-

Table 16.2. *Estimates of* h^2, c^2, t^2, *and* e^2 *at each age*

Variance component	Age (years)						
	1	2	3	4	7	9	10
h^2	.39	.39	.35	.45	.65	.59	.44
c^2	.16	.15	.20	.12	.07	.16	.25
t^2	.02	.24	.15	.15	.14	—	—
e^2	.43	.21	.29	.27	.15	.26	.31

Note: Dashes indicate that this component cannot be estimated at those ages, since twins were not tested.

Table 16.3. *Tests of unique environment development patterns*

Model	Form	$-LL^a$	NPAR^b	χ^2	df^c	p
1	Full model	1874.52	74			
2	Model 1, drop common factor	1875.79	67	2.54	7	>.90
3	Model 1, drop transmission	1875.20	68	1.36	6	>.95
4	Model 1, drop transmission & common factor	1877.84	61	6.64	13	>.90

^{a} log-likelihood function.
^{b} number of free parameters.
^{c} degrees of freedom.

mental processes. These tests appear in Table 16.3. The first test was whether the common factor could be dropped from the model without a decrement in fit. We found that, indeed, a nonshared environmental common factor was unnecessary in explaining these data (Model 2). Next, the transmission parameters were tested and also found unnecessary (Model 3). Finally, the common factor and transmission parameters were tested as a set and, again, were found not to be required to explain these data adequately (Model 4). That is, *environmental influences not shared by siblings do not contribute to continuity in general cognitive ability across ages 1 through 10 and only contribute to change*. It is this model, which contains no nonshared environmental influences accounting for any of the observed covariance among general cognitive ability measures at each age, that we use as the base model for the next set of tests, those of the environmental influences specific to twins (see Table 16.4).

A test of the twin environment common factor indicated that it was not necessary for an adequate model fit (Model 5), nor were the simplex trans-

Table 16.4. *Tests of twin environment development patterns*

Model	Form	$-LL^a$	NPAR^b	χ^2	df^c	p
	Model 4	1877.84	61			
5	Model 4, drop common factor	1878.99	56	2.30	5	>.80
6	Model 4, drop transmission	1879.38	57	3.08	4	>.50
7	Model 4, drop transmission & common factor	1883.97	52	12.26	9	>.15
8	Model 7, drop specifics	1887.29	47	6.64	5	>.20

^{a} log-likelihood function.
^{b} number of free parameters.
^{c} degrees of freedom.

Table 16.5. *Tests of shared-environment development patterns*

Model	Form	$-LL^a$	NPAR^b	χ^2	df^c	p
	Model 8	1887.29	47			
9	Model 8, drop transmission	1893.08	41	11.58	6	>.05
10	Model 9, drop transmission & specifics	1895.00	34	3.84	7	>.75
11	Model 10, drop common factor	1904.19	27	18.38	7	<.02

^{a} log-likelihood function.
^{b} number of free parameters.
^{c} degrees of freedom.

mission parameters (Model 6). Again, as a set, the common factor and transmission parameters could be dropped (Model 7). Lastly, the age-specific variances were dropped from the model (Model 8), yielding a base model with no environmental influences unique to twins and no nonshared environmental influences accounting for the relationship between general cognitive ability measures at each age. That is, *environmental influences shared by twins, over and above those influences shared by ordinary siblings, do not appear to contribute to individual differences in general cognitive ability*. In other words, the twin and sibling correlations can be explained by the same genetic and environmental influences. This provides some justification for combining sibling/adoption data with twin data in a single comprehensive analysis.

Tests of the developmental processes present at the level of the shared sibling environment (which twins also share to the same extent) appear in Table 16.5. Age-to-age transmission was not found to occur at this level of covariance (Model 9). Furthermore, time-specific influences were also

Table 16.6. *Tests of genetic development patterns*

Model	Form	$-LL^a$	NPAR[b]	χ^2	df[c]	p
	Model 10	1895.00	34			
12	Model 10, drop factor	1900.69	27	11.38	7	>.10
13	Model 10, drop transmission	1966.36	28	142.72	6	<.001
14	Model 10, drop specifics	2056.09	27	322.18	6	<.001

[a] log-likelihood function.
[b] number of free parameters.
[c] degrees of freedom.

found unnecessary (Model 10). After these influences were removed from the model, *the common factor at the shared environmental level was found necessary* for an adequate model fit (Model 11). It is the model with only a single common factor at the shared environmental level that is now the base model for subsequent tests of the genetic developmental processes.

Table 16.6 contains the tests of the genetic components in the developmental model. The first test is of the common factor, and it was found to be unnecessary to explain the data (Model 12). The simplex transmission parameters were found necessary, however (Model 13). Finally, new genetic variation at each age, as a set, was also found (Model 14). The final reduced model is, therefore, Model 12, shown in Figure 16.3. In contrast to the shared environment, *genetic influences arise at each age and persist to later ages.*

Figure 16.3 shows the final reduced model, where the observed phenotypic IQ variances were standardized to unity. We see that unique environmental influences, which are time-specific, are the e^2s at each age and are, on average, greater at the younger ages than at the older ages. The loadings from C in the figure are the square roots of the respective c^2 at each age. (That is, the C factor has a variance of unity.) Shared environmental influences appear modest at each age and are perfectly correlated from age to age in this reduced model. The paths from the G_i in the figure to the phenotypes at each age are fixed to unity. Genetic variance at each age is the sum of the new variance appearing plus the variance which is transmitted from the previous age-point. This transmitted genetic variance is the product of the genetic variance present at the previous time-point and the square of the simplex transmission parameter. We see that heritability appears slightly higher at older ages than at younger age points, although it is likely not significantly so. Transmission is moderate from ages 1 to 2 and 2 to 3. The genetic variance present at age 3 is amplified slightly at 4. The magnitude of new genetic variation decreases from ages 2 through 4, with a

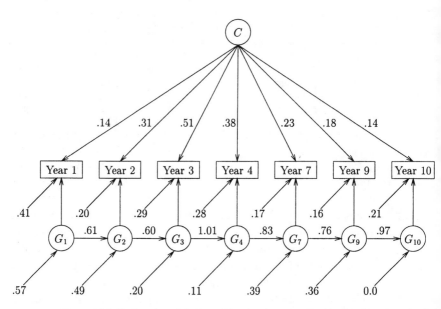

Figure 16.3. Reduced model of cognitive development. Residual variances on the G factors are parameters in Ψ_G. Residual variances loading directly on the phenotypes are parameters in Ψ_E.

rather large increase again at age 7. New variation also appears at age 9 but not age 10.

Assumptions and limitations of our analysis

An assumption of the twin method, and hence of the present analysis, is that MZ twins are not more similar because they are treated differently than DZ twins. That is, we assume equal environments for MZs and DZs. If this assumption is incorrect, we might expect different estimates of h^2 and c^2 obtained from analyses of twins only than from analyses of nonadoptive and adoptive siblings only. This was not found in an earlier analysis of our cognitive data (Cardon et al., 1992b). Furthermore, the effects of perceived zygosity and perceived similarity by the twins themselves and by others were not found to bias twin analyses of cognitive and other data significantly (Scarr & Carter-Saltzman, 1979). In addition, we presently find no evidence of an additional source of environmental similarity for twins over and above that for ordinary nonadoptive and adoptive sibling pairs. This is encouraging, although the parameter estimates are not negligible (even though not statistically significant), and not finding the effect may be an issue of the statistical power of our studies to detect such effects. Nonetheless, we would not go far astray by concluding that additional twin environmental influences are not an important source of variation. If this conclusion is wrong, our estimates of heritability are biased upward slightly.

Another assumption of this analysis is that genetic influences are additive. This assumption is unlikely to be problematic, however, given that no evidence was found for genetic nonadditivity, which would have produced DZ correlations less than half that of MZ correlations. In addition, our analyses assume that parents select a mate randomly with respect to intelligence. However, in the presence of assortative mating, the estimate of heritability obtained from a twin design would be biased downward, but that obtained from a sibling-adoption design would be biased upward. Therefore, our combined design should have offsetting biases and lead to unbiased estimates. Lastly, we assume that correlations and interactions between genetic and environmental effects are of negligible importance. Although such influences may potentially exist and may also be of theoretical interest (e.g., Plomin, DeFries, & Loehlin, 1977), their magnitude cannot be assessed from the data of the present study.

Conclusions from our analyses of cognitive development

These analyses tell an interesting story of the diversity of processes that control the development of general cognitive ability. The nonshared environment, uniquely experienced by individuals, clearly does not drive the developmental process. Its influences are transitory; that is, they are occasion-specific. Since nonshared environmental influences are not correlated over time, we would venture to say that the variance observed at this level is largely measurement error, although presumably not entirely so.

The environmental influences shared by siblings (and twins) in this study were found to be of a global nature. This could be a result of such monolithic factors as socioeconomic status and other relatively constant influences on the family. These environmental influences causing families to differ appear not to drive the developmental process. Although the relative magnitude of the effect may not be the same at each age, the source of influence remains the same throughout early childhood.

The picture at the genetic level is the most interesting. It appears that genes are driving the developmental process, as the simplex model implies. New genetic variation appears at each age, and that variation persists onto later ages. This is characteristic of a truly developmental process. The amount of new variation appears to be decreasing through age 4 but rises substantially at age 7, after the child has completed his or her first year of schooling.

The developmental model based on results from this combined twin and adoption study differs in some respects from that reported by Eaves et al. (1986), based on their reanalysis of published correlations from the Louisville Twin Study. First, their model did not include a common factor across ages for the shared environment, while they did include partial transmission from age to age of shared environmental influences. In practice, these two alternatives were difficult to discriminate in their data, and the robust conclusions about the shared environment that hold for their results as well as ours are that (1) the influence of the shared environment is correlated from age to age, either through transmission of effects or through the influence of a common factor across ages (but not both together), but (2) these correlated influences do *not* increase in their impact on individual differences during childhood.

Second, although both the Louisville Twin Study data and other available genetically informative data sets, reviewed recently by McGue, Bouchard, Iacono, and Lykken (1993), suggest an increase in the impact of genetic influences (i.e., increasing heritability during adolescence), our results do not show this trend for increasing heritability to age 10. However, we should note that the cognitive tests used in our study differed at different ages and include such innovations as telephone testing at ages 9 and 10 (Cardon et al., 1992a). For this reason, we would want to be cautious about overinterpreting our data on this point. What is consistent about our results and others is that genetic influences that manifest themselves at ages as young as 3 or even younger are conserved to a considerable extent in later years. Developmental continuity is largely a function of the age-to-age correlation of genetic influences.

The developmental analysis we have employed clearly shows that there is not a single developmental process determining relative intellectual ability from ages 1 through 10. The three sources of variation we have identified – unique environmental, shared environmental, and genetic influences – each appear to act in a rather different manner, with genetic influences driving both continuity and change.

Future directions for genetic research on cognitive ability

The analyses presented in this chapter resulted in the detection of genetic variation underlying general cognitive ability at each age. However, the studies described cannot tell us anything about where specific genes underlying cognitive ability are located. Future research will likely move toward identifying, or mapping, polygenes, or *quantitative trait loci* (QTLs) as they are now known, the genes that underlie quantitative phenotypes such as cognitive ability.

The development of methodology for the detection and mapping of QTLs has been an active area of research and has led to a number of recent advances (Cardon & Fulker, 1994; Fulker et al., 1991; Fulker & Cardon, 1994; Fulker, Cherny, & Cardon, 1995). Initially, a multiple-regression procedure for use on selected samples was proposed (Fulker et al., 1991). This procedure is based on Haseman and Elston's (1972) method of sib-pair analysis but extended for use on selected samples by application of the DeFries and Fulker (1985, 1988; LaBuda, DeFries, & Fulker, 1986) regression method for analysis of twin and sibling data. The method could also be used to examine multiple genetic markers simultaneously. While this method could detect linkage of a quantitative trait to a marker locus, it could not provide an estimate of the putative QTL's location. Subsequently, interval mapping procedures, also based on Haseman and Elston and DeFries and Fulker methodology, which allow for mapping of QTLs using sib-pair data (although it could be extended to any size kinship) were developed (Fulker & Cardon, 1994; Cardon & Fulker, 1994; Fulker et al., 1995). These methods have been applied to DZ twins and siblings selected for extreme deficits in reading ability and provided strong evidence for a major gene on chromosome 6 in the HLA region (Cardon et al., 1994).

Of necessity, a QTL will only be found if genetic variation is detectable using methods of the type presented in this chapter. Developmental genetic analyses can determine which phenotypes are the best candidates for QTL analysis. Furthermore, such analyses, by pointing to the changes and continuities in genetic architecture, can suggest at which points in the developmental process we should focus. Methods such as these and related methods of association, discussed by Plomin (chapter 3, this volume), should help us to understand better the genetic mechanisms that underlie cognitive ability and other behavioral traits.

Acknowledgments

Our work was supported in part by Grants HD-10333, HD-18426, and HD-19802 from the National Institute of Child Health and Human Development (NICHD), by Grant MH-43899 from the National Institute of Mental Health, by Biomedical Research Support Grant RR-07013-25 from the National Institutes of Health, and by a grant from the John D. and Catherine T. MacArthur Foundation. S.S.C. was supported in part by a postdoctoral fellowship from the Natural Sciences and Engineering Research Council of Canada.

Note

1. Those twins tested by the MALTS were assessed at 14 months rather than at 12 months of age.

References

Bayley, N. (1969). *Manual for the Bayley scales of infant development.* New York: Psychological Corporation.

Benson, J. B., Cherny, S. S., Haith, M. M., & Fulker, D. W. (1993). Rapid assessment of infant predictors of adult IQ: Midtwin–midparent analyses. *Developmental Psychology, 29,* 434–47.

Bentler, P. M. (1989). *EQS structural equations program manual.* Los Angeles: BMDP Statistical Software.

Bouchard, Jr., T. J., & McGue, M. (1981). Familial studies of intelligence: A review. *Science, 212,* 1055–9.

Cardon, L. R., Corley, R. P., DeFries, J. C., Plomin, R., & Fulker, D. W. (1992a). Factorial validation of a telephone test battery of specific cognitive abilities. *Personality and Individual Differences, 13,* 1047–50.

Cardon, L. R., Smith, S. D., Fulker, D. W., Kimberling, W. J., Pennington, B. F., & DeFries, J. C. (1994). Quantitative trait locus for reading disability on chromosome 6. *Science, 265,* 276–9.

Cardon, L. R., & Fulker, D. W. (1994). The power of interval mapping of quantitative trait loci, using selected sib pairs. *American Journal of Human Genetics, 55,* 825–33.

Cardon, L. R., Fulker, D. W., DeFries, J. C., & Plomin, R. (1992b). Continuity and change in general cognitive ability from 1 to 7 years of age. *Developmental Psychology, 28,* 64–73.

CERN (1977). *MINUIT: A System for Function Minimization and Analysis of the Parameter Errors and Correlations.* Geneva, Switzerland: European Organization for Nuclear Research.

Cherny, S. S., & Cardon, L. R. (1994). General cognitive ability. In J. C. DeFries, R. Plomin, & D. W. Fulker (Eds.), *Nature and nurture during middle childhood* (pp. 45–56). Cambridge, MA: Blackwell.

Cherny, S. S., Fulker, D. W., Emde, R. N., Robinson, J., Corley, R. P., Reznick, S. J., & DeFries, J. C. (1994). A developmental-genetic analysis of continuity and change in the Bayley Mental Development Index from 14 to 24 months: The MacArthur Longitudinal Twin Study. *Psychological Science, 5,* 354–60.

DeFries, J. C., & Fulker, D. W. (1985). Multiple regression analysis of twin data. *Behavior Genetics, 15,* 467–73.

DeFries, J. C., & Fulker, D. W. (1988). Multiple regression analysis of twin data: Etiology of deviant scores versus individual differences. *Acta Geneticae Medicae et Gemellologiae, 37,* 205–16.

DeFries, J. C., Plomin, R., & Fulker, D. W. (1994). *Nature and nurture during middle childhood.* Cambridge, MA: Blackwell.

DeFries, J. C., Vandenberg, S. G., McClearn, G. E., Kuse, A. R., & Wilson, J. R. (1974). Near identity of cognitive structure in two ethnic groups. *Science, 183,* 338–9.

DiLalla, L. F., Thompson, L. A., Plomin, R., Phillips, K., Fagan III, J. F., Haith, M. M., Cyphers, L. H., & Fulker, D. W. (1990). Infant predictors of preschool and adult IQ: A study of infant twins and their parents. *Developmental Psychology, 26,* 759–69.

Eaves, L. J., Long, J., & Heath, A. C. (1986). A theory of developmental change in quantitative phenotypes applied to cognitive development. *Behavior Genetics, 16,* 143–62.

Emde, R. N., Plomin, R., Robinson, J. A., Reznick, J. S., Campos, J., Corley, R., DeFries, J. C., Fulker, D. W., Kagan, J., & Zahn-Waxler, C. (1992). Temperament, emotion, and cognition at 14 months: The MacArthur Longitudinal Twin Study. *Child Development, 63,* 1437–55.

Fulker, D. W., & Cardon, L. R. (1994). A sib-pair approach to interval mapping of quantitative trait loci. *American Journal of Human Genetics, 54,* 1092–103.

Fulker, D. W., Cardon, L. R., DeFries, J. C., Kimberling, W. J., Pennington, B. F., & Smith, S. D. (1991). Multiple regression analysis of sib-pair data on reading to detect quantitative trait loci. *Reading and Writing: An Interdisciplinary Journal, 3,* 299–313.

Fulker, D. W., Cherny, S. S., & Cardon, L. R. (1993). Continuity and change in cognitive development. In R. Plomin & G. E. McClearn (Eds.), *Nature, nurture, & psychology* (pp. 77–97). Washington, DC: American Psychological Association.

Fulker, D. W., Cherny, S. S., & Cardon, L. R. (1995). Multipoint interval mapping of quantitative trait loci, using sib pairs. *American Journal of Human Genetics, 56,* 1224–33.

Haseman, J. K., & Elston, R. C. (1972). The investigation of linkage between a quantitative trait and a marker locus. *Behavior Genetics, 2,* 3–19.

Hewitt, J. K., Eaves, L. J., Neale, M. C., & Meyer, J. M. (1988). Resolving causes of developmental continuity or "tracking." I. Longitudinal twin studies during growth. *Behavior Genetics, 18,* 133–51.

Humphreys, L. G., & Davey, T. C. (1988). Continuity in intellectual growth from 12 months to 9 years. *Intelligence, 12,* 183–97.

Jöreskog, K. G., & Sörbom, D. (1989). *LISREL 7: A Guide to the Program and Applications* (2nd ed.). Chicago: SPSS, Inc.

Kent, J., & Plomin, R. (1987). Testing specific cognitive abilities by telephone and mail. *Intelligence, 11,* 391–400.

LaBuda, M. C., DeFries, J. C., & Fulker, D. W. (1986). Multiple regression analysis of twin data obtained from selected samples. *Genetic Epidemiology, 3,* 425–33.

Lange, K., Westlake, J., & Spence, M. A. (1976). Extensions to pedigree analysis: III. Variance components by the scoring method. *Annals of Human Genetics, 39,* 485–91.

McCall, R. B. (1979). The development of intellectual functioning in infancy and the prediction of later IQ. In J. D. Osofsky (Ed.), *The handbook of infant development* (pp. 707–41). New York: Wiley.

McGue, M., Bouchard, T. J., Jr., Iacono, W. G., & Lykken, D. T. (1993). Behavioral genetics of cognitive ability: A life-span approach. In R. Plomin & G. E. McClearn (Eds.), *Nature, nurture, & psychology* (pp. 59–76). Washington, DC: American Psychological Association.

Neale, M. C. (1994). *Mx: Statistical modeling* (2nd ed.). Department of Psychiatry, Medical College of Virginia, Box 710 MCV, Richmond, VA 23298.

Neale, M. C., & Cardon, L. R. (1992). *Methodology for genetic studies of twins and families, NATO ASI Series.* Dordrecht, The Netherlands: Kluwer Academic Press.

Phillips, K., & Fulker, D. W. (1989). Quantitative genetic analysis of longitudinal trends in adoption designs with application to IQ in the Colorado Adoption Project. *Behavior Genetics, 19,* 621–58.

Plomin, R., Campos, J., Corley, R., Emde, R. N., Fulker, D. W., Kagan, J., Reznick, J. S., Robinson, J., Zahn-Waxler, C., & DeFries, J. C. (1990). Individual differences during the second year of life: The MacArthur Longitudinal Twin Study. In J. Columbo & J. Fagan (Eds.), *Individual differences in infancy: Reliability, stability, & predictability* (pp. 431–55). Hillsdale, NJ: Erlbaum.

Plomin, R., & DeFries, J. C. (1985). *Origins of individual differences in infancy: The Colorado Adoption Project.* Orlando, FL: Academic Press.

Plomin, R., DeFries, J. C., & Fulker, D. W. (1988). *Nature and nurture in infancy and early childhood.* Cambridge: Cambridge University Press.

Plomin, R., DeFries, J. C., & Loehlin, J. C. (1977). Genotype–environment interaction and correlation in the analysis of human behavior. *Psychological Bulletin, 84,* 309–22.

Plomin, R., Emde, R. N., Braungart, J. M., Campos, J., Corley, R., Fulker, D. W., Kagan, J.,

Reznick, J. S., Robinson, J., Zahn-Waxler, C., & DeFries, J. C. (1993). Genetic change and continuity from fourteen to twenty months: The MacArthur Longitudinal Twin Study. *Child Development, 64*, 1354–76.

Scarr, S., & Carter-Saltzman, L. (1979). Twin method: Defense of a critical assumption. *Behavior Genetics, 9*, 527–42.

Spearman, C. (1904). General intelligence, objectively determined and measured. *American Journal of Psychology, 14*, 201–93.

Terman, L. H., & Merrill, M. A. (1973). *Stanford-Binet intelligence scale: 1972 norms edition.* Boston: Houghton Mifflin.

Wechsler, D. (1974). *Manual for the Wechsler Intelligence Scale for Children–Revised.* New York: Psychological Corporation.

Wilson, R. S. (1972a). Similarity in developmental profile among related pairs of human infants. *Science, 178*, 1005–7.

Wilson, R. S. (1972b). Twins: Early mental development. *Science, 175*, 914–17.

Wilson, R. S. (1978). Synchronies in mental development: An epigenetic perspective. *Science, 202*, 939–48.

Wilson, R. S. (1983). The Louisville Twin Study: Developmental synchronies in behavior. *Child Development, 54*, 298–316.

17 Intelligence, language, nature, and nurture in young twins

J. Steven Reznick

The relation between intelligence and language is complex. If intelligence is defined as the ability to solve problems, generate creative ideas, recognize emergent categories and patterns, or simply as processing power or speed of neural conduction, then language can be viewed as a *secondary process* in that it merely serves to communicate the workings of the intelligent mind. This language-independent conceptualization of intelligence can be contrasted with a language-dependent perspective, in which verbal skill per se is an integral aspect of intelligence. From the language-dependent perspective, intelligence includes abilities such as knowledge of word meanings, mastery of grammar and syntax, and skilled articulation of ideas.

Verbal and nonverbal skills co-occur in normal adults, and efforts to separate them are problematic: Most verbal tasks tap underlying nonverbal skills, and tasks that seem nonverbal may be influenced by verbal mediation of instructions or ongoing processes. However, from a developmental perspective, separating verbal and nonverbal skills is obvious and necessary. In the first year, most human infants begin to babble, use gestures, and understand and say a few words, but these activities pale in comparison to the infant's interest in object manipulation, locomotion, and visual exploration. Our terminology for the first year of life evokes this relative lack of language – the word *infant* is from the Latin *infans*, meaning "incapable of speech." Infants do communicate, but their linguistic investment seems relatively minor in comparison to their other abilities and to the prodigious advances in language that characterize the second year.

There is a rich complexity to language, and any effort to describe language based on simple parameters is obviously incomplete. Bates, Dale, and Thal (1995) contrast various strategies for specifying components of language and, on the basis of data from a normative sample as well as from abnormal groups, argue that the distinction between expressive and receptive language is more psychologically real than alternative specifications of linguistic substructure, such as vocabulary versus grammer. The distinction between *expressive* and *receptive* ability is clearly relevant for describing the

483

emergent lexicon. Expressive language increases dramatically during the second year for most children. Checklist-based parental reporting of expressive language indicates that 12-month-olds produce an average of 10 words and that 24-month-olds have an average productive vocabulary of 310 words (Fenson et al., 1993), but there are vast individual differences (Bates et al., 1994; Fenson et al., 1994). Receptive language is more difficult to assess, but most theorists believe that children know referents for many more words than they can say. Parents have some indication of the words that their child knows, and the norms cited suggest that 12-month-olds comprehend an average of 80 words, which is eight times more words than reported for productive vocabulary. By extrapolation, the comprehension vocabulary at 24 months is extremely vast. These changes in vocabulary occur gradually for some infants, but many experience a burst or surge in production and comprehension vocabulary at around 18–20 months (Goldfield & Reznick, 1990; Nelson, 1973; Reznick & Goldfield, 1992).

The emergence of language abilities during the second year has implications for theories of intelligence and for efforts to differentiate the effects of nature and nurture. My goal in this chapter is to use data from the MacArthur Longitudinal Twin Study (MALTS) for twins assessed at 14, 20, and 24 months to address two questions about language and intelligence in the second year of life. First, what is the relation between language and intelligence during this time of rapid change in language? Second, are the effects of nature and nurture comparable for expressive and receptive language?

Infant intelligence was assessed for the MALTS cohort using the Bayley Scales of Infant Development (BSID) (Bayley, 1969). This instrument is generally regarded as the best general measure of early development currently available (Sattler, 1990). Its most widely used version has a Mental Scale composed of 163 items arranged by tenths of months. Raw scores are converted into a Mental Development Index (MDI), which is a normalized standard score based on a national stratified sample of normal infants and children. Language was assessed using a modified version of the Sequenced Inventory of Communication Development (SICD) (Hedrick, Prather, & Tobin, 1975). This standardized assessment of general communicative performance has separate tests of expressive and receptive language skills and combines observer report with parent report. In order to provide a broader index of language ability, I also adopted a strategy used by Dale, Bates, Reznick, and Morisset (1989) and sorted items on the BSID into subsets with theoretical, rather than statistical, coherence. For this chapter, I am particularly interested in items that require expressive language skills (e.g., naming objects) and items that require receptive language skills (e.g., pointing to named pictures). Finally, to provide a more objective measurement of receptive language ability, children were tested on a cued-target procedure in which comprehension was inferred from behavior. Children saw pairs of slides of objects, and their fixations to the slides were recorded. After an initial salience assessment phase, the child was cued with a label for one of the objects. Receptive knowledge was inferred from an increase in fixation to the named object. Researchers have used this technique to explore various aspects of receptive language ability and have discovered improvement in word comprehension from 8 to 20 months and acceptable reliability and validity (Golinkoff, Hirsh-Pasek, Cauley, & Gordon, 1987; Hirsh-Pasek, Golinkoff, Fletcher, deGaspe Beaubien, & Cauley, 1985; Naigles, 1990; Reznick, 1990; Reznick & Goldfield, 1992).

The present approach to questions about early language is potentially problematic to the extent that results about twin language do not generalize to nontwin language. Twin language is unique in several ways. One aspect of the twin situation is that each twin receives less individually directed parental speech (Conway, Lytton, & Pysh, 1980; Lytton, Conway, & Sauve, 1977; Stafford, 1987; Tomasello, Mannle, & Kruger, 1986). A second aspect is that twins often participate in three-way conversations in which they may communicate with either the parent or the co-twin (Savic, 1980). Research on twin versus nontwin language is sparse, particularly regarding comparisons during the second year for children speaking English. When quantitative differences emerge, they often reflect auxiliary processes such as a tendency to complete each other's utterances (see Savic, 1979, for twins learning Serbo-Croatian) or syntactic or semantic adaptations to twin status such as misuse of plurals or pronouns (Malmstrom & Silva, 1986). There is a traditional belief that twin language is developmentally delayed (Davis, 1937; Day, 1932; McCarthy, 1954), but this effect may be due to factors not necessarily unique to the twin situation, such as low birth weight or relative birth position (Matheny & Bruggemann 1972). Moreover, detailed analysis of twin language reveals some domains in which twins are more advanced than singletons. For example, twins acquire the ability to use "I" relatively quickly (Savic, 1980). Thus, the presence of a same-aged sibling certainly alters the linguistic environment, but twin language is neither abnormal nor "retarded." Unique aspects of the twin linguistic environment necessitate some caution in interpreting language data from twins in comparison to singletons, but they do not preclude twin analysis as a tool for contrasting genetic and environmental effects on language acquisition. Broadly conceived, the twin situation for learning language is a special case of the multisibling family context, albeit one in which there is no spacing between siblings.

Description of the present study

MALTS is an ongoing investigation of the behavioral genetics of cognitive, social, emotional, and temperamental aspects of behavior being conducted at the Institute for Behavioral Genetics at the University of Colorado. Same-sex twin pairs with relatively high birth weight (96% weighed 1,700 grams or more) and normal gestational age (34 weeks or more) have been observed at 14, 20, and 24 months using an extensive battery described elsewhere (see Plomin et al., 1990, for details). Additional assessment of this cohort at older ages is underway. The ethnic distribution of the 408 participating families is 88.5% European American, 9% Hispanic American, and 2.5% African American. The vast majority are two-parent families with both parents relatively old and well-educated.

Earlier reports have examined a broad array of variables for subsets of the MALTS cohort cross-sectionally at 14 months (Emde et al., 1992) and longitudinally at 14 and 20 months (Plomin et al., 1993) or have focused on the domains of temperament or emotion (e.g., DiLalla, Kagan, & Reznick, 1994; Robinson, Kagan, Reznick, & Corley, 1992). In this chapter, I report on intelligence and language measures assessed at 14, 20, and 24 months, and I expand the sample to include the entire MALTS cohort that was tested in this age range.

Testing took place in the home and in the laboratory. The first home visit occurred between 13.5 and 15.5 months, the second between 19.5 and 21.5 months, and the third between 23.5 and 26 months. Laboratory visits occurred within this same interval, usually from 1 to 2 weeks after the home visit.

Two female examiners visited each home. Each examiner tested one child, but the tests were conducted simultaneously with the mother seated within view of each twin. On the first home visit, examiners were randomly assigned to each twin, but on subsequent visits to the same home, the examiner was assigned to the twin that she had not previously tested. At 20 and 24 months, testers were blind to each twin's previous performance except for the SICD, for which previous test results were used to determine the first item to be administered.

Intelligence and language testing began after various procedures described elsewhere. Each twin was seated in a high chair, but a few children sat in a booster seat or in the mother's lap. The mental development component of the BSID (Bayley, 1969) was administered according to standardized procedures. Items from the expressive and receptive components of the SICD were administered concurrently with the BSID to take advantage of the significant overlap between the two measures. Due to time constraints, administration of the SICD was not complete.

SICD items using parent report were gathered at the end of the home visit, but preliminary inspection of the data revealed marked differences in twin resemblance for parent-reported and observer-reported items. Specifically, there were extremely large MZ and DZ twin correlations for the parent-reported scores, which is problematic because it may indicate that parents have difficulty distinguishing between the language accomplishments of their twins, at least as assessed in the SICD. To avoid this potential source of error, we adopted the conservative strategy of eliminating all parent-reported items from the expressive and receptive language components.

Word comprehension tests were administered in the laboratory. Each twin was tested separately with the mother present. A single examiner tested both twins but was not aware of the zygosity classification or any other relevant data. The child sat in a high chair, 27 inches from two rear-projection screens placed side by side, 9 inches apart. The mother was seated next to the child or, in rare cases, held the child on her lap. An experimenter watched through a small window above the screen and pushed a toggle switch to indicate fixation to the slide on the left or right. A computer cumulated the fixation to each side on each trial.

Each word-comprehension trial had two phases. In the preprompt phase, a pair of slides, each depicting a different object, was visible for 8 sec while the experimenter recorded fixation to the slides. The slides went off for 1 sec and then reappeared in exactly the same locations. The experimenter, who was visible through an opening above the slides, but blind to the location of specific slides, then asked the child "Do you see the ——?, Where is the ——?, Look at the ——," substituting a word that named one of the slides. The slides remained visible for 8 sec while the experimenter prompted the child and recorded fixation. Following a 3-sec intertrial interval, the procedure was repeated with another slide pair and another word prompt. Each child saw a counterbalanced presentation of 15 word-comprehension pairs composed of words selected to be either easy, moderately difficult, or extremely difficult to comprehend (on the basis of norms available from previous research).

Establishing composite variables

Table 17.1 contains a list of the SICD items retained for the expressive component and the percentage of subjects passing each item at each age. Table 17.1 also lists the BSID items that were considered to be relevant to assessment of expressive language and the percentage of subjects passing each item. To form an expressive language composite, relevant items on the BSID or SICD that had either 100% or 0% correct at a particular age were

Table 17.1. *Sequenced Inventory of Communication Development (SICD)
and Bayley Scales of Infant Development (BSID) items in the expressive
and receptive composites*

		Percentage correct		
Expressive language		14 mo	20 mo	24 mo
SICD Expressive scale				
22a	Name picture of baby	5	54	82
22b	Name picture of shoe	3	54	82
22c	Name picture of ball	4	54	81
29a	Answer question "no"	NA	NA	22
29b	Answer question "yes"	NA	NA	45
31a	Answer question "shoes"	NA	NA	24
BSID Expressive items				
79	Vocalize 4 syllables	98	100	100
85	Say "da-da" or equivalent	95	100	100
101	Jabber expressively	80	97	98
113	Say 2 words	26	90	98
116	Use gestures to make wants known	82	99	100
124	Name 1 object	8	67	90
127	Use words to make wants known	5	69	91
130	Name 1 picture	6	63	89
136	Say sentence of 2 words	2	35	74
138	Name 2 objects	0	29	70
141	Name 3 pictures	0	29	72
146	Name 3 objects	0	15	53
149	Name 5 pictures	0	12	52
Receptive language				
SICD Receptive scale				
6b	Respond to "come here"	89	96	98
9	Respond to "don't touch!"	54	71	86
10	Respond to gesture and "give it to me"	85	97	99
11a	Indicate referents "cup, spoon, shoe"	48	92	98
12a	Respond to "get the car"	68	91	98
12b	Respond to "Give it to me"	26	60	83
12c	Respond to "Put it on the paper"	23	69	89
13a	Indicate referent "ears"	5	54	79
13b	Indicate referent "eyes"	17	76	93
13c	Indicate referent "hair"	7	69	92
13d	Indicate referent "mouth"	9	70	93
13e	Indicate referent "nose"	16	76	92
14a	Respond to "sit down"	49	87	95
15a	Indicate referent "socks"	10	69	90
15b	Indicate referent "tree"	6	61	87
15c	Indicate referent "bear"	11	76	92
15d	Indicate referent "chair"	5	66	89

Table 17.1. *(cont.)*

Expressive language		Percentage correct		
		14 mo	20 mo	24 mo
15e	Indicate referent "key"	7	64	90
15f	Indicate referent "box"	5	55	84
16	Respond to "where's mama"	55	90	97
17a	Respond to "put block on box"	1	13	32
17b	Respond to "put block in box"	2	57	79
17c	Respond to "put block beside box"	0	1	4
17d	Respond to "put block under box"	0	7	35
18	Respond to "bye-bye"	25	52	73
19a	Respond to "put 1 spoon in box"	NA	6	12
20a	Respond to "let me see you walk"	NA	52	82
21a	Indicate referent "what mom cooks on"	NA	NA	36
21b	Indicate referent "what you wear on feet	NA	NA	42
21c	Indicate referent "what you read"	NA	NA	21
23	Indicate referent "big, little"	NA	NA	25
BSID	*Receptive items*			
89	Respond to verbal request	97	100	100
94	Respond to "no no!"	89	99	100
117	Indicate referent for article of clothing	25	83	95
126	Respond to command re: doll	17	83	96
128	Indicate referents for body parts on doll	6	76	95
131	Find 2 hidden objects identified by name	20	58	71
144	Indicate referent for 2 objects	1	43	78
152	Indicate referent for 3 objects	0	24	60
158	Respond to 2 prepositions	0	15	46
162	Respond to "put 1 block on paper"	0	3	8
163	Respond to 3 prepositions	0	3	20

Note: NA = not administered at that age.

eliminated, and percentage correct was calculated across the remaining items. Cronbach's alphas (Cronbach, 1951) for the final set of items in the expressive composite were .64 (N of items = 12), .86 (N of items = 14), and .87 (N of items = 16) at 14, 20, and 24 months, respectively.

The SICD Receptive Scale includes items designed to measure the child's awareness, discrimination, and understanding of language. To meet the goals of the present study, items were retained that measured the understanding of single- or multiword utterances. These items were combined with comparable items from the BSID to form a composite representing receptive language. Table 17.1 lists the relevant items and the percentage of subjects passing each item. Items that had either 100% or 0% correct at a particular age were eliminated. Cronbach's alpha for the remaining items in each set was .79 (N of items = 30), .89 (N of items = 37), and .86 (N of items = 40) at 14, 20, and 24 months, respectively.

The dependent variable available from the laboratory word-comprehension procedure was how long each child looked at each stimulus. Derivation of a score reflecting receptive language ability required several steps. Inspection of the data revealed some word-comprehension trials on which performance was ambiguous because the infant had not fixated both stimuli. Fixation to just one stimulus could indicate a marked preference for that stimulus but not necessarily relative to the other stimulus. A conservative criterion was adopted in which word-comprehension trials were eliminated if fixation to either side before the word prompt was 0 sec or if fixation to both sides after the word probe was 0 sec. This caused the loss of 20%, 15%, and 11% of trials at 14, 20, and 24 months, respectively.

Percentage fixation to the target slide before the experimenter's prompt for that target was subtracted from percentage fixation to the target slide after the prompt. This calculation indicates whether the child showed an increase in fixation to the target relative to its initial salience. Following the guidelines established in previous research (Reznick, 1990; Reznick & Goldfield, 1992), an increase of 15% was accepted as the criterion indicating comprehension of each word. A word-comprehension score was created by dividing the number of words comprehended by the number of word tests available. The modal denominator for this ratio was 15 but was lower for some children because of exclusion for low fixation (described earlier) or because the child refused to participate. Word-comprehension scores computed from three or fewer trials were considered unreliable and so were treated as missing. This constraint eliminated word-comprehension data for 2.5%, 1.2%, and 1.2% of children at 14, 20, and 24 months, respectively.

The laboratory-based measure of receptive language was related to the observer-reported component but was not merely redundant, with significant correlations of r (656) = .19, $p < .01$, r (570) = .37, $p < .01$, and r (606) = .26, $p < .01$ at 14, 20, and 24 months, respectively, between word-comprehension scores computed from laboratory measurement and receptive language ability computed from items from the BSID and the SICD. These relatively low correlations for concurrent validity suggest that these measures may be capturing either different aspects of receptive language (e.g., breadth of vocabulary versus knowing the names of common objects) or may differ in sensitivity to language ability. Based on the theoretical similarity of the measures, a final receptive language composite was computed as the average percentage correct on BSID receptive items, SICD receptive items, and word-comprehension tests. The receptive language composite was set to missing for any subject lacking two of these measures, which caused a loss of data for 3.3% of the children.

Zygosity

Zygosity of the twins was determined through aggregation of independent tester ratings on the similarity of eight physical attributes across age. The attributes were selected on the basis of the diagnostic rules developed by Nichols and Bilbro (1966). When the features of the pair of twins were rated consistently as highly similar (scores of 1 or 2 on a 5-point scale), the classification of MZ was made. When two or more features were only somewhat similar (a score of 3) or if one feature was not at all similar (score of 4 or 5), the classification of DZ was made. Zygosity of twin pairs was rated primarily at five age points currently assessed in this longitudinal study (14, 24, 36, 48, and 60 months), and a few twin pairs were assessed at 20 months. At 14, 20, 24, and 36 months, twin zygosity was independently rated by two home and two laboratory testers. Two home testers rated zygosity at later age points. At the time of this writing, the modal number of zygosity ratings per subject is 18. Twin zygosity was considered unambiguous if there was 85% agreement of the MZ or DZ classification across all testers at all ages, on the basis of a minimum of four ratings. A subset of the twin pairs with ambiguous ratings had their zygosity determined through an evaluation of genetic markers from blood samples; seven pairs were classified as MZ, and two pairs as DZ. The current zygosity distribution for the 408 MALTS families is: 210 MZ pairs, 177 DZ pairs, and 21 ambiguous pairs.

The relation between language and infant intelligence

"'When *I* use a word,' Humpty Dumpty said in rather a scornful tone, 'it means just what I choose it to mean – neither more nor less'" (Carroll, 1872).

Researchers who use the term *infant intelligence* are often more laissez-faire in their approach to meaning. Traditional tests define infant intelligence implicitly as the extent to which an infant possesses assorted age-appropriate abilities or capacities. For example, Bayley's pioneering California First Year Mental Scale (Bayley, 1933) contained items to measure behaviors such as response to sound, imitation, visual discrimination, memory, problem solving, language comprehension, and language production. Age-based norms have been used to order these items to reflect the skills that are present for most infants at each age. However, given the fact

Table 17.2. *Correlations among variables at each age*

	SICD expressive	SICD receptive
14 months		
BSID Mental Development Index	.49[b] (779)	.48[b] (779)
SICD Expressive		.38[b] (788)
20 months		
BSID Mental Development Index	.70[b] (682)	.67[a] (682)
SICD Expressive		.58[b] (703)
24 months		
BSID Mental Development Index	.73[b] (694)	.69[b] (694)
SICD Expressive		.57[b] (703)

Note: Values are Pearson product-moment correlations with degrees of freedom in parentheses.
[a] $p < .05$.
[b] $p < .01$.

that items are arranged to reflect age-associated abilities, the configuration of items that infants should pass changes with age, and individual differences at each age can reflect relative performance on distinctly different abilities. For example, between 4 and 12 months, 4% of the items are about expressive language (e.g., says "da-da" or equivalent), 4% of the items are about receptive language (e.g., responds to verbal request), and 92% of the items are about various nonverbal abilities (e.g., rings bell purposively, looks at pictures in a book, holds crayon adaptively). After 12 months, the configuration changes, with 17% of the items about expressive language, 14% of the items about receptive language, and 68% of the items about nonverbal abilities.

If the MDI score becomes increasingly based on linguistic items, then its relation with measures of language should be informative. As indicated in Table 17.2, MDI scores were highly correlated with expressive and receptive language at each assessment. Moreover, the correlation between MDI and language increased significantly between 14 and 20 months for expressive scores, $z = 6.62$, $p < .01$, and for receptive scores, $z = 5.76$, $p < .01$. Note that this is not merely an effect of item overlap. Correlations between the

Table 17.3. *Contemporaneous and developmental regression models predicting MDI*

Step	Variable Entered	Partial R^2	F	DF
Contemporaneous models				
Predicting 14-month MDI from 14-month variables				
1	Verbal expressive	0.24	250.71	1, 647
2	Verbal receptive	0.10	123.15	1, 646
Predicting 20-month MDI from 20-month variables				
1	Verbal expressive	0.50	669.52	1, 532
2	Verbal receptive	0.10	176.28	1, 531
Predicting 24-month MDI from 24-month variables				
1	Verbal expressive	0.53	779.32	1, 581
2	Verbal receptive	0.11	218.89	1, 580
Cross-age models				
Predicting 20-month MDI from 14-month variables				
1	Verbal receptive	0.12	94.81	1, 580
2	Verbal expressive	0.05	40.52	1, 582
Predicting 24-month MDI from 14-month variables				
1	Verbal receptive	0.15	120.76	1, 580
2	Verbal expressive	0.02	19.73	1, 582
Predicting 24-month MDI from 20-month variables				
1	Verbal receptive	0.36	354.95	1, 518
2	Verbal expressive	0.05	50.42	1, 516

Note: All F values in this table are significant at $p < .01$.

MDI score and language measures calculated without the inclusion of BSID items (i.e., scores based on just the SICD expressive items, the SICD receptive items, or the word-comprehension test) were lower than correlations that included overlapping items, but the developmental pattern was the same: The correlation between MDI score and each BSID-independent language score increased significantly from 14 to 20 months.

A *stepwise regression model* was used to determine which aspect of language was the most influential predictor of MDI at each age. As indicated in Table 17.3, expressive language accounted for a large percentage of the variance in BSID at 14 months, and an even larger percentage at subsequent assessments. Receptive language accounted for a statistically significant percentage of variance, but the R^2 was relatively low and was constant across age. This pattern of contemporaneous effects can be contrasted with a developmental analysis in which expressive and receptive language at time 1 is used to predict the MDI score at time 2. As indicated in Table 17.3, expressive language was not particularly potent in this cross-age analysis, and the receptive language score was the stronger predictor of subsequent MDI score for each cross-age comparison. However, in a full model that included cross-age as well as contemporaneous predictors, the contemporaneous effects were potent (particularly for expressive language), but the R^2 values for cross-age prediction to 20 and 24 months based on receptive language were vanishingly small (albeit, statistically significant), and cross-age prediction based on expressive language dropped out of the model entirely.

In traditional accounts, early intelligence is a moving target, defined based on the skills that emerge at particular ages. Specifically, the items used to assess intelligence in the first year are primarily nonverbal but, during the second year, become increasingly oriented toward the ability to use and understand language. From this perspective, it is not surprising that assessment of IQ based on BSID or other tests fits a simplex model in which assessments are highly correlated with neighboring points in time but are not highly predictive of later performance (Humphreys & Davey, 1988). Indeed, this also explains why tasks that assess early emerging specific abilities such as recognition memory, speed of processing, or the ability to sit still tend to be better long-term predictors of intelligence than general indices based on intelligence tests per se (Bornstein & Sigman, 1986; Colombo, 1993; Fagan & Singer, 1983).

The present results suggest that the MDI is particularly sensitive to expressive language and that the percentage of MDI variance accounted for by expressive language increases with age. It is tempting to interpret this finding methodologically: The MDI contains more items that reflect expressive language, and because expressive language is either present or absent, scoring of expressive items can be relatively straightforward. The strong relation between the MDI and the expressive measure, as compared to the receptive measure, could reflect a greater precision of measurement. However, this methodological interpretation is less credible given the cross-age models in which receptive language is the stronger predictor of subsequent MDI. This latter finding is consistent with the claim by Bates, Bretherton, and Snyder (1988) that individual differences in language have unique long-term significance at different points in development. It is unclear whether accomplishment in receptive language in the second year is merely diagnostic of later performance or plays a causal role, but further work to explore this relation is clearly warranted.

Effects of nature and nurture on expressive and receptive language

Individual differences in language acquisition during the second year have been investigated for two decades, but we still know relatively little about genetic influences on early language development. Heritability reflects the extent to which variance in behavior can be attributed to genetic influence. Many speech and language disorders in older children such as developmental dyslexia and stuttering appear to be heritable (DeFries, 1985; Howie, 1981; Lewis, Ekelman, & Aram, 1989; Lewis & Thompson, 1992; Pennington & Smith, 1983; Tallal, Ross, & Curtiss, 1989), but these data do not address the question of genetic influences in normal children at the dawn of language.

Adoption studies

The data from adoption studies suggest some genetic influence on early expressive language. Hardy-Brown, Plomin, and DeFries (1981) used a full adoption design to contrast the effects of heredity and environment on language in 1-year-old children. Children's *communicative performance* (defined as the first principal factor across a variety of expressive measures such as vocalization, gesture, imitation, and phonological ability) was affected by aspects of the behavior of their adoptive mother but more so by the cognitive abilities of their birth mother, suggesting genetic influence on expressive language in the first year. Hardy-Brown and Plomin (1985) reanalyzed these data in comparison with nonadoptive control families and replicated the finding that infant communicative competence is significantly related to general cognitive ability of the birth mother and not the adoptive parents. This additional analysis also revealed that some maternal language variables (e.g., the tendency to imitate the infant's vocalizations) were related to infant communicative competence in both adoptive and nonadoptive homes, but others (e.g., the frequency of question sentences) were only related in nonadoptive homes. These different patterns of correlation suggest that some aspects of the home environment may be mediated by genetic factors shared by parent and child. Thompson and Plomin (1988) assessed language in 2- and 3-year-old adoptees using a composite measure of language that included expressive and receptive items from the SICD, but 60% of the items were expressive. Their analysis revealed both environmental and genetic effects, with the latter effect increasing with age.

Twin studies

Twin studies are a second tool for exploring genetic effects on language. Twin analyses of heritability use MZ twin pairs, who are genetically identical, and DZ same-sex twin pairs, who share half of the genetic material passed to them from their mother and father. The estimate of genetic (hereditary) variation (labeled h^2) can be obtained by doubling the difference between the MZ and DZ correlations. Environmental factors, which include all nongenetic influences, are represented in this model as shared and nonshared components. Shared, or common, environmental variation (labeled c^2) is the twin resemblance that is not explained by hereditary resemblance. It is conceptualized as the contribution to individual variation in a trait from the experiences that both twins share (e.g., prenatal influences, parents' education level, or common child care experiences). The other component of environmental influence is the residual environmental variance or nonshared environment (labeled e^2). This component comprises environmental influences that contribute to individual uniqueness and also error of measurement.

Twin studies suggest a genetic effect on expressive language in preschool-aged children. Mittler (1969) administered the Illinois Test of Psycholinguistic Abilities to 4-year-old monozygotic (MZ) and dizygotic (DZ) twins. Subtests that measure expressive language (e.g., the Vocal Encoding subtest in which the child is asked to describe simple objects, or the Auditory-Vocal Automatic subtest, which measures inflectional aspects of grammar) revealed significant heritable effects. Subtests that should be sensitive to receptive language (e.g., the Auditory Encoding subtest, which assesses the child's ability to understand the spoken word) suggest no heritable effects. Mather and Black (1984) tested preschool twins (mean age = 4.5 years) on standardized measures of vocabulary comprehension, semantic knowledge, morphology, syntax, and articulation (see Mather & Black, 1984, for a detailed description of the specific measures) and found (1) significant heritability for comprehension, and (2) environmental influence for the other measures. However, note that by 4 years of age, it is hard to separate language-comprehension skill from more general intellectual abilities: The Peabody Picture Vocabulary Test used by Mather and Black to measure comprehension correlates highly with broad-based measures of IQ. In a detailed reanalysis of these data, Locke and Mather (1989) discovered that MZ twin pairs were significantly more likely to mispronounce the same sounds on an articulation test than were DZ twin pairs, who in turn shared more errors than children who were unrelated. Finally, Matheny and Bruggemann (1972) used the Templin-Darley Screening Test of Articulation with twins 4 to 8 years of age and found greater MZ similarity than DZ similarity but only for boys.

These studies suggest a genetic effect on expressive language in preschool-aged twins, but we know little about genetic and environmental effects on early language, particularly for receptive skills. Benson, Cherny, Haith, & Fulker (1993) administered the SICD at 5, 7, and 9 months to MZ and DZ twins and found consistent positive correlations with midparent general intelligence (measured using the WAIS–R) for the expressive scale and less consistent results for the receptive scale. This suggests a genetic effect on expressive language in infancy, but during a period that many researchers consider prelinguistic.

Present analysis

Estimates of heritability (h^2) and the proportion of variance due to shared environmental influences (c^2) were computed for the MALTS sample using the method described by DeFries and Fulker (1988) and extended by Cyphers, Phillips, Fulker, and Mrazek (1990). The following model was fitted to the data from MZ and DZ twin pairs simultaneously:

$$P_1 = B_1S + B_2P_2 + B_3R + B_4P_2R + A$$

where P_1 is twin 1's score, S is the sex of the twin pair ($S = 0$ for girls, and $S = 1$ for boys), P_2 is twin 2's score, R is the coefficient of relationship of the twins ($R = 1.0$ for MZ twins, and $R = 0.5$ for DZ twins), P_2R is the product of the coefficient of relationship and twin 2's score, and A is the regression constant. In this model, B_1 is the partial regression of one twin's score on sex and removes variance due to sex from the other parameter estimates. B_2, the partial regression of one twin's score on the co-twin's score, estimates c^2; B_3 is a correction term for mean differences between MZ and DZ twins; and B_4 estimates h^2. Twin data were entered twice to permit the scores of each twin to serve in turn as outcome value and as predictor. Standard errors for the regression coefficients were then adjusted by the square root of $2N - k - 1$ divided by $N - k - 1$ (with $N =$ the number of twin families, and $k =$ the number of regression predictors) to correct for the doubled sample size before performing significance tests. Twin contrast effects and nonadditive genetic influence can result in a c^2 that is negative or an h^2 that exceeds 1.0. Estimates of h^2 and c^2 in the present model were constrained to lie between their theoretical lower and upper bounds of .0 and 1.0 by excluding h^2 or c^2 terms that were negative and fitting a reduced regression model (Cherny, DeFries, & Fulker, 1992). When either parameter was constrained, the remaining parameter was calculated as a weighted average of the individual MZ and DZ regression coefficients (rather than as a difference between MZ and DZ regressions), resulting in a more reliable parameter estimate with a markedly reduced standard error.

Table 17.4 lists the MZ and DZ intraclass correlations for each measure at each age, the associated estimates of heritability and environmental influence, and their standard errors. One-tailed t-tests were used to evaluate the statistical significance of each parameter relative to its standard error.

Table 17.4. *Twin intraclass correlations and modeled effects at each age*

	Expressive				Receptive			
	Twin correlation		Modeled effect		Twin correlation		Modeled effect	
Age	MZ	DZ	h^2	c^2	MZ	DZ.	h^2	c^2
14 months	.38	.37	.01	.35[a]	.53	.46	.14	.38[a]
	204	172	(.19)	(.15)	204	172	(.18)	(.15)
20 months	.76	.63	.25	.49[a]	.71	.59	.23	.45[a]
	181	156	(.16)	(.13)	183	153	(.16)	(.14)
24 months	.79	.60	.38[a]	.40[a]	.66	.61	.09	.56[a]
	179	155	(.15)	(.13)	178	157	(.17)	(.14)

Note: All twin correlations are statistically significant, $p < .01$. Number below correlation is degrees of freedom. Parameter of .00 indicates use of constrained model. Number below parameter is standard error corrected for double entry.
[a] For modeled effects, $p < .05$, one-tailed.

Expressive language

Twin correlations for expressive language were moderate at 14 months and were comparable for MZ and DZ twins. Twin correlations were notably larger at 20 months, with increased differences between MZ and DZ correlations at 20 and 24 months. This pattern suggests a strong effect of shared environment on expressive language across the second year. Additionally, the increasing difference between MZ and DZ twin correlations suggests the emergence of a genetic effect on verbal expressiveness late in the second year.

The timing of the emergence of a genetic effect on expressive language is problematic. MZ and DZ twin pairs differ at 20 and 24 months only, but expressive language is rare at 14 months and increases dramatically later in the second year. Thus, the lack of a genetic effect at 14 months could reflect the difficulty of measuring expressive language when the vocabulary is small. However, this complaint is blunted by the significant twin correlations for expressive and receptive language at 14 months and the fact that expressive language is significantly correlated with receptive language and with MDI scores. This would not be the case if measurement of expressive language at 14 months were inadequate. Therefore, the present results suggest that late in the second year, genetic factors begin to play a significant role in the sophistication of the child's expressed vocabulary and the processes that make children more or less talkative.

Children learn to speak the language that they hear, so environmental effects on language development are not surprising. The genetic effect on expressive language, whatever its exact timing, is more intriguing. Locke (Locke, 1990; Locke & Mather, 1989) has argued for the innateness of phonological development, and previous research with adoptees indicates that communicative performance in the first year defined across measures such as vocalization, gesture, imitation, and phonological ability is more highly correlated with the birth mother than with the adoptive mother (Hardy-Brown et al., 1981; Hardy-Brown & Plomin, 1985). Communicative competence measures at 2 and 3 years also show genetic as well as environmental effects (Thompson & Plomin, 1988). The present study bridges the gap between these reports and suggests genetic influence on expressive language during the second year as well.

The present index of expressive language reflects both the range of the child's expressive vocabulary and also the child's willingness to communicate in the presence of an unfamiliar examiner. The latter condition is particularly salient in the present context because an unwillingness to talk in an unfamiliar situation is related to a heritable temperamental disposition toward shyness and fearfulness (Robinson et al., 1992) and is more likely among MZ twins (DiLalla et al., 1994). Further research is needed to separate the competence component of expressive ability from the more general performance component that emerges in the context of the present measurement procedures. Also, it will be important to determine whether an unwillingness to talk affects the magnitude of the twin correlation.

The presence of a same-aged sibling certainly alters the linguistic environment. However, if one assumes that twin language is essentially normal in most respects, the present data highlight the need to disentangle genetic and environmental influences on language development. For example, the present data are compatible with the hypothesis that expressive parents will have expressive children. Without a genetic analysis, it would be tempting to conclude that the environment's role is causal through the mechanism of expressive parents providing their children with a rich linguistic environment or, perhaps, of expressive children evoking a rich linguistic environment from adults. For example, Hampson and Nelson (1993) find differences between the mothers of children who are early or late talkers and note two directions of effect – maternal differences could enable some children to talk earlier, or mothers might differ as a result of responding differentially to their child's language ability. The present results suggest a third possibility. Genetic mechanisms that affect individual differences in adult speech and are transmitted to offspring may also promote individual differences in the child's expressive language and, hence, early versus late talking. Additional work is needed to explore these possibilities, but as Hardy-Brown and Plomin (1985) have suggested, twin analysis allows a correction to correlational studies that confound environmental influence with genetic influence.

Receptive language

The index of receptive language used in this study is derived from the child's response to questions and thus reflects knowledge of word reference but also the child's compliance with a request to point to, hand over, or look at a particular stimulus. Significant agreement across the various questions about receptive language, and between the SICD items and the laboratory-based, word-comprehension test, boosts confidence that a linguistic interpretation of individual differences is reasonable. But, it must be noted that even if receptive knowledge is being assessed, it is primarily knowledge of nominals and not other aspects of the lexicon.

Twin correlations for receptive language indicated significant shared environmental effects at 14, 20, and 24 months. These results suggest that throughout the second year, receptive vocabulary is primarily influenced by factors in the shared environment such as the quality and quantity of the language that the child hears or perhaps by differences in parental style, such as preferred mode of interaction (e.g., tendency to ask questions) or choice of activities (e.g., naming games, book reading). The present data do not allow specification of the environmental mechanisms that affect receptive language, but subsequent work on this topic may be warranted because of the potential usefulness of early interventions that promote language competence. Indeed, it is interesting to note that Whitehurst et al. (1991) report a home-based intervention that accelerates vocabulary skills for children with expressive language delay but does not decrease the likelihood of later phonological problems.

Comparing expressive and receptive language

The present data are also relevant to efforts to understand the relation between expressive and receptive language. It is generally accepted that individual differences in receptive and expressive language are related. Receptive language is present earlier (Ingram, 1974), but estimates of ability in receptive and expressive language tend to be correlated for observation (Nelson, 1973) or parent report (Reznick, 1990; Tamis-LeMonda & Bornstein, 1990), and there is evidence for periods of marked acceleration in both domains for many children (Reznick & Goldfield, 1992). By the end of the second year, these two aspects of vocabulary appear to be in close alignment (Goldin-Meadow, Seligman, & Gelman, 1976). However, the findings of synchrony between expressive and receptive language can be contrasted with work by Behrend (1988) and others indicating asynchronous development in these two domains. The present data suggest one basis for asynchrony between expressive and receptive language during the second year: differential sensitivity to genetic and environmental influence. For example, in a recent report, DeBaryshe (1993) finds that home story-reading practices are more strongly related to receptive than to expressive language skills. One interpretation of this finding is that environmental factors affect receptive language, but variance in expressive language is more difficult to account for because it is also affected by genetic mechanisms. This may also explain why individual differences in receptive language are strong predictors of subsequent MDI.

Acknowledgments

The results reported in this chapter emerged through collaboration among a group of investigators, including R. Corley, J. C. DeFries, R. N. Emde, D. Fulker, and J. Robinson at the University of Colorado; J. Campos at the University of California at Berkeley; J. Kagan at Harvard University; R. Plomin at the Pennsylvania State University; and C. Zahn-Waxler at the National Institutes of Mental Health. This research was supported by the John D. and Catherine T. MacArthur Foundation through its Research Network on Early Childhood Transitions.

We thank the families who contributed their time and effort, as well as the many research assistants at the University of Colorado, Harvard University, Yale University, and The Pennsylvania State University who were involved in data collection, behavioral coding, and data management.

References

Bates, E., Bretherton, I., & Snyder, L. (1988). *From first words to grammar: Individual differences and dissociable mechanisms.* New York: Cambridge University Press.

Bates, E., Dale, P. S., & Thal, D. (1995). Individual differences and their implications for theories of language development. In P. Fletcher & B. MacWhinney (Eds.), *Handbook of child language.* Oxford: Basil Blackwell.

Bates, E., Marchman, V., Thal, D., Fenson, L., Dale, P., Reznick, J. S., Reilly, J., & Hartung, J. (1994). Developmental and stylistic variation in the composition of early vocabulary. *Journal of Child Language, 21,* 85–123.

Bayley, N. (1933). *The California first year mental scale.* Berkeley: University of California Press.

Bayley, N. (1969). Manual for the Bayley scales of infant development. New York: Psychological Corporation.

Behrend, D. A. (1988). Overextensions in early language comprehension: Evidence from a signal detection approach. *Journal of Child Language, 15,* 63–75.

Benson, J. B., Cherny, S. S., Haith, M. M., & Fulker, D. W. (1993). Rapid assessment of infant predictors of adult IQ: Midtwin–midparent analyses. *Developmental Psychology, 29*, 434–47.

Bornstein, M. H., & Sigman, M. D. (1986). Continuity in mental development from infancy. *Child Development, 57*, 251–74.

Carroll, L. (1872). *Alice's adventures in Wonderland and through the looking glass.* Middlesex England: Penguin.

Cherny, S. S., DeFries, J. C., & Fulker, D. W. (1992). Multiple regression analysis of twin data: A model-fitting approach. *Behavior Genetics, 22*, 489–97.

Colombo, J. (1993). *Infant cognition: Predicting later intellectual functioning.* Newbury Park, CA: Sage.

Conway, D., Lytton, H., & Pych, F. (1980). Twin–singleton language differences. *Canadian Journal of Behavioral Science, 12*, 264–71.

Cronbach, L. J. (1951). Coefficient alpha and the internal structure of tests. *Psychometrika, 16*, 297–234.

Cyphers, L. H., Phillips, K., Fulker, D. W., & Mrazek, D. A. (1990). Twin temperament during the transition from infancy to early childhood. *Journal of the American Academy of Child and Adolescent Psychiatry, 29*, 392–7.

Dale, P. S., Bates, E., Reznick, J. S., & Morisset, C. (1989). The validity of a parent report instrument of child language at 20 months. *Journal of Child Language, 16*, 239–49.

Davis, E. A. (1937). *The development of linguistic skill in twins, singletons with siblings, and only children from age five to 10 years.* Minneapolis: University of Minnesota Press.

Day, E. J. (1932). The development of language in twins. I. A comparison of twins and single children. *Child Development, 3*, 179–99.

DeBaryshe, B. D. (1993). Joint picture-book reading correlates of early oral language skill. *Journal of Child Language, 20*, 455–61.

DeFries, J. C. (1985). Colorado Reading Project. In D. B. Gray & J. F. Kavanagh (Eds.), *Biobehavioral measures of dyslexia.* Parkton, MD: York Press.

DeFries, J. C., & Fulker, D. W. (1988). Multiple regression analysis of twin data: Etiology of deviant score versus individual differences. *Acta Geneticae Medicae et Gemeologiae, 37*, 205–16.

DiLalla, L. F., Kagan, J., & Reznick, J. S. (1994). Genetic etiology of behavioral inhibition among two-year-old children. *Infant Behavior and Development, 17*, 401–8.

Emde, R., Campos, J., Corley, R., DeFries, J., Fulker, D., Kagan, J., Plomin, R., Reznick, J. S., Robinson, J., & Zahn-Waxler, C. (1992). Temperament, emotion, and cognition at 14 months: The MacArthur Longitudinal Twin Study. *Child Development, 63*, 1437–55.

Fagan, J. F., & Singer, L. T. (1983). Infant recognition memory as a measure of intelligence. In L. P. Lipsett & C. K. Rovee-Collier (Eds.), *Advances in infancy research* (Vol. 2). Norwood, NJ: Ablex.

Fenson, L., Dale, P., Reznick, J. S., Thal, D., Bates, E., Hartung, J. P., Pethick, S., & Reilly, J. S. (1993). *The MacArthur Communicative Development Inventories: Users guide and technical manual.* San Diego: Singular Press.

Fenson, L., Dale, P., Reznick, J. S., Bates, E., Thal, D. J., & Pethick, S. J. (1994). Variability in early communicative development. *Monographs of the Society for Research in Child Development, 59* (5, Serial No. 242).

Goldfield, B. A., & Reznick, J. S. (1990). Early lexical acquisition: Rate, content, and the vocabulary spurt. *Journal of Child Language, 17*, 171–83.

Goldin-Meadow, S., Seligman, M. E. P., & Gelman, R. (1976). Language in the two-year-old. *Cognition, 4*, 189–202.

Golinkoff, G. M., Hirsh-Pasek, K., Cauley, K. M., & Gordon, L. (1987). The eyes have it: Lexical and syntactic comprehension in a new paradigm. *Journal of Child Language, 14*, 23–45.

Hampson, J., & Nelson, K. (1993). The relation of maternal language to variation in rate and style of language acquisition. *Journal of Child Language, 20*, 313–42.

Hardy-Brown, K., & Plomin, R. (1985). Infant communicative development: Evidence from adoptive and biological families for genetic and environmental influences on rate differences. *Developmental Psychology, 21,* 378–85.

Hardy-Brown, K., Plomin, R., & DeFries, J. C. (1981). Genetic and environmental influences on the rate of communicative development in the first year of life. *Developmental Psychology, 17,* 704–17.

Hedrick, D. L., Prather, E. M., & Tobin, A. R. (1975). *Sequenced inventory of communication development.* Seattle: University of Washington Press.

Hirsh-Pasek, K., Golinkoff, G. M., Fletcher, A., deGaspe Beaubien, F., & Cauley, K. (1985, October). *In the beginning: One-word speakers comprehend word order.* Paper presented to the Boston Child Language Conference, Boston.

Howie, P. M. (1981). Concordance for stuttering in monozygotic and dizygotic twin pairs. *Journal of Speech and Hearing Research, 24,* 317–21.

Humphreys, L. G., & Davey, T. C. (1988). Continuity in intellectual growth from 12 months to 9 years. *Intelligence, 12,* 183–97.

Ingram, D. (1974). The relationship between comprehension and production. In R. L. Schiefelbusch & L. L. Lloyd (Eds.), *Language perspectives: Acquisition, retardation, and intervention.* Baltimore: University Park Press.

Lewis, B. A., Ekelman, B. L., & Aram, D. M. (1989). A family study of severe phonological disorders. *Journal of Speech and Hearing Research, 23,* 713–24.

Lewis, B. A., & Thompson, L. A. (1992). A study of developmental speech and language disorders in twins. *Journal of Speech & Hearing Research, 35,* 1086–94.

Locke, J. L. (1990). Structure and stimulation in the ontogeny of spoken language. *Developmental Psychobiology, 23,* 621–43.

Locke, J. L., & Mather, P. L. (1989). Genetic factors in the ontogeny of spoken language: Evidence from monozygotic and dizygotic twins. *Journal of Child Language, 16,* 553–9.

Lytton, H., Conway, D., & Sauve, R. (1977). The impact of twinship on parent–child interaction. *Journal of Personality and Social Psychology, 25,* 97–107.

Malmstrom, P. M., & Silva, M. N. (1986). Twin talk: Manifestations of twin status in the speech of toddlers. *Journal of Child Language, 13,* 293–304.

Matheny, A. P., Jr., & Bruggemann, C. (1972). Articulation proficiency in twins and singletons from families of twins. *Journal of Speech and Hearing Research, 15,* 845–51.

Mather, P. L., & Black, K. N. (1984). Hereditary and environmental influences on preschool twins' language skills. *Developmental Psychology, 20,* 303–8.

McCarthy, D. (1954). Language development in children. In L. Carmichael (Ed.), *Manual of child psychology* (pp. 492–630). New York: Wiley.

Mittler, P. (1969). Genetic aspects of psycholinguistic abilities. *Journal of Child Psychology and Psychiatry, 10,* 165–76.

Naigles, L. (1990). Children use syntax to learn verb meanings. *Journal of Child Language, 17,* 357–74.

Nelson, K. (1973). Structure and strategy in learning to talk. *Monographs of the Society for Research in Child Development, 38* (1–2, Serial No. 149).

Nichols, R. C., & Bilbro, W. C. (1966). The diagnosis of twin zygosity. *Acta Geneticae Medicae et Statistica, 16,* 265–75.

Pennington, B. F., & Smith, S. D. (1983). Genetic influences on learning disabilities and speech and language disorders. *Child Development, 54,* 369–87.

Plomin, R., Campos, J., Corley, R., Emde, R. N., Fulker, D. W., Kagan, J., Reznick, J. S., Robinson, J., Zahn-Waxler, C., & DeFries, J. C. (1990). Individual differences during the second year of life: The MacArthur Longitudinal Twin Study. In J. Colombo & J. W. Fagan (Eds.), *Individual differences in infancy: Reliability, stability, prediction* (pp. 431–55). Hillsdale, NJ: Erlbaum.

Plomin, R., Emde, R. N., Braungart, J. M., Campos, J., Corley, R., Fulker, D. W., Kagan, J., Reznick, J. S., Robinson, J., Zahn-Waxler, C., & DeFries, J. C. (1993). Genetic change and continuity from fourteen to twenty months: The MacArthur Longitudinal Twin Study. *Child Development, 64*, 1354–76.

Reznick, J. S. (1990). Visual preference as a test of infant word comprehension. *Applied Psycholinguistics, 11*, 145–65.

Reznick, J. S., & Goldfield, B. A. (1992). Rapid change in lexical development in comprehension and production. *Developmental Psychology, 28*, 406–13.

Robinson, J. L., Kagan, J., Reznick, J. S., & Corley, R. (1992). The heritability of inhibited and uninhibited behavior: A twin study. *Developmental Psychology, 28*, 1030–7.

Sattler, J. M. (1990). *Assessment of children* (3rd ed.). San Diego: Sattler.

Savic, S. (1979). Mother–child verbal interaction: The functioning of completions in the twin situation. *Journal of Child Language, 6*, 153–8.

Savic, S. (1980). *How twins learn to talk*, New York: Academic Press.

Stafford, L. (1987). Maternal input to twin and singleton children: Implications for language acquisition. *Human Communication Research, 13*, 429–62.

Tallal, P., Ross, R., & Curtiss, S. (1989). Familial aggregation in specific language impairment. *Journal of Speech and Hearing Disorders, 54*, 167–73.

Tamis-LeMonda, C. S., & Bornstein, M. H. (1990). Language, play, and attention at one year. *Infant Behavior and Development, 13*, 85–98.

Thompson, L. A., & Plomin, R. (1988). The Sequenced Inventory of Communication Development: An adoption study of two- and three-year-olds. *International Journal of Behavioral Development, 11*, 219–31.

Tomasello, M., Mannle, S., & Kruger, A. (1986). Linguistic environment of 1- to 2-year-old twins. *Developmental Psychology, 22*, 169–76.

Whitehurst, G. J., Fischel, J. E., Lonigan, C. J., Valdez-Menchaca, M. C., Arnold, D. S., & Smith, M. (1991). Treatment of early expressive language delay: If, when, and how. *Topics in Language Disorders, 11*, 55–68.

18 Sources of individual differences in infant social cognition: Cognitive and affective aspects of self and other

Sandra Pipp-Siegel, JoAnn L. Robinson, Dana Bridges, and Sheridan Bartholomew

The term *intelligence* has historically referred to individual differences in the assessment of school-related abilities. Recently, however, the term has been broadened to include individual differences in a number of different domains. Academic intelligence, for example, is not correlated with measures of practical intelligence, defined as the ability to solve problems that arise in natural, nonschool settings (Wagner & Sternberg, 1986). Broadening the definition even further, Gardner (1983) suggested seven different types of intelligence, including traditional academic domain, (e.g., logical-mathematical intelligence), bodily kinesthetic intelligence, musical intelligence, and social intelligence (both intrapersonal and interpersonal). The domain of intelligence that is the focus of this chapter is infants' social cognitions regarding self and other.

How do infants construct understandings of themselves and others? A long theoretical tradition suggests that infants create the self by differentiating self from nonself (Bretherton, 1985; Mahler, Pine, & Bergman, 1975; Stern, 1985). In a review of current theories of self-development, Brownell and Kopp (1991) suggest that with development, infants establish boundaries that define the self compared to others and objects. The core function of self is "to define, locate, demarcate the world from a consistent perspective by organizing, integrating and representing experiences from that vantage point" (p. 288). By creating boundaries between self and other, the construction of the self is hypothesized to parallel construction of the other (Baldwin, 1899).

The process of differentiating self from other may follow a different developmental course in twins compared to singletons. Twins provide a special context in which to understand self-development, since it is presumed that twins have more difficulty in creating a separate identity (Ainslie, 1985; Schave & Ciriello, 1983). In general, fraternal twins tend to resemble each other, and, of course, identical twins have the same physical characteristics. In addition, twins begin to crawl and walk at a similar time (Ainslie, 1985) and share the same sleeping and feeding problems (Brown, Stafford, & Vandenberg, 1967). Perhaps as a function of the physical and behavioral similarity and the task of parenting two children at the same time, there is considerable consistency in parental child-rearing strategies. Especially in the period of infancy, children travel as a unit in the sense that they are less likely to be separated from one another because of the need for intensive supervision from their caregivers at this age. Twins therefore experience shared environments as well as shared genes. The experience of creating boundaries between self and the co-twin, thus, may provide different challenges for twins than singletons.

Research focused on singleton infants' understandings of self and other stems from two different traditions in developmental psychology. Following Piaget (1936/1952), one focus has been on discovering developmental sequences of how infants construct the cognitive self. In this approach, investigators describe normal development, and all infants are assumed to follow universal developmental sequences of self-knowledge. Research from this tradition, for example, has shown that toddlers are capable of first detecting marks on the self, then naming the self, and then labeling the self's gender (Bertenthal & Fischer, 1978; Lewis & Brooks-Gunn, 1979; Pipp, Fischer, & Jennings, 1987). The developmental sequence does not differ significantly between twins and singletons, although identical twins have been reported to self-recognize later than fraternal twins and singletons (Zazzo, 1975, 1979).

Following psychoanalytic and attachment theory, a second focus has been on the individual differences in infants' constructions of the self. Starting from the framework that infants construct the self in relations with others (Bretherton, 1985; Sroufe & Fleeson, 1986), it is assumed that differences in self-development are due to variations in how infants are treated by caregivers. Security of attachment, for example, is related to variations in interactions between infant and mother (Ainsworth, Blehar, Waters, & Wall, 1978), and it has been reported that infants who are securely attached to their caregivers had more complex self-knowledge than infants who were insecurely attached (Pipp, Easterbrooks, & Harmon, 1992). Yet maternal interactions in general differ between twins and singletons. Caregiving resources that are normally allocated to one infant must cover the needs of two infants. Infant twins show a preference for interacting with their mother more than the co-twin, and yet each individual twin experiences less interaction with the mother (Savic, 1980). The quality of infant–mother interactions is also different in infant twins, with twins experiencing more directive statements (Tomasello, Mannle, & Kruger, 1986), less sustained mutual attention (Clark & Dickman, 1984), less consistent discipline

(Lytton, 1980), and greater involvement with father (Ainslie, 1985). In spite of these differences, however, the proportion of twins who are securely attached to their mothers is equal to that of singletons, with securely attached infants being parented by mothers who are more sensitive to their infants' needs independent of twin status (Goldberg, Perrotta, Minde, & Corter, 1986).

While there has been emphasis on normative development and individual differences due to variations in interactions with caregivers, very little theoretical or empirical work has been directed to other sources of individual differences that may influence infants' self-development. This may be because the assumptions underlying current research preclude consideration of other sources. Because normative developmental sequences have been observed in advantaged and disadvantaged contexts, Schneider-Rosen and Cicchetti (1991) hypothesized that self-development is canalized; self-development occurs in the context of many different environments and therefore is relatively insensitive to variations in the environment, suggesting a strong genetic component. Infants who have no experience with mirrors, such as the Bedoins, for example, readily self-recognize when first presented their mirror image (Priel & deSchonen, 1986). Alternatively, in theories that focus on how the self is created from differences in interactions between infants and caregivers, assumptions are made that the environment is the causal agent in observed differences in infants' self-development. In this model, infants' self-development is assumed to be primarily relational and therefore largely environmentally determined. In one theoretical approach, thus, an influence of genetic inheritance is hypothesized, while in the other theoretical approach, the environment is hypothesized to be the critical factor in self-development. In both approaches, the nature–nurture question is assumed and not tested.

Other sources of individual differences, however, may influence infants' self-development. Infant temperament, cognitive ability, acquisition of language, and the timing of motoric ability are all possible influences that may alter the pathways of acquiring knowledge of self and others. To take just one example, temperament may influence how infants learn about themselves and others. Shy or inhibited children either may follow a *different* developmental pathway than other children or *slower* acquisition of self-knowledge. Behavior inhibition has been shown to be heritable (Robinson, Kagan, Reznick, & Corley, 1992); thus, one source of individual differences in self-development may include heritable influences.

In this chapter, we address the question of heritable and environmental influences on infants' knowledge of self and other. Heritability is a descriptive statistic that represents the portion of observed variability that is due to genetic differences among individuals. Whatever proportion of variance is left over is attributed to environmental influences as well as measurement error. Moreover, environmental influences can be of the shared or nonshared variety. Shared environmental influences include any aspects of twins' life that are common to both. These may include going to the same school, sharing the same friends, or living in an environment that encourages athletics. Twins may thus be similar because they share both genes and environments. Nonshared environment refers to those aspects that are unique to each twin. These may include freak accidents that only one twin experiences, differential parental treatment to each twin, and differences in how each twin experiences the twin relationship (Vandell, 1990).

Our approach to determining the heritable and environmental influence on infants' knowledge of self and other will be to examine infants' self-development in monozygotic and dyzygotic twins. Monozygotic (MZ) twins share 100% of genetic information, and dyzygotic (DZ) twins share an average of 50% genetic information. Heritability is demonstrated by showing that behavior is more similar for MZ than DZ twins. In the studies described in this chapter, we base our conclusions on twins studied in the MacArthur Longitudinal Study (Emde et al., 1992; Plomin et al., 1993). We focus on identical and fraternal twins at the end of the second year of life because self- and other-knowledge emerges at this age.

Infants' self-development is not unitary but instead encompasses a number of domains, such as self-recognition (Lewis & Brooks-Gunn, 1979) and role-taking ability (Brownell & Carriger, 1990; Zahn-Waxler, Chapman, & Cummings, 1984). And, in fact, differences are observed in how infants know themselves and others as a function of domain of knowledge (Pipp et al., 1987, 1992). Infant self-development also consists of cognitive and affective components. The etiology or sources of individual differences that underlie self-development may differ with the various aspects of self-development. In the context of the nature–nurture question, it may be that the underpinning of some aspects of self-development are due more to genetic than environmental sources. An alternate way to frame this question is to ask whether all levels of environmental influences are similar across different components of infants' self-development. In this chapter, we examine heritable and environmental factors of cognitive and affective indices of infants' self-development in two different domains: self-recognition and role-taking ability.

Mirror behavior

What is the developmental process of understanding that one's image in the mirror is a reflection of the self? Because one aspect of *self* includes recognition of images of the bodily self, numerous studies examine infants' mirror behaviors, and, in fact, infants exhibit a wide range of behaviors when placed in front of the mirror. One of the most widely studied behaviors is self-recognition, or the ability to detect a spot of rouge on the nose via the mirror reflection of the face. A second group of behaviors that infants exhibit are exploratory behaviors and emotional responses to the mirror image.

Self-recognition

Naming the self or pointing to their nose when a spot of rouge is seen on the nose in the mirror are generally considered to be adequate markers of *self-recognizing* ability. Stating one's own name when looking at one's mirror image is one unambiguous marker of self-recognition in singleton infants. At 18 months, 6% of infants named the self, while 40% of infants named the self at 24 months (Pipp-Siegel, Foltz, Heinbaugh, & Norton, 1994). Additionally, the ability to touch one's rouge-marked nose shows that the infant associates his or her own face with the face in the mirror and is able to compare an internalized image of one's own face (without rouge) with the image in the mirror of a face with rouge (Lewis & Brooks-Gunn, 1979; Mitchell, 1993). Across a number of studies, it has been shown that approximately 30% of the infants passed the rouge task at 18 months, while at 24 months, from 50% to 70% of infants passed this task (Amsterdam, 1972; Bertenthal & Fischer, 1978; Lewis, Brooks-Gunn, & Jaskir, 1985; Lewis, Sullivan, Stanger, & Weiss, 1989; Pipp, Easterbrooks, & Brown, 1993; Schneider-Rosen & Cicchetti, 1984, 1991).

It has been a matter of speculation whether twins recognize the self in a similar manner as singletons. It may be more difficult for twins to establish a separate identity than singletons (Rutter & Redshaw, 1991), especially in the case of self-recognition. Furthermore, it may be a somewhat more difficult process for MZ twins than DZ twins to recognize themselves, since MZ twins' appearance is identical. Zazzo (1975, 1979) reported that the developmental sequence of mirror behaviors did not differ when twins *as a group* were compared to singletons. However, he observed that identical twins were somewhat delayed when compared with singleton children, while fraternal twins were not significantly different from singletons. A difference in the percentage of infants who showed rouge-directed behavior between the singletons and twins would suggest that the individuation process – as indexed by the rouge-task – is slowed in comparison with singletons, replicating Zazzo (1975, 1979). We asked whether heritability is an important influence on infants' mirror behavior. To the extent that MZ twins' behavioral repertoires are more similar than DZ twins', we may infer a role for heritability on individual differences. To the extent that MZ and DZ twins' repertoires are both highly similar, common environmental experience is assumed to play an important role in individual differences.

In order to explore the effect of zygosity on infants' self-recognition, data from twins and singletons were observed (Robinson, Pipp, & Bartholomew, 1994). The sample of MZ and DZ twins was drawn from families who participated in the MacArthur Longitudinal Twin Study (MALTS) at 20 and 24 months of age. In summary, the final sample consisted of 306 20-month-old twins, 157 MZ and 149 DZ; and 337 24-month-old twins, 165 MZ and 172 DZ, equally divided by gender. In the MALTS sample, all twins participating in the study were selected preferentially for high birth weights (>1,700 gm) relative to gestational age, and 62% had birth weights greater than 2,500 gm. Over 90% were Caucasian, the remaining families being primarily Hispanic. Participating parents were slightly older (30 years) and somewhat better educated (14.5 years) than the average Colorado parent of a newborn (28 years old and 12.5 years of education). The singleton sample was obtained from two samples of children who had participated in previous studies (Pipp et al., 1992, 1993). These infants were similar demographically to the twin sample, with the exception that no infants were of low birth weight or less than 38 weeks' gestation. Twenty-eight 20-month-old infants and 36 24-month-old infants had been administered the rouge task in these studies.

For the MALTS sample, self-recognition was assessed by calling the child to the mirror for 30 sec. The child was then led away from the mirror, and a dot of rouge was surreptitiously placed on his or her nose by the examiner, who claimed to need to wipe crumbs from the child's face. The forehead, chin, and cheeks were briefly wiped with a washcloth before the cloth wiped the nose and left a spot of rouge. The examiner then invited the child to come to the mirror for an additional 30 sec. In the singleton sample, self-recognition was assessed by surreptitiously placing rouge on the nose and leading the child to the mirror. The rouge task was always the first task given in a series of featural-knowledge tasks administered to the toddlers. In order to ensure comparability between the studies, only the first 30 sec postrouge were coded for both the MALTS and the singleton samples.

A comparison of the frequencies of self-recognition for the three groups is shown in Table 18.1. Being a twin may have delayed the onset of self-recognition and embarrassment to a minimal degree. A lower percentage of twins touched their rouge-marked nose (approximately 38% averaged across the two ages) than did singletons (approximately 54% across the two ages). These differences, however, did not attain statistical significance.

Table 18.1. *Percentage of infants exhibiting self-recognition (touching the rouge-marked nose or naming self in front of mirror) for singletons, and MZ and DZ twins at both ages*

Age	Singletons	DZ twins	MZ twins
20 months	54	38	36
24 months	56	36	45

Being an identical twin did not seem to further delay the onset of self-recognition compared to fraternal twins, as no significant effects were found as a function of zygosity.

Affective responses to the mirror

Controversy exists about the meaning of passing the rouge task. Lewis et al. (1989) describe this self-recognizing ability as a cognitive index of a referential self. Loveland (1986), however, suggested that self-recognition – as measured by the rouge task – consists of a number of component skills, not just a cognitive representation of self. The mirror is a perceptual stimulus, and the ability to detect and touch a mark of rouge on the nose (with the use of the mirror) requires the infant to coordinate information about visually guided reaching in the mirror and various aspects of self-knowledge. The mirror image of self without rouge on the nose may also evoke a variety of responses. Responses to the self-image may represent either a cognition about the self or an emotional reaction to viewing the self or both.

Infants show a wide range of behaviors when presented with their mirror image. Three additional behaviors were studied because of their use in prior research: social behavior, avoidance, and embarrassment. Some of the earliest behaviors that have been observed developmentally are *social behaviors*, including social play directed at the mirror image (e.g., smiling, talking to the image). These behaviors have been reported to be less frequently observed after 18 months (Amsterdam, 1972; Schulman & Kaplowitz, 1976). Amsterdam suggested that social behaviors cannot be used as evidence to infer that infants exhibit self-knowledge, because the infant may be reacting to the mirror image as if it were another child. Infants from 14 to 24 months universally exhibited *avoidance* or a resistance to looking at the mirror (Amsterdam, 1972; Schulman & Kaplowitz, 1976). In general, these behaviors have not been meaningfully interpreted when observed and are seldom reported by other investigators (e.g., Bertenthal & Fischer, 1978; Lewis et al., 1989). A less extreme form of turning from the mirror image is coy behavior and *embarrassment*. Lewis et al. (1989) demonstrated that most toddlers who showed embarrassment in front of the mirror were also able to self-recognize. They inferred that self-recognition is a necessary cognitive component for self-conscious emotions such as embarrassment. However, embarrassment has been reported to occur infrequently (in less than one-fourth of children) between 18 and 24 months of age (Lewis et al., 1989; Robinson et al., 1994; Schneider-Rosen & Cicchetti, 1991).

Frequencies of mirror behaviors. Marking infants' noses with rouge may change infants' affective responses to their mirror image. The addition of rouge on the infant's nose, for example, may change the amount of social behavior, avoidance, or embarrassment observed in front of the mirror. The previously cited findings may be a result either of finding oneself in front of a mirror or may be specific to the presence of rouge on the nose. Specifically, embarrassment or avoidance may be elicited by the mirror image because feelings are generated by noting the rouge-marked nose or by noting the mirror image itself. Robinson et al. (1994) reported strong similarities in the frequencies of these pre- and postrouge behaviors at both ages. In general, affect displayed when gazing at one's mirror image seems to be a function of observing the mirror image and not a function of marking the nose with rouge.

No significant differences in the frequencies of behaviors were obtained between MZ and DZ twins, and so twin behavior was presented collapsing over zygosity. Table 18.2 presents the percentage of infants who showed social, avoidant, and embarrassed behavior after the rouge was applied for the singleton and the twin sample at 20 and 24 months of age. Most infants exhibited social behavior at both ages, while a relatively small number of infants showed avoidance or embarrassment. The percentage of infants who exhibited self-recognition (presented in Table 18.1) was between the percentage obtained for social behavior and avoidance or embarrassment. Comparisons between twins and singletons showed no difference between the frequencies of behaviors, with the exception that twins exhibited embarrassed behavior postrouge more frequently (approximately 15% averaged across the two ages) than did singletons (approximately 5% averaged across the two ages). The most common behaviors when presented with a mirror image, then, were social behavior and self-recognition. In comparison, affective responses to the mirror were relatively less common.

Although less frequent, when emotions were expressed, they were strongly linked to infants' self-recognition behavior. Specifically, the type of

Table 18.2. *Percentage of infants exhibiting social, avoidant, and embarrassed mirror behaviors in the postrouge condition for each age*

Behavior	Age	Twins % (N)	Singletons % (N)
Social	20 months	63.7 (195)	67.9 (19)
	24 months	78.2 (262)	69.4 (25)
Avoidant	20 months	20.6 (64)	17.9 (5)
	24 months	15.2 (51)	16.7 (6)
Embarrassed	20 months	13.4 (41)	3.6 (1)
	24 months	17.3 (58)	8.3 (3)

emotion that infants expressed when viewing their mirror images before the rouge was applied predicted the incidence of self-recognition after the rouge was applied (Robinson et al., 1994). Collapsed across age, a very low percentage of avoidant infants went on to self-recognize (14%) compared to infants who initially expressed social behavior (44%), who in turn were less likely to self-recognize than infants who initially expressed embarrassment (71%).

A strong hypothesis of developmental contingency has been put forth by Lewis et al. (1989); that is, self-recognition (signaled by touching the rouge-marked nose or naming the self) is a necessary precursor to the display of self-conscious emotion, such as embarrassment. In Lewis et al. (1989), 25% of children showed wariness or embarrassment in the prerouge mirror condition (their table 4, p. 153). Of these, a larger number of infants were nose touchers (73%) than nontouchers (27%). This empirical support for the developmentally contingent relationship between embarrassment and self-recognition was based on a small cross-sectional sample. Our larger sample (Robinson et al., 1994) replicated Lewis et al.'s (1989) findings. We wish to emphasize, however, that embarrassment is a relatively low frequency occurrence. One developmental pathway to embarrassment may be through self-recognition, but only for some toddlers and not others. Additionally, the ability to demonstrate embarrassment may reflect individual differences in emotional responding independent of the ability to self-recognize.

Table 18.3. *Rates of twin concordance for mirror behaviors*

	20 months		24 months	
Behavior	MZ	DZ	MZ	DZ
Social	.76	.86	.80	.85
Avoidant	.25	.28	.30	.15
Embarrassed	.36	.46	.23	.23
Self-recognizing	.35	.35	.51	.48

Concordance rates. Chi-square tests of association showed that MZ and DZ twins showed similar levels of occurrence of all of the behaviors at either age. Because the data are nominal in nature, indices of heritability and common environment cannot be calculated in the traditional manner. Several options were available for assessing the similarity of behavior within twin pairs. In a multiple-response situation such as this, where a child may perform one or more of several possible behaviors, twin similarity may be assessed by comparing behaviors one at a time, or the entire repertoire or patterning of observed behaviors may be the basis for comparison. Further, when behaviors can only be present or absent, comparison of behaviors is often judged based on their presence; this is commonly called a *disease model* of twin concordance. However, we view the absence of possible behaviors (such as touching the rouge-marked nose) as a potentially salient similarity. Thus, similarity of behavioral repertoire was judged both on the presence as well as the absence of behaviors across the entire behavioral repertoire.

First, we represented concordance according to the disease model, where the twin in the twin pair who shows evidence of the behavior is identified as the proband. In twin pairs where both twins exhibited the behavior, both are represented as probands. Using this method, we found that twin similarity was very high for the frequently occurring social behaviors and was low to moderate for the other three mirror behaviors. (See Table 18.3.) In no instance did MZ twins differ significantly from DZ twins in the level of concordance. It should be noted, however, that developmental patterns may be discerned in Table 18.3. The size of the concordance rates for social behaviors was quite large and did not change from 20 to 24 months, suggesting that social behavior in front of a mirror has reached the upper bound. Concordance rates for self-recognition, in contrast, increased from 20 to 24 months. Twenty-month-old infants are in the process of acquiring the ability to self-recognize, so these behaviors are relatively infrequent and perhaps unstable between twins. At 24 months, the stability of behavior increases and so therefore the concordance rate also increases.

In order to assess similarity of the entire behavioral repertoire within twin pairs, a single index of concordance was created within mirror conditions. Each dyad was given a score ranging from 0 to 3 reflecting the number of mirror behaviors both twins expressed or did not express within the prerouge condition at each age (where possible behaviors were social behavior, avoidance, and embarrassment) and from 0 to 4 within the postrouge conditions (where touching the nose was added as a possible behavior). Hence, a dyad was given one point for each behavior where both twins either did or did not perform the behavior; no point was given when the behavior was exhibited by one twin and not the other.

Concordance was quite high during prerouge conditions: 48.3% and 40.6% of dyads were concordant for all three behaviors at 20 and 24 months, respectively. An additional 16.1% and 18.8% of dyads were concordant for two of the three behaviors at each age; two-thirds of dyads were thus highly similar in the behavior profiles they displayed prerouge. Addition of the rouge mark to the nose to twins' behavioral repertoire resulted in only slightly lower twin concordance: 46.8% of dyads were concordant for three or four behaviors at 20 months, and 49.1% at 24 months. These concordance rates did not differ significantly between MZ and DZ twin pairs, suggesting that common environmental experiences, but not heritability, are an important source of individual differences in mirror behavior.

Individual differences in mirror behavior

Exploring the reflected image is a complex skill where each opportunity in front of the mirror may provide new affordances for exploration of this unique perceptual field (Loveland, 1986). Different aspects of the mirror image may be salient for the child at different times. On one occasion, a child may be intrigued with the reflected movement of the limbs in the mirror, on another occasion the detection of rouge or emotional reactions to the reflection of self. In our sample, somewhat more than 50% of the infants who exhibited self-recognition at 20 months also did so at 24 months, with affective behaviors showing less stability across age. Differential salience may be a function of change in developmental capacity to respond to mirror images, both emotionally and cognitively. More transient responses to the mirror may also be elicited both as infants' attention focuses on different aspects of the mirror and/or children select particular behavioral responses.

These data suggest the importance of integrating individual difference and normative developmental approaches to understanding infant behavior. Children showed a strong tendency to touch rouge on the nose after viewing their reflection in the mirror, yet the tendency is influenced by individual differences in infants' emotional responses to the mirror (e.g., avoidance and embarrassment).

The high rate of twin concordance suggests that common environmental experiences of the twins play an important role in the acquisition of exploratory strategies in front of the mirror. When an opportunity to explore the mirror image occurs for one twin, it is likely to occur for the other as well. In addition, any direct tuition that parents may engage in with one twin is likely to also occur with the other. In many respects, 2-year-old twins travel as a unit, sharing many environmental opportunities. (See Vandell, 1990, for an excellent review.) While it is not surprising that common environment influences mirror behaviors, it is somewhat surprising that heritability appears to play no role in mirror behaviors at this age. In later sections, we examine whether other components of infants' behavior influence mirror behaviors, such as temperament and cognitive ability.

Role-taking abilities

Self–other differentiation has also traditionally been assessed through infants' ability to role play (Brownell & Carriger, 1990; Zahn-Waxler et al., 1984). In older children and adults, role taking is a type of perspective taking that involves the ability to take the role or perspective of another and to understand another's thoughts, motives, and feelings (Iannotti, 1985). Young infants, however, are unable to differentiate self from other in the domain of action. Piaget (1936/1952) suggested that with development, infants become decentered by being able to take the self as one object among many. In the transition from sensorimotor to preoperational thought, infants begin to separate objects from their actions upon them, and this process leads them to differentiate self and self's actions from objects and people. In role taking, inability to take another's point of view implies that infants have not differentiated self from other since in order to pretend to be someone else, a child must be able to maintain his or her representation of self while enacting the role of the other (Fein, 1981).

The ability to take the role of another develops dramatically during the second year of life. In general, three developmental steps are described in the second year (Fenson & Ramsey, 1981; Watson & Fischer, 1977, 1980). Brownell and Carriger (1990) describe the developmental sequence as follows. (1) At about 12–15 months of age, infants are able to act upon the self and, for example, can pretend to feed themselves. Infants see themselves as universal agents but are unaware of others' causality. (2) At 15–20 months, infants begin to be able to act on inanimate objects. In this phase, infants still understand the self as a universal agent but differentiate others as passive recipients from self's actions. (3) At 20–24 months, infants begin to "conceive of others as active, autonomous agents in relation to their own actions, independent of the child's actions or wishes" (Brownell & Carriger, 1990, p. 1165). These three steps are labeled, respectively, *self-as-actor*, *passive agency*, and *active agency* (Watson & Fischer, 1977).

Infants in the MALTS sample were presented tasks that assessed passive and active agency. In the passive-agent task, infants were asked to feed a bunny. Two active-agent tasks were given: The first required infants to make the bunny feed itself, and the second required infants to make the bunny hop. Additionally, a third phase was added in order to examine infants' capacities to take the role of the other. For this phase, two additional tasks were modeled in which infants were asked to pretend to be a bunny by hopping to a carrot and eating it and, in the last step, by pretending to be any other animal they wished (e.g., a dog). Bridges (1993) reported that 87.2% of children fit the sequence described, with reliable scalability. Because the sequence scaled, the highest task passed was used in our analyses.

At both ages, the distribution of the highest task passed ranged from none to all five tasks passed. Statistically stable differences were reported between 20- and 24-month-old infants. At 20 months, infants passed an average of 1.6 tasks (with a dual mode of passive and active agent), and at 24 months, infants passed an average of 2.0 (mode: active agent). These findings replicate others (Brownell & Carriger, 1990; Watson & Fischer, 1977), who showed in singleton samples that the modal shift from 18 to 24 months is from passive to active agency. By inference, then, twins are not delayed in their ability to act as an agent on others when compared with singletons. Thus, the capacity to differentiate self from other in twins is not affected, at least when assessed by an agency sequence.

Additionally, the highest step passed in role-taking capacity was generally related to mirror behaviors. At both ages, increases in role-taking ability were associated with social behavior ($r_{20\,mo} = .17, p < .01; r_{24\,mo} = .23$, $p < .001$) and embarrassment ($r_{20\,mo} = .22, p < .001; r_{24\,mo} = .23, p < .001$). Two age-specific findings, however, were also obtained: Role-taking behavior was reliably related to self-recognition only at 20 months ($r = .18, p < .005$), and decreases in role taking were significantly associated with avoidant mirror behavior only at 24 months ($r = -.16, p < .01$). It is currently unclear as to the reason for the age-related findings, although other factors may mediate the expression of phenotypic behavior at one age and not the other.

Calculation of the indices of the heritability and common environment

Table 18.4. *Twin correlations and estimates of heritability (h²) and common environment (c²)[a] for role-taking ability at 20 and 24 months of age*

	Correlations			
Age	MZ	DZ	h^2	c^2
20 months	.66[b]	.46[b]	.30	.27
24 months	.50[b]	.12	.45	.00

[a] Calculated according to the multiple-regression technique described by DeFries and Fulker (1985).
[b] $p < .001$.

for role taking was based on the regression model of DeFries and Fulker (1985). These analyses revealed that infants' role-taking capacities were related to both heritability and common environmental influences at 20 months and only to heritability at 24 months, as shown in Table 18.4. In contrast to individual differences in self-recognition that were associated with environmental influences, moderate heritability was a factor in infants' role-taking ability at both ages and the only factor at 24 months. Role taking is often seen as a cognitive ability, and it may be that heritability of role-taking ability is related to cognitive ability. Dramatic spurts in language development are also observed during these ages, and both cognition and language may be related to role-taking ability. To explore this hypothesis, in the next section we examine the relation of self-recognition and role-taking ability with individual differences in cognitive and language development as well as temperament.

Self-development related to cognition, language, and temperament

Infants' developing capacities to differentiate self from other were shown to be influenced by environmental factors for self-recognition and heritable and environmental factors for role-taking ability. In the second year of life, reorganization of infants' capacities occurs in a number of different domains, and we believe that understanding infants' self-development in the context of these other changes in infants' abilities will better enable us to understand both direct and indirect influences of other capacities on self-development.

Previous publications have detailed twin resemblance for the MALTS sample for a wide array of measures (Emde et al., 1992; Plomin et al., 1993), and in this chapter we examine the relation between self-development and three important domains of infant development: cognition, language and temperament. A widely used, standardized cognitive test of mental abilities is the Bayley Scales of Infant Development's subscale of the mental development index (Bayley, 1969). If cognitive abilities are central or primary in the ability to self-recognize and role take (as has been presumed), then we might expect children who self-recognize and show more advanced role-taking ability to score higher on cognitive tests than those who do not. Additionally, language development is at a peak during the second year, with a salient increase in both naming and imitation. During this age period, the toddler's capacity to map the word and the action or object increases. Since self-recognition requires the ability to map the mirror image and the self's facial features and/or name, it seems reasonable to propose that children with more advanced spontaneous and imitative language abilities will be more likely to self-recognize than children with less spontaneous language. Additionally, since role-taking ability is assessed through the elicited imitation paradigm, it may be that the capacity to imitate language would also be related to role-taking abilities. Categorization of spontaneous and imitative language use was assessed by the Sequenced Inventory of Communicative Development (SICD) (Hedrick, Prather, & Tobin, 1975), following Kubicek, Emde, and Schmitz (1994).

Because parents observe their infants over much longer periods of time and in a number of different situations, parental reports of their infants' temperament may provide a stable estimate of infant behavior. The subscales of shyness and sociability were used from the Colorado Child-hood Temperament Inventory (CCTI) (Rowe & Plomin, 1977) and were averaged across parents. In general, small but significant relations were found between the parental temperament measure of shyness and our measures of embarrassment and avoidance to the mirror image. Infants who were embarrassed in front of the mirror were reported by their parents as being less shy at 20 ($r = -.22$, $p < .01$) and 24 months ($r = -.15$, $p < .01$). Avoidance of the mirror image was related to increased parental reports of shyness, but only at 20 months ($r = .16$, $p < .01$). Parental reports of sociability were unrelated to embarrassment or avoidance.

A different pattern of heritable and environmental influences was found for each of the measures of cognitive ability, language development, and temperament. Table 18.5 presents estimates of heritability and common environment for each three of the measures in the MALTS sample. Bayley MDI is influenced both by heritable and common environmental factors as 20 and 24 months (Cherny et al., in press). Kubicek et al. (1994) reported that common environment accounted both for language scores at 20 months and for spontaneous language at 24 months, although 24-month-old imita-

Table 18.5. *Estimates of heritability (h²) and common environment (c²) for infant cognitive ability, language, and temperament at 20 and 24 months of age*

	20 months		24 months	
	h^2	c^2	h^2	c^2
Bayley MDI[a]	.34	.46	.39	.43
SICD Language[b]				
Imitative	.08	.79	.40	.39
Spontaneous	.00	.59	.08	.60
Temperament				
Shyness	.38	.00	.38	.00
Sociability	.42	.00	.59	.00

[a] These estimates derive from MALTS data and were first reported by Cherny, Fulker, Emde, Robinson, Corley, Reznick, Plomin, & DeFries (1994).
[b] These estimates derive from MALTS data and were first reported by Kubicek, Emde, & Schmitz (1994).

tive language was described by both heritable and common environmental influences. Temperament subscales of shyness and sociability were heritable, with no evidence of common environment. The only developmental change in heritable and environmental factors from 20 to 24 months in this sample, then, was observed in imitative language.

In general, self-recognition and role-taking ability were related to different components of cognitive ability, language development, and temperament, as shown in Table 18.6. Similar patterns of association were found in relation to expressions of self–other differentiation both affectively and cognitively, with one exception. While all indices were related to imitative language ability and shyness, for the more cognitive indices of self-development (e.g., self-recognition and role-taking ability), individual differences in cognitive ability were added as a source of individual differences in self-development. It should be noted that the intercorrelations of cognition, language, and temperament with role-taking ability were stronger than the intercorrelations for self-recognition. This may be due to the relative instability of our index of self-recognition, the rouge task. We have suggested elsewhere that more stable assessments of infants' capacity to self-recognize can be obtained by a series of tasks designed to examine infants' recognition of the self (Pipp et al., 1992, 1993).

Infant avoidance and embarrassment in front of the mirror were related

Table 18.6. *Correlations of self-recognition and role-taking ability with cognition, language, and temperament at 20 and 24 months of age*

	Bayley MDI	Language		Temperament	
		Imitative	Spontaneous	Shyness	Sociability
Role-taking skill					
20 months	.19^c	.39^c	.28^c	−.15^b	.05
24 months	.18^c	.48^c	.15^b	−.16^c	.06
Self-recognition					
20 months	.02	.15^b	.15^b	−.19^c	.10
24 months	.13^a	.21^c	.05	−.18^c	.14^a
Mirror avoidance					
20 months	−.05	−.17^c	−.06	.16^c	−.10
24 months	−.10	−.15^b	−.10	.09	.00
Mirror embarrassment					
20 months	.07	.14^a	.17^c	−.22^c	.09
24 months	.07	.12^a	−.04	−.15^b	−.02

$^a p < .05.$
$^b p < .01.$
$^c p < .001.$

to parental reports of shyness and imitative language, but the direction was opposite for these two affective responses to the mirror. Avoidant infants were less likely to exhibit imitative language skills at both ages and more likely to be rated as shy by their parents at 20 months. Infants who were embarrassed when shown their mirror reflection were more likely to demonstrate more imitative (at both ages) and spontaneous language (at 20 months) and were less likely to be considered shy by their parents at both ages. Avoidance may retard infants' acquisition of skills, while embarrassment may reflect more advanced self-development.

While the phenotypic correlations reported were significant, they were also quite low. As a result, we are limited in the inferences we can make about genetic and environmental covariance in these behaviors. Nonetheless, there are common patterns for some variables and more substantial correlations for others, and we consider these next.

Individual differences in self–other differentiation

If the sources of individual differences in infants' self-development are unitary, the same phenotypic relation between self-development and the other domains of development would be expected. If, however, the sources are multidetermined, phenotypic relations between self-development and other domains would be predicted to vary. Similar sources of individual differerences influenced the phenotypic expressions of self–other differentiation in self-recognition and role-taking abilities. Individual differences between infants in temperament and in cognitive and language ability were related to the level of role-taking ability and to the ability to recognize the self in the mirror. Infants' affective responses to their mirror images were associated with imitative language ability and shyness but were not related to cognitive ability.

Two important and common sources of individual differences in self-development include imitative language ability and shyness. Imitative language ability significantly influenced all measures of self-development in these data. Increased role-taking ability, self-recognition, and embarrassment were related to increased imitative language, and shyness was related to decreased imitative language. The associations between these variables are modest, with the exception of role-taking ability. Common to these tasks was infants' ability to imitate, and the more substantial associations may be a function of this commonality. Role-taking ability is assessed through an elicited imitation technique, while the scale of language ability focuses on toddlers' imitation of verbal productions. One future research direction will be to determine whether role-taking ability is related to language ability per se, or to the more global ability to imitate. Alternatively, both spontaneous- and imitative-language measures were significantly associated with role-taking ability at both ages, and it may be that rapid developmental changes at these ages influence the acquisition of both language and role-taking ability. From these data, it is impossible to determine, of course, whether a third developmental factor influences both language and role taking or if one domain influences the acquisition of the other domain (e.g., development in language ability *causes* development in role taking).

Parental reports of shyness were also associated with all self measures. While the associations are significant and consistent across measures, the magnitude of association was modest. The modest correlations make it unlikely that heritable and common environment covariation would be detected. Theoretically, however, temperament is a very important predictor of infants' performance abilities. This dimension is missing from most theoretical formulations of self-development, with the exception of Kagan (1981). Shyness may result in different developmental pathways in acquiring self–other differentiation. Our data suggest that shyness slows infants' ability to differentiate self from other. Toddlers who are shy are less likely to act on others, and as a consequence, this temperament variable may retard self-development. In addition to delaying the acquisition of self-development, shyness may also alter the quality of self-development. If shyness reduces the amount of contact that toddlers have with others, the character of the self may be more isolated compared with infants whose self is created by more frequent and robust interactions with others. Nonshy children may incorporate their increased interactions with others in their self-definition more than shy children. It should be noted that our assessments of self–other differentiation occurred in a strange place with strange people for brief periods of time, to paraphrase Bronfenbrenner (1979). As such, we increased the probability of eliciting shy behavior. In future work, it will be important to determine whether the same pattern of results is obtained when toddlers are in more familiar environments.

Comparisons between MZ and DZ twins revealed a difference in the genetic and environmental influences of our two measures of self–other differentiation. While self-recognition was associated with environmental sources of individual differences, it should be noted that infant abilities that are known to have both heritable and environmental determinants were correlated with infants' self-recognition. Role-taking ability was related to genetic and environmental factors at 20 months, but only to genetic factors at 24 months. Phenotypic associations, however, were found between role taking and various measures that included genetic, environmental, or both types of sources. It will be important in future research to assess the genetic and environmental covariances among abilities that together serve as constructs that underlie self-development.

Two caveats should be noted. We do not wish to imply that self-development is related only to the constructs measured in this chapter. Self-development is composed of a number of different domains, and a wide array of constructs will be needed to articulate fully the development of the self. One such construct that should be added is motor development, as represented by infants' increasing ability to crawl or walk. The ability to crawl should have large influences on infant sense of autonomy (Mahler et al., 1975), which in turn alters qualities of the relationship between infants and mothers (Biringen, Emde, Campos, & Applebaum, 1995). A second caveat is that our sample consisted of twins. Vandell (1990) reviewed a number of studies that show that twins' environments are dissimilar to singletons. Although we found no significant difference between twins and singletons in the percentage of infants who showed various mirror behaviors, there was some suggestion that fewer twins self-recognized than singletons. These findings, in conjunction with those reported by Zazzo (1979), may indicate that a small amount of variance in self-recognition may be accounted for by twin status, perhaps reflecting the behavior of a small group of twins. Only a few twins, for example, in our sample named their own twin when shown their mirror image. More importantly, qualitative differences in self-development may exist for twins. As a result, the relation between indices of self–other differentiation and other abilities such as cognition, language, and temperament may be different for twins and sin-gletons.

In summary, it has been hypothesized that twins are faced with different challenges in creating an identity compared to singletons. If the self is created from differentiating from significant others, the twin environment is constructed such that twins must separate from both parents and the co-twin. With development, the nature of the self-concept becomes more detailed and abstract (Damon & Hart, 1988). Changes in self-development may be reflected in twin's relationship with each other. Vandell (1990), for example, reported developmental changes in twins' relationships with the co-twin. In infancy, twins are less likely to interact with the co-twin and singleton peers than are singletons, although twins become increasingly more interested in each other from 6 to 24 months of age (Vandell, Owen, Wilson, & Henderson, 1988). Grade school is the golden age of twinship, with twins preferring and spending more time in the company of the co-twin compared to singleton siblings. Finally, adolescent twins in general begin to separate from one another, with a marked increase in conflict. The in-creased closeness in twins in grade school and the increased conflict in adolescence may, in part, reflect changes in genotypic development that underlie social cognitive ability.

The self develops throughout the life span, and it may be that heritability estimates will increase with development. Plomin (1986) reports that – contrary to conventional wisdom – the pattern of data reveals that heritabil-ity increases with age. For example, heritability of mental ability increases from less than 20% in infancy to 40% in childhood to about 50% in adoles-cence. And, in our data, heritability estimates of role-taking ability in-creased with age, while estimates of common environment decreased. Plomin (1986) suggests that increasing heritability could be due to increased genetic variance, due in part either to "the influence of genes expressed only later in development or to the amplification of genetic effects ex-pressed early in development" (p. 326). In the second year, spurts in lan-guage ability, the advent of representational thought, and the increase in self-conscious emotions dramatically alter infants' construction of self. Changes in the phenotypic expressions of the language, cognitive and emo-tion that together are associated with self-development may reflect changes in genotypic development.

Gardner (1983) posited that intelligence should be assessed across a number of domains, including musical, bodily-kinesthetic, and social intel-ligence, as well as the traditional academic ones. In the domain of social intelligence, we found that in infancy, cognitive and affective aspects of self–other differentiation were related to estimates of heritability, common environment, or both depending upon the domain of self-development.

Gardner differentiated two aspects of social intelligence; one focuses on intelligence regarding self, and the other on intelligence about other people. Controversy exists regarding which domains of self and other are differentiated at birth and which undergo a process of differentiation (Pipp, 1990; Stern, 1985). An important strategy for determining estimates of heritability and common environmental influences of social intelligence during this age period should include measures of self- and other-development *and* the underlying abilities that together constitute those components underlying infants' developing differentiation of self from other.

Acknowledgments

The results reported in this chapter were made possible through the collaboration of a group of investigators that also included J. C. DeFries and D. Fulker at the University of Colorado; R. Emde at the University of Colorado Health Sciences Center; J. Campos at the University of California at Berkeley; J. Kagan at Harvard University; R. Plomin at the Pennsylvania State University; J. S. Reznick at Yale University; and C. Zahn-Waxler at the National Institute for Mental Health and was supported by a grant from the John D. and Catherine T. MacArthur Foundation. The authors wish to thank the families who generously contributed their time to this project, John DeFries and Elena Grigorenko for comments on the manuscript, and the research assistants and students at IBG who were involved in data collection, behavioral coding, and data management.

References

Ainslie, R. (1985). *The psychology of twinship*. Lincoln, NE: University of Nebraska Press.

Ainsworth, M. D. S., Blehar, M. C., Waters, D., & Wall, S. (1978). *Patterns of attachment*. Hillsdale, NJ: Erlbaum.

Amsterdam, B. (1972). Mirror self-image reactions before age two. *Developmental Psychology, 5*, 297–305.

Baldwin, J. M. (1899). *Social and ethical interpretations in mental development*. New York: Macmillan.

Bayley, N. (1969). *Manual for the Bayley Scales of Infant Development*. New York: Psychological Corp.

Bertenthal, B. I., & Fischer, K. W. (1978). The development of self-recognition in the infant. *Developmental Psychology, 14*, 44–50.

Biringen, Z., Emde, R. N., Campos, J. J., & Applebaum, M. I. (1995). Affective reorganization in the infant, the mother and the dyadic relationship as a function of walking onset. *Child Development, 66*, 499–514.

Bretherton, I. (1985). Attachment theory: Retrospect and prospect. In I. Bretherton & E. Waters (Eds.), Growing points of attachment theory and research. *Monographs of the Society for Research in Child Development, 5* (Serial No. 209), 3–35.

Bridges, D. L. (1993). *The association between the development of role-taking and empathy in children*. Unpublished senior honors thesis, University of Colorado, Boulder.

Bronfenbrenner, U. (1979). *The ecology of human development.* Cambridge, MA: Harvard University Press.

Brown, A. M., Stafford, R. E., & Vandenberg, S. (1967). Twins: Behavioral differences. *Child Development, 38,* 1055–64.

Brownell, C. A., & Carriger, M. S. (1990). Changes in cooperation and self-other differentiation during the second year. *Child Development, 61,* 1164–74.

Brownell, C. A., & Kopp, C. B. (1991). Common threads, diverse solutions: Concluding commentary. *Developmental Review, 11,* 288–303.

Cherny, S. S., Fulker, D. W., Emde, R. N., Robinson, J., Corley, R. P., Reznick, J. S., Plomin, R., & DeFries, J. C. (1994). A developmental-genetic analysis of continuity and change in the Bayley Mental Development Index from 14 to 24 months. *Psychological Science, 5,* 354–60.

Clark, P. M., & Dickman, Z. (1984). Features of interaction in infant twins. *Acta Geneticae Medicae et Gemellololgiae, 33,* 165–71.

Damon, W., & Hart, D. (1988). *Self-understanding in childhood and adolescence.* New York: Cambridge University Press.

DeFries, J. C., & Fulker, D. W. (1985). Multiple regression of twin data. *Behavior Genetics, 15,* 467–73.

Emde, R. N., Plomin, R., Robinson, J., Reznick, J. S., Campos, J., Corley, R., DeFries, J. C., Fulker, D. W., Kagan, J., & Zahn-Waxler, C. (1992). Temperament, emotion and cognition at 14 months: The MacArthur Longitudinal Twin Study. *Child Development, 63,* 1437–55.

Fein, G. (1981). Pretend play in childhood: An integrative review. *Child Development, 52,* 1095–118.

Fenson, L., & Ramsey, D. (1981). Decentration and integration of the child's play in the second year. *Child Development, 52,* 1028–36.

Gardner, H. (1983). *Frames of mind: The theory of multiple intelligences.* New York: Basic Books.

Goldberg, S., Perrotta, M., Minde, K., & Corter, C. (1986). Maternal behavior and attachment in low-birth-weight twins and singletons. *Child Development, 57,* 34–46.

Hedrick, D. L., Prather, E. M., & Tobin, A. R. (1975). *Sequenced inventory of communication development.* Seattle: University of Washington Press.

Iannotti, R. J. (1985). Naturalistic and structured assessments of prosocial behavior in preschool children: The influence of empathy and perspective taking. *Developmental Psychology, 21,* 46–55.

Kagan, J. (1981). *The second year: The emergence of self-awareness.* Cambridge, MA: Harvard University Press.

Kubicek, L., Emde, R. N., & Schmitz, S. (1994). *Temperament and language in the transition to early childhood.* Manuscript in preparation.

Lewis, M., & Brooks-Gunn, J. (1979). *Social cognition and the acquisition of self.* New York: Plenum.

Lewis, M., Brooks-Gunn, J., & Jaskir, J. (1985). Individual differences in visual self-recognition as a function of mother–infant attachment relationship. *Child Development, 49,* 1247–50.

Lewis, M., Sullivan, M. W., Stanger, C., & Weiss, M. (1989). Self development and self-conscious emotions. *Child Development, 60,* 146–56.

Loveland, K. (1986). Discovering the affordances of a reflecting surface. *Developmental Review, 6,* 1–24.

Lytton, H. (1980). *Parent–child interaction.* New York: Plenum Press.

Mahler, M. S., Pine, F., & Bergman, A. (1975). *The psychological birth of the infant.* New York: Basic Books.

Mitchell, R. W. (1993). Mental models of mirror-self-recognition: Two theories. *New Ideas in Psychology, 11*, 295–325.

Piaget, J. (1952). *The origins of intelligence in children* (M. Cook, Trans.). New York: International Universities Press. (Original work published 1936.)

Pipp, S. (1990). Sensorimotor and representational internal working models of self, other and relationship: Mechanisms of connection and separation. In D. Cicchetti & M. Beeghly (Eds.), *Topics in transition in development: Self development* (pp. 243–64). Chicago: University of Chicago Press.

Pipp, S., Easterbrooks, M. A., & Brown, S. R. (1993). Attachment status and complexity of infants' self- and other-knowledge when tested with mother and father. *Social Development, 2*, 1–14.

Pipp, S., Easterbrooks, M. A., & Harmon, R. J. (1992). The relation between attachment and knowledge of self and mother in one- to three-year-old infants. *Child Development, 63*, 738–50.

Pipp, S., Fischer, K. W., & Jennings, S. (1987). The acquisition of self and mother knowledge in infancy. *Developmental Psychology, 23*, 86–96.

Pipp-Siegel, S., Foltz, C., Heinbaugh, H., & Norton, J. (1994). *Contextual effects on toddlers' acquisition of self and mother knowledge.* Manuscript submitted for publication.

Plomin, R. (1986). *Development, genetics and psychology.* Hillsdale, NJ: Erlbaum.

Plomin, R., Emde, R. N., Braungart, J., Campos, J., Corley, R., Fulker, D. W., Kagan, J., Reznick, S. J., Robinson, J., Zahn-Waxler, C., & DeFries, J. C. (1993). Genetic change and continuity from 14 to 20 months: The MacArthur Longitudinal Twin Study. *Child Development, 64*, 1354–76.

Priel, B., & deSchonen, S. (1986). Self-recognition: A study of population without mirrors. *Journal of Experimental Child Psychology, 41*, 237–50.

Robinson, J., Kagan, J., Reznick, J. S., & Corley, R. (1992). The heritability of inhibited behavior: A twin study. *Developmental Psychology, 28*, 1030–7.

Robinson, J., Pipp, S., & Bartholomew, S. (1994). *What does behavior before a mirror reflect?* Manuscript in preparation.

Rowe, D. C., & Plomin, R. (1977). Temperament in early childhood. *Journal of Personality, 41*, 150–6.

Rutter, M., & Redshaw, J. (1991). Annotation: Growing up as a twin: Twin-singleton differences in psychological development. *Journal of Child Psychology and Psychiatry Allied Disciplines, 32*(6), 885–95.

Savic, S. (1980). *How twins learn to talk.* London: Academic Press.

Schave, B., & Ciriello, J. (1983). *Identity and intimacy in twins.* New York: Praeger Press.

Schneider-Rosen, K., & Cicchetti, D. (1984). The relationship between affect and cognition in maltreated infants: Quality of attachment and the development of visual self-recognition. *Child Development, 55*, 648–58.

Schneider-Rosen, K., & Cicchetti, D. (1991). Early self-knowledge and emotional development: Visual self-recognition and affective reactions to mirror self-images in maltreated and nonmaltreated toddlers. *Developmental Psychology, 27*, 471–8.

Schulman, A. H., & Kaplowitz, C. L. (1976). Mirror-image response during the first two years of life. *Developmental Psychobiology, 10*, 133–42.

Sroufe, L. A., & Fleeson, J. (1986). Attachment and the construction of relationships. In W. Hartup & Z. Rubin (Eds.), *Relationships and development.* Hillsdale, NJ: Erlbaum.

Stern, D. N. (1985). *The interpersonal world of the infant.* New York: Norton.

Tomasello, M., Mannle, S., & Kruger, A. C. (1986). Linguistic environment of 1- to 2-year-old twins. *Developmental Psychology, 22*, 169–76.

Vandell, D. L. (1990). Development in twins. In R. Vasta (Ed.), *Annals of Child Development* (Vol. 7) (pp. 145–74). London: Jessica Kingsley Publishers, Ltd.

Vandell, D. L., Owen, M. T., Wilson, K. S., & Henderson, V. K. (1988). Social development in infant twins: Peer and mother–child relationships. *Child Development, 59,* 169–77.

Wagner, R. K., & Sternberg, R. J. (1986). Tacit knowledge and intelligence in the everyday world. In R. K. Wagner & R. J. Sternberg (Eds.), *Practical intelligence* (pp. 51–83). New York: Cambridge University Press.

Watson, M. W., & Fischer, K. W. (1977). A developmental sequence of agent use in late infancy. *Child Development, 48,* 828–36.

Watson, M. W. & Fischer, K. M. (1980). Development of social roles in elicited and spontaneous behavior during the preschool years. *Developmental Psychology, 16,* 483–94.

Zahn-Waxler, C., Chapman, M., & Cummings, E. M. (1984). Cognitive and social development in infants and toddlers with a bipolar parent. *Child Psychiatry and Human Development, 15,* 75–85.

Zazzo, R. (1975). Des jumeaux devant le miroir: Questions de methode. *Journal de Psychologie, 4,* 389–413.

Zazzo, R. (1979). Des enfants, des singes, et des chiens devant le miroir. *Revue de Psychologie Appliquee, 29,* 235–46.

Zazzo, R. (1982). The person: Objective approaches. In W. Hartup (Ed.), *Review of child development research* (Vol. 6) (pp. 247–90). Chicago: University of Chicago Press.

Part IV

Integration and conclusions

19   Nature vs. nurture: The feeling of *vujà dé*

Earl Hunt

Benjamin Franklin said that nothing is certain except death and taxes. But that was in his day. Modern-day psychologists have the Superbowl and the nature–nurture debate. The Superbowl begins with great publicity and proceeds to a lop-sided victory by one side, after which the winning players graciously tell reporters how well the losers played. So it is with the nature–nurture debate. Nature wins, 48–6, and then the winners say that, well, some of those environmentalist arguments were very good tries, albeit a trifle misguided.

I have a feeling of *vujà dé*, a term you will not find in your French dictionary because I just made it up. *Vujà dé* is the uncanny feeling that I do not ever want to get caught here again. Others may agree. Just before the chapters in this book were written, the American Psychological Association published a volume called *Nature, Nurture, and Psychology* (Plomin & McClearn, 1993). Sound familiar? An article coauthored by one of the contributors to this volume (McGue, Bouchard, Iacono, & Lykken, 1993) began by saying,

it appears that the issue has been resolved. . . . Over 90% of [those responding to a survey] agreed that IQ was, at least in part, heritable [p. 59].

and closes with

Psychology appears ready to move beyond the acrimony that has marked the past century of debate on the nature–nurture issues [p. 74].

Now Bidell and Fischer (chapter 7, this volume) say that the basic technique of behavior genetics, partitioning the variance in measures of behavior into components associated with heritability, environment, and their interaction, is fundamentally flawed. At the same time, Bouchard (chapter 5, this volume) says about the opposition to behavior genetics that

all the criticisms that have been brought against these studies fall into two classes: (1) pseudo-analyses, quantitative analyses giving the appearance of legitimate scientific analysis of the data but actually committing fundamental errors, or, (2) pseudo-arguments, verbal arguments that when put into quantitative form are shown either to be incapable of explaining the facts they purport to explain and that are inconsistent with large bodies of existing evidence.

One of the nice things about acrimony in psychology is that if you missed it the first time, you can always watch the reruns.

Sarcasm aside, science is not a sporting event, and hopefully psychology is more than a television series. I was quite serious when I said that I do not ever want to be caught here again. The chapters of this book provide a reasonably good summary of the current state of the nature–nurture debate. In this essay, I look at what that state is and ask why it continues. I conclude that the most productive thing to do will be to move on beyond the debate to consider new issues. Few will disagree with this aphorism.

What is more contentious is my conclusion that the people on the nativist side of the debate have developed and utilized concepts that are likely to advance our scientific understanding of variations in human intellectual capacity but that those on the nurture side of the debate have not. I believe that if they wish to do so within the framework of science, they will have to experience a sea-change in their views and, especially, in their willingess to submit their ideas to the rigor and impersonality of good science. It is quite possible that many of those who wish to take what Miller (chapter 9, this volume) calls the *culture perspective* on mental competence may wish to do this, not because they are inherently bad scientists, but because they wish to make a humanistic rather than a scientific argument.

Since "being a scientist" carries with it a certain cachet, my contention that the culture perspective is not scientific could be taken as an attack. This is not necessarily true, although it can be true in specific situations. Science is only one of several possible ways of understanding the world. Because scientific reasoning requires a commitment to precise definition and impersonal, objective verification, it is not always the best way to approach a question, including the question of which of several possible scientific approaches to emphasize! Humanistic arguments, by contrast, are subjective interpretations of personal experiences and arguments based on selected examples of situations in the world. The recipient of a humanistic argument should be convinced if he or she finds the proposer's arguments consonant with the receiver's own beliefs and experiences. One of the components of a humanistic argument might well be an appeal to a scientific analysis, but a scientific analysis could not contain a humanistic argument as one of its components.

Several philosophers of science have noted that scientific analyses are contained within world views about the sort of scientific questions that should be investigated. This is particularly the case in the social sciences and most certainly when we come to the study of intelligence. The observation that people differ in their mental capabilities is hardly new. The oldest reference to intelligence that I have been able to find is in the *Odyessy*, but I would not be at all surprised if someone calls my attention to an older one. To go further, we have to say what is going to count as an ability and to what sort of things ability is to be related. Choices at this level of generality are what we mean by world views and are essentially humanistic enterprises. For instance, Howard Gardner's (1983) influential book *Frames of Mind* contains the proposal that talents for bodily expression, such as ballet dancing, should be considered part of intelligence. Virtually all modern psychometricians disagree. This particular debate is a humanistic argument about the concept of intelligence rather than a scientific argument about the systematic organization of objectively recorded data.

Some world views generate useful scientific analyses, while others do not. Sometimes the outcome of the scientific enterprise will be central to the continuation of the world view, while in other cases it will not. This is because different world views assign different priorities to scientific evidence. To take an example highly relevant to this book, Darwin's theory of evolution, a scientific concept, evolved from the rationalist world view that dominated European thought in the late 18th century. Rationalism would not have survived without successful science because rationalism, itself a humanistic enterprise, made science central to its argument. During the 20th century, the theory of evolution has come under attack by fundamentalist Christians, whose religious world view leads them to believe in supernatural creation of the species. Creationist beliefs have utterly failed to generate adequate scientific accounts of evolutionary phenomena, but this has done little harm to the world views of religious fundamentalists because they have always made personal faith rather than objective verification their criterion for truth.

I doubt that there are many religious fundamentalists among the authors of the chapters in this book! On the other hand, there are different varieties of rationalism and, for that matter, different varieties of evolutionary thought. In modern American and Western European science, the evolutionary world view has led to an emphasis on physical aspects of evolution, including human evolution, and eventually behavior genetics. There has been little formal application of evolutionary ideas to social organization. By contrast, Karl Marx used evolutionary concepts to explain social organization. This led to world views that stressed ways in which an individual human's capacities could be fit into an evolving social organization. Grigorenko and Kornilova (chapter 14, this volume) describe how the Marxist view of evolution, in its Soviet interpretation, led to some excellent psychological research on the development of human thought, while almost totally disregarding human behavior genetics. We can see this echoed in the chapters of this volume. Consider how Jensen (chapter 2, this volume) speculates about the way in which an individual's genetic heritage shapes his or her environment compared to Vygotsky's emphasis upon the importance of arranging a society that moves children to the edge of their proximal zone of development. Both approaches deal with the interface between society and the individual, but they look at causal mechanisms on different sides of that interface. There is a similar difference of emphasis when we compare books on "How to talk to your child" to recent American studies of language development that emphasize how independent language development is from its social surroundings.

Neither world views that emphasize individual malleability nor views that emphasize genetic determinism are wrong. They both have their place, and they both can lead to good scientific analyses. The choice between these world views determines the sort of science one does. Scientific arguments within one world view cannot be opposed by arguments for an alternative world view. On the other hand, when both world views lead to scientific analyses dealing with the same phenomenon, we can make a meaningful contrast between the two analyses and, by implication, between the two world views, using scientific rules for the analysis of evidence.

My argument is going to be that we are not at this stage because the behavior geneticists have advanced from the world-view stage to the stage of scientific analysis, while the advocates of cultural dominance are still making humanistic arguments for their world view. This conclusion leaps out when you apply two criteria that are central to science – stating a theory clearly and assembling data that support it. By these criteria, one has to be impressed by the behavior geneticists' view of intelligence as an attribute of human variation that is largely under genetic control. Most of the cultural arguments presented in this book, and elsewhere, do not begin to challenge this. Some of the arguments presented by the culturists, if you will, have focused on changes in the average level of intellectual behavior. Everyone acknowledges that mean levels of performance and variations from the mean can be controlled by different variables. Other culturist arguments focus on alleged deficiencies in the behavior geneticists' acceptance of intelligence test scores as an appropriate operational definition of what we mean by intelligence. Sometimes this objection is developed within the framework of normal science, where the trait is redefined and alternative measures are offered. In other cases, though, the discussion of "the greater meaning of intelligence" moves outside of science to rely on humanistic and almost existentialist arguments. This is certainly a legitimate method of inquiry, but, as I have argued elsewhere (Hunt, 1991), we want to be clear about what the ground rules for the inquiry are.

Theoretical clarity

Science progresses by developing and evaluating theories. Therefore, the theories must be stated clearly and unambiguously. In fact, clarity is a hallmark of the distinction between scientific theory and folk knowledge.

Behavior genetics is on firm ground here. The behavior-genetics approach begins with the very well established idea that intelligence is, to some unknown degree, controlled by genetic inheritance. It follows from this that the relative amount of genetically determined intelligence common to two people will be determined by their shared genetic material. Because quantitative genetic theory focuses on shared variations in traits, the behavior geneticist attends to interindividual variations in intelligence rather than the absolute amount of its expression. It is not too far off the mark to say that the behavior geneticist is interested in correlations between the grades of relatives in mathematics classes but is not too concerned about the class average. The well-developed mechanics of mathematical models of inheritance are then applied to partition the variance in measurement into genetic, environmental, and gene–environment correlations. All the pure genetic approaches in this volume (Bouchard; Cherney, Fulker, & Hewitt; Loehlin, Horn, & Willerman; Pipp, Robinson, Bridges, & Bartholomew; Plomin; Reznick; and Scarr) take this approach. Kinship studies, adoption studies, and especially contrasts between monozygotic (MZ) and dizygotic (DZ) twins are exceptionally informative. In practice, of course, observations have to be taken in quasi-experimental settings, because scientists cannot control the key independent variables. (Indeed, our humanist side chills at the thought that they might!) But this is a detail, albeit an important one. The conceptual basis of the theory is straightforward, and the relation between the theory and the experimental observations is clear.

Behavior geneticists are much less clear about what *the environment* is. They can afford this unclarity because, for them, the environment is that vast whatever that is hopefully distributed randomly with respect to genetic variation. Since the behavior geneticists are not fools, they know that their hopes are not realized, and so they provide, in their models, for a way of measuring the size of the effects of environmental and gene–environmental interaction, relative to pure genetic effects (note again the emphasis on variation). They can do this without specifying just what the environment is.

It seems reasonable to require those who emphasize culture to be as clear about their central concepts of culture and environment as the geneticists are about heredity and kinship. This requirement is not met. Gordon and Lemons (chapter 11, this volume) write

Culture is the omnibus construct that refers to the circumstances of development and performance. Culture has many definitions and connotations that are often confused and used interchangeably. We use *culture* in several senses in this chapter. . . .

Gordon and Lemons go on to say that they use the term *culture* in three ways: as the sum of the ways in which a person is raised (including physical culture), as the set of performances and beliefs of a particular set of people, and as the cultural identity adopted by the individual. These are very different things, would be measured in different ways, might respond to different independent variables (unless the definition itself includes all conceivable independent variables!), and may influence different sorts of behaviors in different ways.

Gordon and Lemons do not represent an isolated case. Consider the definition by Miller (chapter 9, this volume).

> The term *cultural psychology* is applied in this chapter as an umbrella concept to characterize a loosely defined emerging tradition, characterized by a variety of shared assumptions and goals.... Not all theorists whose work is considered as within this tradition ... necessarily share all of the assumptions identified with this perspective.

On occasion, vagueness is hidden by references to mathematical concepts such as the notion of a system:

> Twentieth-century science has taught us that living processes as multileveled, self-organizing, integrative systems.... By *integrative systems*, we mean self-organizing systems in which the parts do not and cannot exist prior to or separately from one another or from the whole in which they participate [Bidell & Fischer, chapter 7, this volume].

The authors then inform us that the human body contains interactive systems, and so does society. I am not sure we can credit 20th-century science for this observation, because John Donne (1572–1631) said it better; "No man is an island...."

I wish that I could be guilty of being facetious, but I am not. Saying that people interact with each other and that we construct our own environment is all very true, but truisms do not constitute a scientific theory. John Donne intended to speak in generalities. After all, he was a poet. Many of the "culture counts" authors in this volume, and elsewhere, are speaking at Donne's level, although not with his skill.

The situation is not improved by using the words of science without restricting those words to the semantics appropriate for scientific discussion. Bidell and Fischer argue vehemently that science deals with systems. Indeed it does, but the term *system* has a much more specific meaning in science than the vague definitions they give. A system is a set of interconnected variables, such that, in the ideal situation, the value of any one variable is uniquely determined once the values of the other variables are known. In some cases, there is an added notion that system states vary with time, so that prediction is possible. For instance, Newton's physics can be used to predict the position of the planets at some future time, given knowledge of their current positions.

If prediction can be perfect the system under study is called a *closed* system – that is, one in which no variables outside the system have any influence on variables within it. In practice, no system is completely closed (the planets are not), but in a manageable science, we come close. The philosophic point to be made, however, does not have to do with prediction. It has to do with clarity. The geneticist's concept of the genome meets a scientific criterion for clarity because it can be used to describe experiments in an ideal system. To take the most striking example, and one that is built upon by several papers in this book, monozygotic (MZ) twins share 100% of their genes, while dizygotic (DZ) twins share 50%. If environment has no effect (the ideal case), then the relative extent to which genetically controlled behaviors should covary over MZ or DZ twins is known exactly. Any actual study, then, can begin by contrasting the observed situation to the ideal case. Being able to proceed this way is an essential aspect of science.

The culturists do not even define the variables in their ideal system. All we know is that humans are systems, cultures are systems, societies are systems, and they all interact. In a sense, this is true. Real engineering and scientific system theory have a way of handling this, with the concept of an *open* system. In an open system, some of the variables are influenced by extrasystem, unmeasured variables. Because we cannot trace the effects of unmeasured variables, we have no way of entering them into causal arguments. However, the amount of such influence can be measured, relative to the extent to which the intrasystem variables are dependent on each other. That is what statistics is used for, and the statistical analyses associated with behavior-genetic models are very good examples of how to deal with open systems that are derived from abstract closed systems. The culturist positions defined in this volume do not approach this clarity, because they do not offer a model for measurement and functioning of their ideal (closed) system.

Mixing sports metaphors shamelessly, to win in baseball you have to score. But you can't score unless you can get to first base. Having a clearly stated theory is the scientific analog of getting to first base. Some of the hitters on the cultural team (notably Bidell & Fischer; Ceci, Rosenblum, de Bruyn, & Lee; Gardner, Hatch, & Torff; Gordon, Lemons, & Miller; all this volume) go down swinging. They state world views that we ought to keep in mind when we look at the behavior-genetic approach to intelligence. They do not present an alternative theory sufficiently precise to grapple with using the methods of science.

There is ample evidence for psychological theories that are actually world views masquerading as theory. The writings of Piaget and Freud could be characterized this way. Since it would be hard to argue that Piaget and Freud have not had an influence, there must be some sense in which the culturists' arguments have legitimacy. And there is: Piaget, Freud, and the current culturists all depend on humanistic argument. A humanistic argument appeals to us because it makes us aware of broad classes of influences on our life, without spelling out the specifics of application. Indeed, in many cases, the humanist's audience is urged to find its own interpretation of the principles being enunciated. It is sensible to argue about the right interpretation of the writings of Freud and Piaget, and the culturists in this book, while it would be ludicrous to have the same argument about the writings of Newton, Einstein, Feynmann, or the behavior geneticists.

Much of the clash between culturist and behavior-geneticist arguments presented in this book, and elsewhere, is not a contest between scientific models; it is a clash between scientific models and world views. That is the reason that some of the chapters seem to be talking right by each other.

Theory evaluation by data analysis

Scientific theories are supposed to describe data. This is of necessity a two-step process. Theories state relationships between theoretical variables, but all that we can observe are relationships between operational definitions of theoretical variables. Jensen (chapter 2, this volume) acknowledges this when he refers to the intelligence test as an indicator of a trait rather than the trait itself. In the classic sciences, such as physics, the distinction seems to be a quibble over words. Physicists record scale readings, not mass, but the two can be considered synonymous because there is a clearly understood relation between the concept and the way the measurement is taken.

The relationship between intelligence test scores and intelligence as a theoretical variable is quite a bit murkier than the relation between scale readings and mass. To an astonishing degree, though, the behavior-genetic chapters in this volume (and elsewhere) simply ignore this point. They implicitly accept Boring's (1923) famous argument that intelligence is what the intelligence test measures. The culturists can raise valid questions about this approach. Most culturist attacks on IQ tests do not question the heritability coefficients associated with the tests; they question the adequacy of general intelligence (g or IQ) as a conceptual description of human cognition.

There are two aspects to this attack. One is the claim that even within the psychometric world, the theory of general intelligence is inadequate. There is a good deal of support for this position. The variations in the behaviors tapped by intelligence tests cannot be summarized by a single dimension. At the least, one needs a hierarchical model, in which we distinguish between the application of specific knowledge (including linguistic knowledge), abstract reasoning ability, and spatial-visual reasoning (Carroll, 1993; Horn, 1985). However, the first two of these factors are substantially correlated. With some minor qualifications, especially about male–female differences in spatial-visual reasoning, the genetic models for general intelligence apply reasonably well to models of the factors that make up *g* (Cardon & Fulker, 1993; Pedersen, Plomin, Nesselroade, & McClearn, 1992).

In fact, very few culturist objections to the behavior-genetics position complain about overly simplified summarizations of psychometric research. The attack is directed at the validity of the psychometric data themselves. Psychometric tests are said not to capture the true meaning of intelligence. As I have already pointed out, this objection is at best a rear guard action unless it is accompanied by a workable definition of what the real meaning of human intelligence is. Absent such a definition (which we certainly are), we fall back into an uninformative clash between world views and models. However, there is a way to make this objection more scientific, not by trying to specify the world view of the culturist but by looking more closely at the world view that generated psychometric testing itself.

The core belief of psychometric studies of intelligence is that there exists some mental capacity (or capacities) of the individual that makes a difference in his or her dealings with the physical and social world. This was explicit in the views of Galton, Binet, and Spearman. The general point was made far earlier. Consider the different ways in which Achilles and Odysseus dealt with adversity. In our more prosaic world, where would psychometric tests be if they did not correlate with socially important behavior? A legitimate scientific objection to the psychometric viewpoint is that these tests predict only very limited aspects of human behavior.

The facts are otherwise. The usefulness of psychometric tests to predict school behavior is well known. Studies of psychometric tests used in industrial and military settings have shown, over and over again, reliable and socially important prediction of workplace performance (Hartigan & Wigdor, 1989; Hunt, 1995a; Hunter, 1986). It has been estimated that the correlation between the traits underlying the tests and the traits underlying overt measures of workplace performance is somewhere between .4 and .6. This has to be qualified somewhat by distinguishing between whether a person is learning to do a job or performing a job that has already been learned (Hunt, 1995b). That is a quibble, however. Psychometric tests are often the best predictors that we have of success in both school and the workaday world. What is more, they are nonnegligible predictors of failures in everyday life, such as slipping to poverty status or requiring aid for dependent children (Herrnstein & Murray, 1994).

The argument that the behavior that underlies psychometric tests has *nothing* to do with human competence cannot be maintained. Bouchard is quite right to complain that the studies that have maintained this are technically deficient. That remark certainly includes all the studies with which I have familiarity and that are cited by the culturists in this volume.

A weaker argument against the genetic-psychometric position is that there are important aspects of cognitive behavior that are not measured by psychometric tests and, by implication, not tied to genetic variability. The available data are quite clear on this point. There is lots of room for explanations beyond psychometric prediction of virtually every aspect of human affairs. For instance, no one would argue against Engeström, Engeström, and Kärkkäinen's (chapter 15, this volume) proposition that people have to fit into social systems, because most human intellectual work is done by teams of people. Social systems often operate by both explicit and implicit rules. Sternberg (chapter 12, this volume) usefully points out that people do not automatically know what such implicit rules are, and he alludes to data showing that specific instruction on the unspoken rules can help children to make their adjustment to the modern school. These controls over the effectiveness of human mental effort are certainly outside the range of conventional psychometric evaluation.

What can the behavior geneticists respond to this? Their strongest response is, of course, that the data show that they have a solid hold on the psychometric data, where heritability estimates vary from .4 to .8 (the precise value does not really matter), and this alone is substantial. They can also point out, with justification, that saying that other things matter is hardly a challenge to the behavior-genetic position until the challenger says precisely what these other things are.

And when this is done, one of two things often happens. Sometimes the ability proposed as being an unevaluated aspect of intelligence turns out to have some degree of genetic determination. Pipp-Siegel et al.'s (chapter 18, this volume) example of genetic determinancies of individual differences in awareness of the views of others is an excellent case in point. In other cases, the variable that, on logical grounds, seemed so very important may turn out, on closer and more objective examination, to have very little influence on behavior. Note the importance of *close* and *objective* examination. Humanistic arguments that the atmosphere in a child's home must influence intellectual development are convincing, but Scarr's (chapter 1, this volume) demonstration that children's intellectual development can be predicted as well by genetic variables alone as by genetic and home environmental measurements combined raises a scientific flag of caution.

Of course, unless every bit of the variance in behavior has been accounted for, there always might be something else important, and that something else might have nothing to do with genetics. It might be that the home-environmental measures considered by Scarr were inadequate. Being yet more general, Ceci et al. (chapter 10, this volume) argue that heritability estimates show only the effects of environmentally actualized genetic potential. It is certainly true that if people were raised in unspecified different environments, there might be different unspecified gene–environment interactions. There is no way that such arguments can be dealt with within the framework of science. As Bouchard (chapter 5, this volume) points out, you can always account for anything by appeal to unmeasured variables! Presentations such as those of Ceci et al. do serve as useful reminders that there is nothing in psychometric or behavior-genetic theories that rules out the discovery of some yet unknown genetic potential, in the sense that our theories of the physical world rule out the discovery of perpetual motion machines and faster-than-light travel. Such an argument only becomes interesting when the culturist points toward specific areas that ought to be investigated.

There is one point at which the cultural argument does begin to indicate where the psychometric–behavior-genetic investigation is limited. Several authors (notably Miller, chapter 9, this volume) have objected that the psychometric view of intelligence is limited because it has only been validated in North American and Western European cultures. This is not quite true; there has been research on intelligence tests in the industrialized nations of Asia. Nevertheless, modern psychology's view of cognition, both in psychometrics and in cognitive psychology, is very much a product of industrial and high-technology civilization.

How serious is this situation? One way of interpreting the culturist position is: That psychometric concept of intelligence is not valid because it has not been shown to apply to every conceivable present and past culture. In fact, the culturists sometimes claim counterexamples. These tend to be of two types. One is a demonstration that some cognitive operation that is required for performance on tests (e.g., abstract reasoning) is not valued in a particular culture or in a particular setting within a culture. The other is a demonstration that some group of people who exhibit low scores on a minimally adapted intelligence test, such as a poor translation of an English language test, can demonstrate substantial reasoning if tested in a culturally appropriate way.

Such demonstrations provide useful cautions for those who want to force a naive adoption of the testing of industrial civilization into a nonindustrial world, but they are certainly not crucial evidence in discussions of the nature of intelligence. The objection confuses the operational definition of intelligence with the underlying concepts. I readily concede that one cannot simply apply Western testing concepts to all other cultures.[1] For that matter, the laboratory procedures of cognitive psychologists do not export very well either.

What may export, though, are the core beliefs that psychologists have developed about cognition. Concepts such as fluid and crystallized intelligence, verbal competence, spatial-visual reasoning, the distinction between long-term and working memory, and the ability to control attention appear, from their definitions, to be applicable to the species even though they have been developed by an industrial-technological civilization. Certainly the same thing can be said of the concepts of genetics. We are quite sure that DNA works the same way on Park Avenue and the Andaman Islands. The link between genetic composition and the conceptual underpinnings of cognitive performance (e.g., working memory) has yet to be established, but, as Plomin (chapter 3, this volume) points out, progress is being made in linking specific genes to specific mental capacities.

Proponents of a cultural position may object that I am distorting their position, for they certainly never meant that different cultures were biologically different in any important way. The problem I have at this point is that it is difficult for me to figure out just what the culture advocates do mean. (I again invite the reader to extract more than a world view meaning from some of the quotations given at the outset of this chapter.) To carry the discussion forward, I now offer an argument that I think could be interpreted as a cultural view and show that it leads to a program of research that would be entirely compatible with both the cultural and the psychometric view of intelligence.

My argument begins with two fairly obvious points. The first is that cultures vary in the value that they place on certain types of mental capacities. Highly specific reading deficits are a substantial problem for the individual involved in our culture but are of no consequence in a nonliterate society.[2] My second point is that mental capacity depends partly on acquired knowledge. The acquisition of this knowledge is determined by the way that a person's culture stores and transmits knowledge and the person's ability to extract information presented in culturally approved manners.

Now let us consider what these propositions mean for the investigation of intelligence. We could first define a set of "culturally appropriate cognitive challenges for some culture" (note the careful avoidance of the word *test*, solely to be culturally correct) and observe a group of individuals meeting them. This would result in a matrix of person-by-situation scores. We could then (horrors!) factor-analyze the matrix. I suspect that this program, with suitable embellishments, would produce something very much like our modern psychometric concepts of verbal ability, visual-spatial ability, and even general intelligence! Furthermore, I suspect that an analogous program in cognitive psychology would rediscover working memory, long-term memory, procedural and declarative knowledge, and the like.

There are certain points at which the resulting program would probably not mirror what we have found in our present, admittedly industrial-technological psychological research. The factor loadings that connected the culturally appropriate challenges to the underlying dimensions of variability would certainly not be the same. Consider, for instance, the psycho-metric distinction between verbal and spatial-visual reasoning. We make this distinction in our culture, but in most situations, we emphasize verbal intelligence. Manuel de Juan-Espinosa, a Spanish psychologist, studied the Fang, a forest-dwelling tribe in Equatorial Guinea. Juan-Espinosa (personal communication) reports that the Fang regard questions about the nature of intelligence as sensible and that they distinguish between verbal and visual-spatial reasoning. They weigh these dimensions of the mind differently than we do, for getting lost in the jungle is a far more serious problem for the Fang than for contemporary Americans or Europeans, but the principle is the same. The dimensions of intellectual variation that humans recognize may be invariant across cultures, even though the importance assigned to each dimension may vary.

Geary (1995) has suggested a general principle to deal with this sort of situation. Consistent with psychometric views (cf. Carroll, 1993), Geary points out that human cognitive abilities can be grouped into a few broad categories, such as verbal, spatial-visual, and general reasoning abilities, with specialized abilities, such as reading and dealing with static or moving objects, grouped under each category. Geary proposes that all cultures deal with the primary abilities in basically the same way but that there may be a good deal of cultural variation in the treatment of specialized abilities. Ideas like this could lead to a coherent scientific view of individual differences in human mental capacity that had both general and specific cultural components. I am sure that other proposals could be developed within a scientific framework. They are what we need rather than further railing against psychometrics because of differences in world views.

In closing this section, I would like to make one point that often seems to be overlooked. Suppose that investigations of the sort that I have described were carried out, and it were to be discovered that our current concepts of intelligence are completely tied to the highly technological, developed societies. Such a finding would be interesting, but it certainly would not be cause for abandoning our interest in intelligence, as used in our own society. While the high-technology societies do not encompass all the globe, depending on how you count them, they collectively constitute between one and two billion people. Regularities in a society this big should not be ignored. Furthermore, unless the technological societies blow themselves up, it is likely that they, not what Miller (chapter 9, this volume) refers to as traditional societies, represent the way of the future. India and China seem far more interested in developing a technological society than North America and Europe are interested in returning to a village-based economy. This does not make the high-technology society any more or less human than the many different traditional societies. It does make the study of intelligence, *as defined by the high-technology society*, a topic of considerable scientific and practical interest.

The relative influences of cultural and biological variables upon intelligence

Many of the foregoing comments have been directed at what could be called the *syntax of science* – rules that have to be followed in order to qualify a discussion as scientific. The semantics of scientific theories connect the theories to the world. How well do the cultural and biological explanations handle the data?

It has been amply demonstrated that behavior-genetic theories do very well. We may quibble about how much of the variability in intelligence, as defined by psychometric theory, is under genetic control, but it is clear that a lot of it is. What are the cultural effects? Here it is important to distinguish between two aspects of culture: the physical and the social environment.

It has often been claimed (e.g., by Jensen, chapter 2, this volume) that the data show that nonshared, between-family environments do not have much influence on intelligence. This contrasts with certain other positions that lead us to expect such influences. Both animal and human models have shown that variables in the psychical environment can indeed have dramatic influences on neural development (Wahlsten & Gottlieb, chapter 6, this volume). However, it is not clear that these findings have any relevance at all to human intellectual development, within the range of environments encountered in our own society. In spite of difficulties of interpretation, epidemiological studies are likely to be of much more help than laboratory studies, because we are concerned not with what might happen but with what actually does happen in our society. Unfortunately, the evidence is mixed.

Nutrition is probably the most heavily studied environmental variable, but the evidence for an effect on intelligence is at best weak. It is well known that extreme and prolonged impoverishment, especially in infancy, is debilitating, although humans are surprisingly resilient when faced with brief deprivations. Eysenck and Schoenthaler (chapter 13, this volume) present some data suggesting that there is enough variation in nutrition in modern Western society to make a difference in children's cognitive behavior. Given the complexity of interpretations of epidemiological data, it would perhaps be best to maintain an open mind until further reviews of this work are presented in the more specialized literature.

On the other hand, there are other variables in the physical environment that clearly do make a difference to cognitive functioning, on a population basis. Alcohol use is a good example, although for some reason this drug, whose use is very sensitive to cultural conditions, is seldom cited as an environmental effect. That is surprising, because it is quite well known that heavy alcohol consumption can lead to debilitating forms of memory loss (Korsakoff's syndrome), to the point at which the individual must be institutionalized. Fetal alcohol syndrome, which is a concomitant of excessive drinking by pregnant women, is a well documented, diagnosable form of mental retardation and is the result of culture-sensitive drinking practices. Less dramatic reductions in intelligence have been associated with lower levels of alcohol consumption and with exposure to other agents (e.g., atmospheric lead). Although causation is extremely hard to assess in the latter cases due to confoundings between social variables and exposure to the damaging agent, the effects of extreme exposure seem well documented.

The influence of the social environment on intelligence is much less clear. This is a point at which the vagueness of the culturists' definition places them at a disadvantage, because it is not clear what effects are supposed to occur! Gardner, Hatch, and Torff (chapter 8, this volume) argue for a *symbol system approach* to intelligence but provide no data in support of their position, nor is it clear what data would be appropriate! Miller's paper (chapter 9, this volume) makes the same point. The position seems to be that every person constructs his or her own view of reality, and that this view is conditioned by the culture. In one sense, this is very true; how could it be otherwise? Furthermore, there are many examples we can give to prove the point. But do these examples, or illustrations of these principles, force any reconsideration at all of our ideas about the basic dimensions of human mental capabilities? I do not think so. To illustrate my point, I will resort to an unabashedly humanistic method of argumentation – analysis of a historical incident. (Once again, there is nothing wrong with humanistic arguments. The error is in confusing them with science!)

Schoolchildren learn that when the Spanish conquistador Hernan Cortes landed in Mexico, the Aztecs thought that he was a god. In fact, there was considerable controversy in the Aztec leadership. One faction, led by the Aztec "emperor" Montezuma, believed that Cortes might be the return of the god Quetzalcoatl. This argument was not just based on superstition. Certain aspects of Cortes' behavior, such as the date and location of his arrival and his destruction of temples associated with Aztec gods other than Quetzalcoatl, were consistent with the Quetzalcoatl legend. Compounding the situation, when Cortes was asked if he was Quetzalcoatl he intentionally gave an ambiguous answer. However, another faction among the Aztecs had a different theory. Montezuma's brother, Cuitlahuac, argued that the Spanish were a dangerous band of military adventurers arriving from some unknown but potentially hostile nation. Montezuma won the argument, Cortes (who was militarily vulnerable when he first landed) won the war, and Mexicans speak Spanish today.

Gardner et al. and Miller would presumably argue that Montezuma was intelligent because he developed a symbolic interpretation of Cortes that was consistent with the Aztec view of the world. The Aztecs themselves disagreed; they eventually killed Montezuma and put Cuitlahuac on the throne (Thomas, 1993). Cuitlahuac and his successor, Cuahtemoc, displayed great ingenuity in redesigning traditional Aztec methods of war to cope with Spanish tactics and weaponry. Cortes himself was a genius at adopting Spanish capabilities to unexpected situations. This is beginning to sound very much like the theory of crystallized and fluid intelligence, with Montezuema using crystallized reasoning and Cuitlahuac using fluid reasoning, almost 400 years before psychometrics was born.

Stating that different cultures develop different symbolic interpretations of the world does not force us to rethink our concepts of intelligence one bit. It is, of course, true that humans are social animals and that a great deal of our cognitive performance is group performance. Therefore, studying the way in which intelligence is drawn out is likely to be a highly useful enterprise. But the enterprise will have to be disciplined. Anecdotes about group performance (e.g., Engeström et al., chapter 15, this volume) are only a small step beyond truisms. What is needed are careful analyses of how social demands interact with personal characteristics. For instance, in the educational world, there is much evidence for an aptitude × treatment interaction; people of lower intelligence generally profit most from structured learning environments, while people of higher intelligence do better when they are allowed to exercise some initiative in a learning situation (Snow & Yalow, 1982). There is also a great deal of lore but, insofar as I can find, precious little data, to the effect that it helps to fit a training situation to a person's learning style. There is room for a research program on cultural effects on intelligence, but it will have to be a disciplined program that distinguishes between the ways in which our mental abilities arise and the ways in which they are shaped and utilized.

The interactionist position: Where does the future lie?

Psychometric measures of intelligence evaluate something that is important in our society (Herrnstein & Murray, 1994; Hunt, 1995a; Hunter, 1986). As the chapters in this volume show, the behavior-genetics model accounts for a good bit of the variability in psychometric intelligence. When the question is posed as Nature vs. Nurture, the debate looks like a stomping match between Godzilla and Bambi. For those who did not see the movie, Godzilla wins. Yet we all love Bambi. Why do people keep looking for cultural explanations of intelligence?

The sociologist Christopher Jencks (1992, p. 95) says that people who distrust behavior-genetic models do so for two reasons – because the models stop at statistical association without explaining the causal link between genome and behavior, and because some people who hold a culture-first world view believe that if the genetic models are correct, social controls over behavior are not likely to work. The latter conclusion is particularly threatening to people who have a deep philosophic commitment to social engineering to produce a more perfect society. In its extreme, such a belief leads one to deny that there is any immutable aspect to individual intelligence. This can even lead to suppression of research on the topic, as apparently happened in the Soviet Union (Grigorenko & Kornilova, this volume).[3]

Jencks (1992) goes on to point out that the conclusion about social control is fundamentally erroneous, because genes do produce their effects through interactions with the environment. Virtually all the authors in this volume agree. However, they differ greatly in their approach to how they deal with the issue.

The approach of statistical behavior genetics has been to measure the effect and let it go at that. Jensen (chapter 2, this volume) is a good example. He observes that the major gene × environment interaction in intelligence is associated with within-family variation rather than between-family variations in environments. Period. Full stop. Somewhat the same thing is said in a different way by Scarr (chapter 1, this volume) when she observes that parental rearing styles seem to have little influence over children's intelligence. The absence of an interaction is important in any treatment of interactions in general.

What we want to do, though, is to go beyond statistical associations to trace out the causal links between genes and behavior. As a matter of logic, these causal links will have to proceed outward, from the presence of a gene to an understanding of its physical effects, and then to an understanding of how these physical effects interact with environmental pressures. Some of our present practices for preventing mental retardation are justified in exactly this way. Our society has virtually eliminated mental retardation associated with improper metabolism of phenylalanine (PKU) because we have developed inexpensive, nonintrusive screening techniques to identify affected individuals, and because we have developed effective interventional therapies once a person at risk is identified. The advent of powerful techniques for genetic screening holds out at least the possibility of studying much subtler variations in the development of cognitive power, perhaps extending into the normal range (Plomin, chapter 3, this volume). As yet, this is a promise rather than an accomplished fact. Such studies can be buttressed by developmental studies that show where in the life cycle various genetic controls begin to exert their influence. The work of Reznick and of Cherny, Fulker, and Hewitt (chapters 17 and 16, respectively, this volume) on the unfolding of genetic influences over the life span is a good example. But we want to be careful about what such studies tell us. They tell us when genetics begins to exert its influence – but not how.

This issue becomes particularly important when we consider how social influences might interact with genetic ones. During the 1970s, there was considerable social debate over the finding that men with an extra Y chromosome (XYY) tended to be large and of moderately low intelligence and to have a history of criminal activity including violence. The relationships were far too low to justify intervention on the individual level, but on a population basis, there might be legitimate cause for concern. In retrospect, this debate shows how little we can learn if all we know is the statistical pattern of associations between genetic composition and behavior. We need to know how these statistical interactions arise, at both the physical and social level. Jencks' point that we are not going to be satisfied with genetic explanations until we can trace out the causal chain is correct. Unfortunately, this is likely to be a very complicated process. Often the answers will be far too tentative to point clearly toward the appropriate social interventions, either as control devices or as a way of drawing out genetic potential.

Gene–environment effects are indeed real. Culturists write as if they can be used to equalize genetic variance. In general, though, social interventions have not done this, in part because we do not know the relevant causal paths, and in part because interventions aimed at the amelioration of possible but not certain defects, such as those associated with the XYY genotype, raise serious issues about individual liberties. American society insists on a very high standard of proof before we intervene to correct what an individual might do, and most of us agree that this is a good thing. It does limit our ability to control gene–environment interactions that may produce deficient cognition. And as for extracting genius from the genes? Frankly, we do not know how to do this.

This brings us to Jencks' second point – that culturists (he says "liberals," but he was speaking in a political context) distrust genetic explanations because they perceive a genetic trait as immutable. Of course, this is not true, as behavior geneticists have repeatedly said. Culturists are fond of citing such examples, too. It should be pointed out, however, that these examples generally illustrate effects on mean performance rather than selective effects at the high or low end of intelligence. The fact that Japanese-Americans are taller than their Asian ancestors is a frequently used example. (Modern Anglo-Saxons are taller than their forefathers, too, as a visit to a museum with suits of armor will show!) I would like to close with a better example, dealing with intelligence itself.

Flynn (1987) has called attention to an interesting phenomenon. A variety of comparisons suggest that over the last few decades, intelligence test scores have been rising in the developed world. Flynn's particular conclusion can be questioned, because it was based on comparisons of samples taken in different countries, over fairly long periods of time, using noncomparable sampling methods. Nevertheless, the general point is certainly true. In 1995, the average American had greater cognitive capabilities than he or she had in 1895, for a very simple reason. The level of schooling was much higher than it had been a century earlier. This resulted in more than just acquisition of a cognitive skill, since literacy has been shown to be associated with a general improvement in (or willingness to engage in) abstract reasoning. This does not mean that the genetic potential for American intelligence changed (disregarding immigration for the moment!). It does mean that there was greater realization of our genetic potential. We now live in a society that is more conducive to cognitive development. We certainly are not perfect, but we are better than we used to be.

In conclusion

Genetics counts. If we insist on treating genetic and cultural explanations of intelligence as a stomping match, then the behavioral geneticists are the stompers and the proponents of cultural effects are the stompees. Twin, kinship, and adoption studies have more than proven the point that genetic variance in intelligence is substantial. It is not clear why we need to tie down the numbers to the second or third decimal point.

What we do need to do is to go beyond statistics to investigate the mechanisms of genetic action, including the interaction between genetic predisposition and the environment. No one inherits a genetic compulsion to watch television or innate knowledge of the answers to intelligence-test questions. We inherit blueprints for constructing proteins; all else follows from interactions! What we need to do is to understand the causal pathways, at both the molecular genetics and the psychological-social ends of the continuum between genes and behavior. We will not be able to do this until those who want to argue that culture counts are far more specific than they have been to date about what culture is and how it counts.

Notes

1. Miller (chapter 9, this volume) and others who emphasize cultural explanations repeatedly refer to *traditional* cultures. I think this is an unfortunate term for several reasons. First, any culture, including our own, has traditions. Second, I know of no proof that there exists any core of beliefs or practices that are shared by all cultures outside the industrial-

techologically oriented ones. This set of cultures is defined by an absence of industrial-technological practices rather than by shared practices of their own.
2. The medieval emperor Charlemagne is said to have been unable to learn to read. He could still rule, because reading was a highly specific trait required in only a few situations, even by rulers. A modern U.S. President could not be dyslexic because there is simply too much written material to be examined.
3. It has been suggested that some opponents of genetic explanations of intelligence do not believe, themselves, that the validity of a genetic explanation justifies abandonment of eforts to improve cognition by educational and social means. Instead, they oppose public discussion of genetic models on the grounds that uninformed or misguided people will leap to the conclusion that genetic effects are immutable and, hence, withdraw support from political efforts in which the culturists believe. Elsewhere (Hunt, 1995b) I have referred to this attitude as *Mokita*, which is a New Guinea word that means "Truth we all know but agree not to talk about." I do not believe that it is either scientifically justifiable or politically feasible to try to solve problems by pretending that the scientific data are anything other than what they are.

References

Boring, E. G. (1923, June 6). Intelligence as the tests test it. *New Republic*, 35–7.

Carroll, J. B. (1993). *Human cognitive abilities: A survey of factor analytic studies.* Cambridge: Cambridge University Press.

Cardon, L. R., & Fulker, D. W. (1993). Specific cognitive abilities. In R. Plomin & G. E. McClearn (Eds.), *Nature nurture & psychology.* Washington, DC: American Psychological Association.

Flynn, J. R. (1987). Massive IQ gains in 14 nations. What IQ tests really measure. *Psychological Bulletin, 101*(2), 171–91.

Gardner, H. (1983). *Frames of mind.* New York: Basic Books.

Geary, D. C. (1995). Reflections of evolution and culture in children's cognition. *American Psychologist, 50*(1), 24–37.

Hartigan, J. A., & Wigdor, A. K. (Eds.). (1989). *Fairness in employment testing: Validity, generality, minority issues, and the General Aptitude Test Battery.* Washington, DC: National Academy Press.

Herrnstein, R. J., & Murray, C. (1994). *The Bell Curve. Intelligence and class structure in American life.* New York: The Free Press.

Horn, J. L. (1985). Remodeling old models of intelligence. In B. B. Wolman (Ed.), *Handbook of intelligence. Theories, measurements, and applications* (pp. 267–300). New York: Wiley.

Hunt, E. (1991). Some comments on the study of complexity. In R. J. Sternberg & P. A. Frensch (Eds.), *Complex problem solving* (pp. 383–96). Hillsdale, NJ: Erlbaum.

Hunt, E. (1995a). *Will we be smart enough: Cognitive changes in the coming workforce.* New York: Russell Sage Foundation.

Hunt, E. (1995b). The role of intelligence in modern society. *American Scientist, 83*(4), 356–68.

Hunter, J. E. (1986). Cognitive ability, cognitive aptitudes, job knowledge, and job performance. *Journal of Vocational Behavior, 29*, 340–62.

Jencks, C. (1992). *Rethinking social policy: Race, poverty, and the underclass.* Cambridge, MA: Harvard University Press.

McGue, M., Bouchard, T. J., Iacono, W., & Lykken, D. T. (1993). Behavioral genetics of cognitive abilities: A life-span perspective. In R. Plomin & G. E. McClearn (Eds.), *Nature, nurture, & psychology* (pp. 59–76). Washington, DC: American Psychological Association.

Pedersen, N. L., Plomin, R., Nesselroade, J., & McClearn, J. E. (1992). A quantitative genetic analysis of cognitive abilities during the second half of the life span. *Psychological Science, 3*, 346–53.

Plomin, R., & McClearn, G. E. (Eds.). (1993). *Nature nurture & psychology.* Washington, DC: American Psychological Association.

Snow, R. E., & Yalow, E. (1982). Education and intelligence. In R. J. Sternberg (Ed.), *Handbook of human intelligence.* Cambridge: Cambridge University Press.

Sternberg, R. J. (1990). *Metaphors of mind: Conceptions of the nature of intelligence.* Cambridge: Cambridge University Press.

Thomas, H. (1993). *Conquest: Montezuema, Cortès, and the fall of old Mexico.* New York: Simon & Schuster.

20 Unresolved questions and future directions in behavior-genetic studies of intelligence

Irwin D. Waldman

Introduction

Most people seem to feel that the "nature versus nurture controversy," as it is commonly and unfortunately phrased, is either passé or settled with regard to intelligence. Indeed, Hunt (chapter 19, this volume) was sufficiently moved by the topic to coin a new term, *vujà dé*, which loosely translated means "get this away from me, I do not ever want to see it again." Rather than representing a lone dissenting voice, it is my sense that Hunt speaks for the masses, or at least for sizable contingents of researchers who occupy very different positions on the relevant issues. Given this, is there justification for the current collection of chapters on the hereditary and environmental bases of intelligence?

If one views the causal basis of intelligence as a battle or controversy (viz., nature *versus* nurture), then I agree that there is little justification, as is the case if one views the disentanglement of genetic and environmental influences as an irrelevant or impossible goal. Contrary to the views of many social science researchers, including a number of contributors to this volume, I believe that characterization of the unique and conjoint contributions of genetic and environmental influences to intelligence is a worthwhile and attainable goal. Behavior-genetic designs have been used – successfully, in my opinion – to characterize broad components of variance corresponding to genetic and environmental influences on intelligence. Hence, I view the chapters in this volume on behavior-genetic studies of intelligence as a welcome summary of recent research, made all the more useful by the fact that this is a literature characterized by quick progress and rapid developments. Nonetheless, the estimation of broad components of variance corresponding to genetic and environmental influences on intelligence has its limits, as sharply pointed out by those contributors to this volume who adopt a more cultural perspective on intelligence.

Given the presentation or review of behavior-genetic data on intelligence in many of the chapters and the culturalist critiques of the same in other chapters, I intend to take a somewhat different tack. Rather than a comprehensive review of genetic and environmental influences on intelligence or a thorough methodological critique, I wish to focus on some important unresolved questions and fruitful future directions in behavior-genetic studies of intelligence. Despite recent advances, many central issues regarding the development of intellectual abilities (broadly construed) remain poorly understood. In this chapter, I hope to highlight some of the more promising trends in recent behavior-genetic studies of intelligence, as well as suggest some interesting pathways that have heretofore received only cursory exploration.

These directions worthy of further study fall loosely under a number of topics. The first concerns what I would call the *generalizability and specificity of findings* regarding genetic and environmental influences on intelligence. The second involves genotype–environment correlation and interaction, examples of how genetic and environmental influences – which are often assumed to be additive in nature – can act conjointly or dependently. The third covers the application of behavior-genetic methods to the study of intelligence in an explicitly developmental context. The fourth concerns the use of behavior-genetic designs to investigate issues in the measurement and construct validity of cognitive abilities. The fifth and final topic explores the nature of behavior-genetic models for examining causal influences. Although the primary focus in this chapter is on the utility of behavior-genetic designs for understanding the causes and development of intelligence, I highlight where relevant some points raised in the chapters that focus on culturalist critiques of intelligence. A brief overview of important findings from recent behavior-genetic studies of intelligence sets the stage for discussion of these future directions.

An overview of findings from recent behavior-genetic studies of intelligence

What have we learned from recent behavior-genetic studies of intelligence? First, data from many different behavior-genetic studies can be combined in model-fitting analyses (Chipuer, Rovine, & Plomin, 1990; Loehlin, 1989) to arrive at estimates of genetic and environmental influences on general cognitive ability. These models allow familial correlations from a wide variety of relationships (e.g., identical twins reared apart, parents and their biological and adoptive children, cousins) to be analyzed simultaneously and facilitate the exploration of more complex genetic and environmental influences than do simple comparisons of correlations. Second, genetic influences on general cognitive ability appear to be substantial, with heritability estimates (h^2) ranging from .4 to .8. This means that about 40–80% of the variability in general cognitive ability among individuals is due to genetic differences among them. Third, for reasons that remain obscure, heritability appears to be higher for *direct* estimates – those based on the familial correlation for individuals who *were not* reared together (e.g., between reared-apart twins) – than for *indirect* estimates – those based on differences in familial correlations for individuals who *were* reared together (e.g., between identical and fraternal twins reared together). Fourth, nonadditive genetic influences (i.e., dominance and epistasis) as well as additive genetic influences appear to be important for general cognitive ability when the effects of assortative mating are considered (Chipuer et al., 1990).

Interesting findings from behavior-genetic studies of intelligence are not limited to genetic influences; in fact, some of the most interesting findings are those pertaining to environmental influences. First, shared environmental influences (c^2) on general cognitive ability – aspects of the environment that family members experience *in common* that contribute to their similarity in cognitive ability – account for about 10–40% of individual differences. Second, nonshared environmental influences (e^2) on general cognitive ability – aspects of the environment that family members experience *uniquely* that make them different in cognitive ability – account for about 10–30% of individual differences (it should be noted that part of this variance, perhaps 10–15%, is due to measurement error). Third, shared environmental influences on general cognitive ability appear to differ in predictable ways among different types of relatives (Loehlin, 1989). Specifically, shared environmental influences appear to decline in magnitude for the following types of relative: twins (35%) > siblings (22%) > parents and their children (20%) > cousins (11%) (Chipuer et al., 1990). Fourth, shared environmental influences on intelligence appear to decline from childhood through adolescence to adulthood. Fifth, although there is little evidence that genetic and environmental influences interact for cognitive ability, there is some evidence that these influences may be substantially correlated (Loehlin & DeFries, 1987).

The generalizability and specificity of findings from behavior-genetic studies of intelligence

Given the foregoing summary, it is reasonable to ask two broad questions. The first is: How generalizable are the findings from recent behavior-genetic studies of intelligence? That is, what populations and what aspects of intelligence do these findings adequately describe? The second is: What do we know about specific genetic and environmental influences on intelligence? That is, what can we say beyond general and anonymous variance components about the development and causes of intelligence?

Most of the data in behavior-genetic studies of intelligence have come from predominantly white samples that, while ranging substantially in socioeconomic status (SES), have not represented well the extremes. Hence, although we have considerable data with which to make inferences regarding genetic and environmental influences on intelligence in the white, largely middle-class population, our knowledge of these influences in minority and lower SES populations is, with some exceptions, fairly limited (e.g., Moore, 1986; Scarr & Weinberg, 1977; Scarr, Weinberg, & Waldman, 1993). Future studies should examine genetic and environmental influences on intelligence in African-American and other minority populations, as well as in the lower-SES white population, to address the generalizability of findings from behavior-genetic studies on intelligence in the white, predominantly middle-class population.

Just as we can address the generalizability of findings from behavior-genetic studies of intelligence in terms of diverse samples of individuals, we can address generalizability with regard to diverse aspects of cognitive ability. Most findings from behavior-genetic studies of intelligence are pertinent to genetic and environmental influences on some index of general cognitive ability, or g. As such, we know relatively little regarding genetic and environmental influences on mental abilities, whether broad or narrow, beyond g.

Let me illustrate the benefits of behavior-genetic studies of more specific mental abilities with just one example here, as I explore this theme in greater depth later. Horn (1988) reported results from a behavior-genetic analysis conducted with McArdle (McArdle, Goldsmith, & Horn, 1981) of eight mental ability measures corresponding to Cattell's higher-order factors of fluid and crystallized ability (G_f and G_c, respectively). Cattell (1941) hypothesized that variability in G_f is due primarily to genetic influences, whereas variability in G_c is due primarily to environmental influences. In contrast, results of these analyses suggested that G_f and G_c were equally heritable ($h^2 = .59$ for each factor) and that the genetic influences on each of these factors were largely independent (shared $h^2 = .14$). These findings suggest that genetic influences play as great a role in G_c as in G_f but that the genes underlying each trait are more different than they are similar. Although the sample in this analysis was small, consisting of only 48 MZ and 63 DZ pairs, these results illustrate the potential for interesting findings that lies in the realm of behavior-genetic analyses of mental abilities beyond g.

The second set of questions concerns the specificity of genetic and environmental influences on intelligence. Although behavior-genetic studies of intelligence have provided good estimates of general genetic and environmental variance components, at present we know little regarding the specific genetic loci or environmental characteristics that underlie intelligence.

Nonetheless, we do have a few good leads, a number of which are described in the chapters in this volume.

Plomin (chapter 3, this volume), for example, describes the search for specific DNA markers that are related to high versus low cognitive ability a search in which he and his colleagues have blazed the trail. In contrast to previous molecular genetic approaches that concentrated on relatively rare medical conditions involving often marked cognitive deficits (e.g., PKU, Alzheimer's disease, fragile X syndrome), Plomin and his colleagues have focused on genetic loci that may be important for cognitive ability within the normal range. After describing the advantages and disadvantages of linkage and association methods for detecting quantitative trait loci (QTLs), Plomin summarizes the findings from initial association analyses (Plomin et al., 1994), which suggested that two DNA markers were associated with high versus low cognitive ability in two different samples. Each of these markers accounted for at least 2% of the overall variance in IQ, which underscores the likelihood that QTLs found to be associated with intelligence (and many other human characteristics) will in general be of relatively small magnitude, consistent with their characterization in the multifactorial polygenic model that underlies most behavior-genetic analyses. Despite the preliminary nature of these findings and the need for independent replication, these findings represent an exciting beginning in the search for specific genetic loci associated with cognitive abilities. This search should be aided significantly by analytic techniques based on familial regression methods that allow the inclusion of quantitative measures (Cardon et al., 1994; Fulker et al., 1991), as well as those that combine molecular and quantitative genetic analytic strategies (e.g., Boomsma, 1995).

Given the plethora of environmental measures that have been correlated with IQ, it may seem odd to assert that we know little regarding the specific environmental characteristics that underlie the development of intelligence. Nonetheless, environmental measures often have a substantial genetic component (Plomin & Bergeman, 1991), because they frequently are indirect manifestations of individuals' characteristics (e.g., intelligence, personality), which are themselves genetically influenced. Unfortunately, there are only a few examples of the inclusion of environmental measures in genetically informative studies of intelligence. It is only by including specific environmental characteristics in behavior-genetic designs that one can truly discover the degree to which they represent *environmental* influences on intelligence.

A simple version of such an analysis is provided by Loehlin, Horn, and Willerman (chapter 4, this volume), who examine the effects on adoptee's IQ of both birth mother's IQ and adoptive family's SES and show that the latter has only very small effects. A better way to examine putative environmental influences is to incorporate specific environmental measures into biometric model-fitting analyses, as these authors did with an index of SES in a previous publication (Loehlin, Horn, & Willerman, 1989). Analyses of this sort would facilitate inferences regarding the causal role of any particular environmental characteristic by quantifying the percentage of shared or nonshared environmental variance (c^2 or e^2) explained by its corresponding specific environmental measure. These environmental measures could represent family environmental characteristics traditionally studied by developmentalists – such as SES, child-rearing styles, and parental warmth and control – or those less typically researched. The latter could include indices of the biological environment, such as nutrition (Eysenck & Schoenthaler, chapter 13, this volume) and the "physical microenvironment" (Jensen, chapter 2, this volume), which may include pre- and perinatal environmental influences. There also are aspects of the social and cultural environment that are less typically studied, such as novel variations in education programs (Sternberg, chapter 12, this volume) or cultural changes that vary geographically or temporally (such as those mentioned by Grigorenko & Kornilova, chapter 14, this volume). The take-home message is that the causal effects on intelligence of any hypothesized environmental characteristics can be evaluated by including them as variables in the biometric models used in contemporary behavior-genetic analyses.

Genotype–environment interaction and correlation for intelligence

The concept of genotype–environment interaction has long held strong appeal to developmental psychologists as a description of how genetic and environmental influences combine to influence developmental outcomes, not just for intelligence but for a wide variety of human traits. Despite its appeal as a model for human development, as well as the frequency with which it is found in the animal behavior-genetics literature, genotype–environment interaction may be considered the bane of human behavior-geneticists' existence. Many behavior geneticists have devoted considerable effort to the search for genotype–environment interactions for human characteristics, only to come up empty handed. This is well illustrated in recent publications from the Colorado Adoption Project (e.g., chapter 9 in Plomin, DeFries, & Fulker, 1988), in which the number of statistically significant interactions was actually *less* than that expected by chance.

Before discussing this further, however, it may be useful to clarify the use of the term *genotype–environment interaction*, especially because it differs markedly from the concept of *interactionism* discussed in some of the other chapters. Behavior geneticists typically use the term *genotype–environment interaction* to refer to the moderation of genetic influences on a trait by differences in environmental circumstances. In other words, if genetic predispositions for a trait are actualized to a greater extent in some environments than in others, this would be evidence of genotype–environment interaction. The oft-made critique of behavor-genetics research – that heritability estimates can differ considerably in different populations – is an example of genotype–environment interaction, despite the fact that its proponents often may not be aware of this. Because it is a symmetrical concept, genotype–environment interaction also describes the phenomenon whereby environmental influences have differential impact depending on individuals' genotypes. In other words, individuals may derive differential benefit or harm from some aspect of the environment depending on their genetic predispositions. When used in these ways, the term *genotype–environment interaction* captures well the concept of *reaction norm* while minimizing confusion with the concept of *reaction range*, which has a distinct meaning (Wahlsten & Gottlieb, chapter 6, this volume).

There are a number of potential explanations for the dearth of genotype–environment interactions in human behavior-genetics research on intelligence. The first is simply that genotype–environment interaction may not be important for individual differences in intelligence in most populations, given the ranges of intelligence, genotypes, and environments represented in these populations. Given the extant data, this may be the most plausible explanation, but there are plausible alternatives. A second explanation is distinct from but related to the first. Almost all investigations of genotype–environment interaction for intelligence have used adoption designs, which are less than ideal for this purpose because the range of adoptive environments may not include the very part that is most important for detecting interactions (cf. Waldman, Weinberg, & Scarr, manuscript in preparation). It may be the case that genotype–environment interactions for intelligence, if they exist, are not linear and constant throughout the range of genotypes and environments but rather nonlinear and local in their effects, being found especially at the extremes. Such a possibility, which was advanced and illustrated graphically by Turkheimer and Gottesman (1991), may have little chance of being detected in most adoption studies. This is because the environmental influences thought to impede intellectual development (e.g., poverty, abuse, deprivation, and harsh or unresponsive parenting) are likely to be unrepresented in adoptive environments, given that potential adoptive parents tend to be screened out on the basis of such characteristics.

A third explanation for the lack of genotype–environment interaction for intelligence involves statistical considerations. These concern a severe lack of power to detect interactions of the sort that are most likely to emerge in adoption studies – namely, ordinal interactions in which the effects of some environmental characteristic are somewhat greater for individuals with certain genotypes than with others. Detailed explanations of the reasons for this dramatic lack of power are presented elsewhere (McClelland & Judd, 1993; Wahlsten, 1990, 1991); suffice it to say that statistical tests of ordinal interactions can have as little as 16% of the power as tests of their corresponding main effects. Another reason for the lack of power to detect genotype–environment interaction for intelligence in adoption studies is that the magnitude of relations obtained among variables is often fairly modest. For example, correlations of adoptees' IQ scores with their biological parents' IQs or educational levels are often in the .1–.4 range, while correlations with characteristics of the adoptive family or adoptive environment (e.g., adoptive mother's IQ, adoptive family SES) are often in the .0–.4 range (Loehlin, Horn, & Willerman, chapter 4, this volume; Scarr et al., 1993). Hence, genotype–environment interaction for intelligence in adoption studies frequently may involve only small differences in correlations of adoptees' IQ with adoptive family characteristics (e.g., $r = .1$ versus $r = .4$) across levels of biological mothers' IQ (i.e., low versus high). Differences in correlations of this magnitude would be quite difficult to detect and would require samples of considerable size.

Given these difficulties, is there any point in continuing the search for genotype–environment interaction for intelligence? I believe so, but only if it continues in directions that are substantially different from those of past attempts. Cronbach (1991) recently has elucidated a number of novel alternative statistical methods for detecting interactions. To his list I would add the more extensive use of graphical statistical methods, especially exploratory analyses of three-dimensional regression surfaces in which algorithms (e.g., diagonally weighted least squares) are used that allow the surface to reflect the local character of the data. Techniques for implementing such exploratory data analyses have become more accessible given their inclusion in widely available statistical packages (e.g., SYSTAT, 1992).

The most important shift in direction, however, may involve the type of behavior-genetic design rather than the type of analytic method. Two such directional shifts come to mind. First, as suggested earlier, a wider range of environments, and perhaps genotypes, should be sampled in behavior-genetic studies. This is especially true of environments that are more harsh or impoverished, which would be expected to impede the actualization of genetic potential for high intellectual ability. Such environments exist, so the trick is to represent them in behavior-genetic designs. Perhaps the best way to accomplish this is through methodologies in which such environments are oversampled. Adequately sampling environments in this range would facilitate the shift from simply estimating the heritability of IQ to examining its reaction range across diverse environments, as a number of authors have suggested (Bouchard, chapter 5, this volume; Turkheimer & Gottesman, 1991; Wahlsten & Gottlieb, chapter 6, this volume).

Second, genotype–environment interaction for intelligence may be more profitably studied in behavior-genetic designs in which the magnitude of effects is greater than that typical of adoption studies. A good example of this comes from a twin study by Heath et al. (1985), in which the effects of secular changes in educational policy on genetic and environmental influences on educational attainment were examined in a large sample of Norwegian twins and their parents. Increases in educational opportunities over time led to a substantial increase in heritability for educational attainment in younger birth cohorts for male but not female twins, thus supporting the hypothesis that genetic potential for high educational achievement is actualized to a greater extent in more rather than less facilitative environments (Ceci, Rosenblum, de Bruyn, & Lee, chapter 10, this volume; Sternberg, chapter 12, this volume). (The sex difference in moderation of heritability may have been due to greater accessibility of educational opportunities for males than for females in the younger cohorts.) This study examined genotype–environment interaction for educational attainment in a novel way, one that capitalizes on the substantial magnitude of twin correlations for educational attainment (twin correlations across gender and birth cohorts: $r_{MZ} = .82 - .89$, $r_{DZ} = .47 - .75$).

Given the power of the twin design, as well as the greater ease of sampling twins than adoptees, it is surprising that twin studies similar to that of Heath et al. (1985) have not been utilized more in the search for genotype–environment interaction for intelligence. Such studies would not be limited to examining differences in twin correlations across cohorts but could examine a wide variety of contemporaneous environmental characteristics (e.g., SES, neighborhood or school differences, urban versus rural setting) as moderators of genetic influences on intelligence. In light of these advantages, I expect twin designs to gain ascendency over adoption designs in future in investigations of genotype–environment interaction for intelligence.

While one may conceptualize genotype–environment interaction as an example of how genetic and environmental influences are mutually dependent, one can alternatively conceptualize genotype–environment correlation as an example of how genetic and environmental influences act conjointly in producing developmental outcomes. Despite this difference, there are a number of similarities in the behavior-genetic literature on intelligence between genotype–environment correlation and genotype–environment interaction. Similar to the situation for genotype–environment interaction, there have been relatively few investigations of genotype–environment correlation; furthermore, most studies of genotype–environment correlation for intelligence also have used adoption designs. One final similarity is that genotype–environment correlation has held an allure for developmentalists similar to that of genotype–environment interaction. When developmental psychologists write about genetic and environmental influences on developmental outcomes, they often mention these influences as working hand-in-hand; in fact, this is often a core part of the *interactionist* model that many developmentalists espouse. Developmentalists often regard the role of genetic influences on developmental outcomes as guides for the exposure to or active search for relevant environmental influences. It is for these reasons that the consideration of genotype–environment correlation for intelligence is so important.

Developmental behavior geneticists also have considered genotype–environment correlation as an important determinant of developmental outcomes. Plomin, DeFries, and Loehlin (1977) carefully elaborated the biasing effects of genotype–environment correlation and interaction on estimates of heritability and environmental influences within traditional behavior-genetic designs (a somewhat similar quantitative genetic approach to genotype–environment correlation and interaction was offered by Eaves, Last, Martin, & Jinks, 1977). Scarr and McCartney (1983) took a more conceptual approach in formulating a theory of developmental processes based largely on three types of genotype–environment correlation: passive, evocative or reactive, and active. They hypothesized that these three types of genotype–environment correlation shift in influence throughout development, such that the passive type declines while the active type increases, whereas the evocative type may be important throughout the life course.

Despite the paucity of studies designed explicitly to examine genotype–environment correlation for intelligence, sufficient data from five adoption studies were available to afford an analysis of the importance of passive genotype–environment correlation for IQ (Loehlin & DeFries, 1987). Although estimates varied considerably across studies and methods of estimation, the authors concluded that passive genotype–environment correlation may account for as much as 30% of the overall variance in IQ. Despite these promising findings, few subsequent behavior-genetic studies of intelligence have systematically examined genotype–environment correlation (cf. Loehlin et al., 1989; Plomin et al., 1988, chapter 10). Hence, future behavior-genetic studies of intelligence should pursue the lead offered by Loehlin and DeFries (1987) and estimate the variance in IQ due to passive genotype–environment correlation. In addition to following up on these findings in adoption study data, researchers should use methods based on twin designs (see Neale & Cardon, 1992, chapter 17) to estimate genotype–environment correlation for intelligence. Finally, although some techniques have been developed for examining reactive genotype–environment correlation (see Plomin et al., 1977; Plomin et al., 1988, chapter 10), better methods are needed for examining reactive and especially active genotype–environment correlation.

Developmental behavior-genetic analyses of intelligence

Just as behavior-genetic designs can be used to estimate genetic and environmental influences on a trait measured at a single point in time, so can they be used to examine *changes* in genetic and environmental influences on a trait over time. Studies examining such changes have become more and more common due to the burgeoning field of developmental behavior genetics (Plomin, 1986). In this section, I highlight a number of interesting questions for further study that have emerged from the initial developmental behavior-genetic studies of intelligence conducted over the past 15 years.

Given the above section on genotype–environment interaction and correlation, a logical jumping off point for this section involves examining how such influences on intelligence might change throughout development. Relatively little theorizing exists regarding developmental shifts in genotype–environment interaction for intelligence (cf. Wachs & Plomin, 1991). Nonetheless, developmental changes in genotype–environment interaction for intelligence have been found in recent analyses of adoption study data (Waldman, Weinberg, & Scarr, manuscript in preparation). With regard to genotype–environment correlation for intelligence, recall that Scarr and McCartney's theory, described previously, is a theory of developmental processes. There are no studies, however, that have directly tested this theory's predictions concerning developmental changes in the influence of passive, reactive, and active genotype–environment correlation on intelligence or on any other developmental outcomes. The closest to this are analyses from the Colorado Adoption Project (Plomin et al., 1988, chapter 10), but these have not yet been conducted across the full developmental range (i.e., infancy to late adolescence/early adulthood) necessary to test the theory. In addition, a full test of Scarr and McCartney's theory awaits the development of adequate methods for studying active genotype–environment correlation.

One of the most intriguing findings to emerge from a number of recent behavior-genetic studies of intelligence is the tendency for h^2 to increase and c^2 to decrease over the life course. These findings are based on a number of studies, most of which are cross-sectional, of IQ in samples ranging in age from infancy (e.g., Plomin et al., 1988) to late adulthood (e.g., Pedersen, Plomin, Nesselroade, & McClearn, 1992). These results raise a number of interesting issues worthy of further study. First, only a few of the relevant findings regarding this developmental shift in genetic and environmental influences on IQ come from longitudinal studies. In both the Colorado Adoption Project (Fulker, DeFries, & Plomin, 1988) and the Louisville Twin Study (Wilson, 1983), genetic influences increased in magnitude and shared environmental influences decreased in magnitude as children progressed from infancy through adolescence. In addition, in the Texas Adoption Study (Loehlin et al., 1989), genetic influences increased in magnitude and shared environmental influences decreased in magnitude as adoptees progressed from childhood through late adolescence/early adulthood. Nonetheless, these two adoption projects studied family members at only two time points, which is far from ideal for drawing inferences regarding continuity and change (Rogosa, 1988). Hence, we will be able to place greater confidence in the developmental shifts in h^2 and c^2 for IQ as we accumulate further data from longitudinal behavior-genetic studies. A good example of the simultaneous analysis of data from multiple twin and adoption samples to test hypotheses regarding genetic and environmental influences on continuity and change in general cognitive ability can be found in chapter 16 (this volume), by Cherny, Fulker, and Hewitt.

A second issue raised by these findings concerns the meaning of the developmental shifts in h^2 and c^2 for IQ. Essentially, there are several nonmutually exclusive phenomena that could give rise to such developmental change. Such a shift could reflect the activation of new genes that influence cognitive ability as development progresses. Alternatively, this shift is consistent with the different types of environments that a person is exposed to throughout development. In infancy and early childhood, individuals are exposed mainly to *strong* environments that exert significant control over and constraints on their behavior; later in development, individuals are exposed mainly to *weak* environments that impose fewer constraints and may allow genetic predispositions on behavior to be actualized to a greater degree (Snyder & Ickes, 1985). Finally, this developmental shift in h^2 and c^2 for IQ is also consistent with Scarr and McCartney's (1983) theory of the shift in importance from passive to active genotype–environment correlation in influencing developmental outcomes. Each of these explanations is currently a plausible contender; hence, it remains for future behavior-genetic studies to evaluate their relative merits.

Many developmental behavior-genetic studies of intelligence have reported that substantial age-to-age correlations in genetic influences on IQ appear to be the primary source of developmental continuity in intelligence (e.g., Cherny et al., chapter 16, this volume). DeFries and his colleague (DeFries, Plomin, & LaBuda, 1987; Plomin, 1986) performed similar analyses but proposed a novel developmental hypothesis – a *genetic amplification* model – in order to reconcile the findings of considerable genetic stability and dramatic increases in heritability for cognitive ability from infancy to adulthood. According to this hypothesis, many of the same genes influence cognitive ability from infancy to adulthood, but they have increasing influence as development progresses. Examples of how such a mechanism might work could include genetic influences on aspects of neuronal development (e.g., the number of axonal projections, the degree of myelination, the complexity of dendritic connections) that have progressively cumulative effects on cognitive ability throughout development due to their influence on intermediate cognitive phenotypes, such as attentional mechanisms and the perception of similarities and differences among stimuli, that are critical to intellectual development (DeFries et al., 1987). The notion of genetic amplification as a source of cognitive stability from infancy to adulthood is a provocative one that deserves further examination in behavior-genetic studies.

Finally, future developmental behavior-genetic studies of intelligence might profit by adopting a growth-curve approach (Willett & Sayer, 1994) to capturing development. In this approach (e.g., McArdle, 1986; Waldman, DeFries, & Fulker, 1992), genetic and environmental influences on the parameters of individual growth curves for cognitive ability (e.g., intercept and slope) are estimated rather than examined in a piecemeal way at each discrete age. The growth-curve approach to developmental behavior-genetic analyses of intelligence has certain advantages and disadvantages relative to the analytic methods (e.g., simplex models) that have been used to date. Although a description of these is beyond the scope of this chapter, growth curve approaches to developmental behavior-genetic analyses may reveal novel insights regarding genetic and environmental influences on the development of intelligence.

Behavior-genetic designs as tools for examining the construct validity of intelligence

As noted previously, the majority of behavior-genetic studies of intelligence – with some recent exceptions – have focused on general cognitive ability, or *g*. In this section, I discuss behavior-genetic approaches to intelligence that move beyond *g* to examine genetic and environmental influences on specific mental abilities, discuss other cognitive abilities outside of the mainstream that appear to be fruitful for behavior-genetic analysis, and offer suggestions for the role that future behavior-genetic studies can play in investigating the construct validity of intelligence.

As most researchers know, there have been two broad traditions in the literature on mental abilities. The first, initiated largely by Spearman, posits a single general cognitive ability factor, g, as the source of most of the substantively interesting variance that is shared among mental ability measures, diverse as they may be. The second, initiated largely by Thurstone, posits a (relatively small) set of higher-order ability factors as the source of this shared variance. To an overriding extent, behavior geneticists have cast their lot with the former approach, focusing mainly on the genetic and environmental influences underlying indices of g.

Recently, a number of attempts have been made in the psychometric literature to reconcile these two approaches to the structure of mental abilities. One of the most promising of these attempts is a hierarchical model of mental abilities proposed by Gustafson (1984) in which specific ability measures load on a series of Thurstonian first-order factors, which in turn load on a series of second-order factors most characteristic of Horn (1988) and Cattell's perspective, which in turn load on a single general factor, g.

Just as behavior-genetic analyses have been conducted on a number of specific mental ability measures, so can they be conducted on the factors at various levels of the hierarchical model. One can pose a number of interesting and novel questions regarding genetic and environmental influences on a broad range of mental abilities using this approach. For example, one can examine whether crystallized ability, G_c, is influenced by genetic factors independent of those that influence g or fluid ability, G_f, thus attempting to replicate the finding by McArdle et al. (1981) cited earlier. One might also hypothesize that some first-order factors have residual genetic influences after the genetic influences on their higher-order factor are accounted for, whereas others do not. In an interesting recent paper, Cardon and Fulker (1994) combined the hierarchical approach to mental abilities with developmental behavior-genetic analytic strategies using longitudinal data from the Colorado Adoption Project. Future behavior-genetic studies should examine genetic and environmental influences on specific mental abilities beyond g, as well as conduct behavior-genetic analyses of mental ability factors within the context of a hierarchical model such as Gustafson's (1984).

In addition to further investigation of specific mental-ability measures that are within the psychometric tradition, behavior geneticists should journey farther afield to examine genetic and environmental influences on cognitive measures that are outside the mainstream. A number of the chapters in this volume that analyze data from the MacArthur Longitudinal Twin Study are good examples of this. Pipp-Siegel, Robinson, Bridges, and Bartholomew (chapter 18, this volume) conducted behavior-genetic analyses of infant social cognition, specifically behavior and affective responses indicative of self-recognition and of role-taking ability. Reznick (chapter 17, this volume) conducted behavior-genetic analyses of the development of expressive and receptive language in the first 2 years of life. A noteworthy feature of both of these studies was the inclusion of measures of general cognitive ability. This allowed the authors to investigate the role of genetic and environmental influences on general cognitive ability in accounting for genetic and environmental influences on infant social cognition and early language development.

Behavior geneticists also should consider studying genetic and environmental influences on measures representing alternative conceptualizations of intelligence, such as those corresponding to Sternberg's triarchic theory (chapter 12, this volume) (e.g., measures corresponding to analytic, creative, and practical processes), or measures that index Gardner's (Gardner, Hatch, & Torff, chapter 8, this volume) multiple symbolic domains of intelligence (e.g., measures corresponding to musical thinking or spatial reasoning). Behavior-genetic analyses of alternative measures of cognitive abilities also can be extended beyond the confines of a Western conceptualization of intelligence. Grigorenko and Kornilova (chapter 14, this volume) describe the differences in the conceptualization of cognitive abilities within Soviet/Russian psychology from that in the West. Behavior-genetic studies of measures drawn from these alternative conceptualizations of intelligence could shed considerable light on the influences underlying their development. Incorporating standard psychometric measures of intelligence along with measures from these alternative conceptualizations in such studies would go a long way toward resolving issues of similarity and difference among these approaches to intelligence, as common and distinct genetic and environmental influences across the disparate measures could be revealed.

The nature of behavior-genetic models for examining causal influences on intelligence

Implicit in several contributors' chapters is a theme that I want to address explicitly in this final section. This concerns the nature of behavior-genetic models for examining causal influences on intelligence, or for making causal inferences regarding the development of intelligence. Although a number of contributors (e.g., Wahlsten & Gottlieb, chapter 6, this volume) criticize behavior-genetic models on several fronts, in this section I choose to highlight some of the important features of these models that make them especially useful for testing hypotheses regarding the causal influences underlying the development of intelligence or other traits.

This theme emerges most strongly in Scarr's paper (chapter 1, this volume), in which she pits Behavior-Genetic Theory against Socialization Theory in explaining the causes of children's intelligence and academic achievement. It should be noted that some behavior geneticists (e.g., Dolan & Molenaar, in press) would not consider behavior genetics to be a theory of development per se but rather a methodology for testing predictions from different developmental theories. Scarr focuses mainly on the genetic side of behavior-genetic models (i.e., on the fact that they include genetic as well as environmental influences as potential causes, whereas socialization theories focus exclusively on environmental influences). In her comparisons, behavior-genetic models come out ahead, consistently providing better fits to the observed data than socialization theories. This should come as no surprise, for these results illustrate a signal advantage of behavior-genetic models – namely, their comprehensiveness. By including broadly construed genetic and shared and nonshared environmental influences, behavior-genetic models are capable of making more comprehensive statements regarding causal influences than competing models. Behavior-genetic models are also highly flexible, allowing for the inclusion of more complex causal mechanisms such as genotype–environment correlation and interaction, causal mediation or indirect causation, and specific indices of genetic and environmental influences.

Additional important features of behavior-genetic models emerge in other chapters. Bouchard (chapter 5, this volume) raises a number of these features. His description of the criticisms of studies of identical twins reared apart (i.e., MZAs) as being either pseudoanalyses or pseudocriticisms highlights the importance of the quantitative specifiability of causal influences in behavior-genetic models. By fully specifying in quantitative form the causal influences contained within their models, behavior geneticists make public their assumptions regarding the influences that are included and excluded from their models. As such, these assumptions and specifications may be examined and tested by others and thus subjected to the risk of falsification. Bouchard demonstrates how most critiques of MZA studies, when put in specifiable quantitative form, do not stand up to close scrutiny. Bouchard also draws the parallel between behavior-genetic models and meta-analytic methods in their focus on replicability and generalizability of findings across studies. Behavior-genetic models also are similar to meta-analytic methods in their emphasis on effect sizes and their variability (viz., on the magnitude and precision of parameter estimates) in addition to statistical hypothesis tests. For example, considerable attention is focused on the *estimates* of genetic and environmental influences (e.g., h^2, c^2, and e^2) and on their ranges in behavior-genetic models, in contrast to the simple hypotheses that these parameters differ significantly from zero. These features are illustrated well in the behavior-genetic analyses of intelligence mentioned previously

(Loehlin, 1989; Chipuer et al., 1990) in which familial correlations for IQ were combined across many different types of relationships, as well as in similar behavior-genetic analyses of personality (Loehlin, 1992). These models also allow one to examine the sensitivity of estimates of genetic and environmental influences to various assumptions concerning issues of sampling and measurement (Horn, Loehlin, & Willerman, 1982).

It seems that our conceptualization of genetic and environmental influences on intelligence, and on many other developmental outcomes more generally, must be altered in response to the findings of recent behavior-genetic studies. In making inferences regarding causation, it is useful to distinguish between factors that are part of the *causal field* or background context, which thus are judged to be of little causal import, and those that represent a *difference-in-a-background* (Einhorn & Hogarth, 1986; Mackie, 1974), which thus are judged to be of strong causal import. Psychologists' traditional conceptualizations of the causes of most developmental outcomes appear to have considered genetic influences as part of the causal field on which environmental influences acted as differences-in-a-background. In the light of recent behavior-genetic findings, it seems that just the opposite may be a more accurate depiction. In this sense, the interactionists are correct, as there is no doubt that genetic influences, which differ among individuals in ways that effect differences in their intellectual development, act on an environmental background context representing the causal field. Interestingly, this environmental-background context may be more similar functionally across individuals than was previously thought, at least within the predominantly white, largely middle-class population that has been most typically represented in behavior-genetic designs. The causal import of genetic and environmental influences on intelligence in ethnic minority and lower-SES populations remains in large part to be elucidated.

Acknowledgments

Preparation of this chapter was facilitated by support from NIMH Grant #R03-MH53450-01. The author thanks Elena Grigorenko, Julia Hough, Scott Lilienfeld, Dick Neisser, and David Rowe for their helpful comments on an earlier draft of this chapter.

References

Boomsma, D. I. (1995). Using multivariate methods to detect pleiotropic Quantitative Trait Loci. *Behavior Genetics, 25*, 257–8.

Cardon, L. R., DeFries, J. C., Fulker, D. W., Kimberling, W. J., Pennington, B. F., & Smith,

S. D. (1994). Quantitative Trait Locus for reading disability on chromosome 6. *Science*, *266*, 276–9.

Cardon, L. R., & Fulker, D. W. (1994). A model of developmental change in hierarchical phenotypes with application to specific cognitive abilities. *Behavior Genetics*, *24*, 1–16.

Cattell, R. B. (1941). Some theoretical issues in adult intelligence testing. *Psychological Bulletin*, *38*, 592.

Chipuer, H. M., Rovine, M. J., & Plomin, R. (1990). LISREL modeling: Genetic and environmental influences on IQ revisited. *Intelligences*, *14*, 11–29.

Cronbach, L. J. (1991). Emerging views on methodology. In T. D. Wachs & R. Plomin (Eds.), *Conceptualization and measurement of organism–environment interaction*. Washington, DC: American Psychological Association.

DeFries, J. C., Plomin, R., & LaBuda, M. C. (1987). Genetic stability of cognitive development from childhood to adulthood. *Developmental Psychology*, *23*, 4–12.

Dolan, C., & Molenaar, P. C. M. (in press). A note on the scope of developmental behavior genetics. *International Journal of Behavioral Development*.

Eaves, L. J., Last, K. A., Martin, N. G., & Jinks, J. L. (1977). A progressive approach to non-additivity and genotype–environmental covariance in the analysis of human differences. *British Journal of Mathematical and Statistical Psychology*, *30*, 1–42.

Einhorn, H. J., & Hogarth, R. M. (1986). Judging probable cause. *Psychological Bulletin*, *99*, 3–19.

Fulker, D. W., Cardon, L. R., DeFries, J. C., Kimberling, W. J., Pennington, B. F., & Smith, S. D. (1991). Multiple regression analysis of sib-pair data on reading to detect quantitative trait loci. *Reading and Writing: An Interdisciplinary Journal*, *3*, 299–313.

Fulker, D. W., DeFries, J. C., & Plomin, R. (1988). Genetic influence on general mental ability increases between infancy and middle childhood. *Nature*, *336*, 767–9.

Gustafson, J.-E. (1984). A unifying model for the structure of intellectual abilities. *Intelligence*, *8*, 179–203.

Heath, A. C., Bevg, K., Eaves, L. J., Solaas, M. H., Corey, L. A., Sundet, J., Magnus, P., & Nance, W. E. (1985). Education policy and the heritability of educational attainment. *Nature*, *314*, 734–6.

Horn, J. L. (1988). Thinking about human abilities. In J. R. Nesselroade & R. B. Cattell (Eds.), *Handbook of multivariate psychology*. New York: Academic Press.

Horn, J. M., Loehlin, J. C., & Willerman, L. (1982). Aspects of the inheritance of intellectual abilities. *Behavior Genetics*, *12*, 479–516.

Loehlin, J. C. (1992). *Genes and environment in personality development*. Newbury Park, CA: Sage.

Loehlin, J. C. (1989). Partitioning environmental and genetic contributions to behavioral development. *American Psychologist*, *44*, 1285–92.

Loehlin, J. C. & DeFries, J. C. (1987). Genotype–environment correlation and IQ. *Behavior Genetics*, *17*, 263–77.

Loehlin, J. C., Horn, J. M., & Willerman, L. (1989). Modeling IQ change: Evidence from the Texas Adoption Project. *Child Development*, *60*, 993–1004.

Mackie, J. L. (1974). *The cement of the universe: A study of causation*. Oxford, UK: Clarendon Press.

McArdle, J. J. (1986). Latent variable growth within behavior genetic models. *Behavior Genetics*, *16*, 163–200.

McArdle, J. J., Goldsmith, H. H., & Horn, J. L. (1981). Genetic structural equation models of fluid and crystallized intelligence. *Behavior Genetics*, *11*, 607.

McClelland, G. H., & Judd, C. M. (1993). Statistical difficulties of detecting interactions and moderator effects. *Psychological Bulletin*, *114*, 376–90.

Moore, E. G. J. (1986). Family socialization and the IQ test performance of traditionally and transracially adopted black children. *Developmental Psychology*, *22*, 317–26.

Neale, M. C., & Cardon, L. R. (1992). *Methodology for genetic studies of twins and families.* Dordrecht: Kluwer Academic Publishers.

Pedersen, N. L., Plomin, R., Nesselroade, J. R., & McClearn, G. E. (1992). A quantitative genetic analysis of cognitive abilities during the second half of the life span. *Psychological Science, 3,* 346–53.

Plomin, R. (1986). *Development, genetics, and psychology.* Hillsdale, NJ: Erlbaum.

Plomin, R. & Bergeman, C. S. (1991). The nature of nurture: Genetic influence on "environmental" measures. *Behavioral and Brain Sciences, 14,* 373–427.

Plomin, R., DeFries, J. C., & Fulker, D. W. (1988). *Nature and nurture during infancy and early childhood.* New York: Cambridge University Press.

Plomin, R., DeFries, J. C., & Loehlin, J. C. (1977). Genotype–environment interaction and correlation in the analysis of human behavior. *Psychological Bulletin, 84,* 309–22.

Plomin, R., McClearn, G. E., Smith, D. L., Vignetti, S., Chorney, M. J., Chorney, K., Venditti, C. P., Kasarda, S., Thompson, L. A., Detterman, D. K., Daniels, J., Owin, M., & McGoffin, P. (1994). DNA markers associated with high versus low IQ: The IQ QTL project. *Behavior Genetics, 24,* 107–18.

Rogosa, D. R. (1988). Myths about longitudinal research. In K. W. Schaie, R. T. Campbell, W. Meredith, & S. C. Rawlings (Eds.), *Methodological issues in aging research* (pp. 171–210). New York: Springer.

Scarr, S., & McCartney, K. (1983). How people make their own environments: A theory of genotype → environment effects. *Child Development, 54,* 424–35.

Scarr, S., & Weinberg, R. A. (1977). Intellectual similarities within families of both adopted and biological children. *Intelligence, 1,* 170–91.

Scarr, S., Weinberg, R. A., & Waldman, L. D. (1993). IQ correlations in transracial adoptive families. *Intelligence, 17,* 541–55.

Snyder, M., & Ickes, W. (1985). Personality and social behavior. In G. Lindzey & E. Aronson (Eds.), *Handbook of social psychology* (3rd ed.). Reading, MA: Addison-Wesley.

SYSTAT (1992). *SYSTAT for Windows: Graphics, Version 5 edition.* Evanston, IL: SYSTAT, Inc.

Turkheimer, E., & Gottesman, I. I. (1991). Individual differences and the canalization of human behavior. *Developmental Psychology, 27,* 18–22.

Wachs, T. D., & Plomin, R. (1991). *Conceptualization and measurement of organism–environment interaction.* Washington, DC: American Psychological Association.

Wahlsten, D. (1990). Insensitivity of the analysis of variance to heredity–environment interaction. *Behavioral and Brain Sciences, 13,* 109–61.

Wahlsten, D. (1991). Sample size to detect a planned contrast and a one degree-of-freedom interaction effect. *Psychological Bulletin, 110,* 587–95.

Waldman, I. D., DeFries, J. C., & Fulker, D. W. (1992). Quantitative genetic analysis of IQ development in young children: Multivariate multiple regression with orthogonal polynomials. *Behavior Genetics, 22,* 229–38.

Waldman, I. D., Weinberg, R. A., & Scarr, S. *Developmental change in genotype–environment interaction for IQ.* Manuscript in preparation.

Willett, J. B., & Sayer, A. G. (1994). Using covariance structure analysis to detect correlates and predictors of individual change over time. *Psychological Bulletin, 116,* 363–81.

Wilson, R. C. (1983). The Louisville Twin Study: Developmental synchronies in behavior. *Child Development, 54,* 298–316.

Name index

Subject index